Introduction to
AEROSPACE VEHICLES

Mahesh M. Sucheendran • Lokanna Hoskoti

Introduction to Aerospace Vehicles
Mahesh M. Sucheendran & Lokanna Hoskoti

Published by White Falcon Publishing
Chandigarh, India

All rights reserved
First Edition, 2024
© Mahesh M. Sucheendran & Lokanna Hoskoti, 2024
Cover/Interior designed by Mahesh M. Sucheendran & Lokanna Hoskoti

No part of this publication may be reproduced, or stored in a retrieval system, or transmitted in any form by means of electronic, mechanical, photocopying or otherwise, without prior written permission from the publisher.

The contents of this book have been certified and timestamped on the Gnosis blockchain as a permanent proof of existence. Scan the QR code or visit the URL given on the back cover to verify the blockchain certification for this book.

The views expressed in this work are solely those of the author and do not reflect the views of the publisher, and the publisher hereby disclaims any responsibility for them.

Requests for permission should be addressed to the publisher.

ISBN - 978-81-19510-45-0

This work is subject to copyright. All rights are reserved, whether the whole or part of the material is concerned, specifically the rights of translation, reprinting, reuse of illustrations, recitation, broadcasting, reproduction in microfilm or in any other way, and storage in data banks. Duplication of this publication or parts thereof is permitted only under the provisions of the Copyright Law in its current version, and permission for use must always be obtained from the authors.

Information contained in this book has been obtained by the authors from sources believed to be reliable. However, neither the publishers nor the authors guarantee the accuracy or completeness of any information published herein. Neither the publishers nor the authors shall be responsible for any errors, omissions, or damages arising out of the use of this information. The work is published with the understanding that the publishers and the authors are supplying information but are not attempting to render engineering or other professional services. If such services are required, the assistance of an appropriate professional should be sought.

The book was prepared using LaTeX typesetting system on *overleaf.com*. Although most of the images featured in the book were captured by the authors at various air shows and aviation museums, open source images and data available at *www.si.edu*, *www.isro.gov.in*, *earthobservatory.nasa.gov* and *commons.wikimedia.org/www.wikipedia.org* were also utilized. Sketches and Info-graphics were prepared using *Microsoft Office* and *Mathworks Matlab*.

The image on the cover and the one before the first chapter display the *Dassault Rafale*, fighter and the vintage aircraft, *T-6G Harvard*, respectively, captured at AeroIndia2023. Beginning of Chapter 2, a full-page image features the *Vostok* space capsule, taken at the Air and Space Museum in Paris. Moving forward, the front images of Chapters 5 and 7 display the *Airbus A321XLR* and the *Lockheed Martin F-35 Lightning*, captured at the Paris Air Show 2023, respectively. Chapter 8 full-page image display the *Ariane 5* rocket. Chapter 9 showcases the *HAL Prachand*, a light Combat Helicopter (LCH) captured at AeroIndia 2023. Lastly, the cover page of Chapter 10 features the *Sol.Ex.* aircraft propelled by combined solar and electric propulsion, captured at the Paris Airshow 2023.

TO ALL THE AEROSPACE ENTHUSIASTS

Preface

Many books cover the history of aviation quite extensively. For example, *Flight: The Complete History of Aviation* by *R. G. Grant* is a well-researched book that provides a comprehensive history of aviation. *Introduction to Flight*, written by *J. D. Anderson, Jr.*, gives a fundamental understanding of the engineering aspects of flight vehicles. The technical details of flight vehicles and associated systems can be accessed from various books like *Jane's All the World's Aircraft*, the Internet and other sources. This book aims to bring together in one place the history, engineering aspects, and the system-level understanding of flight vehicles in a manner which is easily understood by a beginner. To the best of our knowledge, there are few books in that category. As mentioned above, the information presented in this book is already available in the open literature and is scattered across multiple sources. We have tried to bring together the relevant information in one place and present it in a manner that is easily understood by an undergraduate student in engineering. To keep the contents simple and easy enough to be understood from a system engineering perspective, the mathematical derivations and theoretical aspects are not covered in detail.

The book is divided into three parts: the first part emphasizes the historical developments of aviation and space flight, the second part gives an overview of the engineering aspects of aerospace vehicles, and the last part gives details of different classes of aerospace vehicles. The first chapter is dedicated to the brief historical evolution of aviation. It is to be mentioned that the history is covered mostly from an engineer's perspective. The second chapter gives a brief history of rockets and space flight. In the second part, which includes chapters three and four, the engineering aspects of flight vehicles are briefly explained. Chapter three brings out the relevant aspects of the flight environment, both outer space and Earth's atmosphere, whereas chapter four covers the fundamental engineering aspects of flight vehicles. The third part gives a system-level understanding of different types of aerospace vehicles. The third part includes chapters on civilian airplanes, military airplanes, missiles, space vehicles and other miscellaneous flight vehicles.

The book is primarily meant for first-year undergraduate students in aerospace engineering to gain an overall understanding of flight vehicles without placing much emphasis on the mathematical and theoretical details. Hence, detailed mathematical derivations, assumptions involved in the derivation of equations, and the range of applicability of different theories and results are not mentioned in this book. Pilots, military personnel, aviation & space enthusiasts will find the information given in this book interesting and useful. In addition, anyone who wants to get a preliminary understanding of the different types of aerospace vehicles will find this book beneficial.

The scope of the book as envisaged by the title is quite broad. We have made significant efforts to cover pertinent topics. However, some of the important aspects related to aerospace vehicles likely have been missed out. For the information presented in this book, we have referred to some of the standard textbooks in the aerospace engineering domain. Still, there is the possibility that some errors could have happened in the process. After significant deliberation, we decided to divide the book into three parts. There can be overlap and repetition of concepts across different chapters. Also, there can be objections that some of the concepts are introduced without sufficient mathematical rigor. Also, there can be objections to how aerospace vehicles can be classified into various types as described in *Part III*. We acknowledge that many examples of flight vehicles can be found that will fit into multiple types or categories. So, the division of flight vehicles as discussed in *Part III* is for convenience purposes only and many exceptions can easily be found. We believe that such decisions taken while writing a book like this can be subjective. However, we are open to receiving any comments and suggestions for a better presentation of the material via email (`vehiclesaerospace@gmail.com`).

Acknowledgments : Writing a book is always an arduous exercise and the same was the case with this book. The time and effort it took to write the book was much more than what we had anticipated. Writing this book would not have been possible without the help and support of many of our colleagues, friends and well-wishers. We acknowledge the following people for reviewing various parts of this book and giving meaningful suggestions that significantly improved the readability of the book.

- Dr. Abhijith Gogulapati, IIT Bombay
- Wg Cdr Abhilash Raghunathan, Indian Air Force
- Dr. Christopher Ostoich, ATA Engineering, USA
- Mr. Dileep Ravindranathan, Vikram Sarabhai Space Centre, ISRO
- Dr. Kumaravel Gnanashanmugham, Vikram Sarabhai Space Centre, ISRO
- Dr. Muralidhar Patkar, Armament Research & Development Establishment, DRDO
- Dr. Naresh Kumar, Defence Research & Development Laboratories, DRDO
- Mr. Nisar Ali, Aeronautical Development Agency
- Dr. Rudrakant Sollapur, Friedrich Schiller University Jena, Germany
- Dr. Sanjay Talole, Research & Development Establishment (Engineers), DRDO
- Dr. Shakti Gupta, IIT Kanpur
- Mr. Swapnil Sapkale, Bharat Dynamics Ltd.
- Mr. Vishal Shirbhate, National Aerospace Laboratories
- Mr. V. Shankar, National Aerospace Laboratories.

The contents of the book are based on a course for undergraduate students put together by the first author in 2017. We thank the numerous students at IIT Hyderabad who have taken the course since then and whose questions, comments and suggestions have played a role in the structure of the book.

Our visits to the Smithsonian's National Air and Space Museum in Washington, D.C., the National Museum of the United States Air Force at Wright Patterson Air Force Base in Ohio, the San Diego Air & Space Museum, National Air and Space Museum of Paris, the Royal Air Force Museum in London, Italian Air Force Museum in Rome, Tokorazawa Aviation Museum in Japan, the Paris Air Show, AeroIndia and Dubai Air Show have helped in improving our understanding of flight vehicles, and we believe it would have played an important role in the narrative followed in this book and in improving the quality of the book.

The first author thanks his wife, Arathy, and their children, Nandhana and Ananth, for being supportive and providing a lively environment at home. The second author sincerely appreciates his wife, Sudha, and their children, Venkat and Vendat, for their invaluable support and cooperation.

IIT Hyderabad
May 2024

Gentle Breezes, Blue Skies & Calm Space
Mahesh & Lokanna

Contents

Preface	v
Introduction	1

I A Brief History of Aerospace Vehicles — 21

1 History of Aviation — 25
- 1.1 Overview — 26
- 1.2 The Beginning — 26
- 1.3 Aviation during World War I — 38
  - 1.3.1 As War Machines — 38
  - 1.3.2 Flying Aces — 43
  - 1.3.3 Technology Innovations — 47
- 1.4 Between World War I and World War II — 48
  - 1.4.1 Maturing of Aviation — 48
  - 1.4.2 Passenger Airlines — 58
  - 1.4.3 Flying Boats — 62
- 1.5 As a Game Changer during World War II — 63
  - 1.5.1 Role of Aircraft in the War — 63
  - 1.5.2 Notable Aircraft during the War — 65
  - 1.5.3 Flying Aces — 71
  - 1.5.4 Technological Evolution — 71
- 1.6 As a Mainstream Technology after World War II — 74
  - 1.6.1 Robust, Reliable and Sophisticated Technology — 74
  - 1.6.2 Modern Jet Airliners — 85
  - 1.6.3 Military Aviation — 93
- 1.7 History of Helicopters — 96
  - 1.7.1 Early Concepts and Precursors — 96
  - 1.7.2 Post-War Innovation and Commercialization — 100
  - 1.7.3 Modern Era and Beyond — 100
- 1.8 USA's Experimental Aerospace Programs — 104
- 1.9 Aviation Accidents — 114
- 1.10 Future of Aviation — 117

2 History of Rocketry and Space Flight — 123
- 2.1 Introduction — 124
- 2.2 Visionaries and Pioneers — 124
- 2.3 Near Earth Space Explorations — 130
  - 2.3.1 Technological Advancements — 130
  - 2.3.2 First Artificial Earth Satellite — 132
  - 2.3.3 The First Human in Space — 135
  - 2.3.4 Space Stations — 136
  - 2.3.5 Space Shuttle — 139
  - 2.3.6 Hubble and James Webb Space Telescopes — 144
- 2.4 Missions to Celestial Bodies — 145
  - 2.4.1 Luna Mission to Moon — 145
  - 2.4.2 Apollo Space Program — 145
  - 2.4.3 Space Missions to Mars — 152
  - 2.4.4 Space Missions to Venus — 153
  - 2.4.5 Voyager Mission to Jupiter, Saturn, Uranus and Neptune — 155
  - 2.4.6 Pioneer Mission to Jupiter and Saturn — 155
  - 2.4.7 NASA's Juno Space Probe to Jupiter — 156
  - 2.4.8 Cassini–Huygens Mission to Saturn — 156
  - 2.4.9 New Horizons Mission to Pluto — 156
  - 2.4.10 NEAR-Shoemaker Mission to Asteroid — 156

		2.4.11	JAXA's Hayabusa Missions	157
	2.5	USA's Experimental Programs		157
	2.6	Current and Future Space Explorations		162
		2.6.1	Lunar Missions	162
		2.6.2	Planetary Explorations	166
		2.6.3	Missions to Sun	167

II Fundamentals of Flight — 171

3 Flight Environment — 173

3.1	Introduction		174
3.2	Gravitational Force		175
3.3	Solar System		175
	3.3.1	Sun	176
	3.3.2	Planets and Natural Satellites	176
	3.3.3	Dwarf Planets, Comets, Asteroids and Meteors	178
3.4	Earth's Atmosphere		179
	3.4.1	Altitude, Longitude and Latitude	179
	3.4.2	Pressure, Density and Temperature Variation	180
	3.4.3	Real Atmosphere	182
	3.4.4	Aviation Meteorology	183

4 Basics of Aerospace Engineering — 185

4.1	Introduction		186
4.2	Aerodynamics		186
	4.2.1	Basics Definitions	186
	4.2.2	Airfoil	189
	4.2.3	Aerodynamic Forces	190
	4.2.4	Wings	193
4.3	Propulsion		194
	4.3.1	Aspects of Thermodynamics	195
	4.3.2	Reciprocating Engines	197
	4.3.3	Gas Turbine Engines	199
	4.3.4	Ramjet and Scramjet Engines	207
	4.3.5	Pulsejet Engines	210
	4.3.6	Rocket Propulsion	210
4.4	Flight Mechanics		215
	4.4.1	Performance of Flight Vehicles	215
	4.4.2	Principles of Stability	222
4.5	Flight Structures		224
	4.5.1	Material Types	228
	4.5.2	Aircraft Structures	230
	4.5.3	Manufacturing	231
4.6	Orbital Mechanics		234
	4.6.1	Basic Laws	234
	4.6.2	Space Vehicle Trajectories	235

III Types of Aerospace Vehicles — 239

5 Civilian Airplanes — 242

5.1	Introduction		243
5.2	Commercial Airliners		247
	5.2.1	Basic Terminologies	248
	5.2.2	Parts of Airliners	248
	5.2.3	Notable Airliners	254
	5.2.4	Identifying the Airliners	261
	5.2.5	Supply and Services of Airliners	265
	5.2.6	Airliner Operations	268
5.3	Safety and Comfort in Commercial Aviation		270
	5.3.1	Aviation Safety Management System (SMS)	271
	5.3.2	Airliner Noise	276
5.4	Business Aviation		277
	5.4.1	Business Aircraft	278
5.5	Airports		282
	5.5.1	Airport Types	283
	5.5.2	Airport Services	284
5.6	Governing Bodies		287
5.7	Civil Aviation in India		287

6 Military Airplanes — 291

6.1	Introduction		292
	6.1.1	Military Airplane Generations	292
	6.1.2	Next Generation Military Airplanes	296
6.2	Capabilities and Performance Parameters		298
	6.2.1	Maximum and Minimum Speeds	298
	6.2.2	Range and Endurance	299
	6.2.3	Climb Performance	299

6.2.4	Turning Performance	300
6.2.5	Takeoff and Landing	300
6.2.6	Stealth Technology	301
6.2.7	Airborne Armaments	301

- 6.3 Roles of Military Airplanes — 302
  - 6.3.1 Aerobatics — 302
  - 6.3.2 Attack — 306
  - 6.3.3 Bomber — 309
  - 6.3.4 Fighter and Interceptor — 312
  - 6.3.5 Reconnaissance — 312
  - 6.3.6 Tanker — 314
  - 6.3.7 Trainer — 316
  - 6.3.8 Transport — 317
  - 6.3.9 Multirole Airplanes — 321
  - 6.3.10 Airborne Early Warning & Control (AEW&C) System — 326
- 6.4 Military Aviation in India — 328

7 Missiles — 331
- 7.1 Introduction — 332
- 7.2 Brief History of Missile Technology — 332
- 7.3 Basic Concepts — 333
- 7.4 Missile Subsystems — 338
  - 7.4.1 Airframe — 338
  - 7.4.2 Propulsion System — 341
  - 7.4.3 Navigation, Guidance and Control — 343
  - 7.4.4 Warheads — 346
  - 7.4.5 Missile Launchers — 347
- 7.5 Types of Missiles — 347
  - 7.5.1 Based on Launch Mode — 347
  - 7.5.2 Based on Target — 347
  - 7.5.3 Multiple Independently Targetable Reentry Vehicles (MIRVs) — 350
  - 7.5.4 Missile Defense System (MDS) — 350
- 7.6 Some Examples — 353
- 7.7 Major Missile Manufacturers — 356
- 7.8 Treaties & Agreements — 357
- 7.9 Missile Programs in India — 357

8 Space Vehicles — 362
- 8.1 Introduction — 363
  - 8.1.1 What is Space — 363
- 8.2 Types of Space Vehicles — 364
  - 8.2.1 Satellites — 364
  - 8.2.2 Lunar and Interplanetary Vehicles — 367
  - 8.2.3 Launch Vehicles — 367
  - 8.2.4 Space Stations and Space Shuttles — 370
- 8.3 Satellite Launching Stations — 370
- 8.4 Types of Trajectories — 373
  - 8.4.1 Orbital Parameters — 374
- 8.5 Spacecraft Maneuvers and Trajectories — 376
- 8.6 Space Debris — 382
- 8.7 Space Agencies — 382
- 8.8 Space Laws — 384
- 8.9 India's Space Program — 384

9 Helicopters — 390
- 9.1 Introduction — 391
- 9.2 Working Principle — 392
  - 9.2.1 Components of Helicopter — 392
  - 9.2.2 Maneuvering — 394
- 9.3 Helicopter Rotor Configurations — 398
  - 9.3.1 Single Main Rotor with Tail Rotor — 399
  - 9.3.2 Tandem Rotors — 400
  - 9.3.3 Coaxial Rotors — 400
  - 9.3.4 Intermeshing Rotors — 400
  - 9.3.5 NOTAR (No Tail Rotor) — 401
  - 9.3.6 Tilt Rotor — 401
  - 9.3.7 Compound Helicopters — 401
- 9.4 Autogyros — 402
- 9.5 Examples of Helicopters — 404
  - 9.5.1 Older Helicopters — 404
  - 9.5.2 Advanced Helicopters — 406
  - 9.5.3 Future Vertical Lift (FVL) Programs — 414
- 9.6 Helicopters in India — 415

10 Miscellaneous Flight Vehicles — 419
- 10.1 Aerostats — 420
  - 10.1.1 Air ships — 421
  - 10.1.2 Balloons — 423
- 10.2 Autogyros — 423
- 10.3 Convertiplanes — 425

	10.3.1	Tiltrotors	426	10.7.2 Floatplanes	430
	10.3.2	Tiltwings	426	10.7.3 Amphibious Aircraft	431
	10.3.3	Stopped Rotors	426	10.8 Uninhabited Aerial Vehicles	431
10.4	Gliders		426	10.8.1 Micro Aerial Vehicles	432
10.5	Ground-Effect Vehicles		428	10.8.2 Quadcopters	434
10.6	Jet Packs		429		
10.7	Seaplanes		430		
	10.7.1	Flying Boats	430		

Index   **437**

About the Authors   **441**

Introduction

About the book

This book's organization is designed to offer readers an overall understanding of the history, engineering principles, and system level details about the diverse array of aerospace vehicles that have shaped human exploration of the skies and beyond. The book presents a detailed exploration of aerospace technology, spanning from its historical roots to the cutting-edge innovations of today. Divided into three distinct parts, the book aims to provide readers with a holistic understanding of aerospace vehicles, their evolution, underlying principles, and diverse applications.

The first part, *A Brief History of Aerospace Vehicles*, sets the stage by chronicling the remarkable journey of human ingenuity in conquering the skies and venturing into space. Through its two chapters, *History of Aviation* and *History of Rocketry and Space Flight*, readers gain insight into the contribution by the pioneers in the field, breakthroughs, challenges, and triumphs that have shaped the aerospace landscape. From the early days of aviation to the exploration of outer space, this section highlights the pivotal moments that have paved the way for modern aerospace advancements.

The second part, *Fundamentals of Flight*, delves into the core engineering principles that underpin aerospace technology. Through chapters such as *Flight Environment* and *Basics of Aerospace Engineering*, readers are introduced to the basics of flight, including aerodynamics, propulsion systems, structures, flight dynamics and orbital mechanics. This section lays the groundwork for understanding the technical aspects of aerospace design and operation.

The final part, *Types of Aerospace Vehicles*, offers an in-depth exploration of various categories of aerospace vehicles and their roles. The six chapters in this section cover a wide spectrum of vehicles, ranging from *Civilian Airplanes* used for commercial transportation to *Military Airplanes* designed for defense purposes. *Missiles* discusses guided projectiles, while *Space Vehicles* delves into spacecraft designed for space exploration, research, and space applications and *Helicopters* explores the realm of rotary-wing aircraft. The last chapter encompasses a diverse array of vehicles, including balloons, gliders, Micro Aerial Vehicles (MAVs), and Uninhabited Aerial Vehicles (UAVs).

In essence, the book offers readers a journey through the history, principles, and diverse array of aerospace vehicles that have revolutionized the way humans interact with the skies and outer space. It is designed to cater to both newcomers to the field and those seeking a broader understanding of the intricacies and advancements in aerospace technology. By combining historical context, technical insights, and practical applications, the book provides a well-rounded exploration of one of humanity's most awe-inspiring frontiers.

For *students* studying aerospace engineering or related fields, the book serves as an essential foundation. The historical overview in the first part provides context to the evolution of aerospace technology, while the basics of aerospace engineering chapter in the second part offer a solid grounding in key concepts. The third part's exploration of different aerospace vehicles broadens students' understanding of real-world applications and potential career paths.

Pilots, both aspiring and seasoned, can benefit from the book's comprehensive overview of

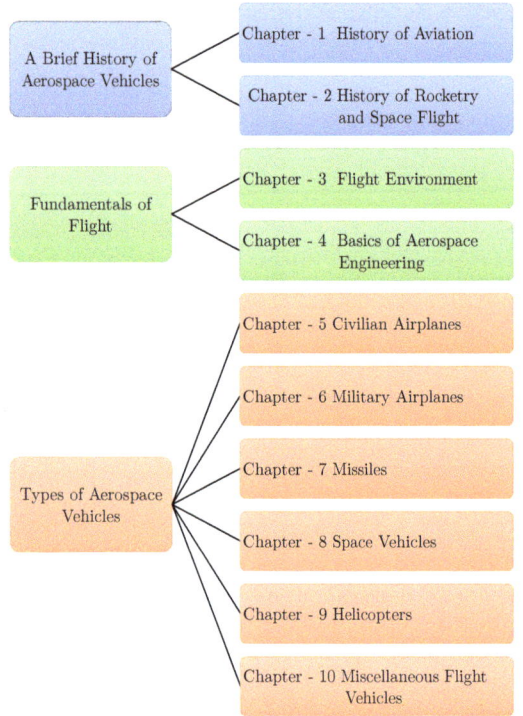

Organization of the book.

ples of aerospace technology. The detailed chapters on different aerospace vehicles can serve as a reference for engineers involved in designing and testing aerospace vehicles.

Aerospace enthusiasts passionate about aviation and space exploration may find the book a treasure trove of information. The historical chapters cater to their curiosity about the beginnings of aerospace achievements, while the comprehensive coverage of various aerospace vehicles in the third part satisfies their desire to delve into the technical details of different aircraft and spacecraft. Historians interested in the evolution of aerospace technology will appreciate the first part of the book. The coverage of the history of aviation and space technology provides an overview of how humanity's achievements have shaped the field. The book's structure also allows historians to gain insights into the development of different types of aerospace vehicles over time.

Aerospace Sector

The sector plays a crucial role in shaping the modern world and is characterized by its significant size and consistent growth over the years. It can be seen that all the developed countries have a strong presence in the sector. Aerospace sector, in which aerospace/aeronautical engineers are actively involved, can be broadly categorized into the following distinct groups:

- Aerospace Industry
- Airlines
- Armed Forces
- Space Agencies
- Aerospace Infrastructures
- Research Institutions
- Regulatory Bodies
- Technical Societies

These groups form the ecosystem that propels the aerospace sector forward, with aerospace engineers and policy makers serving as the driving force behind technological advancement, innovation, and

aerospace vehicles. The historical insights in the first part provide context for the evolution of flight, and the part on fundamentals of flight deepens their understanding of the mechanics and principles behind aviation. Additionally, the coverage of different flight vehicle types can enhance their knowledge of the broader aerospace landscape.

Defence personnel interested in aerospace technology and its history will find the book particularly informative. The historical chapters offer insights into the development of military aviation and space exploration, while the detailed exploration of military airplanes, missiles, and space vehicles in the third part provides a comprehensive understanding of the role played by aerospace vehicles in the military context. *Scientists* working in aerospace research and development can benefit from the technical insights provided in the book. The sections on flight environment and aerospace engineering offer valuable information on the challenges and princi-

the achievement of remarkable milestones in both aviation and space exploration.

Aerospace Industry

The industry is involved in designing, developing, manufacturing and testing cutting-edge aerospace products. They are also involved in the maintenance, repair and operations of aerospace systems. These products range from aircraft and spacecraft to components that drive innovation and elevate the industry's technological standards. The aerospace industry encompasses three main categories of activities: (1) aeronautics, which includes aircraft, propulsion systems, and infrastructure and equipment; (2) astronautics, which include satellite launch vehicles, satellites, and space vehicles; and (3) missile development for defense applications. The aeronautical industry constitutes a major part of the total activity.

The industry contributes significantly to the global economy. The global aerospace and defense sector is projected to grow from $\sim$ \$ 800 billion in 2022 to $\sim$ \$ 1070 billion in 2027 (\$ 1 billion $\sim$ ₹8,300 Crores). The presence of armed conflicts accelerates the growth of the defense sector. The projection underscores the resilience and dynamism of the global aerospace and defense market, driven by the civil aviation boom in countries like India and China, increasing global security concerns, and the pursuit of new frontiers in space exploration.

Based on the revenue generated, *The Boeing Company*, *Lockheed Martin Corporation*, *RTX Corporation*, *Northrop Grumman Corporation*, *GE Aerospace (A subsidiary of General Electric)*, *General Dynamics Corporation*, *Airbus SE*, *BAE systems plc* and *Safran S. A.* are among the top leading aerospace and defense manufacturers worldwide.

In 2022, the USA played a significant role in the global aerospace export landscape, with a substantial contribution of approximately \$ 102.8 billion in exports. This position firmly established the USA as the foremost country in the realm of aerospace exports, owing to its hosting of prominent aerospace

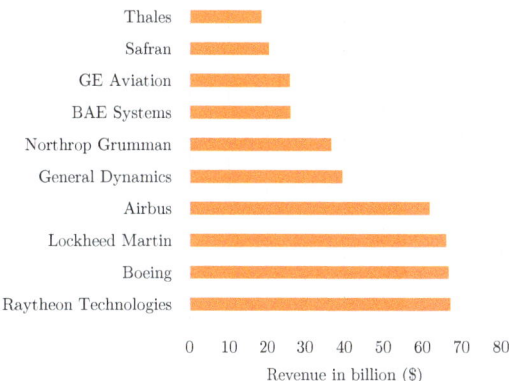

Revenue of leading aerospace and defense manufacturers (2021).

industry giants such as *RTX*, *Boeing* and *Lockheed Martin*.

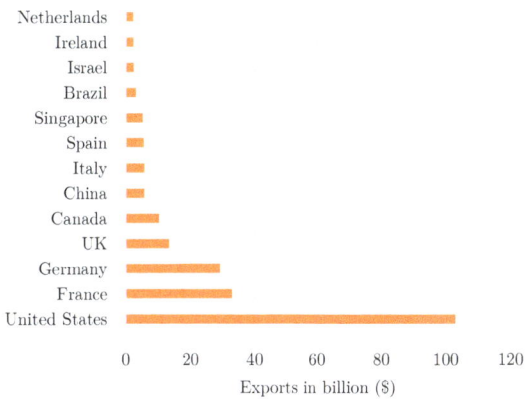

Aerospace exports of selected countries (2022).

Airlines

Airlines rely on the expertise of aerospace engineers to ensure the safe and efficient transportation of both people and goods. Aerospace professionals contribute to aircraft design, maintenance, and the implementation of advanced technologies that enhance passenger safety and comfort.

Air transport plays a key role in global economic development. In a normal year (before COVID-19), the aviation industry supported 87.7 million jobs worldwide and generated \$3.5 trillion economy which is 4.1 % of global GDP (Aviation would rank 17th in size by GDP if it were a country). The aviation industry is truly a global industry connecting all parts of the world seamlessly with ∼ 1,500 commercial airlines operating ∼ 33,000 commercial aircraft from 3780 airports. Also, ∼ 58% of international tourists travel by air.

Airbus and *Boeing* are the two major players in the commercial aircraft manufacturing industry, and they engage in fierce competition in terms of aircraft deliveries. Recently, the battle for market share in the commercial segment was primarily between *Airbus'* A320 family and *Boeing's* 737 series, as well as their wide-body aircraft, including the *Airbus A350* and the *Boeing 787 Dreamliner/777* series.

The number of aircraft deliveries varied from year to year due to factors like market demand, production rates, and unforeseen events, such as the grounding of the *Boeing 737 MAX*.

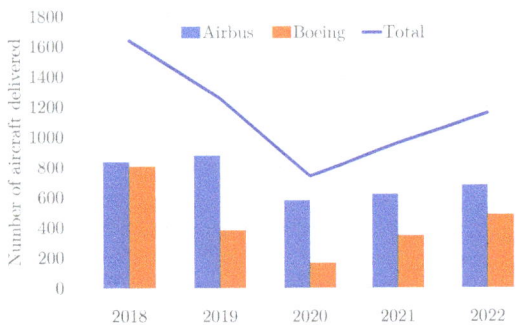

Airbus vs *Boeing* aircraft deliveries 2018-2022

Boeing's 737 series, a twinjet, shorter and more cost-effective aircraft model, has emerged as the company's top-selling product line. Within this series, the *Boeing 737-700*, with an average price of just under $90 million, stands out as one of the more affordable options, while the *Boeing 777-9*, priced around $ 442 million, ranks among the higher-priced models on *Boeing's* product list.

The landscape of the global airline industry is continually evolving, with several prominent airlines standing out as leaders in terms of revenues. These leading airlines are marked by their extensive route networks, strong passenger demand, and innovative strategies.

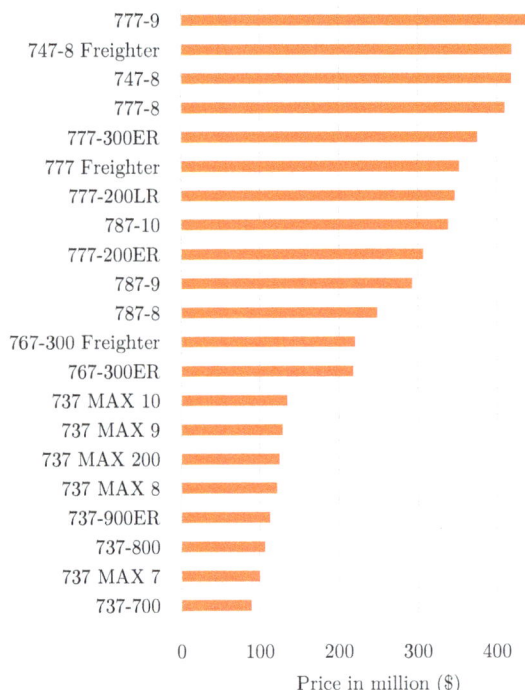

Approximate prices of various *Boeing* models (2022).

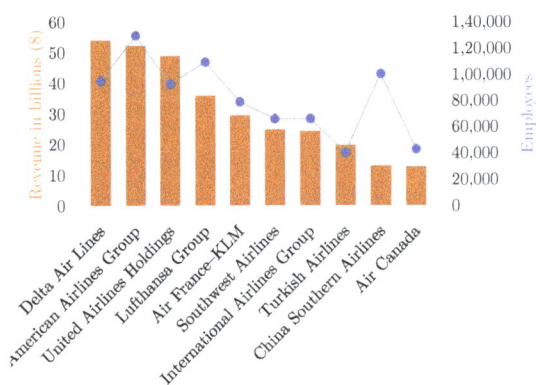

Leading airlines by revenues (2022-2023).

Military Use

Armed forces are discerning clients of the aerospace sector, demanding advanced technologies for their defense and surveillance needs. Aerospace engineers work closely with these entities to develop state-of-the-art fighter jets, helicopters, missiles, satellites, reconnaissance aircraft, and various other types of vehicles and related systems.

In recent years, global military expenditure has

experienced a consistent increase, surging to $ 2.24 trillion by 2022. Contributing factors to this escalation include the outbreak of the Russia-Ukraine conflict in 2022, along with growing tensions in the South China Sea. Notably, the United States assumed the top position among countries with the highest military spending in 2022, allocating a substantial $877 billion to its defense budget, constituting nearly 40% of the world's total military expenditure for that year. A fraction of the overall military budget is spent on purchase or manufacturing of aerospace vehicles and related systems.

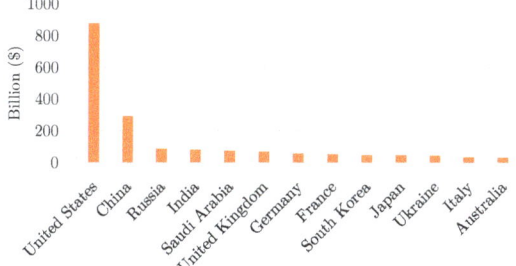

Countries with the highest military spending (2022).

Space Agencies

Space agencies rely on the skills of aerospace engineers to venture into the unknown realms of space. These professionals are at the heart of space vehicle design for commercial, military and scientific purposes. In 2021, the global space economy saw a turnover of roughly $ 469 billion, marking an increase from the $ 447 billion recorded in the preceding year. The paramount sector within the global space economy during 2021 was the commercial space products and services segment, contributing to nearly 48% of the overall turnover.

In 2022, global government spending on space programs reached a historic milestone, totaling around $ 103 billion. Leading the way, the USA allocated nearly $ 62 billion to its space endeavors during that year. Over the past few decades, space agencies have become increasingly significant as they spearhead civilian space programs, space

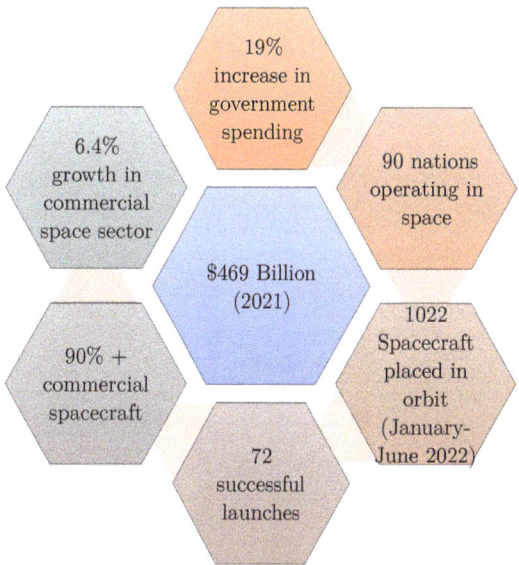

Global space economy 2021

research, and exploration missions. Among these agencies, NASA stands as the most prominent and recognized and a significant portion of its budget is naturally directed toward scientific pursuits and exploration initiatives.

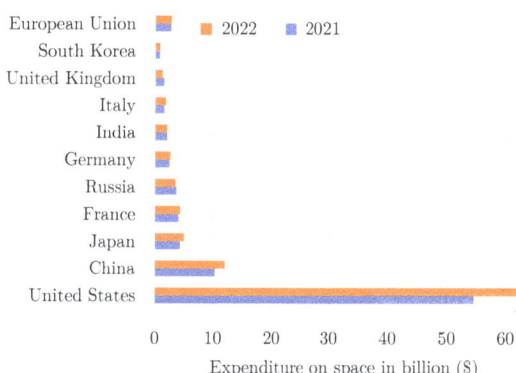

Space budget (2021 & 2022).

The budgets of leading space agencies around the world represent substantial investments in space exploration and research. NASA in the United States allocated over $ 24 billion for various space missions including Mars exploration and the *Artemis* program aiming to return humans to the Moon. The European Space Agency (ESA) had a budget of more than $ 6 billion, supporting a wide range of missions, from Earth observation to interplanetary exploration. China's CNSA continued to increase its budget, surpassing $ 10 billion, focusing on lunar exploration, Mars missions, and the development of its space station. These substantial budgets underscore the importance of space exploration and the growing global interest in advancing our understanding of the cosmos. It is worth noting that budgets for space agencies may vary year by year due to changing government priorities and policies.

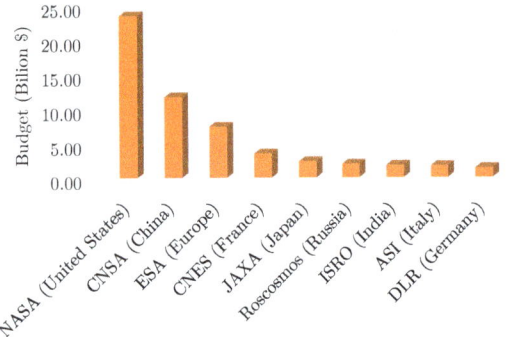

Budget of leading space agencies (2022).

Aerospace Infrastructure

To ensure safe operations for various forms of aviation, including commercial airliners, military aircraft, and space missions, a critical set of infrastructures and human resources is indispensable. The essential infrastructures that support flight operations and space missions encompass airports and air navigation services, particularly relevant for atmospheric flights. In the realm of space transportation, distinct set of facilities, including launch bases, tracking stations, and control and surveillance centers, are vital to ensure the safe and effective execution of endeavors beyond Earth's atmosphere.

Airports serve as fundamental components of the global transportation network, playing a crucial role in connecting people and moving goods across the world. These multifaceted facilities are more than just transit points; they are intricate ecosystems where countless components come together to facilitate safe and efficient air travel. Airports are designed with passenger convenience and safety in mind, offering a diverse range of services and amenities, from baggage handling and security screening to dining options and retail stores. Beyond passengers, airports also support the logistics and supply chain industries. They are central hubs for the transport of cargo and goods, ensuring that products reach their intended destinations in a timely manner. Airports come in various sizes and capacities, ranging from major international hubs to smaller regional airports. In an era of increasing globalization, airports continue to evolve to meet the growing demands of air travel, with a focus on sustainability, innovation, and safety.

Airports around the world shown by dots. Credits: *https://www.flightconnections.com /openstreetmap.org/copyright*

Research Institutions

Research institutions play a vital role as catalysts of technological advancement in the aerospace industry. Research endeavors are primarily conducted in academic institutions, aerospace companies, and government agencies. Their contributions span a wide spectrum of aerospace disciplines, from aerodynamics, propulsion, materials, avionics, communication, airport infrastructure, air navigation and aviation safety. With access to state-of-the-art laboratories and facilities, researchers at these institutions pioneer breakthroughs that drive the industry forward. Apart from the numerous academic institutions, many agencies or organizations like the *National Aeronautics and Space Administra-*

Top 10 busiest airports based on passenger traffic (2022).

Airport	Passenger (Million)
Hartsfield-Jackson Atlanta International Airport, Georgia, USA	94
Dallas Fort Worth International Airport, Texas, USA	73
Denver International Airport, Denver, USA	69
O'Hare International Airport, Chicago, USA	68
Dubai International Airport, UAE	66
Los Angeles International Airport, Los Angeles, USA	66
Istanbul Airport, Istanbul, Turkey	64
Heathrow Airport, London, United Kingdom	62
Indira Gandhi International Airport, Delhi, India	59
Charles de Gaulle Airport, Paris, France	57

tion (NASA) in the United States, the *Office National d'Études et de Recherches Aérospatiales* (ONERA) in France, the *Deutsches Zentrum für Luft-und Raumfahrt* (DLR), *Centro Italiano Ricereche Aerospaziali* (CIRA) in Italy, *Instituto Nacional de Técnica Aeroespacial* (INTA) in Spain, and *Central Aerohydrodynamic Institute* (TsAGI) in Russia have made remarkable contributions in aerospace research. In India, the *Defence Research and Development Organisation* (DRDO), *Indian Space Research Organisation* (ISRO), *Hindustan Aeronautics Limited* (HAL) and *National Aerospace Laboratories* (NAL) are involved with the research leading to the development of flight vehicles.

Regulatory Bodies

International organizations act as custodians of aerospace jurisprudence, regulating and coordinating the global aerospace industry. Two fundamental supranational organizations: The *International Civil Aviation Organization* (ICAO) and the *International Air Transport Association* (IATA) were formed to ensure the promotion of reliable, efficient, and secure air transportation and develop a comprehensive framework of regulations. Both ICAO and IATA are preeminent organizations at the forefront of the aviation industry, each playing a distinctive role in shaping the global air travel landscape.

ICAO, established as a specialized agency of the United Nations, holds the crucial responsibility of coordinating and regulating international air travel. Rooted in the principles set forth by the Chicago Convention of 1944, ICAO develops and enforces a set of rules and standards that encompass everything from airspace management to aircraft registration and safety protocols. This organization has been instrumental in harmonizing international aviation regulations and promoting the orderly growth of civil aviation worldwide. ICAO's enduring commitment to enhancing air travel safety and efficiency has significantly contributed to the global aviation community's success.

On the other hand, IATA, while also promoting safety and efficiency in air travel, primarily focuses on inter-airline cooperation and advocacy. Founded in 1945, IATA serves as a platform for airlines to collaborate and advance the industry collectively. Its mission involves not only supporting the interests of airlines but also working toward the establishment of secure, reliable, and economically viable air services. IATA plays a crucial role in facilitating seamless passenger and cargo transport across the world, striving to make the global airline network function as smoothly as a single domestic airline. Additionally, IATA works to enhance public understanding of the aviation industry and the considerable benefits it brings to national and global economies. These two organizations, ICAO and IATA, continue to be pivotal in ensuring that aviation remains a safe, efficient, and indispensable mode of global transportation.

In addition to these organizations, a multitude of regulatory bodies and agencies worldwide play crucial roles in ensuring the safety and efficient functioning of civil aviation. Among these, the European Aviation Safety Agency (EASA) holds a pivotal position within the European Union. EASA is responsible for overseeing and harmonizing aviation safety standards, certification processes, and regulatory frameworks across EU member states. Its mission is to maintain high levels of air travel safety and environmental protection while facilitating the seamless integration of new technologies into the European aviation landscape. In the United States, the Federal Aviation Administration (FAA) serves as the primary regulatory body for civil aviation. The FAA establishes and enforces regulations, certifies aircraft and pilots, and manages the nation's airspace to ensure safe and orderly air travel. Its comprehensive oversight encompasses various facets of aviation, from commercial airline operations to general aviation and air traffic control. The Directorate General of Civil Aviation (DGCA) in India serves as a regulatory authority responsible for overseeing civil aviation activities. DGCA sets safety standards, oversees aircraft maintenance and pilot certification, and regulates aviation operations within their respective territories. Another critical player is the European Organization for the Safety of Air Navigation (EUROCONTROL). This organization focuses on air traffic management and the coordination of airspace usage across European states, striving to enhance the safety, efficiency, and environmental sustainability of air navigation services.

Together, these regulatory bodies and authorities work collaboratively to ensure that civil aviation worldwide adheres to stringent safety standards, while also striving to accommodate technological advancements and growing demands in this vital global industry.

Technical Societies

Technical societies like the *American Institute of Aeronautics and Astronautics* and *Royal Aeronautical Society* contribute immensely to the dissemination of knowledge in the domain through their scientific journals and technical conferences. They also provide a platform for networking with aerospace professionals. *The Aeronautical Society of India* promotes the advancement and dissemination of knowledge of aeronautics and space technologies in India.

Classification of Aerospace Vehicles

Aerospace vehicles encompass a broad category of vehicles that operate both within Earth's atmosphere and in outer space. Vehicles operating outside the Earth's atmosphere are referred to as *spacecraft*. However, the demarcation between Earth's atmosphere and space is not well defined and there are no universally accepted standards. "Kármán line", the imaginary surface at an altitude of 100 km above mean sea level is the most accepted boundary that separates Earth's atmosphere from outer space. As per *The Aircraft Act* of 1934 in India, "aircraft" means any machine that can derive support in the atmosphere from reactions of the air. Typically, balloons, helicopters, gliders and airplanes are considered as aircraft. And, rockets, missiles and satellite launch vehicles which may have their trajectory partially or fully in the atmosphere are excluded from the definition of aircraft.

A major portion of aeronautics (aircraft) activity occurs within specific altitude bands: typically below 12 km for civil aircraft and below 25 km for military aircraft. Astronautics (spacecraft), on the other hand, promptly transition into space during launch. Satellites are positioned at varying altitudes; for example, the *International Space Station* orbits at around 400 km, sun-synchronous satellites near 800 km, global positioning satellites at approximately 20,000 km, and geosynchronous satellites at approximately 36,000 km above mean sea level. Ballistic missiles ascend to altitudes ranging from 70 to 90 km before launching their payloads into a ballistic trajectory.

Understanding the atmospheric conditions and space in which aerospace operations take place is crucial, especially in connection with engines. For example, piston-engine aircraft and helicopters primarily operate at lower altitudes and speeds, while jet aircraft can reach higher altitudes and speeds. Beyond the Earth's atmosphere, rocket engines become the propulsion method of choice, as they must carry their own oxidizer. This is in contrast to aeronautical vehicles, which rely on atmospheric air for combustion. Satellites are examples of aerospace platforms that operate beyond the Earth's atmosphere.

Aircraft

In a broad context, an aircraft is a vehicle with the capability to navigate through the air using lift, typically within the Earth's atmosphere. It overcomes the force of gravity through mechanisms of lift (static or dynamic) generated by an airfoil. In some instances, aircraft utilize direct downward thrust from their engines to balance its weight. Examples of common aircraft include airplanes, helicopters, airships, gliders and hot air balloons.

(a) A free-flying hot air balloon.   (b) A helium-filled airship, *Zeppelin NT*.

Lighter-than-air aircraft (Aerostat).

Single-seat glider, *Glaser-Dirks DG-808*.   Paraglider   Hang glider

Heavier-than-air non-power driven aircraft. Gliders typically feature fixed wings with an exceptionally high aspect ratio. Paragliders, on the other hand, employ a flexible wing that takes its shape from the surrounding air pressure. At the same time, hang gliders use generally flexible wings but are supported on a rigid frame which determines its shape.

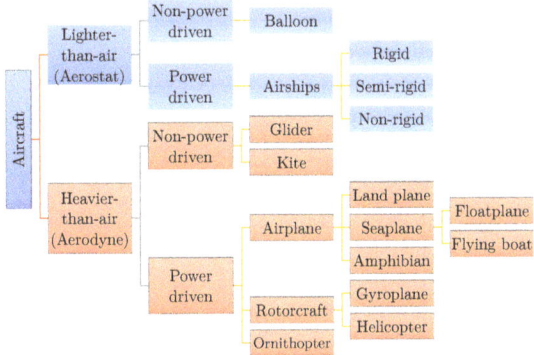

Classification of aircraft.

Based on the way lift is generated, the aircraft can be classified into "lighter-than-air vehicles"(*aerostats*) and "heavier-than-air vehicles" (*aerodynes*). An aerostat remains aloft primarily by utilizing lighter-than-air gases, meaning that the lift is chiefly by its buoyancy in the air. Examples of aerostats are balloons and airships both of which remain aloft by housing a large envelope filled with a gas less dense than the the atmosphere. Heavier-than-air aircraft is used to describe any aircraft whose average density is greater than that of surrounding air. Examples of heavier-than-air aircraft include conventional airplanes, gliders, and helicopters.

Airplanes' characteristics include their speed, altitude of operation, utility, engine type, number of engines, wing-tail configuration, size and weight. It is possible to classify them based on any of these parameters. One common classification criterion is the Mach number, which indicates their speed in relation to the speed of sound. Airplanes can be considered as subsonic or supersonic based on whether the operating Mach number is less than or greater than one, respectively. Additionally, they can be grouped according to their purpose, whether they are designed for civil aviation or military aviation. The type of engines they employ is another crucial factor, as some aircraft use piston engines, while others rely on jet engines or turboprops. Moreover, the number of engines a plane possesses plays a significant role in its classification, distinguishing between single-engine, twin-engine, or multi-engine aircraft. Finally, the number of wings is a defining characteristic, categorizing aircraft into monoplanes, biplanes, or even triplanes based on their wing configurations. Airplanes can be categorized based on their mode of take-off and landing. They can also be categorized based on their maximum take-off mass or payload capacity.

Airplanes are complex machines comprising various components. At the heart of an airplane is the fuselage, the central structure that houses the cockpit, passengers, cargo, and other systems. The wings provide lift and can come in different configurations, such as monoplane, biplane, or more advanced designs like swept wings. Attached to the wings, the ailerons help control the aircraft's roll and flaps augment lift during takeoff and landing, respectively.

Aircraft propulsion is typically achieved through engines, which can be jet engines, turboprop engines, or piston engines, depending on the aircraft type. The tail assembly features horizontal and vertical stabilizers, which help control pitch and yaw. The landing gear consists of wheels, struts, and shock absorbers that support the aircraft during takeoff and landing. Airplanes also have various onboard systems, including avionics for navigation and communication, as well as hydraulic and electrical systems for control and operation. Together, these components enable airplanes to take flight and transport passengers and cargo efficiently.

Helicopters, in contrast to airplanes, possess distinct components that allow them to achieve vertical takeoff and landings and hover in place. At the core of a helicopter is the fuselage, which contains the cabin for passengers or crew, cargo space, and critical systems. The primary lifting component is the rotor system, consisting of a main rotor mounted above the fuselage and a tail rotor, if required, to counteract torque and ensure stability.

Helicopter engines can be turboshaft or piston engines and are responsible for powering the rotor system. Helicopters also feature a complex control system, which includes swashplates that regulate the pitch of the rotor blades. Tail booms and horizontal and vertical stabilizers contribute to stability and control during flight. Landing gear, often including skids or wheels, supports the helicopter during ground operations. Helicopters are versatile aircraft, capable of intricate maneuvers and hovering, making them invaluable in various roles, including transport, search and rescue, and military operations.

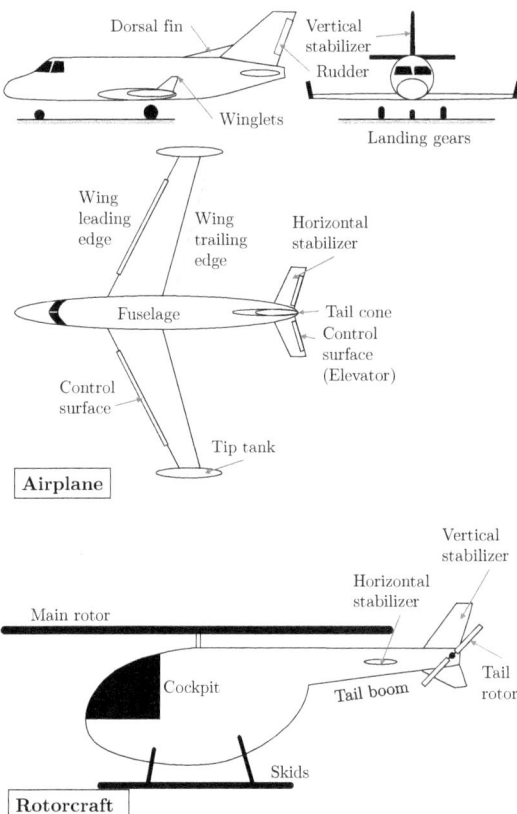

Parts of powered heavier-than-air aircraft (airplane and helicopter).

Wings with airfoil cross-section when swiftly moved through the air, produce lift. They come in a diverse array of shapes and sizes, and their design can be customized to attain specific desired flight characteristics. Wings shapes, including straight, swept, elliptical, or delta configurations are common in the present days. Altering the wing's shape leads to variations in control at different operating speeds, the quantity of lift generated, as well as adjustments to balance and stability. The leading edge and trailing edge of a wing can be straight or curved, or one edge may be straight while the other is curved. Furthermore, one or both edges can be tapered, resulting in a wing that is narrower at the tip than at its root, where it connects to the fuselage. Wing tips themselves can be square, rounded, or even pointed. In contemporary aviation, most aircraft feature a single wing on each side of the fuselage. However, during the early days of aviation, aircraft with two or more wings were commonplace. The angle at which the wings connect to the fuselage can be bend upward (known as dihedral), downward (anhedral), or remain straight.

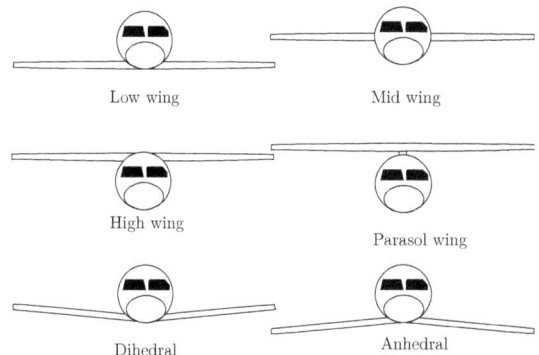

The configuration of the airplane based on the location at which wings are attached to the fuselage

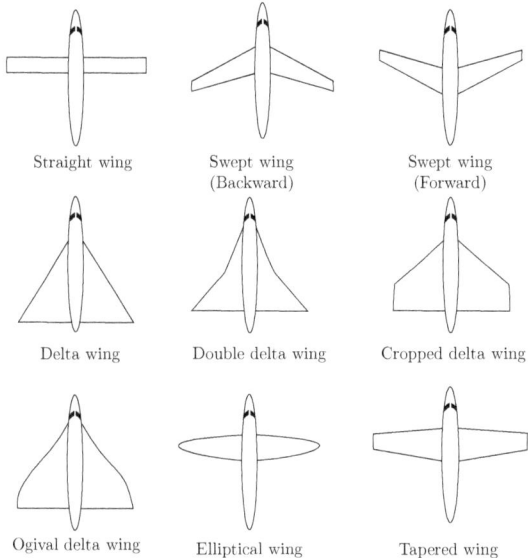

Design of different wing shapes.

Aircraft wings can be affixed to the fuselage at different positions: either atop, midway, or beneath it. These wings may stretch out at a right angle to the fuselage's vertical plane or feature a slight upward or downward inclination. This angle is referred to as the wing's dihedral, and it significantly impacts the lateral stability of the aircraft. In the Figure, typical wing attachment locations and dihedral angles are presented for reference.

Standard wings possess inherent instability in pitch, necessitating the incorporation of a horizontal stabilizing surface to maintain control and stability during flight. This stabilizing surface, often referred to as the horizontal tail or horizontal stabilizer, serves to counteract the destabilizing forces acting on the aircraft. It can take various forms and be positioned either in front and behind the main wing, playing a pivotal role in the aircraft's overall aerodynamic design and control.

The horizontal stabilizing surface, located either forward (canard configuration) or aft of the main wing, effectively mitigates pitch instability. In the canard configuration, the horizontal stabilizer is positioned in front of the main wing, enhancing the aircraft's control and stability by generating additional lift and control authority. Conversely, when situated behind the main wing, it acts as a traditional horizontal tail, stabilizing the aircraft by adjusting the distribution of lift forces and maintaining the desired pitch attitude. This crucial component, along with other control surfaces, ensures that the aircraft maintains its intended trajectory and stability throughout its flight.

Aircraft featuring conventional wing or three-surface configurations typically incorporate a combination of one or more horizontal stabilizers and vertical stabilizers. Among these configurations, the most prevalent design involves a solo vertical stabilizer affixed to the rear fuselage, complemented by horizontal stabilizers on either side.

Attaching the tail to an airplane has several ways, and each method serves specific purposes. The most common tail attachment configurations include the conventional tail, which features a sin-

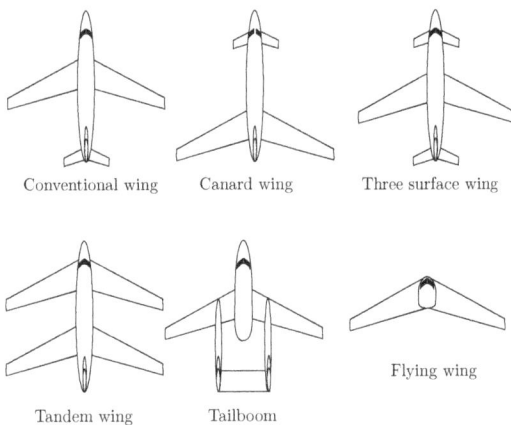

Wing-tailplane configurations.

gle vertical stabilizer with a horizontal stabilizer. In the T-tail configuration, the horizontal stabilizer is mounted at the top of the vertical stabilizer, resembling the letter T. These configurations play a vital role in maintaining proper balance and control during flight. Additionally, unconventional tail configurations such as the V-tail or double tail are employed in certain aircraft to optimize aerodynamic performance and reduce drag. The choice of tail attachment configuration is influenced by factors such as the aircraft's intended mission, design requirements, and aerodynamic considerations, ultimately shaping the aircraft's flight characteristics and behavior in the air.

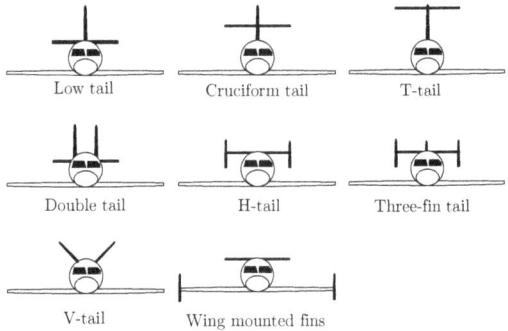

Tailplane configurations.

Most aircraft are equipped with landing gear for safe touchdown on the Earth's surface. Landing gear serves as the interface between the aircraft and the ground, providing support, stability, and shock absorption during these critical phases of flight. It typically comprises a set of wheels, struts, and shock absorbers, although it can vary significantly depending on the aircraft type and mission.

This landing gear can either be fixed in position or designed to retract. It typically comprises two or more main landing gears and supplementary support gear, collectively defining the landing gear configuration.

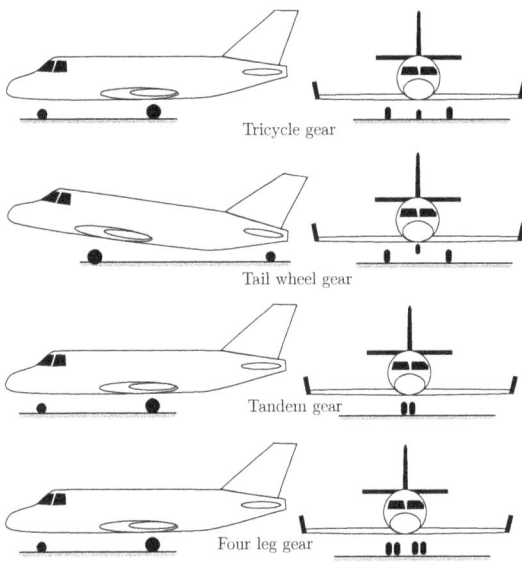

Landing gear configurations.

Propulsion systems are the heart of any flying machine, responsible for generating the necessary thrust to the air resistance and the weight. These systems can vary significantly in terms of design, fuel type, and complexity. Piston engines, turboprop, turbojet, turbofan, ramjet or scramjet and rocket engines are the different types of engines used in flight vehicles. Piston engines, also known as reciprocating engines, are commonly found in smaller general aviation aircraft. They operate on the same principle as automobile engines, using the reciprocating motion of pistons to generate power. The engine power is then used to rotate propeller(s) which provides the thrust. Turbojet engines are the earliest form of jet propulsion. They operate on the principle of sucking in air, compressing it, adding fuel, and expelling the combustion products at high velocity to generate thrust after extracting work through turbine for running the compressor. These engines are common in military aircraft and some older commercial airliners. Turboprop engines combine the principles of a gas turbine (jet) engine with a propeller. They are commonly used in regional and commuter aircraft, providing a good balance between fuel efficiency and speed. Turbofan engines, often seen on modern commercial airliners, use a combination of the traditional jet exhaust and a large fan at the front to produce thrust. They are more efficient than turbojet engines and provide improved fuel economy. Ramjets are air-breathing engines that do not have moving parts. They rely on the forward speed of the aircraft to compress incoming air and mix it with fuel for combustion. These engines are suitable for high-speed applications like missiles. Scramjet, a variant of ramjet engines, is a type of air-breathing engine designed for hypersonic flight. They operate similarly to ramjets but are optimized for high supersonic speeds. Rocket engines are the primary means of propulsion for spacecraft. They can also work in the vacuum of space, where air is not available. These engines carry their own oxidizer and fuel and can operate in a vacuum, making them suitable for operation outside the Earth's atmosphere.

Each of these propulsion systems has its own unique characteristics, advantages, and limitations, making them suitable for different types of flight vehicles and missions. The choice of propulsion system depends on factors such as the flight vehicle's intended use, speed requirements, and operating environment. Aircraft propulsion systems can be categorized into two primary groups: propellers and jets. A propeller functions as a fan that generates thrust by converting rotational motion into forward force. In the early days of aviation, aircraft were powered by piston engines, but nowadays, these engines are primarily employed in lightweight aircraft due to their weight and reduced efficiency at high altitudes. Another propeller type of propulsion is the turboprop, which operates using a turbine engine to drive an aircraft propeller

through a reduction gear. Turboprop engines are particularly efficient within the low subsonic speed range. A jet engine is a propulsion system that generates thrust discharging a high-velocity jet. It encompasses various types, such as turbojets, turbofans and ramjets used in long-haul jet aerospace vehicles. Modern subsonic jet planes favor high-bypass turbofan engines due to their fuel efficiency.

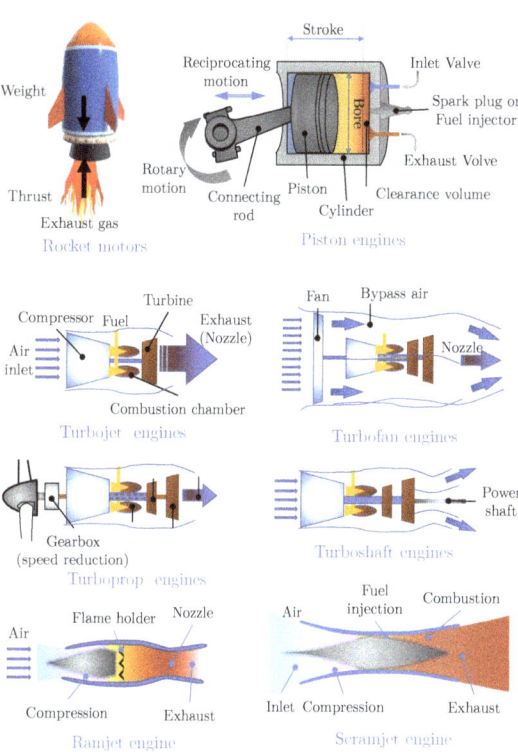

Different propulsion systems.

Aircraft can be classified based on the number of engines they possess. The number of engines plays a significant role in an aircraft's performance, safety, and mission capabilities. It affects factors like power, redundancy, fuel efficiency, and payload capacity.

Helicopters are typically powered by turboshaft or piston engines. Using a gear mechanism, the shaft power is used to drive the rotors. Traditional helicopters employ various mechanisms to counteract the torque force generated by the main rotor, which seeks to rotate the fuselage around the rotor axis. Helicopter rotors come in various configurations, each tailored to specific needs and engineer-

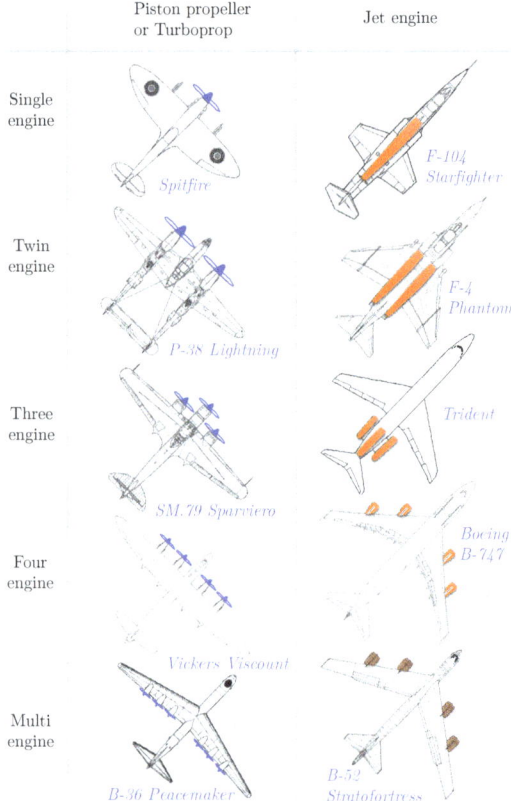

Aircraft classification based on number of engines. (Drawings NOT to scale). Note that *B-36* has 6 radial piston engines and 4 turbojet engines. *B-52* has 8 turbofan engines.

Some helicopters employ unconventional rotor configurations for specific purposes. The NOTAR (No Tail Rotor) system uses a hidden fan within the tail boom to produce a low-pressure area, which counters the main rotor's torque effect. This design reduces noise and maintenance associated with traditional tail rotors. Furthermore, there are compound helicopters, which incorporate auxiliary fixed wings and a rear-mounted propeller in addition to the main rotor. This combination enhances speed and efficiency, enabling the aircraft to achieve both helicopter and airplane-like performance. Each rotor configuration has its unique advantages and limitations, catering to diverse applications in military, civilian, and specialized roles within the aviation industry.

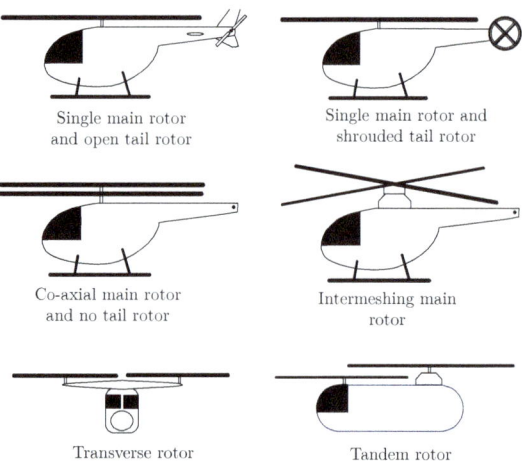

Different helicopter configurations.

ing considerations. The most common rotor configuration is the single main rotor with a tail rotor or anti-torque system. In this design, the primary rotor generates lift and thrust, while the tail rotor counteracts the torque produced by the main rotor's rotation, preventing the helicopter from spinning uncontrollably. Another configuration is the coaxial rotor system, which features two main rotors mounted one above the other, rotating in opposite directions. This design eliminates the need for a tail rotor, simplifying the aircraft's mechanical structure. Additionally, tandem rotors comprise two main rotors mounted in a tandem configuration, typically found in heavy-lift helicopters like the *CH-47 Chinook*. These configurations offer enhanced lift capacity and stability, making them suitable for transporting heavy loads and troops.

Spacecraft

A spacecraft is a specially designed vehicle for controlled flight beyond the Earth's atmosphere, capable of operating with or without a crew. These vehicles serve a wide range of purposes, including communication, Earth observation, meteorology, navigation, scientific exploration, and the transportation of humans and cargo. Notably, they function in the absence of an atmosphere or regions with extremely low atmospheric density, but they must first overcome Earth's gravitational pull to reach space.

The inaugural spacecraft, *Sputnik 1*, was launched by the Soviet Union on October 4, 1957. This historic achievement paved the way for the subsequent deployment of uninhabited Soviet and U.S. spacecraft. Within another four years, on April 12, 1961, the first manned spacecraft, *Vostok 1*, ventured into space, carrying the renowned Soviet cosmonaut *Yury Gagarin*. A few years later, on July 20, 1969, Commander *Neil Armstrong* achieved another significant milestone when he became the first person to set foot on the lunar surface during the *Apollo 11* spaceflight, which marked the first successful landing of humans on the Moon. Chapter 2 of the book is dedicated to the history of the space flight. Further, Chapter 8 of the book covers a wide array of topics related to space technology and explorations, engineering aspects of space vehicles, their types, trajectories through space, orbital parameters, and various maneuvers involved in spaceflight.

The large initial velocity required by spacecraft to carry out their missions is provided by a launch vehicle. This launch vehicle separates from the spacecraft once the mission is accomplished. Typically, the spacecraft is either placed into orbit around Earth or, if it attains sufficient velocity to escape Earth's gravitational pull, continues its journey into space. For guidance and orientation in space, spacecraft are often equipped with small rocket engines. For example, the *Lunar Module*, employed in the *Apollo* program for manned Moon landings, featured rocket engines that enabled a soft lunar touchdown and the safe return of its crew to the lunar-orbiting *Command Module*. The *Command Module*, in turn, possessed sufficient rocket power in its attached *Service Module* to depart lunar orbit for the journey back to Earth.

To operate the equipment carried onboard, spacecraft require an internal source of electrical power. Those intended for extended stays in Earth's orbit typically incorporate solar panels along with storage batteries. The space shuttle orbiter, designed for missions lasting one to two weeks, employs hydrogen-oxygen fuel cells. Deep-space probes, like the *Galileo* spacecraft, which entered Jupiter's orbit in 1995, and the *Cassini* spacecraft launched to Saturn in 1997, are typically powered by small, long-lasting radioisotope thermoelectric generators.

Presently, space hosts various principal categories of uncrewed/robotic spacecraft: flyby spacecraft, orbiter spacecraft, atmospheric spacecraft, lander spacecraft, penetrator spacecraft, rover spacecraft, observatory spacecraft, and communications spacecraft. Robotic spacecraft represent intricately engineered systems, purpose-built to operate in challenging and often harsh environments. Their degrees of complexity and functionality vary significantly, reflecting a wide range of missions and objectives. Spacecraft classification hinges on their designated functions, although alternative classification systems are conceivable, such as distinguishing between powered and unpowered craft or categorizing them as crewed or uncrewed, among other possibilities.

Artificial Satellites are man-made objects deliberately placed into orbit around Earth or other celestial bodies to maintain a long-term presence. They are often referred to as artificial satellites to distinguish them from natural satellites like the Moon. Artificial satellites carry a variety of equipment and subsystems designed to fulfill their designated missions, which can encompass tasks such as data transmission to Earth. The classification of artificial satellites can be based on their mission type (scientific, telecommunications, defense, etc.) or their orbit characteristics (equatorial, geostationary, etc.).

Space Probes are robotic spacecraft deployed for scientific space exploration missions. These probes depart from Earth and venture into space, where they may travel to the Moon, explore interplanetary space, conduct flybys or enter orbits around other celestial bodies, or even journey toward interstellar space. The primary purpose of space probes is scientific research and data collection.

Manned Spacecraft are space vehicles designed

to carry human crew members, typically at least one astronaut. These vehicles can include orbital stations like the *International Space Station (ISS)*. Manned missions focus on research and observational activities, often involving scientific experiments conducted in space.

Space Launchers are vehicles with the primary mission of placing other space vehicles, often satellites, into orbit. Generally, these launchers are not designed for recovery after their mission is complete, with the notable exception of the American space shuttles (*Columbia, Challenger, Discovery, Atlantis,* and *Endeavour*). The space shuttle program operated by NASA from 1981 to 2011 combined rocket launch capabilities, orbital spacecraft, and re-entry spaceplanes. These versatile vehicles were involved in launching satellites, interplanetary probes, conducting scientific experiments, and constructing and servicing the *ISS* during its missions.

Flyby spacecraft perform initial solar system reconnaissance, remaining in continuous solar orbit or escape trajectories without planetary capture. They use instruments to observe passing targets, ideally with panning optical instruments to compensate for target motion. Data is downlinked to Earth, and stored onboard during off-Earth point periods. These spacecraft endure prolonged interplanetary cruises, with stabilization achieved through thrusters, reaction wheels, or continuous spinning. Key examples of flyby spacecraft include *Voyager 1 and 2*, which explored the Jupiter, Saturn, Uranus, and Neptune systems. Other instances include *Mariner 2/4/5/6/7/10*, *Pioneers 10 and 11*, *Stardust*, and *New Horizons*.

Orbiter spacecraft: Orbiters designed for distant planets require substantial propulsion for precise deceleration during orbit insertion. They cope with solar occultations causing power loss and extreme thermal variations. Earth occultations disrupt communication. These spacecraft delve into in-depth planetary exploration. Notable orbiter examples are *Galileo, Messenger, Mariner 9* and *Cassini*. *Galileo*, for instance, successfully orbited Jupiter. Others include *Mars Global Surveyor, Mars Odyssey, and Ulysses*.

Atmospheric spacecraft, engineered for brief data-gathering missions focused on planetary atmospheres, often feature minimal subsystems. These spacecraft may lack propulsion and attitude control systems. They rely on an electric power supply, like batteries, and telecommunications equipment for tracking and data transmission. Scientific instruments on these probes directly measure factors such as composition, temperature, pressure, density, cloud content, and lightning within atmospheres of celestial bodies. Typically, atmospheric spacecraft are delivered to their destinations by other spacecraft. For instance, *Galileo* dispatched its atmospheric probe to Jupiter in 1995, using spin for stabilization during entry. Balloon-equipped atmospheric probes, like those in the Soviet *Vega* missions to Comet Halley, are designed to float in atmospheres. *Huygens*, part of the *Cassini* mission, is a prime example, of exploring Saturn's moon Titan. Examples include *Galileo's Atmospheric Probe, the Mars Balloon, the Vega Venus Balloon, and the Pioneer 13 Venus Multiprobe Mission*.

Lander spacecraft touch down on planets and endure to transmit data back to Earth. Notable examples include the *Soviet Venera landers* on Venus, *JPL's Viking landers* on Mars, and the *Surveyor* series on the Moon. The *Mars Pathfinder*, which landed on Mars in 1997, aimed to study the planet's atmosphere, interior, and soil, also deploying the rover *Sojourner*.

Penetrator spacecraft: Surface penetrators are designed to enter a celestial body's surface, enduring extreme impacts, measuring properties, and transmitting data. Typically, penetrator data is relayed to an orbiter for transmission to Earth. Only a limited number of penetrator missions have been attempted. The *Deep Space 2* penetrators, carried by the *Mars Polar Lander*, struck Mars in 1999 but did not transmit data. The *Deep Impact* spacecraft, launched in 2005, successfully impacted comet 9P/Tempel. Other examples include the *Ice Pick Mission* to Europa and the *Lunar-A Mission* to Earth's Moon.

Rover spacecraft are designed for planetary exploration, equipped with mobility systems for traversing the surface of celestial bodies. They often operate semi-autonomously to make decisions during their missions due to communication delays with Earth. Notable examples include *Sojourner*, *Mars Exploration Rovers*, *Mars Science Laboratory*, Russian *Lunokhod Rovers*, and JPL *Inflatable Rovers*.

Observatory spacecraft remain in Earth or solar orbits to observe distant targets free from atmospheric interference. NASA's Great Observatories program comprises four such spacecraft: the *Hubble Space Telescope (HST)*, *Chandra X-Ray Observatory (CXO)*, *Compton Gamma Ray Observatory (GRO)*, and the *Spitzer Space Telescope (SIRTF)*. While *HST* is still operational, *GRO* completed its mission and was deorbited in June 2000. *CXO*, launched in July 1999, continues to operate, and *SIRTF*, launched in January 2003, is currently operational. The recently launched *James Webb Space Telescope*, will further explore the advantages of space-based observations.

Communication and navigation spacecraft are prevalent in Earth orbit. The *Deep Space Network*'s *Ground Communications Facility* does utilize Earth-orbiting communication satellites to relay data across its sites worldwide. They also rely on Earth-orbiting *Global Positioning System (GPS)* satellites for time synchronization. A prime example is NASA's *Tracking and Data Relay Satellite System (TDRSS)*, which supports missions including the *Hubble Telescope*, the *Space Shuttle*, and the *International Space Station*. Other instances include *Milstar, GPS, DirecTV, Globalstar*, and more.

The configuration of space vehicles varies greatly based on their specific mission objectives and requirements. However, it's worth noting that space launchers often share similarities with missiles in their design and purpose, as they are primarily used to propel other payloads into space.

Missiles

Missiles are a special class of aerospace vehicles designed for various military and strategic purposes. A missile is a sophisticated, unmanned, self-propelled, and guided weapon system. Chapter 7 of the book is dedicated to presenting the details of the missile ranging from historical evolution to operating principles and the different types. These weapons can be categorized based on different criteria. Regarding their trajectory, missiles typically fall into one of three categories: cruise, ballistic, or semi-ballistic. A ballistic missile, for instance, follows a sub-orbital ballistic trajectory, aimed at a predetermined target. Its guidance system operates primarily during the initial powered phase of flight, after which it relies on the principles of orbital mechanics and ballistics to follow its course. In terms of their intended targets, missiles can be specialized for a range of purposes, such as anti-submarine, anti-aircraft, anti-missile, anti-tank, anti-radar, and more. These designations are based on the specific threats they are designed to counter.

From a military perspective, missiles can be classified as either strategic or tactical, depending on their intended use and range. However, a more common and widely accepted classification is based on their launch platform and target. Here are some key categories:

Air-to-Air Missiles: Launched from aircraft to engage and neutralize airborne threats.

Surface-to-Air Missiles: Designed to defend against enemy aircraft or incoming missiles, ensuring airspace security.

Air-to-Surface Missiles: Typically launched from aircraft, they are employed to strike ground or surface targets.

Surface-to-Surface Missiles: These missiles provide support for ground forces in surface-based operations.

Aerospace vehicles include both aircraft and spacecraft operate at a wide range of speeds and altitudes, reflecting their diverse missions and design specifications. The speed and altitude of aerospace

Table 1: Various aerospace vehicles, purpose and operating altitude and speed; UAVs-Uninhabited Aerial Vehicles

Aero vehicle	Altitude, km	Speed, km/h
Paper planes	0.01	< 10
Kites, Gliders	1	< 300
Hang-gliders	< 5	< 150
Micro-lights, home-builts	< 5	< 200
UAVs	< 5	< 100
Civil transport	5 to 12	900
Military supersonic	5 to 20	3000
Gyrocopter, helicopter	< 5	< 300
Tactical missiles	5 to 20	3000
Strategic missiles	< 500	20000
Launch vehicles	Space	20000
Spacecraft	Space	100000

vehicles vary significantly based on their purpose, whether it's carrying passengers, conducting scientific research, or exploring the cosmos.

The design of subsonic, transonic, supersonic, and hypersonic vehicles varies significantly due to differences in their operating speeds, aerodynamic challenges, and materials used. The design is merely not based on speed but also on other factors like application (commercial transport or military operation). Subsonic vehicles typically have straight or moderately swept wings and relatively simple, streamlined shapes. They prioritize lift efficiency and low drag at lower speeds. The design emphasizes optimizing lift-to-drag ratios to maximize fuel efficiency and passenger comfort.

Transonic vehicles often incorporate more pronounced wing sweep, aiming to delay the onset of shock waves and reduce drag. The design focuses on managing the onset of shockwaves and improving control surface effectiveness within the transonic speed range. Commercial airliners like the *Boeing 737* feature a tubular fuselage with straight or slightly swept wings and a focus on passenger comfort.

Subsonic vehicle, piston propeller aircraft like *MD-311 Flamant* operates at Mach number < 1

Transonic vehicle, civilian airliners like *A321 XLR* with maximum speed Mach number 0.82

Supersonic vehicle, *Dassault Rafale* operates at maximum Mach number 1.8

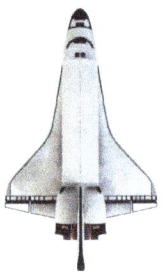
Hypersonic vehicles, *Space shuttle* designed for atmospheric entry at Mach number 25

Hypersonic vehicles, *Apollo space vehicle* designed to return humans from the moon entered the Earth's atmosphere at Mach number 36.

Aerospace vehicles operating at different speed regimes.

Supersonic vehicles adopt more advanced aerodynamic designs with delta or ogival wings to manage shockwaves and reduce drag. They may feature a sharp, needle-like nose and slender fuselage, very thin wings and tail surfaces with very sharp leading edges and a low aspect ratio wing. The design emphasis is on mitigating the effects of supersonic flight, including aerodynamic heating and drag associated with shockwaves. The *Concorde*, a supersonic passenger jet, featured an ogival delta wing design to address supersonic flight characteristics effectively. The *F-104 Starfighter* has a slender, pointed fuselage and swept wings to address flight challenges while maintaining high-speed performance.

Hypersonic vehicles have blunt leading edges and advanced thermal protection systems. They may feature a swept delta wing with a rounded leading edge, and a thick fuselage with a rounded nose. The design primarily focuses on managing the intense aerodynamic heating, structural challenges, and material limitations associated with hypersonic flight. The *Space Shuttle*, during reentry, exemplified hypersonic design, featuring a unique heat shield, ablative tiles, and a highly specialized aerodynamic shape to endure the harsh conditions. The *Apollo* space vehicle designed to return humans from the moon, and to enter the Earth's atmosphere at the extremely high speed of Mach number 36 features a blunt body with no wings.

As there are multiple parameters, characteristics and operating conditions associated with aerospace vehicles, classifying them into various types based on each of the attributes may not be sensible. In this book, especially in part III, we follow the approach of classification based on the utility of the vehicle. This helps in a better understanding of the design and engineering aspects of each class of vehicles. So, aerospace vehicles are classified as civilian airplanes, military airplanes, missiles, space vehicles, helicopters and other miscellaneous types. It is to be noted that this classification is not strictly non-overlapping. Some trivial examples of overlapping usage are: (i) the same aircraft can be used as a civilian airplane or a military airplane, (ii) a space shuttle which can be considered as a spacecraft operates like an aircraft during its return flight to Earth, (iii) a rocket can be customized as a satellite launch vehicle or a missile, and (iv) during a kamikaze attack using an aircraft, it can be considered as a missile. Scrutiny will reveal many notable exceptions like these. So, this classification needs to be taken in a broader sense and is consistent with the vocabulary of the general public.

In the next section, we will introduce some of the terminologies that are used in the aerospace domain. Getting familiar with the vocabulary will facilitate better understanding while reading through the subsequent chapters.

Basic Terminologies

Aeronautics deals with designing and building aircraft that operate in Earth's atmosphere.

Aerospace Engineering is the branch of engineering concerned with the research, design, development, construction, testing, science and technology of aerospace vehicles and allied systems.

Aircraft: As per *The Aircraft Act* of 1934 in India, "aircraft" means any machine which can derive support in the atmosphere from reactions of the air. Typically, balloons, helicopters, gliders and airplanes are considered as aircraft. And, rockets, missiles and satellite launch vehicles which may have their trajectory partially or fully in the atmosphere are excluded from the definition of aircraft.

Airplane (also called **Aeroplane** in British English) is a powered heavier-than-air aircraft, where the lift is generated by the dynamic interaction of air with fixed wings.

Airships (dirigible balloon) are a type of aerostat (lighter-than-air aircraft) equipped with a power plant for propulsion and mechanisms for steering the craft. They achieve buoyancy by enclosing a volume of gas that is less dense than the surrounding air. Airships are classified according to their method of construction into rigid (rigid framework covered by an outer skin or envelope), semi-rigid (has some supporting structure but the main envelope is held in shape by the internal pressure of the lifting gas) and non-rigid types (often called as blimps rely entirely on internal gas pressure to retain its shape during flight).

Astronautics is the science of the construction and operation of spacecraft.

Aviation and Aeronautical Engineering are closely related fields and they are often used interchangeably. It is the design, development, operation, and use of aircraft.

Avionics (aviation electronics) deals with the electronics side of aerospace engineering. Avionics encompass a range of functions on aircraft, such as communication, navigation, and the display and management of multiple systems, each serving specific functions.

Balloon A non-power-driven lighter-than-air aircraft remains aloft due to its buoyancy. A hot air balloon is a balloon that derives its lift from heated air contained within the envelope. A balloon can either drift freely with the wind (free balloons) or be anchored to a stationary point (kite balloons).

Glider: A non-power-driven heavier-than-air aircraft, deriving its lift in flight chiefly from aerodynamic reactions on surfaces that remain fixed under given conditions of flight. Occasionally, an engine can be attached to the glider which then acts like a powered, heavier-than-air aircraft and can have the capability for self-launch or can augment the gliding flight.

Gyroplane (autogyro, gyrocopter or rotaplane) employs an unpowered rotor in the horizontal plane to generate aerodynamic lift through autorotation and conventional engine-driven propeller, similar to that found on fixed-wing aircraft, to supply thrust.

Helicopter is a rotorcraft powered by engine-driven rotors that enable it to achieve vertical take-off and landing, hover in place, and navigate in forward, backward, or lateral directions during flight.

Kite balloon is any non-power-driven structure that is anchored to the Earth, and which is balanced by both buoyancy and aerodynamic lift.

Missile is an unmanned, self-propelled guided weapon system that typically uses rocket propulsion.

Ornithopter: A heavier-than-air aircraft that flies due to the lift generated by flapping wings.

Rotorcraft/rotary-wing aircraft is a heavier-than-air aircraft with rotary wings, which generate lift by rotating around a vertical mast. Rotorcraft include aircraft where one or more rotors provide lift throughout the entire flight, such as *helicopters* and *autogyros*. An aircraft that uses rotor lift for vertical flight but changes to solely fixed-wing lift in horizontal flight is not a rotorcraft but a *convertiplane*.

Spacecraft: Vehicles operating outside the Earth's atmosphere are referred to as spacecraft. Satellites, satellite launch vehicles, space shuttles, space stations and vehicles used for travel to other celestial bodies are considered as spacecraft.

Bibliography

M. S. Arnedo (2017) *Fundamentals of Aerospace Engineering*, http://www.aerospaceengineering.es/. Accessed October 31, 2023.

H. S. Mukunda (2017) *Understanding air and space vehicles* Jain University Press

Global aerospace and defense, *Annual industry performance and outlook* 2023 edition, https://www.pwc.com Accessed October 31, 2023.

Part I

A Brief History of Aerospace Vehicles

Chapter 1

History of Aviation

Contents

1.1		Overview	26
1.2		The Beginning	26
1.3		Aviation during World War I	38
	1.3.1	As War Machines	38
	1.3.2	Flying Aces	43
	1.3.3	Technology Innovations	47
1.4		Between World War I and World War II	48
	1.4.1	Maturing of Aviation	48
	1.4.2	Passenger Airlines	58
	1.4.3	Flying Boats	62
1.5		As a Game Changer during World War II	63
	1.5.1	Role of Aircraft in the War	63
	1.5.2	Notable Aircraft during the War	65
	1.5.3	Flying Aces	71
	1.5.4	Technological Evolution	71
1.6		As a Mainstream Technology after World War II	74
	1.6.1	Robust, Reliable and Sophisticated Technology	74
	1.6.2	Modern Jet Airliners	85
	1.6.3	Military Aviation	93
1.7		History of Helicopters	96
	1.7.1	Early Concepts and Precursors	96
	1.7.2	Post-War Innovation and Commercialization	100
	1.7.3	Modern Era and Beyond	100
1.8		USA's Experimental Aerospace Programs	104
1.9		Aviation Accidents	114
1.10		Future of Aviation	117

1.1 Overview

Compared to the history of other modes of transportation, the history of flight is relatively short. Though the history is short, it is dotted with many significant events highlighting the progress of human beings in the pursuit of a more efficient mode of transportation. 20th century saw tremendous progress being made towards flight through Earth's atmosphere and beyond. Fueled by the two World Wars, aviation was established as a medium of long-distance transport during the first half of the 20th century. After World War II, the Jet Age began, which saw extensive use of jet engines both for civilian and military aviation. Here, the progress in aviation is sketched by highlighting the individual contributions of some of the famous personalities in the field.

The chapter follows a logical sequence, tracing aviation's evolution and impact over history. *The Beginning* section uncovers aviation's origins, key pioneers and their contribution to aviation. *Aviation during World War I* explores the role of aircraft in the war, emphasizing technological advancements in the subsection *Technology Innovations*. The section *Between World War I and World War II* details design breakthroughs and pivotal moments during that period. It also covers various aspects, including pioneering moments, passenger airlines, and the role of aviation in civil and military contexts. Similarly, *As a Game Changer during World War II* underscores aviation's evolution and its significance during the war. Further, the section *As a Mainstream Technology after World War II* highlights aviation's robust transformation, spanning military, civil, and cargo planes.

In order to understand the significant research and development effort in the aerospace domain after World War II, the chapter examines *USA's Experimental Planes*. The aerospace programs of the USA are chosen because they have made significant progress in aerospace technology and the information is available in open literature. *Major Aviation Accidents* chronicles some of the significant accidents involving civilian airplanes and the corrections or modifications made in the aircraft design or aviation operations as a result of the lessons learned from those accidents. In conclusion with *Future of Aviation*, the chapter anticipates some of the innovations in technology that will have a significant influence on the design and operation of future aircraft. The well-structured organization provides a thorough understanding of aviation's transformative journey, covering historical and contemporary facets.

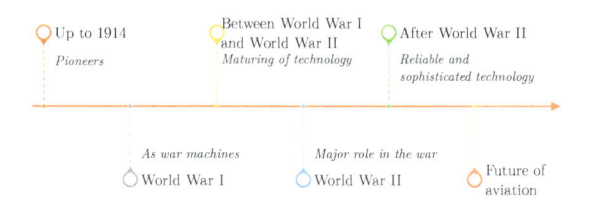

Timeline of the history of aviation.

1.2 The Beginning

Before the history-making flight by the *Wright* brothers on December 17, 1903, at Kitty Hawk in North Carolina, USA, there was a lot of scepticism about whether it will be possible to make a controlled flight using an aircraft that is both powered and heavier than air. Even some of the prominent scientists and policymakers of the time considered the effort to fly as futile and they strongly believed that it would not be possible for human beings to fly. Despite overwhelming scepticism and sometimes outright ridicule even in the newspapers, there were these visionaries who were convinced that it would be possible for human beings to fly. Most of these pioneers worked out of their own conviction and interest, which eventually culminated in the successful flight of *Wright* brothers. They made great personal sacrifices and some of them even met with fatal accidents in their pursuit. We will look at some of these great personalities who laid the foundation for the marvelous field of aviation.

1.2. THE BEGINNING

Leonardo Da Vinci (1452 - 1519), the genius who lived in Italy, drew sketches in his notebooks resembling modern-day flying machines, helicopters, and gliders. He had more than 100 drawings that illustrated his interest in flight. One such sketch is shown here. From these sketches, it seems that he conceptualized a flapping/moving wing for human flight.

Self-portrait of *Leonardo da Vinci*. His notebooks contain sketches and designs of flying machines, leading to his recognition as one of the early pioneers of aviation.

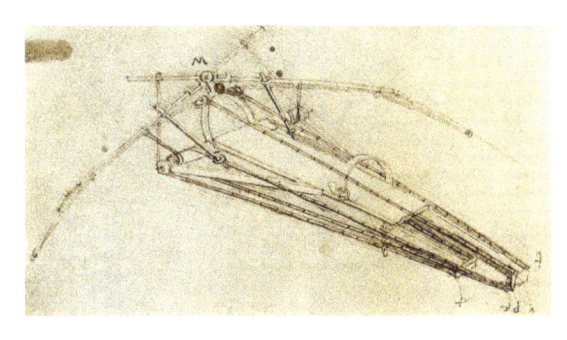

Da Vinci's sketch of an ornithopter, a device meant for achieving flight by flapping of wings like birds.

Joseph-Michel Montgolfier (1740 - 1810) and *Jacques-Étienne Montgolfier* (1745 - 1799), **(Montgolfier brothers)** were French pioneers in developing the hot air balloon and conducted the first untethered flights. They discovered that hot air, when collected inside a large lightweight paper or fabric bag, caused the bag to rise into the air. They made the first public demonstration of this discovery by filling a balloon of a diameter of more than 10 meters with heated air, which rose into the air about 1,000 meters, remaining there for some 10 minutes traveling about two kilometers. Later in 1783, the first manned (two passengers) untethered flight took place over Paris in a *Montgolfier* balloon that sailed to a height of around 900 meters and flew 9 kilometers in about 25 minutes. Balloon flight works on the principle of buoyancy, and to generate the buoyant force, the balloon has to be filled with a lighter gas.

Joseph-Michel Montgolfier (1740 - 1810)
Jacques-Étienne Montgolfier (1745- 1799)

First public demonstration of hot air balloon

A drawing showing the first free ascent of a hot-air balloon over Paris with human passengers on November 21, 1783.

Sir George Cayley (1773 - 1857) was an English engineer, inventor, and aviator. He is called *the inventor of flight* or *the father of aeronautics* as his meticulous observations and findings were to guide later pioneers in aviation. As early as 1799, he

Sir George Cayley (1773 - 1857), pioneered the principles of aerodynamics, including lift and cambered wings, and laid the foundation for modern aeronautics. He is credited with designing the first successful gliders that could carry human passengers.

had grasped the basic issues of heavier-than-air flight, that lift should balance weight and thrust must overcome drag, which should be minimized. He deduced that machines would fly better with cambered wings and began actual gliding experiments in 1808 and had his first manned glider flight in 1853. His flying machine had provisions for pilot-operated three-axis controls and undercarriage wheels. His legacy remains foundational in aviation's progress, inspiring generations of innovators. He published his findings in the *Mechanics'*

Magazine, a prominent scientific journal at that time.

Model of *Sir George Cayley's Aerial Carriage* kept in San Diego Air and Space Museum, San Diego, USA.

Depiction of *Chanute's Katydid* glider on the shores of Lake Michigan at Miller Beach, Indiana. Weighing approximately 15 kg, the frame of *Katydid* glider was made of spruce and the surfaces were made of Japanese silk varnished with pyroxelene.

Octave Chanute (1832 – 1910) was an aviation historian and aeronautical pioneer who designed gliders successfully. One of his favorite gliders was the *Katydid*, which featured multiple wings that could be moved about on the fuselage to facilitate experimentation. A second glider was the

Octave Chanute (1832 – 1910), a French-born American civil engineer and aviation enthusiast who advocated the glider experiments and the biplane configuration. His work greatly influenced the aviation pioneers of his time, including the *Wright Brothers*.

Chanute/Herring biplane that employed Pratt trussing to achieve its considerable strength. He maintained correspondence with most aviation pioneers of the time. He acted as a tireless publicist, collecting information on aeronautics from all corners of the globe and sending out information to all seekers. *Chanute* published a compendium of early aviation experiments in his book titled *Progress in Flying Machines*.

Samuel Pierpont Langley (1834 - 1906) was an American aviation pioneer, astronomer, and physicist who conducted extensive research on aerodynamics and conducted experiments with various flying models. In the realm of aviation, *Langley*'s contributions were pioneering. On May 6,

Samuel Pierpont Langley (1834 - 1906) was the third Secretary of the Smithsonian Institution and a Professor at the University of Pittsburg, USA. His experiments with larger-scale Aerodrome aircraft marked crucial steps in advancing the understanding of powered flight.

1896, the successful flight of an unpiloted, engine-driven, heavier-than-air craft of substantial size by his steam-powered model, the *Aerodrome No. 5*, demonstrated the feasibility of sustained flight. This achievement garnered significant attention and support from the U.S. War Department and private donors, allowing him to pursue the design of larger-scale aircraft. Upon achieving a flight of 1,500 meters on his new *Aerodrome No. 6*, Langley secured a War Department grant of $ 50,000 and an additional $ 20,000 from the Smithsonian for the development of a piloted aircraft. He meticulously devised novel launch techniques, employing a catapult launch, and engineered a robust houseboat scale to accommodate the aircraft's weight. Following an exploration of European engines, his

Langley Aerodrome Number 5 (1896), unpiloted, tandem-wing double monoplane.

Mass	11.3 kg
Dimensions	1.25 m x 4.78 m x 4.17 m
Engine	One-horsepower, single-cylinder steam engine
Drive	Two pusher propellers
Material	Wood, silk, steel tubing, steel wires, copper alloys, cord
Flight	Two flights on May 6, 1896; one of 1,005 m and a second of 700 m at a speed of approximately 40 km/h.

assistant *Charles Manly* conceived a 52 HP radial-cylinder internal combustion engine for the ambitious *Great Aerodrome*. On October 7, 1903, the aerodrome's inaugural launch culminated in a river crash due to structural front wing failure, the result of excessive stress induced by the catapult launch. A subsequent launch on December 9, 1903, after meticulous refurbishments, witnessed the catastrophic collapse of the rear wing and tail during launch. These repeated setbacks led *Langley* to abandon his flight endeavors.

Just a few days after *Langley*'s failed attempt, the *Wright* brothers flew successfully a sustained, heavier-than-air, powered airplane on December 17, 1903, in Kitty Hawk, North Carolina. Although some of his attempts were unsuccessful, his work laid the groundwork for subsequent aviation advancements, and his legacy endures as a trailblazer in the journey toward controlled flight.

Otto Lilienthal (1848 - 1896) was a German aviation pioneer who explored the physical principles governing winged flight. *Lilienthal* designed and built several gliders with which he demonstrated the concept of heavier-than-air flight. After his series of well-publicized experiments, engineers were able to build on his findings and research methods on a course toward developing the world's first heavier-than-air, powered and controlled aircraft. He documented his studies on the aviation of birds in a book titled *Der Vogelflug als Grundlage der Fliegekunst (Birdflight as the Basis of Aviation)*. *Lilienthal* designed and built many gliders based on careful observations of birds, and he flew some 2,000 flights in at least 16 distinct glider types. His work and flights in gliders laid the groundwork for future aviation advancements and influenced other aviation pioneers, including the *Wright brothers*. Tragically, *Lilienthal*'s life was cut short when he sustained fatal injuries during a glider flight in 1896. His dedication to flight experimentation and his innovative spirit left a lasting impact on the field of aviation.

Lilienthal preparing for an Ornithopter flight.

John Joseph Montgomery (1858 – 1911), an American inventor and Professor at Santa Clara University in California, began his experimental flights 20 years before the *Wright Brothers* and is best known for his invention of controlled heavier-than-air flying machines. He is reported to have made and tested his first full-scale glider (span slightly more than 6 m) in 1883 and to have manned it on runs down a long slope in California. During those runs, he was lifted from the ground. He continued to design and test gliders, using curved wings, a stabilizing tailplane, and efforts at control by manipulating wires that extended laterally to brace the wings. On October 31, 1911, while piloting his new

Montgomery and his glider named *Evergreen* which is a monoplane glider with a conventional tail and the pilot seated below the wing. Between October 17-31, 1911, *Montgomery* made more than 50 glides with *Evergreen*.

glider named *Evergreen* at an altitude of under 6 meters, *Montgomery* encountered a stall, followed by a sideslip. The glider then descended, striking the ground on its right wingtip and overturning. Tragically, *Montgomery* sustained a fatal head injury after hitting an exposed bolt and died two hours later.

Wilbur Wright and **Orville Wright**, the American **Wright brothers** and aviation pioneers successfully flew the first powered, sustained, heavier than air and controlled airplane on December 17, 1903. *Wright* brothers owned a bicycle business and the profits from the business were used to conduct aeronautical experiments. The death of the German glider pioneer *Otto Lilienthal* in a glider crash in 1896 was the starting point when they took a serious interest in human flight.

In 1899, the *Wright* brothers wrote to the Smithsonian Institution for information about aviation and introduced themselves to *Octave Chanute* for guidance in their pursuit in aviation. The brothers realized that a successful airplane would require wings to generate lift, a propulsion system to move it through the air, and a system to control the aircraft

Wilbur Wright (1867 - 1912) and *Orville Wright* (1871 - 1948).

in flight. The first *Wright glider*, a biplane featuring 15 square meters of wing area and a forward elevator for pitch control, was tested in October 1900 but did not perform as expected. They believed that the reason for the failure was errors in the experimental data published by their predecessors. Then *Wright brothers* constructed a small wind tunnel in 1901 and tested between 100 and 200 wing designs to gather information on the relative efficiencies of various airfoils and determine the effect of different wing shapes, tip designs, and gap sizes between the two wings of a biplane in an airstream before they constructed the powered aircraft.

With the help of the experimental data from the wind tunnel, the *Wright brothers* improved the aerodynamic and flight control problem to build a successfully powered aircraft. Between 1903 and 1905, the *Wright brothers* made 124 flights, the longest lasting over 38 minutes. In 1904 and 1905 they refined the *Wright Flyer* to increase controllability and stability and by 1906 had a fairly workable and reliable flying machine was available.

Samuel Franklin Cowdery (1867 - 1913) was an American Wild West showman who later became interested in aviation and made significant contributions to aviation in Great Britain. *Cowdery* moved to England as part of a Wild West show. The large kites developed by *Cowdery* were used for observation purposes by the British military and mete-

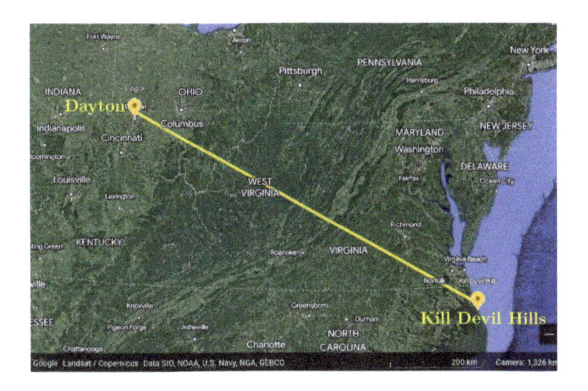

Distance from *Wright brothers*' workshop in Dayton, Ohio, to the flight testing site in Kill Devil Hills of Kitty Hawk, North Carolina is around 1100 km.

In 1900, the brothers consulted with the US Weather Bureau and fellow aviation experimenters to identify suitable locations based on topography, vegetation and wind conditions. They ultimately chose the Kill Devil Hills of Kitty Hawk, North Carolina. This landscape featured expansive shifting sand, prevailing northeasterly winds, large sand dunes, and minimal vegetation. Transporting their aircraft was a significant challenge. Disassembling and packaging the components, they securely packed the flyers into crates. These crates were transported via rail and ship from their Dayton, Ohio, workshop to Kitty Hawk. Amid flight experiments, the *Wright brothers* faced numerous challenges and setbacks. Each test, whether successful or not, yielded valuable insights to refine their concepts. They conducted multiple flight tests with gliders and early flyers before achieving their historic success on December 17, 1903. Their dedication and innovation marked their relentless pursuit of powered flight.

orological department. His flight in the October of 1908 is credited as the first piloted, heavier than air flight in Great Britain. He created various records for range, and endurance using his aircraft. *Cowdery* was killed in an accident on August 7, 1913, when his aircraft came apart due to inherent structural weakness.

Louis Blériot (1872 - 1936), French airplane manufacturer and aviator, made the first airplane flight across the English Channel in 1909. In 1907, both *Louis Blériot* and *Robert Esnault-Pelterie* achieved short flights in tractor monoplanes (powered from the front), a configuration that would soon play a crucial role in the evolution of flight. Between July 1909 and the beginning of World War I, the *Blériot* factory produced more than 800 aircraft, many of them *Type XI* monoplanes or variations of that design. In 1913, a consortium led by *Blériot* acquired the *SPAD* aviation manufacturing company that went on to produce the famous *SPAD S.XIII* in large numbers during World War I.

Alberto Santos-Dumont (1873 - 1932, Brazilian) was an aviation pioneer who captured the imagination of Brazil, Europe and the United States with his balloon, and airship flights and made the first significant flight of a powered airplane in France with his aircraft *No. 14-bis*. A total of 20 dirigibles emerged from *Santos-Dumont*'s workshop between 1898 - 1909. He won the 100,000-franc Deutsch Prize for an 11.3 km flight with his airship No. 6 from the Paris suburb of St. Cloud to the Eiffel Tower and back in less than half an hour. Inspired by stories of what the *Wright brothers* had accomplished in the United States, *Santos-Dumont* designed and flew a series of heavier-than-air flying machines. His major accomplishments include making the first public flight in Europe with a powered, winged aircraft in 1906. He unsuccessfully championed against the use of aircraft for military purposes.

Henri Farman (1874 – 1958) was a French aviation pioneer and aircraft builder. He acquired flight skills using a *Voisin* aircraft and later promptly procured a customized version. He refined the design through a series of instinctive adjustments rather than scientific methods. These initial steps marked the start of a prolific career during which *Farman* adeptly identified and resolved a wide range of aircraft control and structural challenges. *Farman*'s successes as a pilot including a cross-country flight soon made him one of the most famous men in France. *Farman* made aviation history when he won the *Grand Prix d'Aviation* by completing the first 1-km circular flight in Europe on January 13, 1908. He set up an aircraft factory, enjoying immediate success with a box-kite biplane. His brother *Maurice* became a partner in the enterprise and, in 1912, the *Farmans* became France's largest aircraft manufacturer, producing over 12,000 military aircraft dur-

Wright Flyer was the product of an intricate four-year research and development conducted by *Wilbur* and *Orville Wright* beginning from 1899.

Dimensions
- Length 6.4 m
- Wingspan 12.3 m
- Height 2.8 m

Empty mass 274 kg

Loaded mass 341 kg

Material
- Airframe Wood
- Fabric cover Muslin
- Engine Aluminum

Design: Canard biplane with non-wheeled, linear skids as a landing gear. It had a wooden frame covered in muslin fabric.

Powerplant: One 12-horsepower *Wright* horizontal four-cylinder engine driving two pusher propellers via sprocket-and-chain transmission system

Flight: The first flight, piloted by Orville Wright, lasted 12 seconds, covering a distance of 36 m on the sandy dunes of Kitty Hawk on December 17, 1903, traveling 36 m, with *Orville* piloting. The fourth and longest flight of the day covered 255.6 m in 59 seconds at 9.1 m attitude with *Wilbur* at the controls.

Control: Wing warping, a method of flexing the wings, to achieve lateral control and maintain balance during flight.

Impact on Aviation

The 1903 *Wright Flyer*'s success marked a pivotal moment in aviation history and laid the foundation for the rapid advances in aviation technology and industry in the 20th century.

Smithsonian Controversy

The *Wright brothers* had a contentious relationship with the *Smithsonian Institution* over the credit for their achievements. It took years of dispute before the *Smithsonian* officially recognized the *Wright brothers* as the first to achieve a sustained, powered, heavier than air flight.

Preservation

The original *Wright Flyer* is preserved at the *Smithsonian National Air and Space Museum* in Washington, D.C., and is considered one of the most historically significant artifacts in aviation history.

The *Wright brothers* displayed remarkable dedication, ingenuity, and determination in their pursuit of powered flight. Their efforts to transport the delicate flyers weighing 340 kg to Kitty Hawk from Dayton, Ohio (around 1100 km) for flight tests, their unwavering commitment to continue despite challenges, crashes, and setbacks and their deep motivation to achieve their dream laid the foundation for their groundbreaking success on December 17, 1903 which was followed by years of refining the design of their aircraft and conducting numerous tests.

1.2. THE BEGINNING

Britain's first powered airship of *British Army Dirigible No 1, Nulli Secundus*, flew from Farnborough to London in 3 hours 25 minutes on October 5, 1907, with *Cowdery* and his commanding officer Colonel *J. E. Capper* on board.

On July 25, 1909, *Louis Blériot* became the first person to cross the English Channel in a heavier-than-air aircraft. His feat drew great popularity for him and his aircraft, but also coined the phrase *Great Britain is no longer an Island*. The *Blériot XI* was a versatile machine for its day and served as a training aircraft at flying schools all over the world.

ing World War I. The company was taken over by the state in 1936.

Harriet Quimby (1875 - 1912) was an American aviator and writer, widely recognized as the first licensed female pilot in the United States (1911) and the first woman to cross the English Channel. Her feat of crossing the Channel on April 16, 1912, guiding her French *Blériot* monoplane from Dover city in England to Calais in France went largely unnoticed as the Titanic sank a day before the same day. The feat had been first accomplished by *Louis Blériot* in 1909 and subsequently by other pilots, but never by a woman. *Quimby* died in an accident involving her airplane on July 1, 1912, at the Harvard-Boston Air Meet.

Alberto Santos-Dumont (1873 - 1932), a Brazilian aviator, built and flew practical airships and airplanes, demonstrating the feasibility of powered flight.

(a) *Dumont's* first balloon, *Brésil*, in 1898. 6 m in diameter and a mass of 27.5 kg, Brésil has a 113 m^3 hydrogen chamber

(b) *Dumont* going around the Eiffel Tower in *Dirigible balloon* on October 19, 1901.

Henri Farman (1874 – 1958), popularized the use of ailerons, movable surfaces on the trailing edge of a wing solving intricate flight control challenges. He played a pivotal role in advancing the field of aviation.

Gaetano Arturo Crocco (1877 – 1968) was an Italian aerospace engineer and visionary known for his innovative contributions to astronautics. He proposed the concept of a spaceplane in the 1930s, envisioning reusable vehicles capable of reaching space and returning to Earth. *Crocco*'s concepts laid the groundwork for future advancements in space exploration, and his legacy continues to inspire modern aerospace engineering.

Glenn Hammond Curtiss (1878 – 1930) was an

Harriet Quimby (1875 – 1912), her daring spirit and contributions shattered gender norms, inspiring future generations of women to pursue careers in aviation and aviation-related fields.

On October 3, 1908, the inaugural flight of the Italian military airship, *N1*, designed by *Crocco* and his team, took place. Subsequently, the *N1* glided over Rome, sparking excitement among the public and media.

American aviation and motorcycling pioneer. *Curtiss* was responsible for creating innovative aircraft designs, including seaplanes and flying boats. His contributions to naval aviation at time of the United States' entry into World War I were significant, as he developed aircraft capable of taking off and landing on water, which greatly advanced the field of naval aviation. Additionally, *Curtiss*'s innovations in engine design and manufacturing techniques played a crucial role in the development of aircraft engines. His legacy remains as a key figure in the history of aviation and aeronautical engineering.

Arthur Charles Hubert Latham (1883 – 1912) was a French aviator known for his pioneering efforts in early aviation. He made numerous daring attempts to achieve significant aviation milestones, including the first attempt to fly across the English Channel

Glenn Hammond Curtiss (1878 – 1930) was an American aviation and motorcycling pioneer. Notable among his creations are the *Curtiss JN-4 "Jenny"* a versatile biplane, the *Curtiss Model E*, a seaplane and the *Curtiss OX-5 engine*.

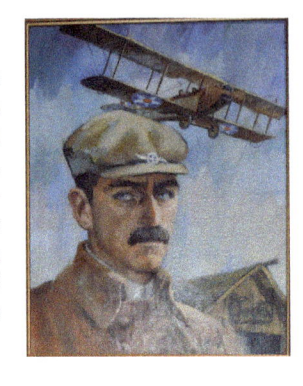

and the first to reach high altitudes. On the 19th and 27th of July, 1909, he attempted to cross the English Channel on *Antoinette IV* monoplane, though he managed to fly a significant distance both times, engine failure forced him to land in the Channel. Despite falling short of some goals, *Latham*'s determination and courage exemplified the spirit of aviation exploration during his era, leaving a lasting legacy that inspired future aviators. *Latham* climbed to an altitude of 1,100 m more than 610 m higher than the previous world record in 1910 (Mourmelon-le-Grand, France). Later in the same year, during the second *Semaine de l'Aviation de la Champagne* at Reims, *Latham* again set a world altitude record of 1,384 m. Further, he set the official World Airspeed Record of 77.5 km/h in his *Antoinette VII*.

Latham over English Channel on second attempt in his *Antoinette IV* monoplane. *Latham* was the first person to attempt to cross the English Channel in an aeroplane.

Eugene Burton Ely (1886 – 1911) was an American aviation pioneer recognized for his historic

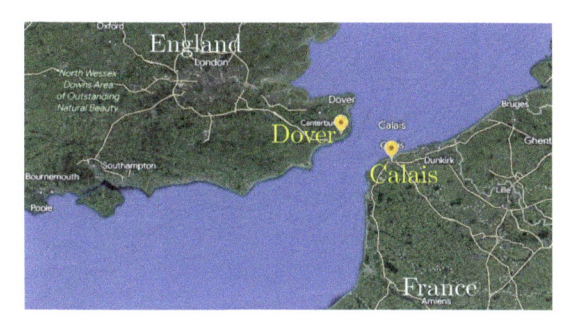

English Channel is a narrow arm of the Atlantic Ocean that separates Southern England from northern France. There are notable flights across the English Channel including first crossing on balloon in 1785 by *Jean Pierre François Blanchard* (France) and *John Jeffries* (US), first crossing on gas balloon in 1906 by *Frank P. Lahm* in the Gordon Bennett gas balloon race, first crossing on heavier-than-air aircraft (*Blériot XI*) in 1909 by *Louis Blériot* (France), first double crossing on *Wright biplane* in 1910 by *Charles Stewart Rolls* (UK), first crossing by a woman on *Blériot XI* in 1912 by (*Harriet Quimby* (US), first crossing on autogyro (*Cierva C.8*) in 1928 by *Juan de la Cierva* (Spanish), first crossing on glider in 1931 by *Lissant Beardmore* (UK), first crossing on helicopter (*Focke-Achgelis Fa 223*) in 1945 by *Helmut Gerstenhauer* (Germany). Calais in France and Dover in England, separated by distance of $\approx 50km$, are pinned in the map indicating *Blériot's* flight starting and ending points.

Eugene Burton Ely takes off from the *USS Birmingham* on November 14, 1910.

First fixed-wing aircraft landing on a warship: *Ely* landing his plane on board the *USS Pennsylvania* in San Francisco Bay on January 18, 1911. Hooks were attached to the plane's landing gear, a primitive version of the system of arresting gear and safety barriers used on modern aircraft carriers.

achievements in early aviation. He made history by conducting the first successful takeoff from a ship and the first landing on a ship, marking significant milestones in naval aviation. These historic events, accomplished using a *Curtiss* pusher biplane, demonstrated the viability of aircraft operations from ships, shaping the trajectory of naval aviation and contributing to the evolution of aircraft carrier technology.

Prominent Scientists

Theoretical studies in aviation are of paramount importance for the aviation industry. These foundational concepts underpin the understanding of various disciplines in aerospace engineering, forming the cornerstone of aerospace design. By delving into the relationship between air pressure, lift generation, and boundary layer behavior, these theories empower engineers to optimize vehicle performance and troubleshoot potential issues. They not only guide the creation of efficient and safe aircraft but also stimulate innovation by inspiring novel designs and configurations. Ultimately, theoretical studies serve as the intellectual catalyst driving the progress of aviation, equipping professionals with the knowledge to navigate the complexities of flight and engineering in an ever-evolving aerospace landscape. The rest of this section presents the contributions by a few scientists which had significant impact in the realm of aerospace engineering.

Daniel Bernoulli (1700 – 1782) was a distinguished Swiss mathematician and physicist of the Enlightenment era. Renowned for his groundbreaking work in fluid dynamics, *Bernoulli*'s eponymous principle elucidated the relationship between fluid velocity and pressure, which helps in explaining many fundamental concepts in Aerodynamics like the generation of lift over a wing. Pitot tube, an instrument used for measuring air speed, also uses the principle.

Daniel Bernoulli (1700 – 1782), formulated the principle explaining the relationship between airspeed and pressure, essential to lift generation in flight.

Leonhard Euler (1707 – 1783), a Swiss mathematician and physicist, made monumental contributions to various fields, including mechanics, fluid dynamics and astronomy which have significant applications in aerospace engineering. Euler's groundbreaking equations in fluid dynamics provided the mathematical foundation for understanding how air interacts with aircraft surfaces. His legacy extends to modern aerospace engineering, where his mathematical insights continue to shape our understanding of mechanics that contribute to advancements in aviation technology.

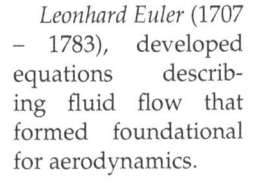

Leonhard Euler (1707 – 1783), developed equations describing fluid flow that formed foundational for aerodynamics.

Claude-Louis Navier (1785 - 1836), a French physicist and engineer, made significant contributions to aviation through his development of the *Navier-Stokes* equations, which describe the motion of fluids. These equations form the basis for understanding the aerodynamics of aircraft and the behavior of airflow around wings, propellers and rotary blades. *Navier*'s work provided essential tools for developing the equations for predicting and analyzing the complex fluid dynamics involved in flight, contributing profoundly to the advancement of aviation science and engineering.

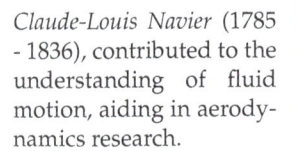

Claude-Louis Navier (1785 - 1836), contributed to the understanding of fluid motion, aiding in aerodynamics research.

George Stokes (1819 - 1903), a prominent British mathematician and physicist, made significant contributions in fluid dynamics. His study helped in understanding the effect of viscous forces due to fluid motion that contributed to the development of the *Navier-Stokes* equations. The equations are solved using the modern approach of Computational Fluid Dynamics which helps in estimating the forces experienced by flight vehicles moving in air.

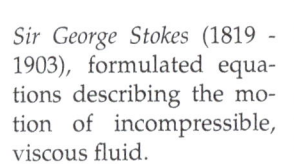

Sir George Stokes (1819 - 1903), formulated equations describing the motion of incompressible, viscous fluid.

Nikolay Zhukovsky/Joukowsky (1847 - 1921), a Russian scientist and engineer, made pioneering contributions to aviation. He formulated the concept of the lifting force produced by an air-

1.2. THE BEGINNING

foil, laying the foundation for modern aerodynamics. *Zhukovsky*'s research also led to the creation of wind tunnels for aerodynamic testing, revolutionizing aircraft design. His work significantly advanced the understanding of flight principles and played a pivotal role in shaping the field of aviation. The *Joukowsky* transform is named after him, while the the *Kutta–Joukowski* theorem is named after him and the German mathematician *Martin Kutta*.

Nikolay Zhukovsky (1847 - 1921), introduced the concept of lift and circulation around airfoils, fundamental to wing design.

Martin Wilhelm Kutta (1867 - 1944), a German mathematician and aeronautical engineer, made notable contributions in numerical analysis and aerodynamics. He formulated the *Kutta–Joukowski* theorem, fundamental in aerodynamics, linking lift to circulation around airfoils. *Kutta*'s collaboration with *Carl Runge* led to the development of *Runge-Kutta* methods for solving differential equations numerically. *Kutta*'s legacy also includes the *Kutta* condition, an aerodynamic principle that helps understand lift generation during flight.

Martin Wilhelm Kutta (1867 - 1944), developed the Kutta condition, a vital component of modern lift analysis.

Ludwig Prandtl (1875 - 1953), considered "father of aerodynamic theory", was a German physicist and professor celebrated for his pioneering contributions to fluid dynamics and aerodynamics. His boundary layer theory explaining the behavior of fluid flow near surfaces helped settle the *d'Alembert*'s paradox which is that a body in motion in an inviscid fluid experiences zero drag and is contrary to experimental measurements. His contribution to the development of thin airfoil theory helped to predict he variation of lift with angle of attack for an airfoil. He also contributed to the development of theory for predicting the induced drag and lift from wings of finite size which plays a significant role in practical wing design. The famous *Prandtl-Glauret* correction helps to extend the results from low speed wind tunnels to high subsonic flows. His theoretical contribution towards gasdynamic phenomena like shock and expansion waves helps in the design of high speed flight vehicles and experimental facilities for high speed applications. The prolific contributions by *Prandtl* and his students strongly built the theoretical foundations required for the progress of aerodynamics in the 20th century.

Ludwig Prandtl (1875 - 1953), introduced the concept of the boundary layer.

Theodore von Kármán (1881 – 1963) was a Hungarian-American aerospace engineer of great renown due to his pioneering contributions in the fields of aerodynamics and aerospace engineering. His innovative research in fluid dynamics, turbulence, and supersonic flight laid down fundamental principles that underpin these disciplines. His remarkable achievements led to the establishment of the *'Kármán line'*, a human-defined demarcation for outer space. Serving as a co-founder of the Jet Propulsion Laboratory (JPL) and playing a pivotal role in aeronautical research, *von Kármán*'s legacy extends to his instrumental involvement in propelling rocketry and space exploration, profoundly shaping the course of modern aviation and space science. *Theodore von Kármán* stands as a preeminent theoretician of the 20th century who contributed significantly to the advancement of aerospace technology. He was instrumental in setting up various organisations and institutes like International Union of Theoretical and Applied Mechanics (IUTAM), JPL, AGARD and von Kármán Institute of Fluid Dynamics in Brussels which immensely fostered the growth of Aerospace Engineering.

Theodore von Kármán (1881 – 1963) played a significant role in the advancement of supersonic and hypersonic aerodynamics.

Apart from those mentioned above, there are many scientists, mathematicians, engineers and businessmen who had made direct or indirect contributions which eventually saw the boom in Aerospace technology in the 20th century. Till about the beginning of World War I, the progress was spearheaded by mostly individuals out of their sheer sense of passion and determination. Beginning with World War I, many Governments and militaries started seriously recognizing the utility of aviation. In the next section, we will look at the progress in aviation during World War I.

1.3 Aviation during World War I

Before World War I, the primary use of aircraft was for experimentation, exploration, and limited military reconnaissance. Aviation was in its infancy and airplanes not yet widely utilized for practical applications. Pilots and engineers were focused on improving flight capabilities, understanding aerodynamics, and demonstrating the potential of aviation. Military forces occasionally used aircraft for reconnaissance purposes, gathering information on enemy positions and activities. However, the roles and capabilities of aircraft were still evolving, and they had not yet become a dominant or established tools in warfare or other fields.

1.3.1 As War Machines

The catalyst for the advancement of aviation was the onset of the World War I. Initially, aircraft held a modest role in warfare, but as the conflict progressed, the aviation branch gained prominence within the armed forces. The military swiftly embraced this pioneering technology for reconnaissance, subsequently deploying aircraft to deliver hand-dropped bombs onto ground troops. The era's aircraft predominantly consisted of biplanes, primarily designed for military purposes, including aerial combat. It is noteworthy that these wartime developments were initiated in peacetime. World War I spurred the demand for enhanced speed, altitude, and maneuverability, driving significant enhancements in aerodynamics, structural design, as well as control and propulsion systems.

Reconnaissance Aircraft

In the early stages of World War I, aircraft were primarily used for aerial reconnaissance. They had basic cameras and observers in the aircraft would take photographs or make sketches of enemy positions. Examples of such aircraft include the French *Maurice Farman MF.7* and the German *Aviatik B.I* and *B.II*. They were simple biplanes with limited engine power and open cockpits. An instance when the aerial reconnaissance played an important role is when the Royal Flying Corps aviators provided crucial information about the invading German troops to the British and French armies in the decisive *Bat-*

The *Aviatik B.II* is a German two-seat reconnaissance biplane. The plane has a wing span of 12.35 m, overall mass of 1,090 kg driven by a *Mercedes D.II* 6-cylinder liquid-cooled inline 89 kW engine with 2-bladed fixed-pitch propeller. It has a maximum speed of 100 km/h.

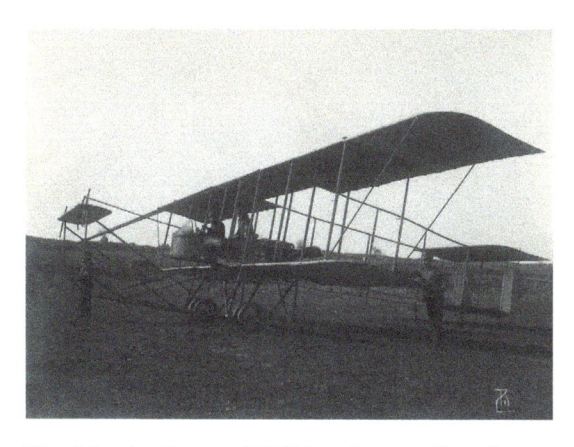

The *Maurice Farman MF.7 Longhorn*, a French two seat biplane used for reconnaissance by the French and British air services in the early stages of the war. The plane has a wingspan of 15.4 m, overall mass of 855 kg and is driven by one *Renault 8C V-8* air-cooled 52 kW piston engine. It has a maximum speed of 95 km/h at a service ceiling of 4,000 m.

tle of the Marne which helped them to plan the counterattack leading to the withdrawal of the enemy troops.

As the War progressed, the engines became more powerful and the aircraft design improved helping them to fly at higher altitudes to avoid interception. For example, the German *Rumpler* that started operation in 1917 could operate as high as 7,300 meters.

Fighter Aircraft

When the World War I started, aircraft were rarely used for aerial fighting. Occasionally, pilots would shoot at each other using pistols. Subsequently, a gun was attached to the front of the plane so that it could shoot down other planes. Machine guns were a logical complement to aircraft. Since most of the aircraft of the time had the tractor configuration with the propeller ahead of the pilot, there was the problem of the bullets from the machine gun hitting the propeller. The issue was solved by the famous aircraft designer *Anthony Fokker* by synchronizing the firing from the gun with the propeller in such a way that the bullets passes in between the propeller blades.

RAF, S.E.-5 Fighter, one of the fastest aircraft of the War and described as the '*Spitfire* of World War I'.

Design	Single-seat British biplane fighter aircraft
Wingspan	8.11 m
Speed	222 km/h
Range	480 km
Engine	110 kW *Wolseley Viper* water cooled V8 engine
Armament	2 x 7.7 machine guns; 4 x 11-kg bombs
Number of aircraft built	∼ 5,200

In the early stages of the war, aircraft were primarily designed for reconnaissance and *dogfighting*, a form of aerial battle between fighter aircraft at close range. These early fighters were characterized by their agility, speed, and maneuverability, which

Table 1.1: Major aircraft manufacturers during World War I

Company (country, year of establishment)	Founder	Notable aircraft
Royal Aircraft Factory (England, 1911)	British Government	S.E. series, B.E. series, F.E. series
Sopwith Aviation Company (England, 1912)	Sir Thomas Sopwith	Camel, Pup, Triplane, $1\frac{1}{2}$ Strutter, Dolphin, Snipe, Cuckoo, Salamander
Bristol Aeroplane Company (England, 1910)	Sir George White	Boxkite, Bristol Fighter
Nieuport (France, 1908)	Édouard Nieuport	Nieuport 11, Nieuport 17
SPAD (France, 1911)	Armand Deperdussin	SPAD S.VII, SPAD S.XIII
Voisin (France, 1906)	Voisin brothers (Gabriel & Charles)	Voisin VII, Voisin Canard, Canon, Voisin III
Albatros-Flugzeugwerke (Germany, 1909)	Walter Huth and Otto Wiener	Albatros D.I, Albatros C.III, Albatros D.III and Albatros D.V
Rumpler Flugzeugwerke (Germany, 1909)	Edmund Rumpler	Rumpler C.I, Rumpler C.IV
Fokker-Flugzeugwerke (Netherlands, 1912)	Anthony Fokker	Eindecker, Scourge, D.VI biplane, Dr.I triplane

Albatros D.Va Fighter, a fighter aircraft with Germany during World War I, has a streamlined fuselage and the propeller spinner smoothly contoured into the nose of the fuselage

Design	Single-seat, German biplane fighter
Wingspan	9 m
Speed	186 km/h
Empty mass	680 kg
Material	Wood and Fabric
Armament	Guns: 2 × 7.92 mm LMG 08/15 machine guns

Sopwith F.1 Camel, among the most significant of all World War I aircraft, shot down ∼ 1,300 enemy aircraft, more than any other Allied fighter in the war.

Design	Single-seat, biplane fighter
Wingspan	8.5 m
Empty mass	420 kg
Engine	1 × 97 kW *Clerget 9B* rotary engine
Number of aircraft built	∼ 5,500

were essential for engaging enemy aircraft in aerial combat. However, as the conflict progressed, it became evident that a new type of aircraft was needed to carry out strategic bombing missions against enemy targets on the ground. This shift in focus led to the development of dedicated bomber aircraft.

Bomber aircraft

Dropping explosives on people and other targets on the ground was conceived as a military use for aircraft. During the war, this rudimentary attack was refined into strategic bombing of enemy infrastructure and tactical bombing of military targets. The earliest bombs were light artillery shells dropped by pilots sitting in open cockpits.

The need for dropping heavier bombs drove the development of bomber aircraft. Bomber aircraft were developed which had the specialized role to deliver ordnance, often targeting strategic infrastructure of the enemy on the ground. Bombers carried a heavier payload of bombs and hence were heavier and less fast and maneuverable compared to fighter aircraft. So, they were often accompanied by fighter aircraft as escort for protection from enemy aircraft. Bomber aircraft from the World War I era include the French *Voisin III* and the German *Gotha G.I* which were modified to carry bombs for attacking enemy targets. By 1917, dedicated bombers like the British *Handley Page Type O* and the German *Gotha G.V* emerged. Compared to their predecessors, they could carry substantial ordnance over longer distances. Compared to *Gotha G.II* which could carry 14 × 10 kg bombs, *Gotha G.V* could carry 14 × 25 kg bombs. Although introduced late in the War in 1918, the British *Handley Page*

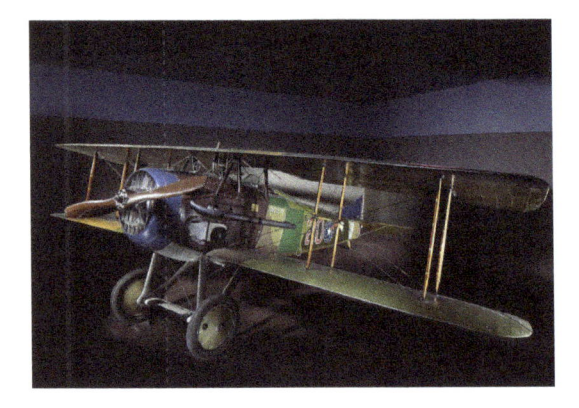

SPAD S.XIII, among the most successful fighters of World War I in a class along with *Fokker D.VII* and *Sopwith Camel*, was flown by some of the legendary aces like *Rickenbacker*.

Design	single-seat, biplane fighter
Wingspan	8.2 m
Material	Wood, Fabric
Number of aircraft built	∼ 8,500

0/400 made a significant impact and helped shape the role of strategic bombing and demonstrated the potential of long-range bombers in modern warfare. In France, they were used for tactical nighttime attacks on targets within German-occupied France and Belgium. The *Gotha G.IV* and *Gotha G.V*, German twin-engine biplane bombers that became operational in the later part of the War, was used by the *Luftstreitkräfte*, the Air wing of the German Imperial Army, for bombing raids on England. It was capable of carrying a substantial bomb load of 350 kg and had a service ceiling of 6.5 km, higher than most of its contemporaries. They had this operational advantage over enemy fighter aircraft of the time and these bombers posed a significant threat to British cities. However, their relatively slow speed (maximum speed 140 km/h) and limited maneuverability made them susceptible to attacks from more agile enemy fighter planes. The *Caproni Ca.3* was an Italian triplane bomber. It was introduced in 1916 and was used by the Italian Army and later by the Royal Air Force. It had a distinctive triple-wing

Fokker D.VII, referred as one of the best fighter aircraft of the World War I, had the unique ability to seemingly hang on its propeller and fire into the underside of enemy aircraft which made it a highly feared combat aircraft. Thick wing section gave the airplane good stall characteristics which enabled the pilot to safely put his airplane into a nose-high attitude.

Design	single-seat, German biplane fighter
Wingspan	8.9 m
Empty mass	700 kg
Material	Wood, Fabric
Propeller	2.8 m (diameter), 22.9 cm (chord)

configuration and could carry a significant payload.

The relatively slow, fragile looking, tiny reconnaissance planes of 1914 had evolved into agile, sturdy, and sophisticated fighter and bomber aircraft towards the end of the War. For example, the British *Sopwith Camel* had become renowned for its aerial fighting capability. On the other hand, the German *Gotha G.V* with a range of 840 km and a bomb carrying capacity of 350 kg had grown into a massive bomber capable of strategic bombing. This remarkable transformation illustrates the rapid advancement of aviation technology driven by the exigencies of war, setting the stage for further aviation advancements in the post-War era.

The *Zeppelin* bomber, designed and manufactured by the German company *Luftschiffbau Zeppelin GmbH*, was a remarkable development during World War I that combined the capabilities both of

Nieuport 28C.1 , the French fighter aircraft went on to serve with an American fighter unit and also made its mark in U.S. aviation history after the war.

Design	single-seat, French biplane fighter
Wingspan	8.2 m
Material	Wood, linen
Engine	1 × 119 kW *Gnome Monosoupape 9N* rotary engine
Number of aircraft built	∼ 300

De Havilland DH-4, the first American-built version of *Geoffrey de Havilland*'s British World War I bomber. On May 13, 1918, *Orville Wright* made his last flight as a pilot in a 1911 *Wright Model B* alongside *DH-4*.

Design	Two-seat, British biplane observation and bomber aircraft
Wingspan	13 m
Material	Wood, Fabric
Number of aircraft built	∼ 6,300

airships and bombers. They had improved range, higher altitude capabilities, and the ability to carry significant bomb loads. The result was a extremely large aircraft capable of carrying out bombing missions on a significantly large scale. However, their large size and slower speeds made them vulnerable to enemy attack. One of the most popular Zeppelin models was the *Zeppelin Staaken R.VI*, which first flew in 1916. Its structure featured an internal framework of lightweight metal and wooden materials covered with fabric. The aircraft had four engines and could accommodate a crew of up to seven or eight personnel.

The German bombing on London using *Zeppelin* airships during the War marked a new chapter in aerial warfare and subjected the civilian population to a new form of terror. These bombings brought the civilian population under the direct attack of the enemy. These raids often occurred during night, taking advantage of darkness at night to reduce the risk of attack by the British forces. The German bombings using *Zeppelins* continued till 1917 when advancements in anti-aircraft guns and aircraft technology had increased the risk of airships and reduced its effectiveness. Though the bombings using *Zeppelins* became obsolete, their role in World War I is an important aspect of the history of aerial warfare.

1.3.2 Flying Aces

The airplanes took to the air with purpose of air-to-air combat and skilled aviators were venerated for their flying acumen and courage in engaging in air combat with their adversaries. A military pilot is considered as a *flying ace* when s/he is credited with shooting down five or more enemy aircraft in aerial combat. Flying aces emerged as prominent figures during the early days of aviation when piloting skills were of paramount importance in aerial warfare.

Handley Page Type O bomber, a British twin-engine biplane developed by *Handley Page Limited*, was one of the significant bombers used during World War I. When built, the *Type O* was among the largest aircraft that had been built in the world.

Crew	4 or 5
Wingspan	30 m
Empty mass	3,856 kg
Powerplant	2 × Rolls-Royce Eagle VIII V-12 water-cooled piston engines, 270 kW each
Propellers	4-bladed wooden propellers
Maximum speed	157 km/h
Range	1,100 km
Payload	Guns: 5 × 7.7 mm *Lewis* Guns and 910 kg of bombs

A *dogfight* between two aircraft engaged in aerial combat is perhaps the most fascinating type of combat. The skill required to operate a fighter aircraft along with the physical and mental strain of a dogfight make the fighter pilots who excel at them truly exceptional. The capabilities and shortcomings of the aircraft also play a significant role in who emerges as the winner. The number of aircraft a single ace pilot can shoot down has steadily decreased because the modern aviation technology have made dogfights rare in recent warfare. From *Erich Hart-*

Voisin Type-8, single-engine, two-seat French World War I pusher biplane bomber aircraft entered service in November 1916.

Wingspan	18.0 m
Empty mass	1,310 kg
Material	Airframe: Wood, Covering: Fabric and Silver finish overall
Engine	*Peugeot 8Aa* V-8 water-cooled pusher-piston engine with 160 kW power.

Zeppelin LZ 38 near its hangar. At the beginning of January 1915, a *Zeppelin* airship bombed London. The bombings continued during the War and the sight of *Zeppelins* caused panic among the general public.

A poster issued to the British public showing the British and German airships and planes so that the public can take shelter if they see an enemy aircraft.

mann, the Nazi fighter pilot credited with shooting down 352 Allied aircraft, to *Giora Epstein*, the Israeli pilot considered the ace of aces of supersonic jet pilots, these pilots are among the most skilled fighter pilots to ever enter a cockpit.

The term, *Flying Aces*, gained prominence during World War I when aerial warfare was still in its infancy. Some famous World War I Flying Aces include *Manfred von Richthofen*, *Eddie Rickenbacker*, and *René Fonck*. The title continues to be used in military aviation contexts to recognize and honor exceptional aerial combat achievements. The term flying ace encompasses not only the skills of the pilots but also the capabilities of the aircraft they flew and the tactics they employed. Flying aces continue to be venerated for their skill, courage, and heroism.

Manfred von Richthofen (1892 – 1918), famously known as *The Red Baron*, is perhaps the most famous

Dogfight scene during World War I, a fast-paced aerial combat between fighter planes, involving sharp turns, evasive tactics, and precise shooting at enemy at close range. Skillful piloting and precise deployment of weapons determine the outcome of these intense aerial battles, playing a pivotal role in achieving air superiority. (Pencil sketch by *Dishant*)

flying ace of all time credited with 80 air combat victories during World War I. He was a pilot for the Imperial German Army Air Service, and had the most number of aerial victories in World War I. In his fabled red *Fokker Dr.I* fighter aircraft, *Richthofen* achieved world wide fame and became a national hero in Germany. He led the *Jasta 11* air squadron which enjoyed more success than any other squad in World War I, particularly in *Bloody April of 1917* battle when *Richthofen* shot down 22 aircraft alone. He eventually commanded the fighter wing formation that became known as the *Flying Circus*. He died at an young age in an aerial combat in 1918.

Eddie Rickenbacker (1890 – 1973): When the United States joined World War I in 1917, Rickenbacker volunteered to join the military. In March 1918, *Rickenbacker* was transferred to the United States Air Service and became a member of the 94th Aero Pursuit Squadron. *Rickenbacker* earned the nickname *Ace of Aces* because he shot down 22 airplanes and 4 balloons during the war. *Rickenbacker* earned numerous awards for heroism during World War I, including the Distinguished Service Cross, the Congressional Medal of Honor, and the French Croix de Guerre. After the war ended, he published his memoirs of his experiences, which he titled *Fighting the Flying Circus* (1919).

Fokker Dr. I, one of the most famous German aircraft of World War I, probably because it was associated with the *Red Baron*.

Design	Single-seat, German triplane fighter
Upper wingspan	7.2 m
Empty mass	406 kg
Max. speed	180 km/h
Range/Service ceiling	300 km/6.1 km
Engine	110hp *Oberursel Ur.II* 9-cylinder rotary piston engine
Guns	2 × 7.92 mm *Maschinengewehr 08* synchronized machine guns
Number of aircraft built	320

Manfred von Richthofen (1892 – 1918).

Thomas Mottershead (1892 - 1917) : A British flying ace during World War I, showcased his exceptional piloting skills against the enemy aircraft. His

Eddie Rickenbacker (1890 – 1973)

remarkable courage in combat led to him being recognized as a recipient of the *Victoria Cross*, the highest and most prestigious award for gallantry in the face of the enemy that can be awarded to British and Commonwealth forces. Tragically, he lost his life in action at the young age of 24 in 1917, leaving behind a legacy of valor and skill in the annals of aviation history.

James McCudden (1895 - 1918), a Royal Flying Corps member, stood as a distinguished fighter ace with 57 victories, becoming Britain's most decorated combatant of World War I. His methodical approach and meticulous aircraft maintenance, honed by his mechanic background, contributed to his successes. He was awarded the Victoria Cross, Distinguished Service Order, the Military Cross and the Military Medal for his exceptional career as a pilot. Tragically, he lost his life due to a flying accident in 1918, ending a remarkable legacy of valor and service.

James McCudden (1895 - 1918), one of the greatest fighter aces of the World War I.

1.3.3 Technology Innovations

World War I witnessed several significant innovations that transformed the nature of aerial warfare and aviation. In a short span of only four years, aircraft designs evolved from simple, fabric-covered frames to more streamlined, aerodynamically efficient, robust flying machines. The development of monoplanes and biplanes with improved wing designs, such as the British *S.E.5* and the German *Fokker D.VII*, led to enhanced maneuverability and overall performance. Advancements in engine technology led to increased horsepower and reliability. Aircraft engines evolved from low-powered rotary engines to more powerful and efficient inline engines. The introduction of engines like the British *Rolls-Royce Merlin* and the German *Mercedes D.III* played a crucial role in enhancing aircraft performance.

Synchronous Machine Gun Designed By Anthony Fokker: Mounting the machine gun in front of the pilot posed the problem that the bullets can hit the propeller and damage it. The solution to this problem was an invention called the *synchronization gear*, which restricted the machine gun operation in such a way that it could only fire in between the propeller blades.

Anthony Fokker (1890 – 1939) was a Dutch aviation pioneer and aircraft manufacturer known for designing and producing innovative aircraft during World War I and the interwar period. *Fokker*'s engineering prowess led to the development of synchronized machine guns that fired through rotating propellers, giving his aircraft a significant advantage in air combat. His aviation company, *Fokker Aircraft*, played a crucial role in the evolution of military and civil aviation, leaving a lasting impact on the aerospace industry. Some famous products of *Fokker Aircraft* are the *Eindecker* monoplane, the *Dr.1* triplane and the *D.VII* biplane.

The Synchronous Machine Gun, designed by the Dutch aviation pioneer *Anthony Fokker* during World War I, was a significant advancement in aircraft armament technology. *Fokker*'s innovation involved synchronizing a machine gun with the aircraft's propeller, allowing the gun to fire through the rotating propeller blades without causing damage. This breakthrough addressed the challenge of firing a machine gun without hitting the propeller, which had been a major limitation in earlier aircraft designs.

Pusher solution was a notable innovation that aimed to resolve the challenge of firing machine guns through a plane's propeller arc without causing damage. This approach involved placing the engine and propeller behind the pilot, along with the machine gun, to eliminate the risk of bullets hitting the propeller blades. The pusher configuration eliminated the need for complex synchronization systems. This simplified maintenance and reduced the risk of mechanical failures. With the machine gun positioned behind the pilot, there was a clear line of sight for firing without the propeller obstructing the bullets' trajectory. This allowed for consistent and accurate targeting of enemy aircraft.

However, the pusher configuration introduced aerodynamic challenges. The engine and propeller at the back created drag, affecting the aircraft's overall performance, speed, and maneuverability. This design also required adjustments to maintain stability. While the pusher configuration protected the propeller, it made the engine more vulnerable to enemy fire, as it was exposed at the rear of the aircraft. Achieving the right balance between the weight of the engine and the pilot's position was crucial for a stable flight. Any imbalance could lead to difficulties in control and handling.

Despite the challenges, the pusher solution provided a viable alternative for firing machine guns and engaging in aerial combat without the complexities of synchronization mechanisms. Aircraft like the British *F.E.2b*, designed for reconnaissance and bombing missions, successfully utilized the pusher configuration. This innovation demonstrated the adaptability and ingenuity of aviation engineers during the early years of air combat.

Facing off against the enemy aircraft in aerial warfare, the necessity for better aerodynamic design, more reliable and powerful engines and sturdy airframe design was felt by the aircraft manufacturers. Compared to the use of struts and wires for the airframe before the start of the War, the streamlined design of the wings and the fuselage during the War resulted in lesser drag and thereby better performance of the aircraft. During the War, the power of the combat aircraft engine used in World War I grew from around 80 hp at the start of the war to a maximum of 400 hp (*Liberty* engine) by 1918. The two main types of power plants during that era were the rotary and in-line water-cooled engines. Rotary engines were lighter and more compact but ran into problems when required to deliver over 150 hp. With their cylinders arranged around a fixed crankshaft they created a gyroscopic effect that made an aircraft difficult to fly, but they worked well on maneuverable *dogfighters* such as the *Sopwith Camel* and *Fokker Triplane*.

World War I saw significant progress on all aspects of aviation. The early days of the War saw only limited usage of aircraft for reconnaissance. Towards the end of the war, airplanes were used extensively for aerial combat, bombing, ground attack and reconnaissance. It became clear for military strategists and policy makers in the Government that aircraft will be an integral part of future wars. The significant increase in demand for aircraft during the War saw the establishment of many aircraft manufacturing companies, some of them are in existence even today. There was an exponential increase in the number of aircraft and pilots available after the War compared to before the start of the War. This set the scene for use of aircraft for many civilian purposes, especially commercial aviation. The scale of aircraft operation saw significant jump during the War which led to the establishment of infrastructure required for aviation like runways and maintenance divisions.

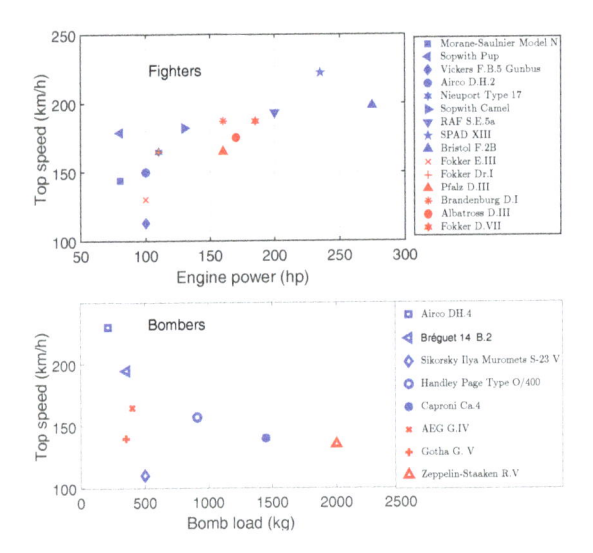

Performance of bombers and fighter aircraft used in World War I. The blue and red colors represent aircraft belonging to Allied Powers (United Kingdom, United States, France, Russia and others) and Central Powers (Germany and others), respectively.

1.4 Between World War I and World War II

The period between the two World Wars saw aviation maturing as a reliable technology for transportation, both for civilian and military purposes. Following the recognition of the potential for aviation during World War I, the interwar period witnessed an intense evolution in aviation technology, design, operations and applications. The era was characterized by remarkable achievements, pioneering endeavors and challenges that paved the way for the modern aviation landscape as we see it today. In this section, we delve into the captivating journey of aviation as it matured to a more reliable technology.

1.4.1 Maturing of Aviation

In the 1930s, the progress in aircraft design that began in the previous decades continued resulting in more reliable aircraft characterized by enhanced performance characteristics. This saw aircraft able to fly faster and longer distances with more passengers in a comfortable environment. The fre-

1.4. BETWEEN WORLD WAR I AND WORLD WAR II

B.E.2 was a single-engine tractor two-seat biplane designed and developed by Royal Aircraft Factory for Reconnaissance

Interrupter mechanism (guns fire through moving propeller blades) first fitted to the *Fokker Eindekker monoplane*

Handley Page 0/400 - Long range British bomber

Sopwith Triplane was a highly maneuverable single-seat, with phenomenal rates of climb and roll.

The *Bristol F.2*, a British two-seat fighter and reconnaissance biplane developed by Frank Barnwell at the Bristol Airplane Company.

Siemens-Schuckert - Single-seat German fighter plane.

Gotha G V - Long range German bomber.

Innovation during WWI

- The pusher solution for machine gun carrying, Ex. *Vickers Biplane 1, F.E.2b* or *Airco DH.1*,
- Machine gun synchronization in tractor: fire the forward-firing armament through spinning propeller without bullets striking the blades. Ex. *Fokker Eindecker*

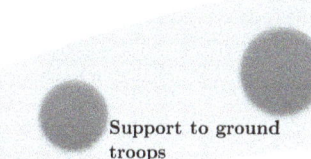

Reconnaissance
- The airplanes would fly above the battlefield and determine the enemy's movements and position.
- Formally aerial photography was introduced early in 1915
- Photographs were either taken by hand or by attaching the camera to the aircraft

Message Streamer
- Without the radio, pilots used drop messages in weighted bags
- Use message streamers to drop

Message streamer used to drop messages to forces on the ground.

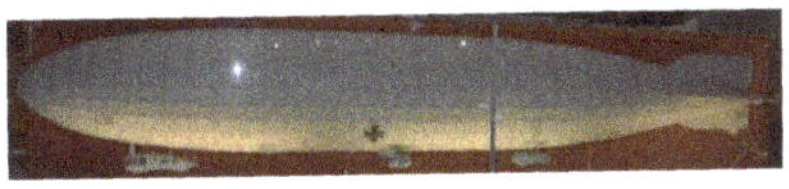

Air-to-air combat
- Initially most aircraft were unarmed but pilots did carry weapons like pistols and grenades
- By 1915, forward-firing machine guns were being fitted onto aircraft
- Interrupter Mechanism – This allowed machine guns to fire through moving propeller blades
- Dog fights

Support to ground troops

Bombing
- Tactical aerial bombing, or the hitting of targets on the battlefield

The Bristol Braemar was a British heavy bomber aircraft developed at the end of the First World War for the Royal Air Force

A Zeppelin is a type of rigid airship. During World War I, the German military made extensive use of Zeppelins as bombers and as scouts, resulting in over 500 deaths in bombing raids in Britain

A flying ace (military aviators credited with shooting down more than five enemy aircraft during aerial combat.

C Type camera used in the WW1 for reconnaissance attaching to aircraft

Manfred von Richthofen (1892 – 1918), known as "The Red Baron" was the highest-scoring ace of the war with 80 victories. He was killed in action in April 1918 and buried by the British with full military honors

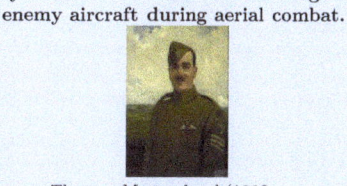

Thomas Mottershead (1892 – 1917) was an English recipient of the Victoria Cross, the highest and most prestigious award for gallantry in the face of the enemy that can be awarded to British and Commonwealth forces. He was the only Non-Commissioned Pilot to be awarded the Victoria Cross

James McCudden (1895 – 1918), recipient of the Victoria Cross, was a British flying ace and one of the highest scoring British fighter pilots with 57 victories. He was killed in a flying accident on 9 July 1918.

Aviation during World War I.

quency of emergency landings and aviation accidents reduced as the technology became more mature. This era witnessed the establishment of monoplanes as the normal wing configuration compared to the high-drag biplanes, signifying a major shift in the aerodynamic design of wings. Another important breakthrough was the widespread adoption of metals, specifically aluminum alloys, for airframe construction. Engines also continued to evolve, both in terms of power-to-weight ratio and reliability. Radial engines saw an increase in power by incorporating an additional ring of cylinders, while the implementation of the *NACA* (National Advisory Committee for Aeronautics) cowling enhanced their efficiency. The efficiency of in-line engines improved significantly with the adoption of ethylene glycol as a coolant. This coolant, with its low freezing and high boiling points, allowed for the design of smaller radiators, reducing both weight and drag on aircraft. This era also saw the beginning of the development of jet engines which markedly transformed the aviation landscape after World War II. Compared to the rudimentary instruments used in the aircraft during World War I, the aviation instruments became more reliable and effective paving the way for night flying, radio communication with the ground, and more effective route planning. This era saw the establishment of aviation as a reliable and comfortable mode of transportation.

Monoplane Configuration

Following *Louis Blériot*'s successful English Channel crossing in 1909 aboard his *Blériot XI*, the monoplane design gained popularity among aircraft designers. Monoplanes marked a significant departure from the biplane configuration, with their streamlined design offering improved aerodynamic efficiency and performance. The development of monoplanes was a pivotal phase in aviation history, contributing to the advancement of aircraft technology and shaping the way for modern aviation. The development of monoplanes faced structural challenges. Early monoplanes experienced the problem of wing folding, a dangerous phenomenon where wings would collapse in flight due to insufficient structural strength. In the aftermath of monoplane crashes in 1912, French military officials suggested strengthening the structure of their *Blériot XI* monoplanes. The War triggered a great deal of innovation by airplane designers, who suddenly had more money and more pressure forcing them to design faster, more maneuverable and more reliable planes. Designers found effective ways to reinforce the wings by fitting additional structures inside the wings that connected them to the fuselage, a design called *cantilevered wings*. *Hugo Junkers* (1859 - 1935), a professor at RWTH Aachen University and founder of the *Junkers Flugzeug-und Motorenwerke AG*, pioneered all-metal construction of airframe and cantilever wings, a significant improvement compared to the braced wings and wire construction common during the early days of aviation. Also, aircraft designers during this era developed a much better understanding of the aerodynamic forces acting on a wing which led to better structural designs. The inter-War period saw significant monoplane development, with all-metal stressed-skin construction becoming the norm, and advances in engine technology further enhancing the performance capabilities of the aircraft.

Airframe Construction

Though there were significant improvements in aircraft range, maximum speed, engine power, load-carrying capacity, and reliability from pre-War to post-World War I, the strut-and-wire biplanes were aerodynamically inefficient. The most innovative airframe designs of the immediate post-war period came from the designers like *Anthony Fokker*, *Hugo Junkers* and *Reinhold Platz*.[1] They developed stronger monoplanes adopting an all-metal (lightweight Duralumin, an alloy of aluminum) construction with a strut-free, single cantilever wing. The metal skin of *Junkers* aircraft was

---
[1] Platz stuck to wood as his prime material, although he used steel tubing for the internal structure of the fuselage. However, wooden construction turned out to be retrograde.

corrugated to give extra structural strength, at the expense of increasing drag.

F.VII-3m (trimotor), a significant monoplane during the early days of aviation.

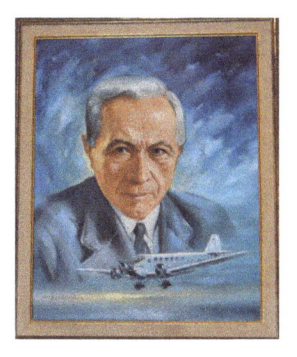

Hugo Junkers (1859 - 1935) was a German aircraft designer who was famous for the all-metal construction for airplanes. His company, *Junkers Flugzeug- und Motorenwerke AG (Junkers Aircraft and Motor Works)*, was among the leading German aircraft manufacturers in the years between World War I and World War II.

By 1920, another German aircraft designer, *Adolph Rohrbach* [2], had worked out the concept of constructing aircraft wings and tail surfaces using metal sheets stretched over box spars. This innovation called the 'stressed skin construction' had the metal skin sharing the load along with traditional framework. Eventually, this proved to be immensely influential in shaping the design philosophies of *Jack Northrop*, *Boeing*, and numerous other manufacturers. His all-metal, multi-engine passenger and cargo aircraft played a pivotal role in the establishment of airlines worldwide, including Germany. Notably, *Hugo Junkers* had previously introduced the *Junkers J4* in 1917, a groundbreaking all-metal airplane crafted primarily from the lightweight aluminum alloy known as duralumin.

Junkers J4 airplane: Engine - *Benz Bz. IV*, 147 kW. Wingspan-16 m; Top speed - 155 km/h; Armament 2 x 7.92 mm machine gun.

Drag Reduction

Numerous research laboratories, including America's NACA, achieved significant understanding of the aerodynamics of aircraft in flight through extensive wind tunnel testing. These advances were subsequently used for better aircraft design. For example, NACA's wind tunnel experiments on air-cooled radial engines revealed the high drag because of the exposed engine cylinder rings. To overcome this, NACA's scientists suggested enclosing the cylinders resulting in a significant decrease in aerodynamic drag (by as much as 60%). This profound engineering insight was pioneered by a team led by *Fred Ernest Weick* (1899 – 1993). [3]

Other innovations that evolved from experimental studies at wind tunnels are:

A fixed undercarriage contributes to a significant 40% of the entire drag acting upon an airplane. This was addressed by enclosing fixed undercarriages

---

[2] Adolph Rohrbach (1889 - 1939), was a German engineer and entrepreneur whose influence on aviation history was profound. He made significant advancements in utilizing duralumin and constructing a series of large monoplane flying boats, aiming to establish passenger travel across the Atlantic Ocean. In 1917, he founded his aircraft manufacturing company, where he focused on groundbreaking flying boat designs. Rohrbach's aircraft earned a reputation for their reliability and exceptional long-range capabilities, setting distance records during the 1920s.

[3] Fred Ernest Weick (1899 – 1993) was an airmail pilot and aircraft designer. He promoted the need for full-scale propeller tests and carried on the first full-scale wind tunnel propeller research at NACA in 1925. He won the 1929 Collier Trophy for his design of the NACA cowling for radial air-cooled engines. He was responsible for the design of *Ercoupe*, the first commercial plane with tricycle landing gear and the design of a plane incapable of spinning. At *Piper*, he guided the design of the initial *Cherokee* and *Pawnee* series.

The *Lockheed Vega*, the most advanced design of its day by *Jack Northrop*, was one of the first airplanes to adopt the NACA cowling, Fokker-influenced high, strut-free single wing and streamlined monocoque fuselage. The aircraft was flown by *Amelia Earhart* in breaking two world records.

in streamlined trousers or by retracting undercarriages which became the standard later.

Variable-pitch and constant-speed propellers were introduced for efficient use of propellers over different flight conditions. Prior to this innovation, propellers setting was optimized for a flight condition, often compromising performance at other conditions. For instance, during his transatlantic flight in the *Spirit of St. Louis*, *Lindbergh* faced challenges due to his propeller being optimized for cruise rather than takeoff, affecting his departure. The ability of the pilot to vary the propeller pitch depending on the flight condition resulted in a more efficient operation. In a constant speed propeller, the pitch of the propeller is automatically changed using a governor mechanism keeping the rotational speed fixed.

Flaps are deployed to alter the wing's shape during take-off and landing. For instance, they enable high-speed aircraft to enhance lift for smoother landings and take-off.

Safety measures were enhanced by affixing de-icers to the front edges of wings. In the beginning, these were inflatable rubber devices that fractured ice upon impact, effectively preventing ice buildup.

Blind Flight Technology

After World War I, aviation's role modified from primarily military applications to include commercial activities involving mail, cargo, and passenger flights. The nascent aviation industry faced operation challenges during night and in adverse weather conditions, such as fog and clouds which obstructed the pilot's view of the ground or other obstacles. A serious challenge for scheduled commercial flights was the difficulty associated with flying in low visibility which posed the problem not only of how to find your way but also of how to avoid hitting any tall obstacles.

Pilots usually flew by relying on their sense of sight. When visibility is reduced due to fog or dense clouds, pilots can easily became disoriented leading to dangerous situations. Throughout the 1920s, there was gradual advancement in cockpit instruments designed to aid pilots when navigating under low visibility conditions. Initially, these mechanical instruments provided altitude, attitude, direction, and airspeed information but lacked the crucial spatial positioning data essential for safe landings. To enable aircraft to navigate through fog or low visibility conditions, the development of accurate instruments and navigation aids became imperative.

Blind flight refers to the navigation and control of an aircraft solely relying on instruments inside the cockpit, rather than on visual cues from outside the aircraft. For the first time, blind flight took place in 1929. Test pilot *James Doolittle*'s biplane was equipped with an altimeter 20 times more accurate than the standard devices in use. To replace the turn-and- bank indicator, an artificial horizon was used. This instrument combined a bar representing the horizon and a small airplane symbol. When the aircraft banked, the horizon bar tilted, and if it changed pitch the bar rose or fell accordingly. *Doolittle*'s primary locating device was a radio. He was equipped to receive instructions from a ground controller and to orient himself on a radio beam.

James Harold Doolittle (1896 – 1993), the first blind flight in September 1929 in a *Consolidated NY-2 Husky* modified with a new generation of flying instrumentation.

Robert Alexander Watson-Watt (1892 – 1973), a Scottish engineer and a pioneer of radar technology, patented the first practical radar system for meteorological applications in 1935. During World War II, radar was successfully used in Great Britain to detect incoming aircraft for interception.

Air Traffic Control: As the number of aircraft increased and flight operations grew complex, the need for a system to manage the air traffic to ensure safe air travel became evident. In the early days of aviation, there was minimal regulation for managing air traffic. Subsequently, airports established control towers from where the operations could be observed and the controllers communicated with pilots using radio signals for safer movement of the aircraft. In the USA, the first air route traffic control center, which directs the movement of aircraft between departure and destination, was opened in Newark in 1935, followed in 1936 by Chicago and Cleveland. After the World War II, with the use of radar and other equipment, the operations of the ATC became more organized and safer.

Jet Engines

From the *Wright brothers' Flyer-I* to the airplanes used in World War II were powered by reciprocating engines that worked by repeatedly injecting a small amount of fuel inside a piston-cylinder configuration and then igniting it. The resulting motion of the piston was used to rotate a propeller, which helped to move the aircraft forward. However, a propeller-driven aircraft has limitations on speed and altitude. The propeller performance degraded significantly as the tip speed of the propeller approached the speed of sound ($\sim$ 340 m/s) due to the occurrence of gasdynamic shock waves. The engineers began to look for engines which could overcome the issue. Therefore, jet engine technology became a more viable option for the pursuit of breaking the sound barrier and reaching new heights.

Hans von Ohain (1911 – 1998) was a German engineer credited with independently inventing and developing the jet engine around the same time as *Sir Frank Whittle*. His work resulted in the first flight of a jet-powered aircraft, the *Heinkel He 178*, in 1939 and the *He S 8A* jet engine propelled the aircraft. In 1947 *von Ohain* came to the United States and became a research scientist at Wright-Patterson Air Force Base in Ohio. *Ohain*'s contributions to aviation technology were instrumental in shaping the future of jet propulsion and modern aircraft design.

Ohain's journey into jet propulsion began during his pursuit of a doctorate at the University of Göttingen, where he formulated his groundbreaking theory in 1933. His dedication to innovation and research led to the awarding of a patent for his turbojet engine concept in 1936. Subsequently, *Ohain* became part of the *Heinkel Company* in Germany, where he further contributed to the evolution of jet propulsion technology. In 1937, he successfully constructed a factory-tested demonstration engine. Notably, his groundbreaking work led to the development of the world's inaugural jet aircraft, the experimental *Heinkel He 178*, which took its maiden flight on August 27, 1939. Building on this advancement, German engine designer *Anselm Franz* [4] engineered an engine tailored for fighter jet applications, leading to the creation of the *Me-262* aircraft by *Messerschmitt*. While being the sole jet fighter operational in combat during World War II, the *Me-262* faced extended periods of inactivity due to its substantial fuel consumption. It was also vulnerable to Allied attacks, earning the reputation as an exposed target.

Whittle W.1X, turbojet engine designed by *Frank Whittle*.

Manufacturer	Power Jets, Ltd. (London)
Designer	Frank Whittle
Thrust	5,516 N at 17,750 rpm
Compressor	Centrifugal (1 stage, 2 entry),
Combustor	10 reverse flow chambers
Turbine	Single-stage axial
Mass	254 kg

During the same time, a similar development of jet engine was underway in England. *Frank Whittle* (1907 - 1996) independently engineered a complete jet engine. *Whittle*'s ingenuity led him to secure a patent for the turbojet engine in 1930, although the engine didn't undergo flight testing until 1941. The British subsequently utilized this groundbreaking engine for another early jet fighter, the *Gloster Meteor*. While this aircraft was employed for homeland defense, its speed limitations prevented its deployment in combat against Germany during World War II.

The United Kingdom shared *Whittle*'s revolutionary technology with the United States, allowing *General Electric (GE)* to manufacture jet engines for America's inaugural jet fighter, the *Bell XP-59*. Building upon *Whittle*'s designs, the British continued to advance jet engine technology. In 1944, *Rolls-Royce* initiated work on the *Nene* engine. *Klimov VK-1*, a reverse-engineered version of the *Rols Royce Nene*, was used by the Soviets to power the *MiG-15* jet fighter.

Air Commodore **Frank Whittle** (1907 – 1996) was a British aviation engineer and inventor whose groundbreaking work led to the development of the jet engine. His innovative concept of the turbojet engine revolutionized aviation by enabling high-speed, efficient propulsion. In 1939, the British Air Ministry placed a contract for the *W.1* engine to be flight tested on the new *Gloster E.28/39* aircraft. The *W.1* flew officially in the *E.28/39* on May 15, 1941. *Whittle*'s contributions transformed air travel, military aviation, and the entire aerospace industry, marking him as one of the key figures in 20th century aviation history.

The end of World War II unveiled significant wartime advancements made in the jet engine technology. Companies like *General Electric*, *Pratt & Whitney* and *Rolls Royce* drew insights from these war-time innovations, amalgamating them with the knowledge of designers like *Whittle* and *von Ohain*. Early jet engines exhibited high fuel consumption, thus prompting a fundamental challenge: the creation of a high-thrust engine which has an economical fuel usage. *Pratt & Whitney* addressed this

---

[4] Anselm Franz (1900 — 1994) was a pioneering Austrian jet engine engineer known for the development of the *Jumo 004*. The *Junkers Jumo 004* powering the *Messerschmitt Me 262* fighter and the *Arado Ar 234* was the world's first production turbojet engine in operational use, and the first successful axial compressor turbojet engine. Around 8,000 units were manufactured by *Junkers* in Germany. After migrating to the USA as part of *Operation Paperclip*, *Franz*'s work on turboshaft designs included the *T53* and the *T55* (turboshaft engines used in helicopters), the *AGT-1500* (a gas turbine engine main power plant of the *M1 Abrams* series of military tanks), and the *PLF1A-2*, the world's first high-bypass turbofan engine.

predicament in 1948 by integrating two engines into one. This innovative design featured twin compressors, each with independent rotation-the inner compressor provided robust compression for enhanced performance. Both compressors were powered by their respective turbines, arranged sequentially. This revolutionary development culminated in the birth of the *J-57 engine or JT3*, a hallmark of postwar aviation. Commercial airliners, including the *Boeing 707* and the *Douglas DC-8*, embraced the *J-57* engine, which debuted in 1953 with the U.S. Air Force, solidifying its place as one of the era's eminent engines.

The *Pratt & Whitney JT3* turbojet engine (1/4 scale model shown in figure) revolutionized air transportation when it entered into service with the *Boeing 707* in 1958. In the 1960s, the turbojet engine was modified into a low-bypass turbofan engine - the *JT3D* by replacing the first three compressor stages with two fan stages. The resulting increase in airflow lowered fuel consumption, noise, and emissions. *JT3D*s became popular and was used on aircraft like *Boeing 707-300*s and *Douglas DC-8*s.

Airmail Flight

An important utility of aviation after the World War I was for delivering mails. After the War, there were many aircraft and trained pilots available. Utilizing these resources for delivering mails across long distances became widespread after the War. This became quite popular because the delivery was much faster compared to that by trains or ships. The first recorded delivery of airmail happened in India in 1911. On February 18, 1911, French aviator *Henri Pequet* (1888 - 1974)[5] achieved this historic feat,

---
[5]Henri Pequet (1888 – 1974) was a pilot in the first official airmail flight on February 18, 1911. He began his aviation career as a balloonist, then joined the aircraft manufacturer, *Voisin*, in 1908, and made his first flight at Hamburg in 1909.

carrying the first officially sanctioned mail transported by airplane. Operating a *Humber-Sommer biplane*, *Pequet* ferried around 6,500 letters and cards during his flight. Covering a distance of 13 km, the flight embarked from an Allahabad polo field, soared over the Yamuna River, and concluded in Naini. Each piece of mail bore a distinctive cancellation showcasing an airplane, mountains, and the inscription *First Aerial Post, 1911, U. P. Exhibition Allahabad*. As the capabilities of aircraft underwent a revolution in range, speed, and payload, the prospects of practical airmail services burgeoned. Regular airmail services were established in various parts of the world in the following decades.

First official airmail flight by airplane, India, 1911.

Use of Parachute

In 1912, aviators initiated experiments with parachutes, a technology that was already extensively utilized by balloonists. The inaugural parachute jump from a powered airplane was executed by Captain *Albert Berry*, an American aviator in Missouri. Captain *Berry*'s feat involved testing a mechanism for carrying and deploying parachutes from an aircraft, which subsequently earned a U.S. patent for its inventors. This pioneering event marked a significant advancement in aviation safety and innovation. However, parachutes were not commonly available for the pilots of World War I. Due to this, pilots were sometimes faced with the grim prospect of jumping to death from their cockpits when the aircraft caught fire. After the World War I, more extensive use of parachutes were seen.

Antony Jannus and Captain *Albert Berry* of the U.S. Army, before their flight in Missouri on March 1, 1912. The parachute was packed inside the inverted cone.

Aerial Refueling

Aerial refueling is a remarkable aviation exercise which involves transferring of fuel from a tanker aircraft to a receiver aircraft while both are in flight. This complex and precise maneuver has had a profound impact on the capabilities of aircraft. Aerial refueling offers a multitude of advantages, extending significantly the operational range and endurance. The concept of aerial refueling dates back to the early days of aviation, but it wasn't until the mid-20th century that practical methods were developed and refined.

The idea of aerial refueling was proposed by *Alexander P. de Seversky*, a pilot in the Imperial Russian Navy, in 1917. After immigrating to the United States, he pursued a career as an engineer in the War Department. In 1921, *de Seversky* secured the first patent for air-to-air refueling. The first practical demonstration of fuel transfer between aircraft involved wing-walker *Wesley May*, aviators *Frank Hawks*, and *Earl Daugherty*. On November 12, 1921, *May* climbed from a *Lincoln Standard* aircraft to a *Curtiss JN-4* airplane, carrying a can of fuel on his back. He poured the fuel into the *JN-4*'s gas tank mid-flight. In essence, the inaugural aerial refueling was primarily a novelty, a daring endeavor born of curiosity and innovation, even though its practicality was limited.

Wesley May climbing from the *Lincoln Standard* to the *Curtiss Jenny* — the fuel-can can be seen strapped to his back as he hangs from the bottom of *Jenny*'s bottom left wing.

Following the audacious feat of the three aviators, a significant breakthrough was achieved in 1929. The US Army Air Corps successfully conducted the first refueling operation with *Question Mark* aircraft, a modified *Atlantic-Fokker C-2A* transport airplane, as the receiver and a *Douglas C-1* aircraft as the tanker. The result was a remarkable record of 150 hours of continuous flight. The advent of aerial refueling has transformed aviation, expanding its horizons and capabilities.

Crossing the Oceans

Crossing of the Atlantic Ocean in an aircraft stands as an iconic event in the history of aviation, marking an important landmark that redefined the possibilities of long-distance flight. This daring achievement showcased not only technological advancements but also the courage of aviators to fly long

The *Atlantic-Fokker X-2A (Question Mark)* stayed airborne for seven days in the January of 1929, refueling 42 times in the air.

distances over the ocean.

The transatlantic flight was challenging because of the large distances, unpredictable weather conditions, and the need for reliable aircraft capable of enduring extended flights. The first successful non-stop transatlantic flight was accomplished by *John Alcock*, a British Royal Navy and later Royal Air Force officer, and navigator Lieutenant *Arthur Brown* in June 1919. They flew in a *Vickers Vimy* biplane, a British heavy bomber aircraft, from Newfoundland, Canada, to Clifden, Ireland in approximately 16 hours.

Charles Lindbergh's [6] historic solo non-stop flight across the Atlantic in May 1927 captured global attention. Piloting the *Spirit of St. Louis*, *Lindbergh* completed the journey from New York to Paris in just over 33 hours, becoming an instant hero and symbolizing the triumph of human ingenuity. His achievement not only earned him international acclaim but also highlighted the potential of aviation for rapid global connectivity.

---

[6] Born in Michigan, *Charles Lindbergh* dedicated his life to aviation. In 1924, he enlisted in the Army, where he completed his pilot training. Following his military service, *Lindbergh* flew regular mail flights between Chicago and St. Louis, honing his flying skills amid all weather conditions. Equipped with his flying experience and unwavering determination, Lindbergh and his team aspired to claim the coveted Orteig prize, a reward of $25,000 to the first aviator of any Allied Country crossing the Atlantic in one flight, from Paris to New York or New York to Paris. The historic flight by *Lindbergh* commenced on May 20, 1927 from New York. After a long non-stop flight of 33 1/2 hours, Lindbergh landed in Paris. This historic flight in his aircraft called the *Spirit of St. Louis* etched his name into the annals of aviation history.

Five years after *Charles Lindbergh*'s solo transatlantic flight, in 1932, *Amelia Earhart* set off from Harbour Grace, Newfoundland in her *Lockheed Vega 5B*. Originally intending to land in Paris, bad weather and mechanical problems forced *Earhart* to instead touch down in Northern Ireland. The 3,260 km long solo flight covering in 14 hours 56 minutes put *Earhart* in the history books as the first female to fly solo across the Atlantic. *Earhart*'s adventure sparked interest in aviation on both sides of the Atlantic and underlined the potential for long-range air transport. Even today, *Earhart*'s legacy continues to serve as a role model for female aviators. Some records from *Earhart* are

- 1928: First woman to be a transatlantic passenger

- 1932: First woman to fly a nonstop transatlantic solo flight

- 1932: First woman to fly solo nonstop across America in a flight from Los Angeles to Newark, New Jersey covering 3,938 km in 19 hours and 5 minutes.

- First solo Hawaii to mainland flight in *Lockheed 5C Vega* aircraft covering 3,875 km.

She tragically went missing while attempting to fly around the world.

One of the most notable transpacific flight occurred in 1927 when *Albert Hegenberger* and *Lester Maitland* together made a historic, non-stop transpacific flight from Oakland, California, to Honolulu, Hawaii crossing the north Pacific Ocean. They piloted the *Bird of Paradise*, a *Fokker trimotor* airplane. This achievement marked the first successful flight across the Pacific Ocean, covering a distance of approximately 3900 km. Their daring endeavor solidified their place in aviation history, demonstrating the feasibility of long-distance flights over vast expanses of water. In 1928, Australian aviators *Charles Smith* and *Charles Ulm* completed the first transpacific flight. Piloting the

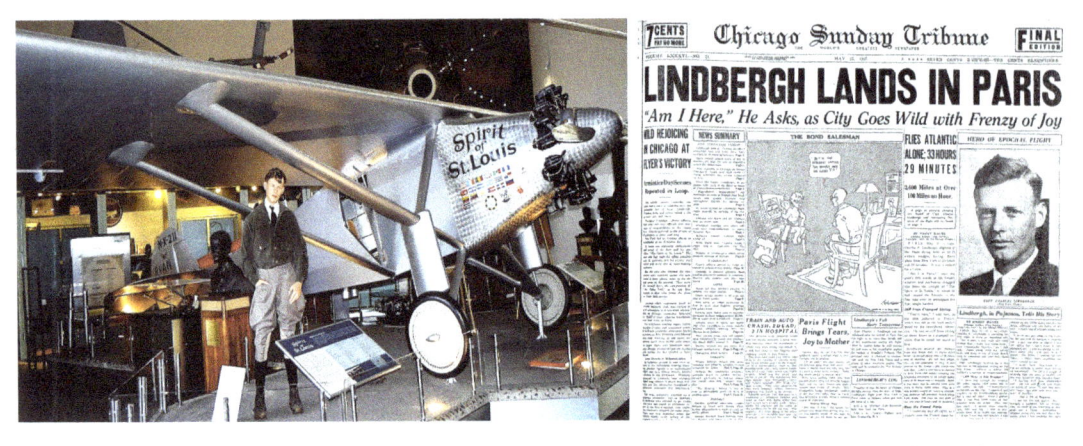

Charles Lindbergh (1902 - 1974), credited with the first solo flight across the Atlantic Ocean.

The *Spirit of St. Louis* was designed by *Donald Hall* under the direct supervision of *Charles Lindbergh*. The silver-colored aircraft had a single radial engine with a high-wing, monoplane configuration.

Wingspan	14 m	Engine	J-5C Wright Whirlwind
Airfoil	Clark Y		166 kW
Empty mass	975 kg	Maximum speed	210 km/h
		Range	6,600 km

Southern Cross, a *Fokker F.VIIb/3m trimotor monoplane*, they embarked on a 16,000 km journey from Oakland, California, to Brisbane, Australia.

1.4.2 Passenger Airlines

Post World War I, passenger air travel underwent rapid progress. The first daily international scheduled air service started in 1919 between Hounslow, London, and Le Bourget, Paris in an *Airco DH.16* aircraft. However, the travel was uncomfortable be-

Following his record-breaking flight, *Lindbergh* returned from Paris to receive a hero's welcome in New York. The welcome parade through New York was witnessed by around four million people. For his amazing achievement, *Lindbergh* was promoted from Captain to Colonel in the US Army Air Corps Reserve and received the first Distinguished Flying Cross.

Amelia Earhart (1897- 1937) was the first female aviator to fly across the Atlantic Ocean.

cause the passengers were exposed to the outside weather, noise and turbulence. The aircraft was forced to land on open field multiple times for re-

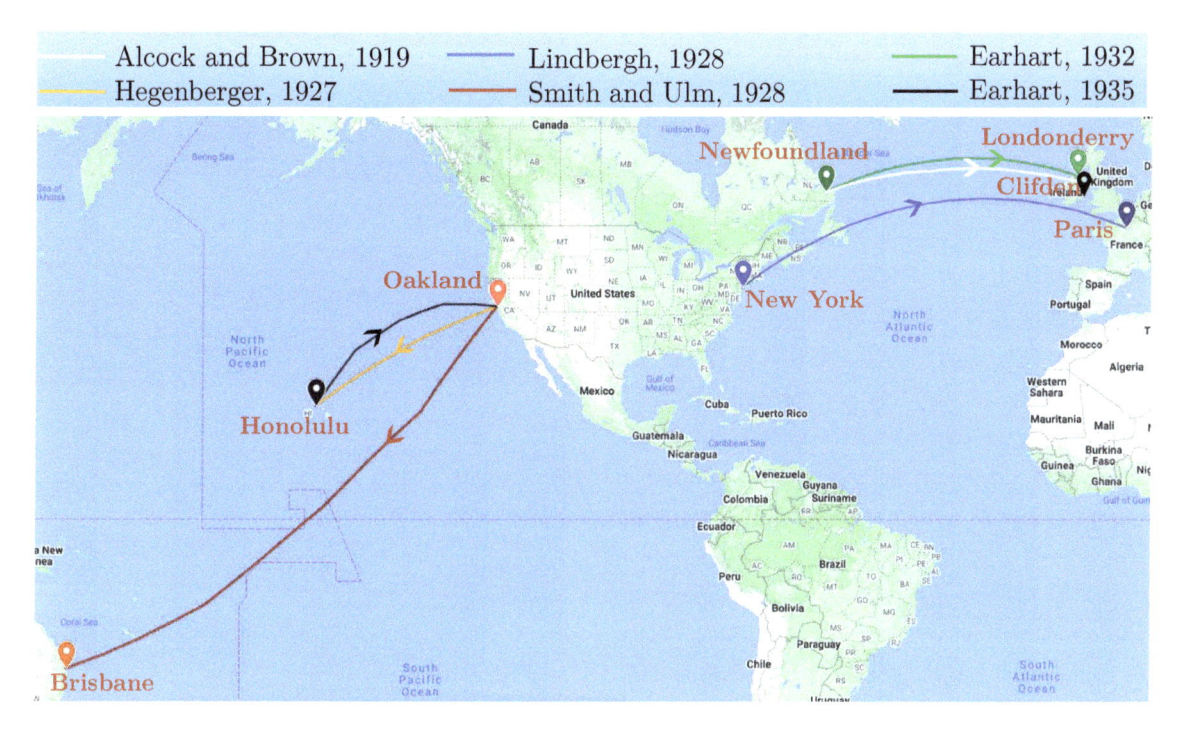

Representative paths taken by various aviators during transatlantic and transpacific flights.

A photograph commemorating the first trans-Pacific flight.

The American-built *DH-4*, manufactured by the Dayton-Wright Airplane Company. Modified *de Havilland DH-4* light bombers were being used as mail planes.

pairs and refueling. The airfare was so high for ordinary citizens to afford.

In the early postwar period, a few companies commenced air service in different parts of the word using bombers from World War I converted as passenger airlines. However, the seating capacity of most these aircraft were limtied to less than ten. For instance, in the *Handley Page O/11*, a converted *O/400 bomber*, two of the passengers had balcony seats in the open bow, combining an astonishing view with total exposure to the outside weather. Further, custom-built passenger aircraft emerged, Britain and France largely stuck with biplanes, such as the *de-Havilland D.H.34*, while the German and Dutch airlines flew more modern monoplanes such as the *Junkers F 13* and *Fokker F.III*.

As commercial aviation continued to advance, the 1920s witnessed the introduction of dedicated aircraft designed explicitly for commercial transportation of passengers. An example being the *Fokker F.II*, introduced in 1920, which featured a cabin capable of accommodating five passengers. However, it was the *Handley Page Type W8*, introduced in 1920 by *Handley Page*, that is often regarded as the first true commercial airliner. With its streamlined design, spacious cabin, and the ability to carry up to 15 passengers, the *W8* set the stage for the development of the modern airliner.

Introduced in 1926, The *Ford Tri-Motor* (nicknamed the *Tin Goose*), one of the first widely-used commercial airplanes designed and built by the *Ford Motor Company*, played a crucial role in changing the prospects for commercial aviation. It could accommodate up to 11 passengers and projected the potential for reliable commercial aviation.

During the 1930s, airlines faced tough challenges. On top of the economic recession, the government supports were reduced. The regulation for air transportation was confusing and people were still skeptical about its use as regular form of transportation. Surviving these challenges required airlines to have better planes that could operate profitably and reliably. Improved navigation and communication equipment were also essential for safety and efficiency. The aviation industry responded by designing advanced, high-performance airliners.

The *B-247* designed by the *Boeing company* became the world's first modern airliner, equipped with innovations like night flying capabilities, twin engines, two-way radios, and safety-enhancing equipment. The aircraft featured an all-metal cantilever wing design, retractable landing gear, and

Ford Tri-Motor, popularly known as the *Tin Goose*, was the largest civil aircraft in USA when it first flew on August 2, 1926. Its all-metal, corrugated aluminum construction and the prestigious *Ford* name made it popular with passengers and airline operators.

Wingspan	23.7 m
Empty mass	2,948 kg
Maximum speed	212 km/h
Engine	3 *Pratt & Whitney Wasps* or *Wright Whirlwind*

pneumatic deicing. With a capacity for 10 passengers, it outperformed the *Ford Tri-Motor* by flying 50 percent faster and completing cross-country trips in less than 20 hours. Its supercharged, air-cooled engines brought efficiency to air travel. The *Boeing 247-D* version introduced pitch control for propellers and wing de-icer boots, further advancing aviation technology.

Douglas company designed aircraft which were faster and more comfortable than the competitor, *Boeing-247*. The first aircraft was *DC-1* which could carry 12 passengers. The design was then stretched to the 14-seater, *DC-2*. *Douglas* went on to dominate airliner production until the jet age. For the *American Airlines*, *Douglas* created a larger version of the *DC-2* with sleeping berths called the *Douglas Sleeper Transport*. The regular version became the famous *DC-3*. First flown in 1936, the 21-passenger *DC-3* became the first airliner that could make a profit without subsidy, and it helped the US airlines in-

1.4. BETWEEN WORLD WAR I AND WORLD WAR II

The *Boeing 247D*, considered the world's first modern airliner, revolutionized air transportation when it entered service with United Air Lines in 1933. It is the first commercial aircraft with night flying capability. Its features included multiple engines, two-way radios, and other equipment that enhanced safety.

Wingspan	22.6 m
Empty mass	4,055 kg
Maximum speed	320 km/h
Engine	2 x *Pratt & Whitney Wasp S1H1-G* 370 kW each

DC-2 (top) and *DC-3* (bottom).

dustry survive cutbacks in government assistance. The *DC-3* revolutionized commercial aviation and by the time the World War II began, three-fourths of American air passengers were traveling in a *DC-3*. *DC-3*s continued to operate many decades after the World War II. The *Douglas C-47* (known as the *Dakota* in Britain), the military transport version of the *DC-3*, was produced in large numbers for military transportation during World War II.

Many flights were uncomfortable for the passengers because of air sickness due to turbulence experienced at lower altitudes. The solution would be to fly above the clouds close to stratosphere. By 1938, *Boeing* developed the four-engined *B-307 Stratoliner*, the first commercial airplane capable of operating at stratospheric height. The *Stratoliner* flew at sufficiently high altitude to reduce the effects of atmospheric turbulence, thereby providing a smoother ride for the passengers. The cabins were pressurized to protect the passengers from the low atmospheric pressure at high altitude, while turbo-superchargers enabled the engines to function efficiently and their performance enhanced by a high-octane fuel. It could carry 33 passengers in great comfort and cruise at 6 km altitude while maintaining a cabin pressure that corresponding to 2.4 km.

Newly graduated hostesses lined up in front of *TWA's Stratoliner*. *Boeing 307 Stratoliner*, an American stressed-skin four-engine low-wing tail-wheel monoplane airliner, was the first airliner with a pressurized fuselage, first flown in late 1938 and entered commercial service in 1940. It had a wingspan of 32.7 m, and had an empty mass of 13,600 kg with a top speed of 400 km/h.

Meanwhile, *Lockheed Corporation* was pushed by the *Trans World Airlines* to develop a competition for the *Stratoliner* resulting in *Constellation (Connie)*, a four-engined airliner capable of transporting people nonstop across the United States. With the World War II, all production of commercial aircraft was stopped and resources were devoted exclusively to building military aircraft. Work on the *Constellation* was halted, and then the aircraft was ordered by the armed forces as the *C-69* military transport aircraft.

Lockheed Constellation with its characteristic dolphin-shaped fuselage, triple tail and popularly known as *Connie*, was a sleek, powerful, and graceful aircraft.

1.4.3 Flying Boats

The flying boats take-off and land on water, which eliminate the requirement for elaborate runways and associated infrastructure. This forms a markedly different approach to aviation compared to the conventional aircraft. Before World War II, the development of flying boats was a significant aspect of development in aviation. These unique aircraft combined the capabilities of airplanes with the versatility of boats, enabling them to take off and land on water. However, this unique feature poses its own challenges. The concept of the flying boat emerged in the early 20th century when aviation was at its infancy. These aircraft are equipped with large, buoyant hulls that allows them to take off and land on water surfaces.

On March 28, 1910, the French aviator *Henri Fabre* performed a water takeoff using an aircraft, known as a *Hydravion*. This achievement drew the attention of aviation figures like *Glenn Curtiss* and *Gabriel Voisin*, who leveraged *Fabre*'s invention to develop their own seaplanes. *Curtiss* flew a series of seaplanes in 1911 and creating the first practical float plane in 1912. In between the World Wars, flying boat design saw significant development. Companies like *Short Brothers* in the UK and *Consolidated Aircraft Corporation* in the US and *Dornier* in Germany produced successful flying boats for both civilian and military purposes.

The German-built *Dornier Do X* with twelve-engines, launched in 1929, was the largest aircraft at that time. It could carry up to 169 passengers for a range of 1,700 kilometers.

The historic *Boeing 314 flying boat* made the inaugural around-the-world flight by a commercial airliner. The flying boat emerged as the natural choice for extended overseas flights. Given that many major global cities are situated along coastlines with accessible ports, the need for costly conventional airfields was minimized. During World War II, flying boats were used for maritime reconnaissance and anti-submarine warfare. However, World War II changed the prospects of flying boat. Wartime necessity prioritized swift movement of people and supplies, leading to widespread production of thousands of four-engine newer and larger planes such as *Boeing B-29 bombers* and *Douglas DC-4s* and *Lockheed Constellation* transports. For their operation, extensive concrete and asphalt runways were constructed globally. As a result of these developments, four-engine heavy airplanes dominated international air travel after the War. With

Boeing 314 flying boat, the first Around-the-World Commercial Flight

widespread runways available, the bulky, inefficient flying boat became obsolete due to its high-drag fuselage.

1.5 As a Game Changer during World War II

During the course of World War II, aviation established its pivotal role in modern warfare. The War saw extensive use of aircraft in most of the battles and events that determined the course of the War. A few such events are the *Battle of Britain* where the British *Royal Air Force* defended against the onslaught of German *Luftwaffe*, the attack on Pearl Harbor by the Japanese Navy Air service which pulled in USA into the War, the *Operation Overlord* where more than a thousand aircraft were used by the Allied forces for airborne assault before the *Normandy landings* and the harrowing atomic bombings of Hiroshima and Nagasaki which eventually led to the surrender of Japan in the War. Both the Axis forces and the Allied forces deployed formidable air fleets, engaged in fierce confrontations both against each other and in support of ground operations. During the War, strategic bombing of civilian areas and industrial infrastructure assumed a scale of operation never seen before in any other war. The War saw aircraft carriers assuming a significant role in shaping the outcome of battles, compelling significant changes in the naval warfare tactics. We will examine the role played by aircraft in some of the watershed events during the War. To aid our understanding, the details some of the aircraft which played a crucial role and the ace pilots who operated them will be looked into.

1.5.1 Role of Aircraft in the War

From reconnaissance and air superiority to strategic bombing and troop transport, aircraft were vital across diverse roles during the War.

Battle of Britain

The *Battle of Britain*, from July to October 1940, witnessed the Royal Air Force and the German *Luftwaffe* fighting for air control over southern England. Famous aircraft like the British *Spitfire*, *Hurricane*, and German *Messerschmitt Bf 109* took center stage in this historic aerial confrontation which marked a critical phase of World War II. The aircraft engaged in relentless dogfights, and the British aircraft were trying to intercept German bombers such as *Heinkel He 111*, *Dornier Do 17*, *Junkers Ju 87*, and *Ju 88*. The use of radar technology allowed the RAF to detect incoming German squad with advanced warning, giving the RAF pilots a strategic advantage. The outcome of the *Battle of Britain* was largely determined by fighter aircraft, as the British Royal Air Force were able to push back the German air attacks.

Dunkirk Evacuation

The evacuation of Allied soldiers from Dunkirk, France during World War II was a pivotal event that showcased the critical role of airplanes in rescue operations. In 1940, as the German forces advanced, around 400,000 Allied troops found themselves trapped on the beaches of Dunkirk. The situation was dire, and the only way to rescue these troops was through a daring and coordinated effort involving naval fleet and a wide range of aircraft.

Aircraft played a vital role in the Dunkirk evacuation, providing essential air cover for the Royal

Naval fleet crossing the English Channel, reconnaissance, and transport capabilities. The RAF launched air attacks against German forces, disrupting their advance and helping the safe evacuation of the troops on the ground. British fighter planes like *Spitfire* engaged in dogfights with German *Luftwaffe* to defend against enemy aircraft and create a safer environment for the evacuation of ships. *Spitfire*s successfully engaged with the *Bf 109*s, *Bf 110*s and *Junkers Ju 87*s inflicting significant damage and contributing to the progress of the evacuation. The fighter Command of RAF lost 87 airmen and more than 100 aircraft to enemy action over Dunkirk. In contrast, the *Luftwaffe* lost 97 aircraft to the RAF, including 28 *Messerschmitt Me 109s* and 13 *Me 110s*. The Dunkirk evacuation rescued more than 338,000 people while the British forces lost around 68,000 soldiers and extensive military equipment and supplies were left behind.

Attack on Pearl Harbor

It was the unexpected air strike by the Imperial Japanese Navy Air Service on the US naval base at Pearl Harbor in Hawaii on December 7, 1941 that pulled in the USA into World War II. The attack involved a two-wave aerial assault from six Japanese aircraft carriers, using more than 350 aircraft to devastating effect. All the US warships present there were either sunk or damaged, more than 180 aircraft destroyed, and around 2,400 American servicemen and women lost their lives. *Mitsubishi A6M Zero* fighters provided air cover and engaged American aircraft in dogfights. The Japanese *Aichi D3A Val* dive-bombers targeted American battleships, cruisers, and aircraft on the ground. The assault was exacerbated by the *Nakajima B5N Kate* torpedo bombers, which unleashed torpedoes against key naval battleships. The devastating attack forced USA to join the War along the Allied powers.

Hiroshima and Nagasaki Bombings

With the War dragging on, the United States dropped the first atomic bomb, *Little Boy*, on Hiroshima city of Japan from the *B-29 Superfortress*, *Enola Gay* bomber on August 6, 1945. With an impact equivalent to approximately 12.5 kilotons of TNT, the bomb wiped out about 13 square km of the city center, resulting in the deaths of around 120,000 individuals within the initial four days following the explosion. The devastating effects of the bomb lasted much longer. The *Enola Gay* travelled a distance of approximately 5000 kms and took 12 hours for the mission from Tinian island in the Pacific Ocean. On August 9, 1945, a second atomic bomb, *Fat Man* weighing around 4,600 kg, detonated over Nagasaki. Originally intended for the city of Kokura, the *B-29 Superfortress*, *Bockscar* bomber carrying the bomb was redirected to Nagasaki due to thick cloud cover. Nevertheless, around 5 square km of the city were completely obliterated, resulting in the loss of approximately 73,000 lives. As on today, these two bombings marked the only instances of nuclear weapons being used in warfare. The devastation caused by these attacks led to Japan's surrender, effectively ending World War II and shaping the new world order after the War.

Operation Overlord

The D-Day landings (Operation Overlord or Battle of Normandy) was the Allied invasion of German-occupied Normandy, France, on June 6, 1944. It involved coordinated amphibious assaults and airborne operations to penetrate into the German occupied territory. The successful landings marked a watershed moment that led to the liberation of Europe from the German control. Transport planes like the *C-47* facilitated troop movement, supply drops, and paratrooper deployments, playing a remarkable role in the D-Day landings. More than 11,000 aircraft participated in the *Operation Overlord*. The fleet that carried out the operation included the American *P-51 Mustangs*, *P-47 Thunderbolts*, British *Supermarine Spitfires*, and *Hawker Hurricanes*. Behind enemy lines, bombers such as *A20 Havocs*, *B-26 Marauders*, *Avro Lancasters*, and *P-38 Lightnings* carried out tactical bombing of enemy targets. The *C-47 Dakotas* transported airborne troops for pre-beach landings, while the British and American gliders like *CG-5 Hadrins* and *A.S. 51 Horsas* transported soldiers inland to strike German territory. Heavy bombers such as the *B-17 Flying Fortress* and the *Lancaster* carried out strategic bombings. Aircraft like the *IL-2 Sturmovik* and the *A-10 Thunderbolt II* provided close air support to ground troops. Planes like the *P-38 Lightning* and *Focke-Wulf Fw 189* played key roles in obtaining crucial information for both sides.

It is quite clear from these battles or events that various aircraft played a multitude of significant roles that eventually determined the outcome of the War. Next, we will examine some of the aircraft that played a major role during the War.

1.5.2 Notable Aircraft during the War

World War II witnessed the deployment of numerous aircraft in large quantities that played out in shaping the course of the conflict. From the fighters and bombers to the reconnaissance and transport planes, this section delves into the notable aircraft that left a lasting mark on aviation history during this eventful War. They not only showcased technological advancements but also demonstrated the military significance of air power during the War.

Supermarine Spitfire (United Kingdom)

The *Supermarine Spitfire*, a British single-seater fighter aircraft, stands as one of World War II's most renowned aircraft. Designed to be an interceptor by *R. J. Mitchell* at the *Supermarine Aviation Works*, its design not only influenced numerous other British planes of the era but also showcased a streamlined fuselage with elliptic wings that enabled maximum speeds of up to 600 km/h. The *Spitfire*'s superior capabilities came to light when the *Hurricane* proved inferior to German fighter aircraft.

The *Supermarine Spitfire (HF. Mk. VIIc)* is a single-seater high-altitude fighter. It features an enclosed cockpit, aluminum monocoque stressed skin construction, and an elliptical, monoplane low-wing.

Wingspan	11.2 m
Empty mass	2297 kg
Maximum speed	600 km/h
Engine	*Rolls-Royce Merlin 45*, liquid cooled V-12, fitted with two-speed two-stage supercharger providing 1,100 kW
Number of aircraft built	∼ 20,300

Hawker Hurricane (United Kingdom)

The *Hawker Hurricane*, a single-seater fighter, exhibited remarkable capabilities and played a crucial role in various combat scenarios during World War II. The *Hurricane* possessed excellent maneuverability and stability, making it suitable for dogfights and aerial combat. Notably, *Hurricanes* played a pivotal role during the Battle of Britain, where they defended British airspace against German air raids.

De Havilland DH 98 Mosquito (U.K.)

De Havilland's Mosquito design emerged in response to Britain's war-time need for an affordable, easily manufacturable, and capable bomber. Con-

Hawker Hurricane Mk. IIC, single seat, low wing monoplane ground attack fighter, is one of the most important aircraft designs in military aviation history. Designed in the late 1930s, the *Hurricane* was the first British monoplane fighter and the first British fighter to exceed 483 km/h in level flight. *Hurricane* pilots fought the Luftwaffe and helped win the Battle of Britain in 1940.

Wingspan	12.19 m
Empty mass	2,606 kg
Powerplant	*Rolls-Royce Merlin XX* V-12 liquid-cooled piston engine
Maximum speed	550 km/h
Range	970 km
Ceiling	11,000 m
Rate of climb	14.1 m/s
Number of aircraft built	∼ 14,500

structed primarily from wood and adhesive, it became the fastest aircraft of its era. With the maximum speed of 668 km/h and agility made it an important aircraft with the British. It had a predominantly wooden frame, earning it the nickname *Wooden Wonder*.

Avro Lancaster (United Kingdom)

The *Avro Lancaster* stands as one of the most famous bombers of World War II. Designed and manufactured by the British aircraft manufacturer *Avro*, the

The *de Havilland DH.98 Mosquito (B Mk.XVI variant)*, a British twin-engined, multirole combat aircraft, made its debut during the World War II.

Wingspan	16.51 m
Empty mass	6,486 kg
Powerplant	*Rolls-Royce Merlin 76/77* V-12 liquid-cooled piston engine, 1,280 kW
Maximum speed	668 km/h
Range	2,100 km
Service ceiling	11 km
Climb Rate	14.5 m/s
Guns	4 × 7.7 mm *Browning* machine guns, 4 × 20 mm *Hispano* cannons
Bombs	1,800 kg
Number of aircraft built	∼ 7700

heavy bomber played a pivotal role in the War, particularly during the nighttime bombing campaigns against Germany.

North American P-51 Mustang (USA)

The *P-51 Mustang*, designed by *North American Aviation*, is renowned as a successful fighter in World War II. Distinguished for its extended range,

1.5. AS A GAME CHANGER DURING WORLD WAR II

The *Avro Lancaster* was a British heavy bomber used during World War II.

Wingspan	31.09 m
Empty mass	16,738 kg
Powerplant	4 × *Rolls-Royce Merlin XX* V-12 liquid-cooled piston engines, 950 kW each
Maximum speed	454 km/h
Range	4,070 km
Number of aircraft built	∼ 7,300
Bombs	6,400 kg

speed, and high-altitude capabilities, it was a game-changer in the War. Introduced to the War in 1942 with the RAF, the *Mustang* swiftly altered the course of the war, competing effectively against the German *Bf 109* and *FW 190* of the *Luftwaffe*.

Boeing B-29 Superfortress (USA)

Boeing's *B-29 Superfortress*, four-engined, propeller driven heavy bomber, was known for its dropping of atom bombs over Hiroshima and Nagasaki in Japan. Apart from dropping the bombs, these aircraft carried out a diverse range of aerial operations, from delivering conventional and incendiary bombs to deploying mines. The *Boeing B-29* is called *Superfortress* due to its exceptional size, and superior capabilities compared to other bombers of its era. The *B-29* had an array of innovative features,

North American P-51D-30-NA Mustang, single-engine, low-wing, long-range fighter, one of the best fighters of World War II. Its combination of speed, range, maneuverability, and firepower gave it great versatility. Its use in all major theaters of the war included long-range high-altitude escort, strafing, and photo reconnaissance. Six *Browning* machine guns were embedded in the wings and it could carry up to 10 rockets or a 230 kg bomb making it one of the most lethal fighters of the War.

Wingspan	11.28 m
Empty mass	3,463 kg
Powerplant	*Packard (Rolls-Royce) V-1650-7 Merlin*, 12-cylinder liquid cooled engine, 1,110 kW
Maximum speed	710 km/h
Range	2,660 km
Service ceiling	12,800 m
Climb Rate	16 m/s
Number of aircraft built	∼ 15,000

such as a pressurized cabin, a centralized fire control system, a set of 12 50-caliber machine guns, and four 2200 hp engines.

Yakovlev Yak-3 (Soviet Union)

The *Yakovlev Yak-3* was a Soviet World War II fighter aircraft renowned for its agility, maneuverability, and exceptional performance. Designed by *Alexander Yakovlev*, its lightweight construction, streamlined design, and a powerful engine made the *Yak-*

Boeing B-29 Superfortress, an American four-engined propeller-driven heavy bomber, was the first bomber to house its crew in pressurized compartments and one of the largest aircraft of World War II.

Wingspan	43.05 m
Empty mass	33,793 kg
Powerplant	4 × *Wright R-3350-23 Duplex-Cyclone* 18-cylinder, air-cooled turbosupercharged radial piston engines, 1,600 kW each
Maximum speed	575 km/h
Range	5,230 km
Service ceiling	9,710 m
Number of aircraft built	∼ 4000

The *Yakovlev Yak-3*, a single-engine, single-seat World War II Soviet fighter. Among the most compact and lightweight combat fighters used by any faction in the war, its remarkable power-to-weight ratio translated into exceptional performance, establishing it as an adept dogfighter.

Wingspan	9.2 m
Empty mass	2,100 kg
Powerplant	1 × *Klimov* VK-105PF2 V-12 liquid cooled piston engine, 960 kW
Maximum speed	646 km/h
Power to mass ratio	0.35 kW/kg
Number of *Yak* series aircraft built	Over 20,000

3 one of the most nimble fighters of its time. The aircraft's excellent aerodynamics coupled with the powerful *Klimov* engine allowed it to excel in dogfights and close-quarter combat. Its top speed of around 655 km/h made it highly competitive against German fighters. The *Yak-3*'s smaller size and lightweight design contributed to its agility and responsiveness, enabling it to outmaneuver opponents effectively.

Messerschmitt Bf 109 (Germany)

The *Messerschmitt Bf 109* was the backbone of the German *Luftwaffe* during the War. With more than 34,000 aircraft produced, it became the most manufactured German aircraft. Renowned ace pilot *Erich Hartmann* achieved an astonishing 352 kills flying this aircraft, earning it the title of the deadliest World War II plane.

Junkers JU 87

Stuka dive bombing was a distinct aerial tactic employed by the German *Luftwaffe* during World War II, primarily using the *Junkers JU 87* dive-bomber, commonly known as the *Stuka*. This technique involved a steep and controlled dive, allowing the aircraft to release the bombs at low altitudes before pulling up. Armed with four 7.9-millimeter machine guns, it carried one 500-kg or one 250-kg bomb beneath its fuselage, along with two 50-kg bombs under each wing. A distinctive feature of

1.5. AS A GAME CHANGER DURING WORLD WAR II

The *Messerschmitt Bf 109 G-6/R3*, a single engine, single seat, low wing, fighter, was produced in larger quantities than any other combat airplane in history except the Soviet *Ilyushin Il-2 Sturmovik*.

Wingspan	9.92 m
Maximum speed	640 km/h
Rate of climb	20.1 m/s
Engine	Daimler-Benz DB 600 series inverted-V, liquid-cooled engine 1085 kW
Number of aircraft built	∼ 34,000

the *Stuka* was the inclusion of wind-driven sirens on its landing gear and cardboard sirens on the bombs. These sirens emitted a distinctive and ominous wailing sound during the dive, creating an unnerving effect on the enemy.

Focke-Wulf FW-190 (Germany)

The *FW-190* featured the advanced *BMW 801* radial engine – an innovative choice for its time. Capable of reaching speeds up to 652 km/h, the aircraft possessed considerable agility and power. Its armament included four *MG 151/20* cannons, which granted it a potent punch, supplemented by two *MG 131* machine guns.

Mitsubishi A6M Zero (Japan)

The *Mitsubishi A6M Reisen*, also known as the *Zero* Fighter, is an embodiment of Japanese air power

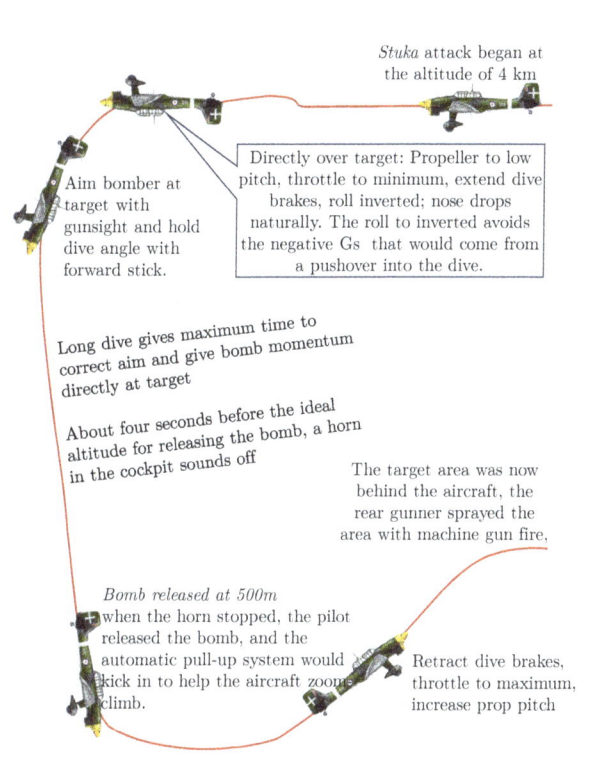

Stuka attack began at the altitude of 4 km

Aim bomber at target with gunsight and hold dive angle with forward stick.

Directly over target: Propeller to low pitch, throttle to minimum, extend dive brakes, roll inverted; nose drops naturally. The roll to inverted avoids the negative Gs that would come from a pushover into the dive.

Long dive gives maximum time to correct aim and give bomb momentum directly at target

About four seconds before the ideal altitude for releasing the bomb, a horn in the cockpit sounds off

The target area was now behind the aircraft, the rear gunner sprayed the area with machine gun fire,

Bomb released at 500m when the horn stopped, the pilot released the bomb, and the automatic pull-up system would kick in to help the aircraft zoom climb.

Retract dive brakes, throttle to maximum, increase prop pitch

Stuka dive bombing: A typical *Stuka* attack began from an altitude of 4 km, descending toward the target at a speed of approximately 480 km/h. Roughly four seconds prior to the optimal bomb release altitude, a cockpit horn signaled the pilot. Upon the horn's cessation, the bomb was released, activating the automatic pull-up mechanism for a rapid zoom-climb. As the aircraft moved away from the target zone, the rear gunner engaged with machine gun fire, suppressing defenders and aiding the subsequent *Stuka* dive bombers in achieving their missions.

during World War II. Co-produced by *Mitsubishi* and *Nakajima*, over 10,000 aircraft were built from 1939 to 1945. The *Zero*'s lightweight design and agility gave it an edge in maneuvers. Characterized by its sleek and lightweight design, the *Zero* boasted impressive agility and a remarkable rate of climb. Its design features included a low-wing monoplane configuration and a fully retractable landing gear. Another notable aspect of the *A6M Zero* was its exceptional range, allowing it to operate at extended distances from its carriers.

Towards the beginning of the War, there were very few airports that could support military operations. Throughout the war, aerodromes were

The *Focke-Wulf FW 190 A-8*, single engine, single seat, ground support fighter, was the only completely successful piston-engined fighter introduced by the German *Luftwaffe*, after World War II started.

Wingspan	10.506 m
Empty mass	3,200 kg
Powerplant	1 × *BMW* 801D-2 14-cylinder air-cooled radial piston engine, 1,250 kW
Maximum speed	652 km/h
Service ceiling	10,350 m
Climb rate	15 m/s
Wing loading	241 kg/m²
Guns	2 × 13 mm synchronized *MG 131* machine guns, 2 × 20 mm *MG 151/20 E* cannons, synchronized in the wing roots, 2 × 20 mm *MG 151/20 E* cannons in mid-wing mounts
Number of aircraft built	Over 20,000

Mitsubishi A6M5 Reisen (Zero Fighter) Model 52 ZEKE, Single-engine, low-wing fighter, emerged as the symbol of Japanese air power during World War II. The two companies–Mitsubishi designed and co-produced with Nakajima–built more than 10,000 Zeros between March 1939 and August 1945.

Wingspan	12 m
Empty mass	1,680 kg
Powerplant	1 x *Nakajima NK1C Sakae*-12 14-cylinder air-cooled radial piston engine, 700 kW
Maximum speed	533 km/h
Range	1,870 km
Service ceiling	10,000 m
Climb rate	15.7 m/s
Number of aircraft built	More than 10,000

rapidly constructed all over the world. Many of these aerodromes opened up for civilian operations after the War. Aircraft companies transitioned from producing a few aircraft intermittently to bulk-producing thousands on assembly lines. Despite the devastating effects of the War, many of the aviation technologies we take for granted today may not have happened otherwise.

The build-up of the necessary infrastructure for aircraft operations for the War allowed for a self-sustaining aviation ecosystem after the War. The War witnessed tremendous growth in the size of American military aviation, from about 2,500 airplanes to nearly 300,000 towards the end of the War. Manufacturers channeled their resources to the civil market after the War was over, with several military units being converted to passenger and cargo units. It can be seen that the World War II planes outperformed those of the World War I era planes. So, the aviation landscape was poised for making air travel commonplace after World War II. Invention of jet engines and the establishment of manufacturing and operation infrastructure during the War ushered in the *Jet age*, where aviation became the common mode of long distance transportation for the general public.

1.5.3 Flying Aces

Flying aces of the World War II were exceptionally skilled pilots who achieved noticeable success in aerial combat. They played a crucial role in shaping the outcome of the aerial battles by dominating the skies and contributing to the air superiority of their respective nations. The achievement of each ace differs in kill scores due to an array of influencing factors including the pilot's proficiency, the capabilities and limitations of their aircraft and opponents', their service duration and combat opportunities.

Erich Hartmann (1922 - 1993), a legendary figure in aviation history, is often regarded as one of the greatest flying aces of all time. *Erich Hartmann* is credited with an unparalleled record of 352 aerial combat victories. Flying the *Messerschmitt Bf 109*, he honed a distinctive approach as a 'stalk-and-ambush' fighter. His tactics included audacious dives through entire enemy formations, showcasing his audacity and unparalleled skill.

Gerhard Barkhorn (1919 - 1983) achieved the distinction of a German military aviator during World War II. A remarkable fighter ace, he is ranked as the second most accomplished fighter pilot in history, with 300 claimed victories as a fighter ace. Piloting the *Messerschmitt Bf 109*, *Barkhorn* flew more than thousand combat sorties to achieve this feat.

Marmaduke Pattle stands out as the foremost ace of World War II within the ranks of the Royal Air Force. Records suggest a kill count exceeding 60. Hailing from South Africa, *Pattle* predominantly served in the Mediterranean Theater and the Western Desert. He piloted both *Gloster Gladiators* and *Hawker Hurricanes*, securing the highest number of victories for both aircraft types. A significant portion of his confirmed kills-approximately 26-came from engagements against Italian aircraft.

1.5.4 Technological Evolution

By the time of World War II, streamlined, all-metal, cantilevered *monoplane* design replaced wood and fabric biplanes of World War I. Although some biplanes remained in service during the War, the design of new aircraft largely tilted towards the clean, unbraced monoplane wing design. The aircraft of World War II were mostly built using lightweight metals such as *duralumin* and began to use *retractable landing gear* and *flaps*, which significantly affects the performance and are essential to modern aircraft. Some of the technological progress related to aviation made at NACA are mentioned below.

U.S. Aviation Research at NACA

Drag cleanup: During World War II, NACA spearheaded efforts for drag cleanup with the aid of windtunnel testing. This entailed refining designs to reduce drag, thereby bolstering speed and efficiency. Utilizing the wind tunnels in its arsenal, NACA assessed drag, offering solutions for reducing drag. An apt illustration of the impact of such an effort is the drag reduction on *Bell P-39 Airacobra*. Through two months of focused effort, its top speed increased from 550 km/h to 630 km/h, a significant achievement which can help when facing-off with enemy aircraft in combat.

Deicing System: Icing is a pilot's nightmare, coating wings and propellers, undermining lift, and increasing drag, often leading to fatal accidents. NACA aimed to counter this threat by devising a heat deicing system for wings by channeling the engine-exhaust-warmed air along the wing's leading edge. This innovation was recognized with the Collier Trophy in 1946.

Trouble-shooting Engine Issues: NACA led pioneering efforts in troubleshooting high-powered piston engine issues at the Aircraft Engine Research

(a) *Consolidated B-24J Liberator* – range: 2,480 km, maximum speed: 478 km/h and bomb load: 1200-3600 kg depending on range

(b) *Junkers Ju 88 A-4* – range: 1,790 km, maximum speed: 470 km/h and bomb load: 1,400 kg

(c) *Boeing B-17G Flying Fortress* – range: 3,200 km, maximum speed : 462 km/h, and bomb load: 7,800 kg

Three most produced bombers of all time used in World War II, American *Consolidated B-24 Liberator* (18,188 built), *German Junkers Ju 88* (15,183 built) and American *Boeing B-17 Flying Fortress* (12,731 built).

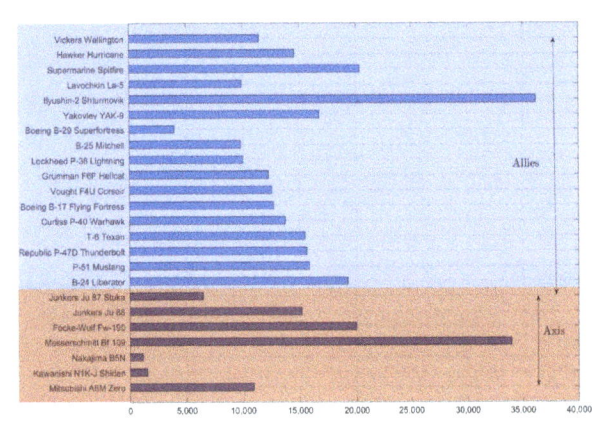

(a) Number of individual aircraft of Allies and Axis powers

Boeing made the B-52 bomber in 1950. It has eight turbojet engines, an intercontinental range, and a capacity of 225 ton.

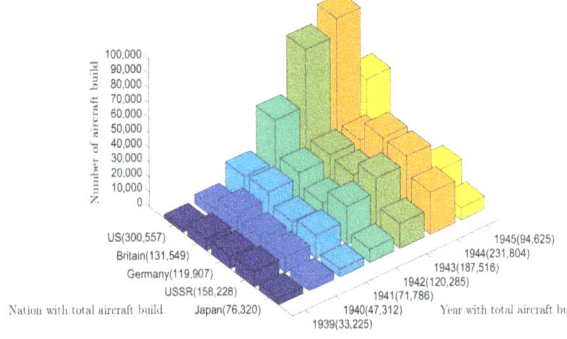

(b) Nations with total number of aircraft

Number of aircraft produced during World War – II

Erich Hartmann (1922 - 1993), a German fighter pilot and the highest-scoring flying ace in the history of aerial warfare.

Laboratory in Cleveland. Engineers examined engines earmarked for military aircraft production, innovating solutions for combustion, heat exchange, and supercharger challenges. With superchargers posing a major problem, NACA developed a centrifugal supercharger. Their work on *Boeing B-17* engines significantly enhanced their altitude and speed through turbo-superchargers.

Research Airfoil: NACA's renowned airfoil research culminated in a game-changing airfoil series introduced in 1940, revolutionizing low-drag wing design during World War II. These airfoils, particularly effective in laminar flow, enhanced aircraft

Marmaduke Pattle (1914 - 1941), a South African-born English World War II fighter pilot and flying ace.

Junkers Jumo 004 B4, designed by *Anselm Franz*, was the first mass-produced, operational turbojet engine. Its features included axial-flow compressors, afterburning, and a variable area exhaust nozzle.

speed and range. NACA's research on low-drag airfoil benefited aircraft like the *P-63 Kingcobra*, *A-26 Invader* and *P-80 Shooting Star*.

Stability and Control: NACA's wartime research on stability, control, and aircraft handling brought out quantifiable standards for plane stability and handling, influencing design across the aviation industry. The NACA also addressed spinning risks of aircraft by testing scaled models in spin tunnels. Research also addressed the phenomenon of uncontrolled dives near the sonic speed.

Design for Water Impact: NACA undertook an intensive initiative to study the load on aircraft upon impacting on water. A full-scale *Consolidated B-24 Liberator* was intentionally ditched in the James River, enabling measurement of impact forces on structural components. Insights gathered from these experiments were shared with aircraft manufacturers.

Seaplane's Porpoising Issue: Seaplanes faced *porpoising* (a bounced landing) during takeoff due to their bulky hulls which posed controllability issues. NACA delved into seaplane hull design and float problems, utilizing tow tanks and an impact basin at Langley Research Center. Study revealed that introducing a *step* on the hull's surface eliminated the problems. NACA also addressed water spray concerns by adding metal strips to the hull, redirecting spray from the propellers and wings.

Jet engine

The jet engine stands as a remarkable innovation in aviation history, maintaining its prominence as the preferred choice for high speed aircraft. Following World War II, jet engines evolved into larger, more efficient forms, becoming the standard form of propulsion for both military and commercial aircraft. Just prior to World War II, on August 27, 1939, the German *Heinkel He178* made history as the first jet-powered aircraft. Yet, the Germans encountered significant challenges and delays in their pursuit of practical military jet development. *Junkers* succeeded in refining their *Jumo 004* into an efficient turbojet engine, supplying it to *Heinkel*, *Arado*, and *Messerschmitt*. Among them, *Messerschmitt*'s *Me 262* emerged as the leading jet aircraft of the era. The pioneering *Me 262*, the world's first operational jet fighter, was equipped with two *Junkers Jumo 004 B* jet engines. This new jet engine technology enabled unprecedented altitude and speed, serving as a foundation for subsequent developments in passenger aircraft worldwide after the War.

Radar

Radio detection and ranging, widely known as *radar*, played an important role during the war in aviation. World War II introduced widespread radar usage, detecting enemy planes up to more than 120 km away, providing early alerts during the *Battle of Britain*. While modern radar has evolved, it remains essential for aviation safety, underpinning the global traffic control network. During the *Battle of Britain*, the British unveiled airplane-detecting radar, while the Germans developed *radio wave navigation techniques*. Both sides ventured into airborne radar, particularly effective for night attacks. In 1944, Britain introduced the *Instrument Landing Sys-*

tem (ILS) for adverse weather landings. These technologies developed during the War were harnessed extensively for making commercial aviation reliable and common place after the War.

1.6 As a Mainstream Technology after World War II

After World War II, aviation transformed into a mainstream technology with profound implications. The rapid advancements in aircraft design, propulsion systems, and infrastructure during the War formed the basis for the post-War era of aviation. As countries shifted from wartime to peacetime priorities and with the cold war looming, the aviation industry found itself at crossroads, navigating new opportunities and challenges. This section delves into this transformation of aviation as a vital component of modern society after World War II. From the boom in commercial air travel to the technological innovations that revolutionized aerospace engineering, this era marked the dawn of a new era where aviation became an integral part of the globalized world.

1.6.1 Robust, Reliable and Sophisticated Technology

Aviation technology evolved from its tottering origins to a robust, reliable, and sophisticated technology that revolutionized the way people and goods traveled across the globe. This period witnessed the use of cutting-edge research and innovative design, resulting in faster aircraft that were not only safer and more dependable, but also provided unprecedented levels of comfort to the passengers. The focus on enhancing safety measures, refining aerodynamics, and optimizing propulsion technologies led to the globalization of air travel, making it more accessible and affordable for both passengers and freight. On the military front also, the conventional performance characteristics improved steadily. The period also saw the introduction of disruptive technologies like stealth and use of advanced electronic and telecommunication systems that transformed military aviation significantly. In this subsection, we look at the advancements in technology in fostering a new era of aviation, where flying became a symbol of speed, convenience, and sophistication, opening up unparalleled opportunities for global connectivity and economic growth.

Pressurized Cabins

Cabin pressurization is the procedure by which conditioned air is pumped into and exhausted out of the cabin of an aircraft to keep the pressure in the cabin at a level equivalent to the pressure between sea level and 8000 feet, called *cabin altitude*. Up to an altitude of 8000 feet, use of supplemental oxygen is not required.

Although popular planes of the era such as the *Boeing 247* and the *DC-3* represented significant advances in aircraft design, they could not fly higher than 10,000 feet because people felt dizzy due to the reduced levels of oxygen at higher altitudes. This was a problem for many passengers and was an inhibiting factor to the industry's growth. The airlines wanted to fly higher, to get above the atmospheric turbulence common at lower altitudes. The problem was addressed with the introduction of the *Boeing 307 Stratoliner* in 1940. This marked the beginning of commercial transport aircraft equipped with a pressurized cabin. With the introduction of the digital electronic cabin pressure control system in 1977, the cabin pressurization became easy and a regular practise. This was followed up in 1979 with the implementation of fully automatic digital cabin pressure control systems. These systems utilized converging nozzle thrust recovery valves, further enhancing the aircraft's ability to maintain optimal cabin pressure.

Retractable Landing Gear

In the 1920s, aircraft designers recognized the significance of reducing drag for better performance. However, the understanding of what contributes

1.6. AS A MAINSTREAM TECHNOLOGY AFTER WORLD WAR II

Boeing's 307 Stratoliner – nicknamed the *Flying Whale* - began flying passengers in pressurized cabin from 1940.

The *Boeing 377 Stratocruisers* introduced with the *Pan American Wolrd Airways* in 1949 was an advanced aircraft for its time. It featured a pressurized cabin, two floors of seating for passengers, and it could carry up to 100 people with luxurious interiors.

to drag and how the shape affects the drag was sketchy. The NACA introduced the Propeller Research Tunnel in 1927, a large wind tunnel at Virginia's Langley Memorial Aeronautical Laboratory. Testing an entire airplane with engine and propeller in the tunnel revealed that the fixed landing gear contributed up to 40 percent of fuselage drag, surprising researchers. To reduce the contribution to drag from landing gear, various approaches were considered. Retracting the landing gear into the aircraft or redesigning fixed landing gear to reduce drag were the most evident options. Aircraft designer *Jack Northrop*, driven by the goal of streamlining for better performance, crafted his

aircraft during the 1930s with streamlined coverings extending from the fuselage. The fixed landing gear with streamlined coverings retained benefits of being lighter, cost-effective, reliable, and easier to maintain than retractable gear. However, despite these advantages, numerous designers in the 1930s adopted the retractable landing gear, prioritizing enhanced performance despite the challenges associated with the choice. Aircraft designers pursued various methods to incorporate retractable landing gear into aircraft fuselages. In certain aircraft, the gear was pulled straight upward, often into cowlings situated behind the engines (seen in *DC*-1, 2, and 3), sometimes allowing parts of the wheels to protrude externally. In smaller aircraft, the struts were folded inwards, positioning the wheels horizontally at the fuselage base, sometimes concealed by a door to further minimize drag. Ensuring secure deployment and locking of the gear was paramount, as a collapsible landing gear upon landing could lead to loss of control or damage.

In the early stages of retractable landing gear, the benefits were not always obvious. This was due to the added weight of the motors and machinery, which offset the drag reduction advantages and required a greater lift from the aircraft. As aircraft speeds increased, the drag penalty associated with fixed landing gear was enormous. Consequently, retractable landing gear became the norm for most high-speed airplanes.

Pilot Seat Ejection System

The pilot's seat ejection system, typically found in military aircraft, is a complex safety feature designed to protect pilots and crew in the event of severe aircraft damage or accidents. These seats incorporate explosive propellant or rocket motors to propel the pilot away from the aircraft, followed by parachute deployment after sufficient clearance. This system can subject pilots to high acceleration. The necessity for ejection seats emerged with the introduction of high-performance aircraft. Traditional methods like opening the cockpit and jumping out

Retractable landing gear of (a) *A340-600* (b) *Boeing 747-400* (c) *Airbus A380*

over the side were dangerous at high speeds. Countries like Sweden and Germany began experimenting with ejection seat concepts. *Saab* tested a gunpowder ejection seat for the *Saab J 21* fighter, while Germany developed a mechanism for the *Dornier Do335 Pfeil*. Over the years, the ejection seats have undergone significant improvements to accommodate higher speeds, altitudes, and various pilot profiles from all over the world. Still, ejection from aircraft using the mechanism continues to be risky because of the high acceleration experienced, the chances of accidental collision with any other object in the vicinity or due to getting into the aircraft plume.

Capt. *Christopher Stricklin* ejects from his *F-16* aircraft with an *ACES II* ejection seat on 14 September 2003 at Mountain Home AFB, Idaho.

G-suit or anti-G suit

A G-suit is a specialized piece of equipment worn by pilots and astronauts to help prevent loss of consciousness during high levels of acceleration or gravitational forces (G-forces).

Pilots who do such high-performance maneuvers, such as sharp turns, rapid climbs or dives, or combat situations, can experience significant G-forces that can cause blood to pool in the lower extremities. During combat situations where pilots need to perform evasive maneuvers to avoid enemy fire, the rapid acceleration and deceleration forces experienced can be extreme. This can lead to a condition called *G-induced loss of consciousness (G-LOC)* or *Black-out*, where the brain is deprived of oxygen and the pilot loses consciousness. Black-out and g-LOC have caused a number of fatal aircraft accidents. In such situations, G-suit is required for the pilot.

The G-suit works by applying pressure to the lower body, typically the legs and abdomen, using inflatable bladders or panels. This pressure helps to counteract the effects of the G-forces, keeping the blood from pooling in the lower body and ensuring a steady supply of oxygenated blood to the brain.

In 1931, Professor *Frank Cotton*, a physiology professor at the University of Sydney, introduced a novel method for determining the human body's center of gravity. This breakthrough allowed for a better understanding of how mass moves within the body when subjected to acceleration. It was observed that 30% of pilot fatalities resulted from accidents, including blackouts. The *Spitfires*, known for their agility in executing rapid turns that produced high g-forces, often led to pilots experiencing blackouts when diving to engage or evade enemy fire. *Cotton*'s insight into the necessity of an anti-gravity suit emerged during the 1940 *Battle of Britain* and resulted in the design of the world's first successful gas-operated anti-G suit.

In the development of g-suits, various designs have been created. Initially, water-filled bladders were utilized around the lower body and legs, followed by later designs incorporating pressurized air to inflate the bladders. These air-filled g-suits proved to be lighter than their liquid-filled counterparts and remain widely used. However, a new g-suit called the *Libelle*, designed by Swiss company *Life Support Systems AG* and German company *Autoflug* specifically for the *Eurofighter Typhoon* aircraft, has introduced a return to liquid as the medium, enhancing performance.

Breaking the Sound Barrier

In the early days of aviation, many believed that it will not be possible to travel faster than the speed of sound. This belief was bolstered by simplified aerodynamic theories that predicted that the drag approaches infinity as the speed of the vehicle approaches the sound speed. An accident involving a *de Havilland DH 108* in 1946 where *Geoffrey de Havilland Jr.* died while flying close to the sound speed reinforced the aviation myth called the *sound barrier*. Breaking the sound barrier or traveling faster than the speed of sound marks the achievement of supersonic flight and the overcoming of a formidable aerodynamic challenge. This milestone not only revolutionized aviation but also represented a triumph of engineering and human courage. (Speed of sound depends on temperature, and both the temperature and speed of sound decrease with altitude up to about 10 km. At 30 °C, the speed of sound is around 349 m/s.)

Owing to the remarkable performance of the *Spitfire*, propelled by its *Rolls-Royce Merlin* engine, some *Spitfire* pilots engaged in high-speed test flights for aeronautical research purposes. These flights yielded valuable insights that ultimately contributed to test pilot *Chuck Yeager*'s later accomplishment of achieving supersonic speeds. One no-

Squadron Leader *Anthony F. Martindale* (left) and other test pilots.

table occurrence in the pursuit of high-speed flight was witnessed during a test conducted at Farnborough, England. In April 1944, Squadron Leader *Anthony F. Martindale* achieved a remarkable feat aboard the *Spitfire* aircraft by breaking the 1,000 km/h (278 m/s) barrier during a high-speed dive. This achievement marked the highest dive speed recorded in a piston-engined aircraft.

During the flight, *Martindale* initiated a dive in his *Spitfire*, simultaneously encountering an unforeseen brake failure that was intended to control the aircraft's speed. As a consequence, the aircraft crossed the speed of 1,000 km/h, resulting in the propeller being torn off. Fortunately, *Martindale*'s survival hinged on the detachment of the propeller, which caused the tail-heavy aircraft's nose to pitch upward, inducing a rapid ascent. After experiencing significant acceleration that rendered him unconscious, *Martindale* regained consciousness to find his *Spitfire* soaring at an altitude of 40,000 feet.

Remarkably, *Martindale* skillfully maneuvered the aircraft back to the base through gliding, emerging from the incident unscathed. This harrowing encounter caused the wings of the *Spitfire* to bend into a swept configuration due to the forces endured. This incident threw light on the wing shape's role in supersonic flight, paving the way for the development of future supersonic aircraft.

Additionally, the incident highlighted limitations in propeller technology, prompting the pursuit of advanced solutions.

This event prompted engineers in the USA to create the experimental aircraft *Bell-X*, which featured an upgraded adjustable stabilizer. This innovation allowed pilots to make instantaneous adjustments to the angle of attack as the aircraft neared the speed of sound. This adjustment enhanced elevator effectiveness and played a crucial role in facilitating supersonic flight.

One of the most iconic breakthroughs occurred on October 14, 1947 when test pilot *Chuck Yeager* flying the *Bell X-1*, a rocket-engine powered aircraft built by the *Bell Aircraft Company* receiving contract from NACA and US Army Air Forces, achieved supersonic flight over the Mojave Desert in the Southwestern United States. This historic achievement not only demonstrated that it was possible to break the sound barrier but also opened the door to a new era of aviation.

The experimental aircraft known as the *Bell X-1 Glamorous Glennis* achieved an impressive speed of 1,127 km/h (equivalent to Mach 1.06) at an altitude of 13 km. This achievement was realized through a unique process, with the *X-1* aircraft being launched from the bomb bay of a flying *Boeing B-29* bomber. The *X-1* program gathered crucial flight data about the transonic and supersonic flight for the Air Force and the NACA, NASA's predecessor.

The *Bell X-1* was followed by even more remarkable advancements in speed with the introduction of aircraft such as the *Lockheed F-104 Starfighter* and the *Lockheed SR-71 Blackbird*. These aircraft not only surpassed the Bell X-1's groundbreaking accomplishment but also demonstrated the continual pursuit of higher speeds and performance in the field of aviation.

The *Lockheed F-104 Starfighter*, first flew in 1954, was designed as a high-speed interceptor and served in various air forces around the world. The *F-104* was capable of reaching speeds in excess

The *Bell X-1* achieved the historic feat by surpassing the speed of sound on October 14, 1947. U.S. Air Force Capt. *Charles Yeager* piloted the *X-1*, reaching a speed of 1,127 km/h (Mach number 1.06) at an altitude of 13,000 m.

Charles Yeager (1923 - 2020), a United States Air Force officer, flying ace, and record-setting test pilot, standing near the Bell *X-1*. He excelled as a *P-51 Mustang* fighter pilot during World War II and earned the credit for downing 11.5 enemy aircraft, with a half credit shared with another pilot in a joint victory. Notably, on October 12, 1944, he achieved "ace in a day" status by successfully shooting down five enemy aircraft during a single mission.

of Mach number 2 far surpassing the *Bell X-1*'s achievement. Additionally, the *F-104* held altitude records, with one aircraft reaching an astonishing height of approximately 31.5 km. Its sleek design, powerful engine, and exceptional climb rate made it one of the fastest and most remarkable aircraft of its time.

The *Lockheed SR-71 Blackbird* is perhaps the most renowned high-speed aircraft in history. Developed during the 1960s, it pushed the boundaries of what was achievable in terms of speed and altitude. The

SR-71 was capable of cruising at speeds in excess of Mach number 3 making it the fastest manned aircraft ever built. It could reach speeds of approximately 3,540 km/h. Its incredible speed allowed it to traverse continents in a matter of hours, making it a vital reconnaissance asset during the Cold War.

Swept Wing

Swept wings are a distinctive wing design characterized by their angled orientation, either backward or rarely forward, from the wing's root rather than a straight sideways direction. This innovative aerodynamic feature significantly impacts an aircraft's performance, particularly at high speeds near or beyond the speed of sound.

While the concept of swept wings had been contemplated by many prior to World War II, it was *Adolf Busemann* (1901 – 1986), a German aeronautical engineer of the early 1940s, who demonstrated their utility. The first practical application appeared on the *Messerschmitt Me 262* jet fighter. Following the war, the United States Air Force swiftly embraced this technology and effectively integrated it into jet aircraft.

This configuration, known as wing sweep, serves to delay the formation of shock waves and the associated rise in aerodynamic drag caused by fluid compressibility near the speed of sound, thus enhancing overall performance. Swept wings find common usage on jet aircraft designed to operate at high speeds. Additionally, swept wings are occasionally employed for reasons such as structural practicality or improved visibility.

A *B-52 Stratofortress* with its noticeable swept back wings.

Grumman X-29 experimental aircraft, an rare example of a forward-swept wing.

Forward-swept wings offer similar advantages to aft-swept wings such as pushing the critical Mach number to higher values and mitigating wingtip stall issues. NASA conducted trials with forward-swept wings on the *X-29*, revealing potential benefits. However, this design also presents challenges like flutter, wingtip stress, and manufacturing complexity. As new composite materials and advanced production techniques emerge, the possibility of forward-swept wings becoming a reality gains momentum.

Delta Wing

A delta wing is a triangular-shaped wing, often with a truncated tip, when viewed from above. It sweeps dramatically back from the fuselage, creating an angle of up to 60° between the leading edge of the wing and the fuselage, and approximately 90° between the fuselage and the trailing edge of

Adolf Busemann (1901 – 1986) proposed the use of swept wings to reduce drag at high speed, at the *Volta Conference* in Rome in 1935.

the wing. Some delta-wing aircraft lack horizontal stabilizers.

Delta wings are well-suited for high-speed flight due to their reduced drag and stall characteristics. The delta shape reduces the drag caused by wingtip vortices, allowing aircraft to achieve higher speeds with less power. This shape also excel in supersonic flight regimes. Their unique shape and sweep angle help delay the onset of gasdynamic shock waves, reducing wave drag and drag divergence. Delta wings often have fewer control surfaces, which can simplify the aircraft's design, leading to lower production and operational costs. The large surface area of delta wings provides good control characteristics, allowing for agile and responsive maneuvering. This is particularly advantageous in combat aircraft and situations requiring rapid changes in direction. However, delta wings are not universally superior. Conventional wings, such as swept wings or variable-sweep wings, have their own strengths, especially in subsonic flight and specific mission profiles. The choice between delta wings and conventional wings depends on the aircraft's intended purpose, design constraints, and overall performance requirements.

Notable modern delta-wing aircraft include the *HAL Tejas*, *J-20*, *Eurofighter Typhoon*, and *Dassault Rafale*. In the 1950s, the delta wing found application in aircraft requiring speed, such as the *B-58 Hustler* and the canceled *XB-70 Valkyrie* bomber. The Soviet Union utilized the delta wing for its *Tu-144* supersonic passenger jet and the renowned *MiG-21*, a widely-used fighter jet during the Cold War. The French adopted the delta wing for the successful *Dassault Mirage III*. The delta wing achieved iconic status through the *Concorde* and the *Space Shuttle*. The *Concorde*'s delta wing enabled its sustained cruising speed of Mach number 2. On the other hand, the *Space Shuttle*'s wing, referred to as a *cranked delta* due to a slight bend near the midpoint of the leading edge, served a different purpose. Initially designed for high cross-range capabilities, the *Space Shuttle*'s large delta wing allowed gliding for thousands of miles sideways during landing. Conventional straight wings could not generate sufficient lift at high speeds and altitudes for such extensive range, making the delta wing an essential choice.

IAF's *HAL Tejas* shows off its delta wing at AeroIndia 2023.

The Whitcomb Area Rule

After the rocket-powered *Bell X-1* successfully broke the sound barrier, the US Air Force sought advanced aircraft capable of supersonic flight. Despite industry engineers proposing their best designs, initial model tests indicated that surpassing this speed barrier was beyond their reach. The challenge lay in the considerable increase in drag as speeds approached the sonic threshold, overwhelming the limited power of jet engines at that time. This impasse was ingeniously resolved by *Richard T. Whitcomb* (1921 – 2009), an aerodynamicist stationed at the National Advisory Committee for Aeronautics (NACA), Langley Research Center in Virginia. The *area rule* suggests that in order to minimize drag buildup as an aircraft nears the speed of sound, the cross-sectional area distribution of the aircraft's fuselage should be kept as smooth and continuous as possible. When an aircraft's fuselage cross-sectional area changes abruptly, shock waves can form along these discontinuities, resulting in increased drag. *Whitcomb*'s innovation, the *area rule*, brought a groundbreaking shift in how engineers approached high-speed drag. This transformative concept influenced the design of many transonic and supersonic aircraft that followed. The data from experimental tests revealed a remarkable 60% reduction in the drag profile near the operational speed of Mach number 1 when the airframe was designed according to the area rule. The *Area rule* and supercritical airfoil, both innovations by *Whitcomb*, helped in the aerodynamic design of aircraft operating at speeds close to the speed of sound. The most common application of the area rule involves shaping the fuselage of an aircraft to resemble a *Coke bottle*. This involves narrowing the fuselage where it intersects with the wing and then gradually expanding it towards the rear. Another instance of application of the rule involves *Convair's YF-102* aircraft which employed a delta wing configuration to diminish drag and achieve supersonic speeds. However, the *Pratt & Whitney J-57* turbojet engine's power fell short in overcoming the transonic drag. *Convair*'s exploration led them to NACA, where *Whitcomb*'s concept gained prominence. The engineers integrated the concept into the *YF-102* airframes, calling the new model *YF-102A*, with significantly improved top speed.

Aircraft Simulator

Airplanes, being highly sophisticated machines, are challenging to operate and demand extensive training that spans hundreds of hours. Aircraft accidents pose risks to both occupants and those on the ground. Consequently, engineers and designers have long sought methods to instruct future pilots how to fly airplanes using ground based facilities thereby reducing the actual number of hours flying in the air without compromising on the skills to be developed. This led to the development of flight simulators, which have evolved into sophisticated tools crucial for training, safety protocol learning,

Richard T. Whitcomb (1921 – 2009) with area-ruled *F-106* aircraft at the retirement in 1991. As a budding mechanical engineer, he joined NACA to conduct tests in the transonic wind tunnel. *Whitcomb* proposed a groundbreaking idea to minimize aircraft drag: maintaining a smooth distribution of area near the wing structure, effectively creating an aircraft with a streamlined fuselage shape resembling a *Coke* bottle. This innovative approach, known as the Area Rule concept, emerged as a potential solution for reducing drag at high speeds.

Convair F-102, American interceptor aircraft, as originally designed, it could not achieve supersonic flight until redesigned with area ruling.

and even aiding in aircraft design and development. Aviation simulators fulfill three key functions. Firstly, they train pilots in essential operational procedures, proving cost-effective and safer compared to actual flights. Secondly, they serve advanced training needs, especially in emergency protocols, which might be too hazardous to practice during real flights or impossible to recreate. Lastly, simulators play a role in the design process for new aircraft.

Following World War II, pilot training became intricate, making use of analog computers, which eventually gave rise to simulators. The *Massachusetts Institute of Technology* embarked on the *Whirlwind project*, a pivotal early computer endeavor that produced the first interactive flight simulator responsive to user input. Its features were subsequently integrated into other simulators. Initially developed for specific aircraft, the *B-47B* flight simulator emerged as the first for a modern jet bomber. The *F-4C* Weapon System Trainer facilitated *F-4 Phantom* crew in mastering operation of missile and radar systems. Military requirements predominantly drove the demand for advanced simulators. By the 1950s, commercial airlines also showed interest in simulators for passenger jets, particularly to simulate in-flight emergencies. The *X-15* program introduced simulator use for practicing flight maneuvers before implementation, contrasting with earlier simulators developed post-understanding aircraft flying characteristics.

Later, in-flight simulators played crucial roles in developing aircraft like the *F-16*, *F-18*, *A-10*, *B-1*, and the *Space Shuttle*. Presently, it's rare for any significant aircraft to undergo production without cockpit simulator testing. Aircraft are simulated extensively, allowing designers to optimize instrument layouts, improve information displays, and assess handling qualities before physical production. This practice prevents costly design errors that might otherwise emerge post-production which can be even difficult to correct.

Fly by Wire

Fly-by-wire (FBW) technology represents a significant advancement in aviation, transforming the way aircraft are controlled and operated. The earliest aircraft were controlled by the pilot using steel cables, pulleys and hydraulic actuators. The use of these mechanical components like pulleys, cranks, tension cables and hydraulic pipes adds much weight to the aircraft and it requires careful routing and maintenance work. Unlike traditional mechanical systems that connect the pilot's controls directly to the aircraft's control surfaces, FBW employs electronic systems to transmit the pilot's inputs to the aircraft's flight control computers, which then interpret and execute these commands through electro-mechanical actuators. This electronic interface offers several benefits and has become a standard feature in modern aircraft, from commercial airliners to military jets.

Key advantages of FBW technology include enhanced safety, improved maneuverability, reduced weight, greater operational flexibility and ease of operation for the pilot. By eliminating physical linkages, FBW systems allow for smoother and more precise control responses. They also incorporate features such as envelope protection and stability augmentation, which prevent the aircraft from entering dangerous flight conditions and improve overall stability.

The initial example of a fully electronic fly-by-wire aircraft without any mechanical or hydraulic backup was the *Apollo Lunar* Landing Research Vehicle, which conducted its inaugural flight in 1964. Subsequently, similar technology was used in aircraft such as the *F-8 Crusader*, *Sukhoi T-4*, and *Hawker Hunter*. Notably, the *Airbus A320* from *Airbus Industries* marked a significant milestone in 1984 by becoming the inaugural commercial airliner to operate with a comprehensive all-digital fly-by-wire control system.

Autopilot

An autopilot is a system employed to steer an aircraft without direct input from the pilot. In its early stages, autopilots were limited to maintaining a consistent heading and altitude during cruise. However, modern autopilot systems have

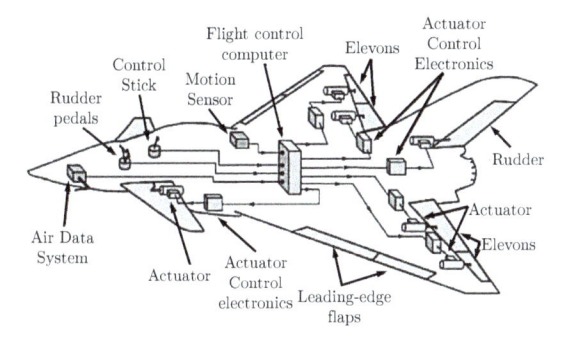

1.6. AS A MAINSTREAM TECHNOLOGY AFTER WORLD WAR II

Basic elements of the Fly-By-Wire (FBW) system. Unlike the mechanical systems that are used to transmit control inputs to the control surfaces, FBW systems convert the movements of the controls by the pilot into electrical impulses. The signals are sent to flight-control computers that reconvert the electrical impulses into instructions for control surfaces. Potentiometers in the control surfaces measure their position and transmit the data back to the flight computer. Once a control surface is in the correct place, the computer freezes the component, ensuring the pilot's commands are followed. The control system not only makes the controls more precise but also means airplanes no longer have to be fitted with the mechanical components like cranks, gears, pulleys and cables which are part of older aviation systems. FBW systems also make the aircraft substantially safer and easier to fly, as the flight computers can be programmed to carry out adjustments to control surfaces automatically. This is largely done using gyroscopes fitted in the aircraft which are connected to the on-board computers. These instruments measure disturbances in pitch, roll and yaw axes and the flight computer compensates for the disturbances.

The *Vought F-8 Crusader*: A single-engine, supersonic, carrier-based air superiority jet aircraft built by *Vought* for the US Navy and for the French Navy. The aircraft was used for proving the feasibility of digital FBW technology.

Avro Canada CF-105 Arrow, an aircraft designed and flown with a FBW system.

advanced capabilities, encompassing control over the entirety of flight operations from takeoff to landing. Typically, modern autopilots are integrated with the Flight Management System (FMS).

Integrated with navigation systems, autopilot software assumes command over the aircraft throughout various flight phases. When coupled with an autothrottle/autothrust system, this software can automatically set the appropriate thrust during takeoff, adjusting it as the climb progresses. As the aircraft ascends, it maintains the proper speed for its weight and surrounding conditions, leveling off at the designated altitude. Simultaneously, the aircraft adheres to the prescribed flight plan route. In case the autothrottle is absent, the pilot must manually make power adjustments based on the flight phase.

Upon entering the descent phase, power settings are modified, guiding the aircraft to descend at the suitable speed and along the specified route. The descent concludes with leveling as stipulated by the flight clearance, leading to the initiation of the approach phase. In instances of an Instrument Landing System (ILS) approach with Autoland capability, the autopilot manages the aircraft's flight path, aligning it with the ILS glide path and localizer. Power adjustments maintain the requisite speed, and the flare is initiated to ensure a secure landing even if the runway remains obscured until the final stage of the approach. Certain aircraft models enable the autopilot to sustain the runway centerline alignment until coming to a stop.

At any juncture during the flight, the pilot can intervene by providing appropriate inputs to the autopilot or FMS. In emergencies, the pilot can disengage the autopilot and take manual control, usually through a conveniently situated switch on the control column. Contemporary aircraft also include switches or throttle positions allowing for swift transition from approach to go-around mode, if necessary. When automatic go-around functions are absent, pilots must manually disconnect the autopilot to execute a missed approach. Effective utilization of automatic systems relies on a comprehensive grasp of equipment capabilities and design principles. Failing to attain this level of understanding has resulted in several fatal accidents.

Stealth Technology

The objective of stealth technology is to make the flight vehicle almost undetectable to enemy. The invisibility to radar is achieved through two broad approaches. Firstly, the aircraft's shape can be altered to deflect the incoming radar signals away from the receiving radar of the enemy. Most conventional aircraft external shape is designed for its aerodynamic performance, inadvertently causing efficient radar reflection. In contrast, stealth aircraft employ flat surfaces and sharp edges, deflecting radar signals at angles upon impact. Secondly, the aircraft's surface can be coated with radar-absorbent materials (RAM) that either trap or absorb radar signals. These materials encompass dielectric composites, metal fibers containing ferrite isotopes, and specialized paint with pyramid-like colonies filled with ferrite-based RAM.

Beyond radar evasion, stealth considerations extend in reducing an aircraft's heat signature to be less observable by enemy missile's heat seekers. Achieving this requires channeling engine exhaust through extended tubes, blending it with cooler external air. Notably, the roots of stealth technology trace back to Nazi Germany during World War II, with the *Horten Ho 229* prototype serving as a precursor. Prominent contemporary instances of stealth aircraft include the *F-117 Nighthawk*, the *B-2 Spirit*, the *F-22 Raptor*, the *J-20*, *F-35 Lightning II* and *Sukhoi Su-57*.

Aircraft designers quantify an aircraft's radar cross-section in terms of *decibel square meters (dBsm)*, drawing an analogy to the radar reflectivity of a specific-sized aluminum sphere. The *B-2* and *F-22 Raptor* exhibit radar signatures akin to an aluminum marble, with the *F-117* slightly less stealthy. The *Joint Strike Fighter*'s signature is comparable to an aluminum golf ball. Intriguingly, an aircraft's size doesn't correlate significantly with its radar cross-section; instead, its shape dictates stealthiness. The *B-1 bomber*, conceived in the 1970s and 1980s, showcases a one-meter-diameter sphere equivalent, while the *B-52 Stratofortress*, from the 1950s, presents an extensive radar cross-section of a 52-meter-diameter sphere. In summary, while size may not be a direct factor, an aircraft's shape remains pivotal in determining its radar cross-section.

The *B-2 Spirit bomber*, the Stealth Bomber, employs cutting-edge stealth technology to make it nearly invisible to radar detection. Its sleek, diamond-shaped design and specialized coatings enable it to operate deep within heavily defended airspace, making it a formidable asset in modern military operations.

The *F-117 Nighthawk*, a Stealth Fighter, was a revolutionary aircraft designed for low-observable stealth technology. It played a pivotal role in the U.S. Air Force's arsenal by conducting precision strikes during its service, including during the Gulf War.

de Havilland DH.106 Comet was the world's first commercial jet airliner. The *British Overseas Aircraft Corporation (BOAC)* launched the first commercial jet service on May 2, 1952 when a *Comet* flew from London to Johannesburg.

1.6.2 Modern Jet Airliners

The advent of the jet engine brought about a revolutionary transformation in air travel. With their remarkable power, jet engines empowered aircraft manufacturers to construct larger, swifter, and more efficient airliners. This technological advancement also facilitated airlines in reducing operational expenses and airfares. Utilizing kerosene as fuel, jet engines generate immense thrust relative to their weight. Consequently, jet-powered aircraft can be designed on a larger scale and attain greater speeds compared to piston-engine counterparts.

The inaugural passenger jet aircraft, the *de Havilland Comet*, was introduced for service in 1952. Although it marked a significant leap in aviation, this early jet-powered aircraft encountered substantial issues. Particularly noteworthy were challenges related to its fuselage, windows, and pressurization system. The resolution to these problems was achieved with the fourth iteration, the *Comet 5*, leading to increased sales. However, by this point, rival aircraft designers had gleaned insights from the initial endeavors and presented competitive alternatives.

A series of tragic incidents involving the *Comet 1*, including the crashes of *BOAC Flight 783* on May 2, 1953, *BOAC Flight 781* on January 10, 1954, and *South African Airways Flight 201* on April 8, 1954, resulted from structural issues. These unfortunate events compelled authorities to ground the entire *Comet* fleet. Following thorough modifications to the airframe, commercial flights of the *Comet* resumed in 1958.

The *Tupolev Tu-104* emerged as a twinjet, medium-range, narrow-body turbojet, marking the Soviet Union's inaugural venture into the realm of jet airliners. The world's introduction to this prototype occurred when the *Tu-104* landed at London Airport on March 22, 1956, revealing the Soviet Union's pioneering achievement. Following the British *de Havilland Comet*, the *Tu-104* became the second jetliner to enter regular service. It held the distinction of being the sole operational jetliner globally from 1956 to 1958, as the *Comet* was grounded due to safety apprehensions, and *Boeing's 707* was still two years away from commencement of service. Commencing its first Soviet domestic jet service in 1956, the *Tu-104* initially accommodated a mere 50 passengers. The enhanced *104A* model went on to establish several commercial records, becoming the most prolific variant among all the *Tu-104*s manufactured.

The *Vickers Super VC10*, a narrow-body, midsized British jet airliner with long-range capabili-

The *Tupolev Tu-104* was the only operational commercial jet-powered aircraft flying between 1956 and 1958.

ties, was crafted by *Vickers-Armstrongs (Aircraft) Ltd* in 1958 as a novel solution for *BOAC*'s extended flight needs. Despite earning high praise from passengers, the *VC10* series struggled to secure significant orders due to its comparatively lower fuel efficiency when compared to its counterparts, the *B707* and *DC-8*. By 1969, *VC10*s were servicing all of *BOAC*'s long-haul routes, yet they were gradually phased out in the early 1970s with the arrival of *Boeing 747*s. The *VC10* often draws comparisons to the larger Soviet *Ilyushin Il-62*, as these two aircraft are the sole airliners to feature a rear-engined quad layout.

Several successors and rivals emerged in the wake of the *Comet*, such as the *DC-8*, *Vickers VC-10*, *Tupolev Tu-104*, and *Boeing 707*. While each of these aircraft held its own unique appeal, the *Boeing 707* stood out in terms of unparalleled success. Although the *B707* was not the initial commercial jet aircraft, it undoubtedly became the first immensely successful one, often hailed for inaugurating the era of jet travel. This achievement solidified Boeing's prominence as a leading civilian aircraft manufacturer and marked the inception of the enduring 7X7 series, which continues to evolve to this day.

Boeing meticulously incorporated various design features based on lessons learned from earlier jet aircraft and valuable customer input. These enhancements included a wider fuselage to accommodate five-abreast seating and enhance cargo capacity, the strategic relocation of engines to underwing pods for enhanced safety in case of fires, and the adoption of improved flap designs and reinforced fuselage structure.

Since the 1950s, there have been fewer fundamental changes to commercial jet-airplane design. The cylindrical fuselage with swept wings has remained the standard airplane configuration. The period saw increasing use of composites compared to metals for manufacturing aircraft components. Turbofan engines have been the primary choice for propulsion but has improved in power and efficiency. Cabin and cockpit technology have similarly improved but are still based on the same designs and concepts.

SuperSonic Transport, SST

The *Concorde*, a remarkable feat of technology and the sole supersonic passenger-carrying commercial aircraft (with the *Tu144* being its Russian counterpart), was a collaborative effort between aircraft manufacturers in Great Britain and France. This delta-wing marvel took to the skies for the first time on March 2, 1969. Capable of reaching a maximum cruising speed of 2,179 km/h (Mach number 2.04 or slightly more than twice the speed of sound), the *Concorde* drastically reduced flight times. On September 26, 1973, it accomplished its maiden transatlantic crossing, and on January 21, 1976, it proudly launched the world's inaugural scheduled supersonic passenger service. *British Airways* commenced flights between London and Bahrain, while Air France offered service between Paris and Rio de Janeiro.

Some of the reasons for its retirement are: The aftermath of the September 11, 2001 attack on the World Trade Center profoundly impacted the aviation industry in the USA, resulting in a sharp decline in air travel. The event led to a significant reduction in passenger demand for flights. The *Concorde*'s operational costs and noise levels were limiting factors for its service. The aircraft's unique design and supersonic capabilities resulted in higher

(a) *Vickers Super VC10*, a mid-sized, narrow-body, long-range British jet airliner, achieved the fastest crossing of the Atlantic by a subsonic jet airliner of ∼5 hours, a record that was held for 41 years, until February 2020 when a *British Airways Boeing 747* broke the record at 4 hours 56 minutes due to Storm Ciara.

(b) The *Ilyushin Il-62*, long-range, narrow-body, Soviet jetliner developed by *Ilyushin* as successor to the *Il-18* with a capacity of almost 200 passengers and crew, was the world's largest jet airliner when first flown in 1963. It was one of four pioneering long-range designs (the others being *Boeing 707*, *Douglas DC-8*, and *Vickers VC10*).

VC10 and *Il-62* were the only airliners to use a rear-engined quad layout.

(a) *Vickers Viscount 701*, a British medium-range first turboprop airliner, was first flown in 1948. The *Viscount*'s pressurized cabin, relatively short landing run, and quiet, smooth turbine engines made it popular with passengers and airlines alike. The last aircraft was delivered in 1964, bringing the total number built to 444.

(b) The *Ilyushin Il-18*, a turboprop airliner that first flew in 1957 and became one of the best-known Soviet aircraft of its era. The *Il-18*'s successor was the long-range *Il-62*. The *Il-18* resembled the *Lockheed Electra* and the *Vickers Vanguard* but with over 550 built, production of the *Il-18* was more numerous than its competitors.

Turboprop Airliners.

expenses and noise concerns, which affected its operational viability. The *Concorde* was predominantly utilized by affluent passengers who could afford its premium price tag in exchange for its unparalleled speed and luxurious service. The high operating costs and limited passenger base made its commercial model unsustainable. For instance, in 1997, a round-trip ticket from New York to London on the *Concorde* cost $8,000 a rate 30 times greater than the most economical option available for the same route. When compared to modern aircraft like the *B-787* and *A350*, the *Concorde*'s fuel consumption would have been significantly higher—around 4 to 5 times greater in terms of seat-mile cost. These various factors collectively played a role in the decision to retire the *Concorde*.

Aircraft Crash: On July 25, 2000, a *Concorde* aircraft traveling from Paris to New York City experienced engine failure shortly after takeoff due to debris from a burst tire. This debris caused a fuel tank to rupture, resulting in a fire that engulfed the aircraft. Tragically, the *Concorde* crashed into a small

Boeing 707-80, better known as the *Dash 80*, is America's first jet airliner which revolutionized commercial air transportation. It made its first flight on July 15, 1954, with built-in features like swept-winged aircraft and powered by four revolutionary new engines that first took to the sky above Seattle. *Boeing*'s 707 was designed for the transcontinental or one-stop transatlantic range. But modified with extra fuel tanks and more efficient turbofan engines, *707-300*s could fly nonstop across the Atlantic. Boeing built 855 *707*s galvanizing its place as a manufacturer of jet airliners.

hotel, resulting in the loss of all 109 individuals on board, including 100 passengers and 9 crew members. Additionally, 4 individuals on the ground lost their lives in the incident.

The Russian-produced *Tu-144* emerged as the world's inaugural commercial supersonic transport aircraft. Its prototype undertook its maiden flight from *Zhukovsky* Airport near Moscow on December 31, 1968–two months prior to *Concorde*. The *Tu-144* possessed larger dimensions, a greater weight, and increased power in comparison to the *Concorde*. However, it lagged behind technologically, with lesser advancements in engine control, inferior aerodynamics, and a less effective braking system. These modest disparities prevented it from becoming a genuine contender in the market.

After the unfortunate *Paris Air Show Crash* in 1973, which occurred during a low-altitude demonstration of rapid acceleration, and a subsequent crash in *Yegoryevsk*, Russia in 1978 during a test flight, *Aeroflot* ceased its passenger services using the *Tu-144*.

During the early 1960s, *Boeing* put forth a propo-

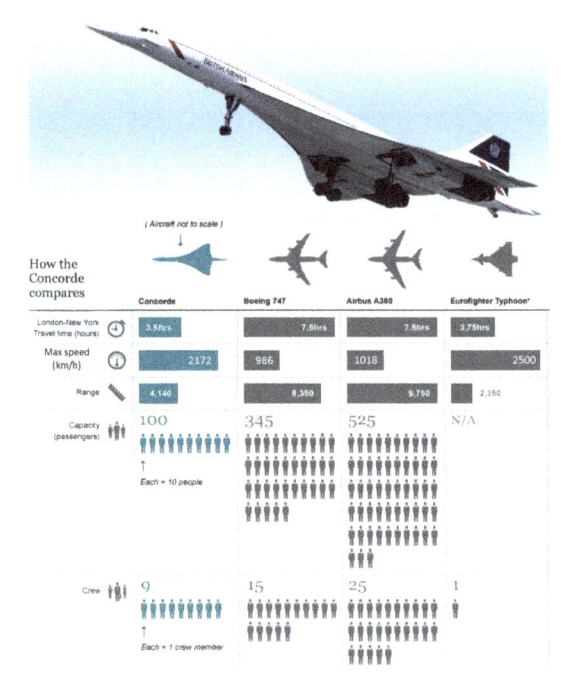

Concorde compared with other aircraft. The *Concorde*, introduced in 1976, stood as the singular and operational supersonic transport of its time. With the capability to transport 100 passengers across the Atlantic in under four hours, its rapid travel came at a significant cost, making its airfares notably expensive. The collective 14 *Concordes* that entered service were procured by the British and French governments for their respective national airlines. Unfortunately, the era of *Concorde* flights came to a close in 2003 following the unfortunate accident involving a *Concorde* in 2000 in Paris.

The Tupolev *Tu-144* is a Soviet supersonic passenger airliner designed by *Aleksey Tupolev* in operation from 1968 to 1999 (Number built: 16).

sition to create an American iteration of a wide-body supersonic passenger aircraft, aiming to compete with the *Concorde* as well as the Soviet *Tupolev Tu-144*. The proposed supersonic airliner was christened the *Boeing 2707* and boasted the potential to ferry around 250 to 300 passengers at speeds nearing Mach number 3, covering a distance of 6,400 kilometers, driven by four *General Electric GE4* turbojet engines. The *B-2707* was designed to incorporate a variable-sweep wing (swing-wing) mechanism, allowing the in-flight adjustment of wing geometry, alongside canards positioned behind the aircraft's nose. Boeing had envisioned conducting flight tests for the supersonic jet around late 1972, with plans for certification and entry into service by 1975.

However, the swing-wing mechanism turned out to be considerably more intricate and weighty than expected. Furthermore, concerns emerged regarding the colossal expenses associated with development and operation, notably high fuel consumption, exorbitant fares, disruptive sonic booms, and other ecological considerations. These mounting challenges became insurmountable hurdles. The convergence of these concerns, coupled with the broader downturn in the aviation industry, contributed to a waning interest in the concept of a supersonic airliner. The government declined further funding in 1971, ultimately leading to the demise of the *B-2707* project.

Boeing vs Airbus

By 1980, *Airbus* positioned as a primary contender against *Boeing*, brought forth a lineup of distinct variants infused with advanced technology and substantial innovation. While *Airbus* embarked on its journey after *Boeing*, it has experienced remarkable achievements in its own right. Serving as a direct rival to *Boeing*'s widely favored narrow-body airliner, the *B-737*, Airbus introduced the *A-320* as its response. Notably, the *A-320* became the pioneer civil aircraft to incorporate fly-by-wire technology, a standard it has upheld since entering service in 1988. Airbus's portfolio encompasses an array of wide-body airliners, including the *A-330* family, the *A-350* family, and the largest airliner *A380*. This comprehensive range is designed to contend with *Boeing*'s counterparts, namely the *B-787*, *B-777*, and *B-747*, further intensifying the competition between the two giants in the aviation industry.

Boeing 737 and Airbus 320

Building on the triumphs of the *707* and *727* models, *Boeing* embarked on the creation of a fresh aircraft, the *B-737*, with the aim of outperforming competitors and capturing a larger customer base. Launched in 1967, the *B-737* introduced several distinctive design features that set it apart from rivals. Notably, it adopted a twin-engine configuration in contrast to the three or four engines seen in competing aircraft, a move that resonated with customers seeking cost efficiency. The aircraft's wider fuselage, accommodating six abreast seating and the handling of standard cargo containers, further contributed to its appeal.

With an impressive track record, the *Boeing 737* stands as the most successful aircraft in terms of sales to date. Despite encountering setbacks with its latest iteration, the *737 MAX*, including two tragic accidents and subsequent grounding, the aircraft remains on course for continued success. As of March 2023, orders close to 15,500 *Boeing 737*s have been placed, out of which 11,377 have been successfully delivered. As a competitor to *B-737*, Airbus devised the *A320* family of aircraft. At one point, the *A320* family even managed to surpass the *Boeing 737* in terms of the number of aircraft orders. Setting new standards, the *A320* introduced a significant incorporation of composite materials in its construction, making it the pioneering narrow-body airliner to do so. These advancements encompassed enhancements in aerodynamics, the incorporation of sizeable winglets, expanded luggage bins, a revamped cabin design, weight reduction measures, and the option to choose between two new engine types, including *LEAP* engines. Notably, the *A320* Family's cockpit shares commonalities with other

(a) *Boeing 737 classic (B737-300)*, the second generation derivative, entered service in December 1984 with high-bypass turbofan engines for better fuel economy and upgraded avionics.

(b) *Boeing 737 Next Generation (737-600)*, the third generation *737*, entered service in 1997 with a redesigned wing with a larger area, a wider wingspan, greater fuel capacity and longer range.

(c) *Boeing 737 MAX (737-800)*, the fourth generation derivative, entered service in 2011 with more efficient *Leading Edge Aviation Propulsion (LEAP)* engines, aerodynamic changes, including distinctive split-tip (Scimitar) winglets, and airframe modifications.

The most sold airliner, the *Boeing 737*.

The Airbus A320 family (A321 XLR is shown in figure), a single-aisle narrow-body, short to medium range commercial passenger twin engine jet airliners, is a direct competitor to the *Boeing 737*.

Airbus fly-by-wire aircraft, solidifying its position within the Airbus lineup.

The giant *Boeing 747*, dubbed the *Queen of the Skies or Jumbo Jet*, is the world's first twin-aisle airplane which revolutionized air travel as it enabled more people to fly farther, faster and more affordably than ever before. Since production began in 1967, more than 1,500 aircraft have entered the market. Originally designed for *Pan American* Airlines to replace the *707*, *B-747* offered significantly lower seat-mile costs and therefore much greater efficiency.

Boeing 747 and Airbus 380

Since its introduction into service in 1970, the *Boeing 747 Jumbo Jet* revolutionized the scope of long-distance passenger flights. Its substantial size allowed for the transportation of three to four times the number of passengers compared to earlier jet airliners, marking the onset of the era of widespread air travel. While the *747* represented a remarkable feat of technology, its widespread success eventually led to a certain normalization of air travel experiences.

By the late 1980s, *Airbus* had firmly positioned itself as a strong contender against *Boeing*, as evidenced by the success of its *A300* and *A320* family aircraft. While advancing its *A330* and *A340* wide-body programs, Airbus set its sights on the 'ultra-high-capacity' segment as a response to *Boeing*'s popular *747 Jumbo Jet* which resulted in the the *A380* program, which capitalized on *Airbus*' collaborative construction approach, involving contributions from various factories across Europe. These components were then brought together at *Airbus*' primary assembly plant in Toulouse, France. More than 250 *A380* aircraft were manufactured. Despite the aircraft's favor among flight crews, passengers, and aviation enthusiasts, its substantial size and fuel-intensive quad-engine design posed limitations on its commercial success. In 2021, *Airbus* officially discontinued the *A380* program due to gradually declining orders for four-engine jets.

The *Airbus A380* is the world's largest wide-body passenger airliner and the only full-length double-decker airliner. *Airbus* initiated studies in 1988, and the project was announced in 1990 to challenge the dominance of the *Boeing 747* in the long-haul market.

Boeing 777 and Airbus 350

Since its introduction in 1989, the *777* series has undergone numerous enhancements, particularly in terms of efficiency. The upcoming *777X* model aims to build upon these improvements and push the boundaries even further. The *777X* offers exceptional fuel efficiency through a range of innovations. These include the utilization of the largest engines ever installed on a civilian aircraft, enhanced with lightweight composite fan technology. The addition of folding wingtips, designed for larger wings to enhance efficiency without limiting airport operations, further contributes to its fuel-saving capabilities. Additionally, composite wing construction and raked wingtips enhance the aircraft's overall aerodynamics. In a similar vein, the Airbus *A350* also boasts impressive fuel efficiency attributes. With 53% of its structure composed of composites, adaptive wings that adjust in-flight to reduce drag, and advanced aerodynamic enhancements to the wing shape, the *A350* sets a high standard.

The *Boeing 787*, introduced many features like

The *Boeing 777X*, launched in 2013, is the latest series of long-range, wide-body, twin-engine jetliners from the *Boeing 777* family. The *777X* features new *GE9X* engines, composite wings with folding wingtips, greater cabin width and seating capacity, and technologies from the *Boeing 787*.

an upgraded fly-by-wire flight control system that translated flight deck crew inputs into electrical signals, moving away from mechanical processes, the traditional hydraulic brake power was replaced with electrical systems, and electro-thermal ice protection blankets were implemented to melt ice on the wings. Interestingly, the aircraft features large-format head-up displays (HUDs) that project flight information onto the pilots' line of sight, allowing them to monitor both the external environment and vital flight data simultaneously, a feature typically found in military combat aircraft.

The *787*'s construction involves extensive use of composites, making up 50% of its weight and 80% of its volume. Carbon fiber-reinforced polymer is employed in the wing design, offering a superior strength-to-weight ratio compared to traditional metals, resulting in significant weight advantages. The aircraft's flexible wings promote improved aerodynamic stability and reduced drag by curving more prominently. The use of hybrid laminar flow control on the leading edges of the vertical and horizontal stabilizers in the *787-9* further refines airflow, reducing drag on the tail. Since the inception of the *Dreamliner* program, over 1,000 *Boeing 787*s have been manufactured. The aircraft's popularity among long-haul operators across the globe

The Boeing 787 Dreamliner.

The Airbus A350, a long-range, wide-body twin-engine jet airliner developed in response to the Boeing 787 Dreamliner.

has been substantial since the first flight of the *787-8* on December 15, 2009. In 2004, *Airbus* unveiled the *A350* aircraft, an extended-range and lighter version derived from the *A330* model. Subsequently, after a thorough evaluation of market demands, *Airbus* introduced the *A350 XWB* (Extra Wide Body) in December 2006, offering a larger and heavier aircraft. A notable feature of the *A350* is its extensive use of carbon-fiber-reinforced polymers, marking a significant departure for *Airbus*. The aircraft is constructed with a blend of materials, comprising 45% lightweight, high-strength composite construction and 55% a mix of low-density aluminum-lithium alloy, steel, aluminum, and titanium. The *A350 XWB* is powered by the *Rolls-Royce*-developed *Trent XWB* engine. The progress in aircraft design for commercial airliners has been marked by remarkable advancements over the years, driven by innovation, technology, and a growing emphasis on efficiency, safety, and passenger comfort. From the pioneering days of jet travel with aircraft like the *Boeing 707* and the *Concorde* to the modern era with the *Boeing 787 Dreamliner* and *Airbus A350 XWB*, the industry has witnessed significant leaps in aerodynamics, materials, propulsion, and avionics. The incorporation of composite materials, improved wing designs, and state-of-the-art engines has led to greater fuel efficiency and reduced environmental impact. Future developments are poised to further reshape the landscape, with a focus on electric and hybrid propulsion systems, increased use of sustainable materials, and advanced aerodynamic solutions to minimize drag and improve performance. These innovations promise to make air travel even more efficient, sustainable, and enjoyable for passengers worldwide.

Cargo Planes

Cargo planes (freighters) play a vital role in global commerce by facilitating the efficient transportation of goods and materials across the world. These specialized aircraft are designed to carry various types of cargo, from perishable goods and consumer products to heavy machinery and oversized cargo that wouldn't fit into passenger aircraft's holds. Cargo planes come in a variety of sizes and configurations, each optimized for different types of cargo and distances.

One of the distinctive features of cargo planes is their ability to handle a wide range of cargo sizes and shapes. Some cargo planes are equipped with large doors, ramps, or even nose-loading capabilities to accommodate oversized and bulky items. This flexibility allows for the transport of goods that wouldn't be feasible on standard passenger flights.

There are two main categories of cargo planes: dedicated cargo aircraft and converted passenger aircraft. Dedicated cargo planes are purpose-built to carry cargo and usually have a larger cargo capacity compared to converted aircraft. These aircraft are designed with features like reinforced floors, specialized loading systems, and efficient cargo hold layouts to optimize space for various types of cargo.

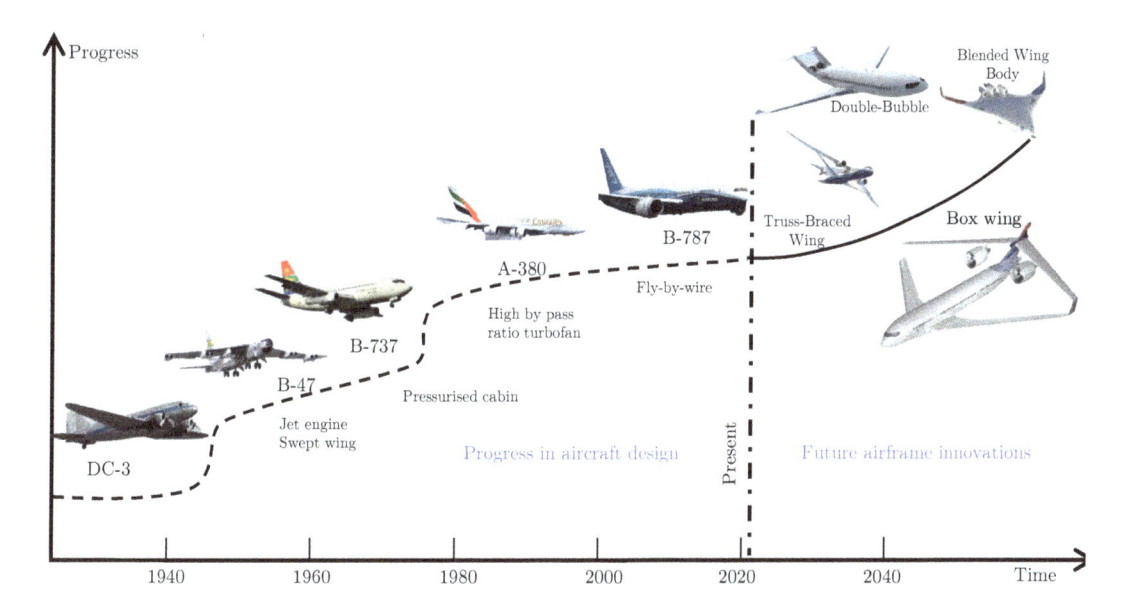

Progress in aircraft design of commercial airliners and future developments.

Converted passenger aircraft, on the other hand, are commercial airplanes that were originally built for carrying passengers but have been modified to carry cargo. This process involves removing seats and reconfiguring the interior to create an open cargo space. While these converted planes may have limitations in terms of cargo capacity and loading efficiency compared to dedicated cargo planes, they offer a cost-effective solution for transporting cargo, especially during periods of high demand.

1.6.3 Military Aviation

After World War II, military aviation technology underwent rapid development, fueled by lessons learned during the conflict and the Cold War tensions.

Jet Propulsion and Supersonic Flight (1950s - 1960s): The 1950s marked the widespread adoption of jet propulsion in military aviation. The introduction of jet engines significantly increased aircraft speeds and performance. The *Bell X-1*, which broke the sound barrier in 1947, paved the way for supersonic flight. The 1960s saw the development of iconic aircraft like the *North American XB-70*

The *Antonov An-225 Mriya* is the biggest strategic airlift cargo aircraft developed in the 1980s. It holds the world record for the largest single-item payload, 210 tons. Only one aircraft in this model was built in the late 1980s by the Soviet Union in what is now Ukraine. *Antonov An-225* is so massive and its landing gear has a total of 32 wheels. Maximum speed: 850 km/h, Range: 15,400 km, Maximum takeoff mass: 640 tons and payload: 250 tons. The aircraft was partially destroyed during the Russian attack on Ukraine in 2022.

Valkyrie, capable of sustained Mach number 3 flight, showcasing the potential of hypersonic flight.

One notable example of innovation came from the United States with the introduction of the *North American F-86 Sabre*. This jet fighter incorporated swept wings, a design feature that allowed for higher speeds and improved maneuverability at transonic and supersonic speeds. The *F-86*'s success was highlighted during the Korean War, where it played a pivotal role in achieving air superior-

The 747-8 Freighter, a cargo version of the passenger *747-8 Intercontinental*, is a heavier plane developed as an alternative to the double-decker *A380* and had its first flight in 2010. Maximum speed : 956 km/h, range : 7,899 km, maximum take-off mass : 448 tons and payload : 132 tons.

The *Lockheed C-5 Galaxy* belongs to the United States Air Force and began production in 1968. There were a total of 131 of these aircraft built. Speed : 869 km/h, range : 11,705 km, maximum takeoff mass : 417 tons and payload : 127 tons

The *Aero Spacelines Super Guppy* is a large, wide-bodied cargo aircraft, *NASA*'s preferred plane for transporting spacecraft and other large objects. The *Super Guppy* is the only airplane to carry a complete *Saturn S-IVB* stage. The first one flew in 1965 and was constructed from the fuselage of a *Boeing C-97 Stratofreighter*, which was modified for larger payloads.

The *Airbus Beluga XL* gets its name not only from its large size but also because it resembles a beluga whale. The aircraft made its first flight on July 19, 2018. There are a total of 6 *Beluga XL*s, which are used for transporting *Airbus* parts, such as fuselages, across multiple *Airbus* production sites. Speed : 737 km/h, range : 4,300 km, maximum takeoff mass : 227 tons and payload: 50 tons

ity against Soviet-built *MiG-15*s, showcasing the importance of aerodynamic advancements in combat. Similarly, the Soviet Union made significant strides in military aviation with the introduction of the *Mikoyan-Gurevich MiG-15*. This jet fighter was known for its swept-wing design and a powerful jet engine, making it a formidable adversary against Western aircraft during the early years of the Cold War. These jets are most often provided with an 'afterburner' for short bursts of enhanced thrust by burning extra fuel in the tail pipe of the engine located after the turbine at the price of higher fuel consumption. Some of these jets have the capability for 'Thrust Vector Control (TVC)', ability to orient the direction of thrust produced by the engines, for improved maneuverability and 'Vertical Take-Off and Landing (VTOL)', where the requirement of a runway can be done away with for take off and landing.

Fly-by-Wire and Avionics : The 1970s and 1980s saw the integration of fly-by-wire technology and advanced avionics systems. The American *F-16 Fighting Falcon* was among the first aircraft to employ fly-by-wire controls, enhancing maneuverability and stability. Avionics innovations led to the introduction of advanced radar, sensor systems, and cockpit displays like the 'Heads-Up Display', enhancing situational awareness and combat effectiveness.

Stealth Technology: The development of stealth

technology revolutionized military aviation. The *Lockheed Martin F-117 Nighthawk* and *Northrop Grumman B-2 Spirit* introduced radar-absorbing materials and innovative designs to reduce an aircraft's radar cross-section, resulting in airplanes with improved survivability in contested environments.

Uninhabited Aerial Vehicles (UAVs): The 21st century witnessed a surge in *UAV*s, more commonly knows as drones, for military applications. These platforms range from reconnaissance drones like the *General Atomics MQ-1 Predator* to armed *UAV*s like the *MQ-9 Reaper*. They offer persistent surveillance, reduced risk to human pilots, and the ability to carry out missions without risking crew lives. *UAV*s have become integral parts of modern air forces, offering extended endurance and reduced risk to human pilots.

Network-Centric Warfare and Connectivity: The integration of information technology and networking into military aviation led to the concept of network-centric warfare. Modern military aviation emphasizes network-centric warfare, where aircraft and assets are interconnected for real-time data sharing and coordination. Aircraft like the *Boeing E-3 Sentry AWACS (Airborne Warning and Control System)* and the *Northrop Grumman E-2 Hawkeye* provide real-time surveillance, command, and control capabilities, allowing for improved situational awareness and coordination. Aircraft like the *F-35 Lightning II* are designed with data fusion capabilities, allowing pilots to access and share information from various sources for improved situational awareness. Electronic measures and electronic countermeasures are frequently used by modern combat aircraft for jamming enemy signals.

Advanced Stealth and Hypersonic Technologies: The development of stealth technology aimed to reduce an aircraft's radar cross-section and increase survivability in contested environments. Ongoing advancements in stealth technology have resulted in aircraft like the *F-22 Raptor* and *F-35 Lightning II*, which incorporate improved stealth characteristics. Hypersonic technology is also a focus area, with experimental aircraft like the *Boeing X-51A* and research on hypersonic weapons promising unprecedented speed and agility.

Autonomy and Artificial Intelligence: Military aviation is exploring the integration of autonomous systems and artificial intelligence (AI). Aircraft equipped with AI can process vast amounts of data, make complex decisions, and assist human pilots in high-stress situations. This technology has the potential to enhance combat efficiency and response times.

Lockheed SR-71 Blackbird, the fastest jet-propelled aircraft in existence. Flying at speeds more than Mach number 3, no reconnaissance aircraft has navigated the world's most hostile airspace with as much impunity as the *SR-71*. The *Blackbird*'s exceptional performance and operational feats positioned it as the epitome of aviation technology advancements during the Cold War era. This twin-engine, two-seat supersonic strategic reconnaissance aircraft boasted an airframe primarily constructed from titanium and its alloys.

The evolution of military aviation after World War II has been marked by a series of transformative innovations. From supersonic flight and stealth technology to UAVs, these advancements have redefined the capabilities and possibilities of modern aircraft. As we look to the future, continued innovation will shape the landscape of military aviation, enhancing operational effectiveness and expanding the boundaries of what is achievable in the skies and beyond.

The *F-104 Starfighter* was a groundbreaking supersonic fighter jet known for its slender, rocket-like design. Introduced in the late 1950s, it achieved remarkable speeds, becoming one of the first operational aircraft to cruise at Mach number 2. Despite its impressive performance, the *F-104* was also known for its challenging handling characteristics.

The *F-15* (referred as *Eagle*) is a highly acclaimed and formidable fighter aircraft that has been a cornerstone of the U.S. Air Force's fleet for decades. Renowned for its exceptional speed, agility, and firepower, the *F-15* has earned a stellar reputation as an air superiority fighter. It continues to serve as a critical asset for US and its allies in safeguarding airspace and is known for its dogfighting capabilities and advanced avionics.

1.7 History of Helicopters

The history of helicopter development is a testament to human ingenuity and persistence in overcoming the challenges of vertical flight. The journey towards achieving this level of flight has been marked by groundbreaking inventions, daring experiments, and continuous refinements. A comprehensive overview of the evolution and milestones in the design and technology of helicopters is pre-

The *F-16* (referred as *Viper*) is a versatile multirole fighter aircraft that has been adopted by numerous air forces worldwide. Known for its agility and adaptability, the F-16 is capable of performing a wide range of missions, including air-to-air combat, ground attack, and reconnaissance. Its ease of maintenance and cost-effectiveness have contributed to its enduring popularity and service across diverse military operations.

The *F/A-18* (referred *Hornet*) is a versatile naval fighter aircraft designed for both air-to-air combat and ground attack missions. It has served as a staple in the U.S. Navy and Marine Corps, as well as several other nations' naval aviation forces. The *F/A-18* is appreciated for its carrier-based capabilities, agility, and capacity to operate from shorter runways, making it a crucial asset in naval aviation.

sented.

1.7.1 Early Concepts and Precursors

The concept of vertical flight and the development of helicopters have a long and fascinating history dating back to ancient times. The concept of vertical flight was explored through various means like Chinese flying toys and ancient Greek devices. *Leonardo*

The *F-22 Raptor* is an advanced fifth-generation stealth fighter aircraft renowned for its unmatched air superiority capabilities. Designed for the U.S. Air Force, it excels in both air-to-air combat and air-to-ground missions. Its combination of stealth, agility, and state-of-the-art avionics makes the *F-22* a dominant force in modern aerial warfare, solidifying its status as one of the world's most formidable fighter aircraft.

The *F-35* developed under the *Joint Strike Fighter* program, is a cutting-edge fifth-generation multi-role combat aircraft designed for a variety of roles, including air-to-air combat, ground attack, and reconnaissance. With different variants for the U.S. Air Force, Navy, Marine Corps, and allied nations, the *F-35* program represents one of the most extensive and ambitious fighter aircraft projects in history. It is known for its advanced technology, stealth capabilities, and the ability to seamlessly integrate with modern military operations, underscoring its status as a pivotal asset in contemporary military aviation.

da Vinci (1452 – 1519), the Italian Renaissance polymath, sketched designs for what is considered one of the earliest concepts of a vertical flight machine.

The *Sukhoi Su-34* is a twin-seat, twin-engine Russian fighter-bomber designed for a wide range of combat missions. With its distinctive duckbill-shaped nose, the *Su-34* is equipped for precision strike capabilities and all-weather, day-and-night operations. It has earned recognition for its versatility, making it a pivotal asset in the Russian military's arsenal.

The *MiG-25* (referred as the *Foxbat*) was a high-speed reconnaissance and interceptor aircraft developed by the Soviet Union during the Cold War. It was one of the fastest aircraft ever built, capable of reaching speeds in excess of Mach number 3. The *MiG-25* played a significant role in reconnaissance and defense, with its speed and altitude capabilities making it a formidable asset in Soviet and later Russian military operations.

His aerial screw or helical air screw drawings depicted a device resembling a modern helicopter's rotor. But, there is no evidence that it was built or tested during his lifetime.

In the 19th century, various inventors and pioneers experimented with early vertical flight machines. Some notable names include *Sir George Cayley (1773 – 1857)*, who designed a model helicopter with counter-rotating blades. *Étienne Oehmichen (1884 - 1955)*, a French engineer, achieved tethered

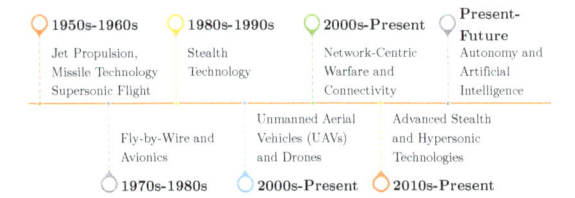

1950s-1960s	1980s-1990s	2000s-Present	Present-Future
Jet Propulsion, Missile Technology Supersonic Flight	Stealth Technology	Network-Centric Warfare and Connectivity	Autonomy and Artificial Intelligence

	1970s-1980s	2000s-Present	2010s-Present
	Fly-by-Wire and Avionics	Unmanned Aerial Vehicles (UAVs) and Drones	Advanced Stealth and Hypersonic Technologies

An overview of some of the key innovations in military aviation after World War II.

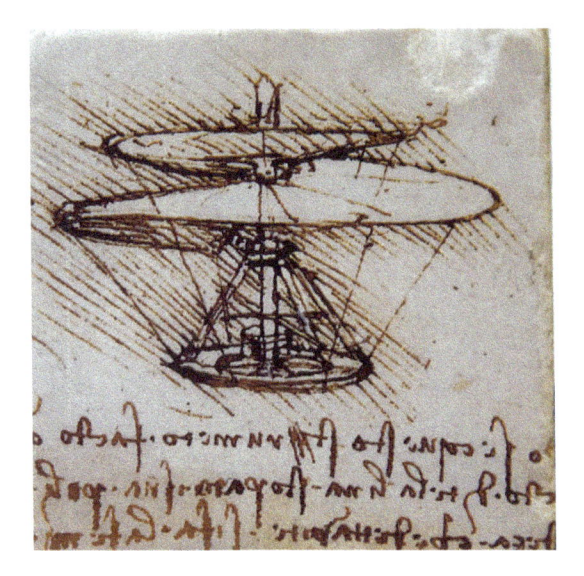

Leonardo da Vinci's sketch showing the design of a vertical flight machine

flights with his steam-powered helicopter. However, it was not until the 20th century that the idea of a practical helicopter began to take shape. The pioneering work of inventors such as *Igor Sikorsky*, *Frank Piasecki*, *Arthur Young*, and *Juan de la Cierva* led to significant advancements in helicopter technology.

In September 1907, *Louis Breguet* (1880 - 1955), a visionary French aviation pioneer, marked a significant achievement by lifting a man with a tethered rotary-winged aircraft for the first time. While the flight reached a height of only 2 feet, it lasted for one minute. The aircraft's stability relied on four assistants physically holding it. Despite later attempts to build two more air vehicles, *Breguet* faced ongoing obstacles due to inadequate power and difficulties in both flight control and landing.

In the same year (1907), French inventor *Paul*

Louis Breguet (1880 - 1955), a visionary French aviation pioneer, played a pivotal role in the evolution of aviation during the early 20th century.

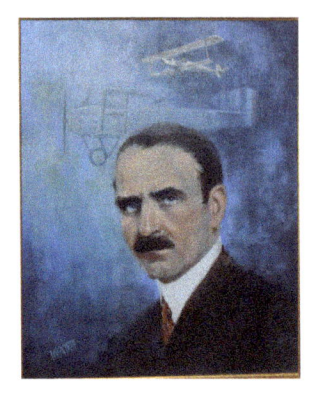

Cornu (1881 - 1944) achieved a short, uncontrolled flight in a twin-rotor helicopter he built. His aircraft, powered by a 24-horsepower engine, briefly lifted off the ground, marking one of the first recorded instances of a manned, controlled helicopter lifting off the ground. During 1916 - 1918,

Paul Cornu (1881 - 1944), a pioneering French engineer, achieved a remarkable feat by creating and piloting the world's first helicopter to achieve sustained flight.

Hungarian-born aerodynamic genius *Theodore von Kármán (1881 - 1963)* helped to design a helicopter called the *Petróczy, Kármán and Žurovec (PKZ-1 and PKZ-2)*, named after three Hungarian and Czech engineers, driven by an electric motor, that proved powerful enough to lift three men. It is notable as one of the earliest known attempts at creating a practical helicopter. In the early 1920s, a Spanish civil engineer, pilot and self-taught aeronautical engineer, *Juan de la Cierva (1895 - 1936)*, became intrigued by the idea of achieving vertical flight with a safer and more stable aircraft than the contemporary helicopters of his time. He developed a concept

1.7. HISTORY OF HELICOPTERS

The *PKZ-2* with a single observer on board.

he called the *Autogiro*, which combined features of both fixed-wing aircraft and helicopters. His pioneering work in the autogyro, a small, flying airplane with a rotor on top, laid the foundation for future advancements in vertical flight technology.

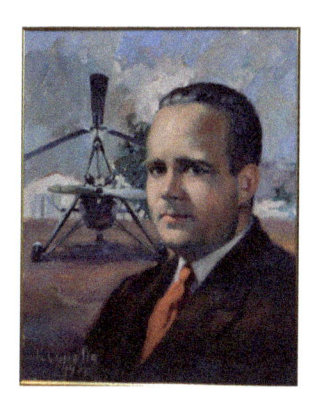

Juan de la Cierva, inventor of the autogyro.

Igor Sikorsky (1889 – 1972), a Russian born American aviation pioneer, made significant contributions to the development of helicopters and fixed-wing aircraft. *Sikorsky* gained prominence with his successful creation of large multi-engine airplanes, including the *Russky Vityaz* and the *Ilya Muromets*. These aircraft marked milestones in aviation history, setting records for passenger capacity, range, and payload. However, *Sikorsky*'s most enduring legacy lies in his pioneering work on helicopters. In the late 1930s, he designed and built the *VS-300*, the world's first successful single-rotor helicopter. This achievement revolutionized vertical flight, laying the groundwork for the modern helicopter industry. His design principles, including the main rotor, tail rotor, and cyclic control, remain integral to helicopter technology to this day. During World War II, *Sikorsky*'s expertise contributed to the development of military helicopters, such as the *R-4*, which became the first helicopter to enter active service with the U.S. military. This opened doors to diverse applications, from medical evacuation to reconnaissance and combat support.

Sikorsky's dedication to aviation extended beyond engineering. He co-founded *Sikorsky Aircraft Corporation* in 1923, which remains a prominent aircraft manufacturer. His relentless pursuit of innovation earned him numerous awards, including the National Medal of Science.

Pioneering Aviation Innovator: *Igor Sikorsky (1889 – 1972)*.

The *Vought-Sikorsky VS-300*, an American single-engine, three-blade rotor, helicopter designed by *Igor Sikorsky*. *VS-300* is considered the first successful single-lifting rotor helicopter.

Sikorsky R-4 in a hover.

1.7.2 Post-War Innovation and Commercialization

During World War II, helicopters started to see military use for reconnaissance, medical evacuation, and other specialized missions. The introduction of gas turbine engines provided helicopters with enhanced power and capabilities. This period also witnessed the development of dual-rotor designs like the tandem-rotor and coaxial configurations, offering increased lift capacity and versatility. Notable examples include the *Sikorsky R-4* and the *Focke-Achgelis Fa 223*. After the war, helicopter technology advanced rapidly, and various companies worldwide began producing civilian and military models. As technology advanced, helicopters began finding applications in civilian domains such as search and rescue, medical evacuation, and offshore operations. The 1950s and 1960s saw the birth of iconic helicopters like the *Bell UH-1 Huey* and the *Boeing CH-47 Chinook*, which became emblematic of rotary-wing operations.

A renowned American aeronautical engineer and entrepreneur, *Frank Piasecki (1919 - 2008)*, made significant contributions to the field of aviation, particularly in the development of tandem-rotor helicopters. His innovative work laid the foundation for the iconic *Boeing CH-47 Chinook*, a versatile and widely-used military transport helicopter. In the 1940s, *Frank Piasecki* co-founded the *Piasecki Helicopter Corporation* (later known as *Vertol Corporation*), along with *Harold Venzie* and other engineers. *Piasecki*'s vision was to create a helicopter with increased lifting capacity and stability by using tandem rotors. The company's first successful tandem-rotor helicopter was the *PV-2*, nicknamed the *Flying Banana* due to its elongated fuselage and shape. The *PV-2* made its maiden flight in 1945 and became the first helicopter in the world to have a fully controllable all-moving stabilizer, which greatly improved its stability during flight. *Piasecki*'s tandem-rotor design provided enhanced lifting capabilities, making the *PV-2* suitable for military and heavy-lift operations. The *Flying Banana*'s successful flights demonstrated the viability of tandem-rotor technology and laid the groundwork for future developments.

In the 1960s, *Piasecki*'s company, *Vertol Corporation*, merged with *Boeing*, forming *Boeing Vertol*. The tandem-rotor helicopters developed by the company eventually evolved into the *Boeing CH-47 Chinook*. The *CH-47 Chinook* became one of the most iconic and versatile military helicopters in history.

An American engineer, inventor, and entrepreneur, *Arthur Young (1905 – 1995)* designed the *Bell Model 47*, which has the distinction of being the world's first commercially successful helicopter. In the 1930s, *Young* joined the *Autogyro Company of America*, where he worked under the guidance of aviation pioneer *Harold Pitcairn*. During this time, *Young* became fascinated with rotor-craft and vertical flight. In the late 1930s, *Young* moved on to work for *Bell Aircraft Corporation*. There, he began working on the design of the *Model 30*, which would later be known as the *Bell Model 47*.

1.7.3 Modern Era and Beyond

The latter half of the 20th century and the beginning of the 21st century brought further advancements in materials, avionics, and automation. Composite materials reduced weight and improved structural integrity, while advanced fly-by-wire systems enhanced control and stability. Additionally, research and development in areas like autonomous flight and electric propulsion are shaping the future of helicopters as greener and more autonomous aircraft.

During the Korean War (1950 - 1953), helicopters played a crucial role in revolutionizing military operations and rescue missions. The conflict served as a significant turning point in the development of helicopter technology and tactics, marking the first extensive and widespread use of helicopters in combat. The Korean War saw the deployment of various helicopter types for different purposes. The *Sikorsky H-19 Chickasaw* (known as the *Sikorsky S-55* in civilian use) was extensively used for troop transport, medical evacuation, and resupply missions. It proved to be a reliable and versatile workhorse, capable of operating in a wide range of environments. The Korean War also witnessed the experimental use of armed helicopters for attack missions. While not yet fully dedicated attack helicopters as we know them today, helicopters such as the *Sikorsky H-19* were fitted with improvised armaments, such as machine guns and rocket pods, to provide suppressive fire against enemy positions.

An American aerospace engineer and entrepreneur, *Charles Kaman*, made significant contributions to helicopter development. In the early 1940s, he founded the *Kaman Aircraft Corporation* and set out to create a new and innovative helicopter design. He envisioned a compact and lightweight helicopter that could offer improved performance and handling compared to existing models. One of his notable creations was the *K-225* helicopter. With the intermeshing rotor system which later became a hallmark of *Kaman helicopters*, *K-225* helicopter was one of *Kaman's* groundbreaking achievements. The *K-225* helicopter, which made its maiden flight in 1947, featured two rotors mounted on separate masts that rotated in opposite directions. This unique configuration eliminated the need for a tail rotor, simplifying the helicopter's mechanical design while providing enhanced stability and maneuverability. Throughout its history, *Kaman Aircraft Corporation* produced several successful helicopter models. The company's helicopters were known for their stability,

Kaman Model K-225, the first turbine-powered helicopter.

versatility, and reliability. Some notable *Kaman* helicopter models include *HOK-1/HU2K-1 Seasprite* – a naval anti-submarine warfare helicopter used by the United States Navy. *HH-43 Huskie*– a search and rescue (SAR) helicopter known for its unique "banana" shape and used by various military and civilian organizations. *K-MAX*– an unmanned cargo helicopter designed for autonomous heavy-lift operations. Today, *Kaman Aerospace* remains an active and successful company in the aerospace and defense sectors, offering a wide range of products and services.

Boeing, primarily known for its commercial and military fixed-wing aircraft, has also made significant contributions to helicopter development. In 1960, *Boeing* acquired the *Vertol Corporation*, a well-known helicopter manufacturer founded by *Frank Piasecki*. *Vertol* was renowned for its tandem-rotor helicopters, including the successful *CH-46 Sea Knight* and the larger *CH-47 Chinook*. *Boeing*'s acquisition of *Vertol* expanded its portfolio to include helicopters, allowing the company to diversify its offerings in the aerospace market. In 1961, fast, cargo-carrying tandem rotor *CH-47 Chinook* helicopter made its maiden flight. The *Chinook*, derived from *Vertol*'s tandem-rotor designs, has been in service since the early 1960s. *Boeing* later developed, *AH-64 Apache*, a highly advanced attack helicopter used by various military forces globally. Originally

CH-46 Sea Knight, *Boeing*'s first tandem-rotor helicopter.

developed by *Hughes Helicopters* (which later became a part of *Boeing*), the *Apache* is designed for combat operations and is equipped with sophisticated avionics, sensors, and weapons systems. The *Apache* is renowned for its lethality and has been continuously upgraded to maintain its effectiveness on the modern battlefield. *Boeing* has been actively involved in the development of unmanned rotorcraft, also known as Unmanned Aerial Vehicles (UAVs) or drones. *Boeing* is actively participating in the U.S. Army's *Future Vertical Lift (FVL)* program, which aims to develop the next generation of advanced vertical lift aircraft.

Sud Aviation company, established in 1957 through the merger of several French aviation firms, played a significant role in helicopter development. *Sud Aviation* was involved in various aircraft projects, but it gained particular recognition for its advancements in helicopter technology, including the development of the *fenestron* which was invented by two French engineers, *Raymond Corvec* and *Charles Hureau*. The fenestron is characterized by a fan-like rotor enclosed within a cylindrical shroud, located at the tail of the helicopter. The design differs from traditional helicopters, which typically have an exposed tail rotor. The company began testing the fenestron on the *Sud Aviation SA 340 Gazelle* helicopter, which was a highly successful utility and scout helicopter. The *Gazelle* was introduced in the late 1960s and became the first helicopter to incorporate the fenestron tail rotor. Following the *Gazelle*'s success, *Sud Aviation* fur-

ther developed and refined the fenestron design. It was later incorporated into several other helicopter models, including the *Sud Aviation SA 360 Dauphin* and the *Eurocopter AS365 Dauphin*, which have since become known for their use of the fenestron tail rotor.

Fenestron on a *Kawasaki OH-1* reconnaissance helicopter.

The development and testing of the NOTAR (no tail rotor) system were significant milestones in helicopter technology. The NOTAR system was developed by *McDonnell Douglas Helicopter Systems* (later acquired by *Boeing*) as an alternative to the traditional tail rotor used for anti-torque and directional control in helicopters. The concept of the NOTAR system was conceived in the 1970s as a response to some of the limitations associated with conventional tail rotors. The key principle behind NOTAR is using a combination of air jets and a variable-pitch fan, known as the fan-in-fin, to produce anti-torque and directional control forces. The first flight testing of the NOTAR system was conducted on a modified *Hughes OH-6A (Loach)* helicopter in the late 1970s. The *OH-6A* was chosen as a test platform due to its small size and agility, making it suitable for conducting early flight trials. Following the successful flight testing, the first production helicopter to feature the NOTAR system was the *MD 520N*, which was a derivative of the *MD 500* series. The *MD 520N*, introduced in the late 1980s, became known for its quiet and smooth operation, making it suitable for various missions, including law enforcement, emergency medical services, and private use. The NOTAR system was later applied to other models, including the *MD 500E* and *MD 600N*.

In 1983, *Bell Helicopter* and *Boeing* were awarded a contract by U.S. Department of Defense to develop the tiltrotor aircraft. As a response, the *Bell-Boeing* developed the *V-22 Osprey* which was a groundbreaking tiltrotor aircraft designed for both vertical takeoff and landing (VTOL) like a helicopter and efficient forward flight like a fixed-wing aircraft. The *V-22 Osprey* is a result of a joint partnership between *Bell Helicopter*, a subsidiary of *Textron Inc.*, and *Boeing*'s defense division. The aircraft's unique tilt-rotor design allows it to combine the capabilities of helicopters and fixed-wing aircraft, making it highly versatile and well-suited for a wide range of military missions.

Throughout the latter half of the 20th century and into the 21st century, helicopter technology has seen significant advancements and innovations, driven by advancements in materials, avionics, aerodynamics, and propulsion systems. These developments have led to improvements in safety, performance, efficiency, and mission capabilities of helicopters. The increased use of advanced composite materials, such as carbon fiber, in helicopter construction has reduced weight while maintaining structural integrity. For instance, the *Airbus H160* incorporates a significant amount of composite materials in its airframe.

The adoption of fly-by-wire technology has improved helicopter handling, stability, and control. Fly-by-wire systems use electronic signals to transmit pilot inputs, reducing the mechanical complexity of control systems and enhancing flight safety. For instance, the *Sikorsky S-76D* features a fly-by-wire flight control system, offering enhanced stability and control. Helicopters have transitioned from traditional analog instrument panels to modern glass cockpit displays. Integrated avionics and digital displays provide pilots with better situational awareness, enhanced navigation aids, and customizable information, improving flight safety and efficiency. For example, the *Leonardo AW139* is equipped with a modern glass cockpit featuring large displays and integrated avionics.

Health and Usage Monitoring Systems (HUMS): HUMS technology has become more prevalent in modern helicopters. These systems monitor various parameters of the helicopter's systems and provide real-time feedback to maintenance crews, enabling proactive maintenance, reducing downtime, and enhancing operational safety. The *Bell 525 Relentless* is equipped with an advanced HUMS that continuously monitors critical components and systems. This technology enables predictive maintenance and, on average, reduces unscheduled maintenance events. Rotor designs have evolved to optimize aerodynamics, noise reduction, and performance. Advancements include improved blade shapes, increased use of composites in rotor blades, and active rotor control systems for reduced vibrations and noise. The *AH-64E Apache Guardian* features improved rotor blades with swept tips, which enhance lift and maneuverability. The *Leonardo AW189* is powered by two *General Electric CT7-2E1* engines, which deliver higher power output and fuel efficiency.

Helicopter engines have seen improvements in fuel efficiency, power output, and reliability. Turbine engines have become more compact and lightweight, enhancing helicopter performance and range.

There is ongoing research and development in electric and hybrid-electric propulsion for helicopters. These technologies aim to reduce emissions, noise, and operating costs while increasing sustainability. The *Airbus CityAirbus* is an all-electric vertical takeoff and landing (eVTOL) aircraft designed for urban mobility.

The integration of autonomous systems in helicopter technology is being explored for various applications, including unmanned cargo delivery, search and rescue missions, and surveillance. Autonomous systems can enhance mission capabilities and reduce pilot workload in certain scenarios. The *Sikorsky S-70M Black Hawk AD* is a modified version of the *Black Hawk* helicopter with autonomous capabilities. It demonstrated the ability to fly autonomously, reduce pilot workload, and perform complex missions with a higher level of precision. The *Eurocopter EC145 T2* incorporates advanced soundproofing materials and an innovative Fenestron tail rotor. These noise reduction measures result in an average noise reduction of 6 dB(A) compared to previous models. Helicopter manufacturers have invested in noise reduction technologies to address environmental concerns and community noise abatement. Improved rotor designs, sound insulation, and active noise cancellation systems have contributed to reducing helicopter noise levels. The *Bell 429 GlobalRanger* is equipped with crash-resistant fuel systems and energy-absorbing seats. These safety features significantly increase occupant survivability in the event of an accident. Helicopter safety has been a focus of technology development. Advancements include crash-resistant fuel systems, terrain awareness and warning systems (TAWS), traffic collision avoidance systems (TCAS), and advanced autopilot capabilities.

Today, helicopters serve a wide range of roles, including search and rescue, transportation, law enforcement, firefighting, military operations, offshore support, and humanitarian aid. They play a vital role in both civilian and military applications, providing rapid and flexible aerial transportation and assistance.

In the previous sections, we looked at the technological evolution of aircraft. In the next section, we look at some of the advanced experimental aerospace programs pursued by the USA which brought out numerous innovative aerospace designs.

1.8 USA's Experimental Aerospace Programs

The United States has a rich history of developing experimental aircraft and spacecraft that push the boundaries of aviation and space exploration. These innovative vehicles played a crucial role in advancing technology, testing new concepts, and expanding human understanding of flight and the cosmos. Aircraft like the *X-1*, which famously broke the sound barrier, and the more recent *X-59 QueSST* (Quiet Supersonic Transport) are examples of vehicles designed to explore supersonic and hypersonic flight capabilities. These experimental aircraft provide valuable data on aerodynamics, propulsion, and materials under extreme conditions. They form the basis for the development of future commercial and military aircraft.

Bell X-1 (1946 - 1951): The *Bell X-1* was a pioneering experimental aircraft designed to explore flight characteristics at supersonic speeds and to study structural and physiological phenomena within the transonic range.

On October 14, 1947, *Bell X-1* became the first aircraft to break the sound barrier by flying faster than the speed of sound. The *Bell X-1* was air-launched from a *Boeing B-29* at an altitude of 7 km. Utilizing its rocket engine, the *X-1* climbed to a test altitude of 21.9 km and achieved a speed of 1,540 km/h (Mach number 1.45).

Bell X-1 (1946 - 1951), the first aircraft to break the sound barrier.

Bell X-1A: aimed to build upon the research goals of the original *X-1*, focusing on higher speeds and altitudes. The *X-1A* was designed with increased fuel capacity, an ejection seat, and a taller, more conventional cockpit canopy. In succession, the *X-1B* was developed for exploratory aerodynamic heating tests and experimental reaction control systems.

On December 12, 1953, the *Bell X-1A* achieved a remarkable Mach number 2.44 (2,604 km/h) at a maximum altitude of 27.7 km. However, during this flight, pilot *Chuck Yeager* faced a perilous situation. The aircraft experienced violent jerks and went out of control due to low dynamic pressure at the high altitude. It plummeted nearly 15.24 km in just 51 seconds, slowing down from 2574 km/h to low speed (around 170-273 km/h). This rapid descent subjected *Yeager* to forces ranging from 8 to -1.5 g, causing a crack in the plane's canopy and his flight helmet. Despite the challenging circumstances, *Yeager* managed to regain control at an altitude of 9.144 km and safely landed the aircraft at Edwards Air Force Base.

Bell X-1A (Skyrocket, 1953-55), the first aircraft to cross Mach number 2.

Bell X-2 (*Starbuster*) was a rocket-powered, swept-wing research aircraft designed to investigate the structural effects of aerodynamic heating as well as stability and control effectiveness at high speeds and altitudes. The *Bell X-2* resembled the *X-1* but featured swept wings and an elongated nose. Captain *Ivan Kincheloe* achieved a remarkable altitude of 38.466 km on September 7, 1956, in the *X-2*, experiencing nearly complete weightlessness. Later on September 27, 1956, Captain *Milburn G. Apt* initiated a dive from the mother ship, which had climbed to an altitude of 22 km. During this dive, the *X-2* reached a speed of Mach number 3.2. However, after sustaining this speed for about 10 seconds, the pilot executed a sharp turn that resulted in a series of rapid rolls. This led to inertial coupling, in which centrifugal forces caused the aircraft to move uncontrollably in all the three axes. Unfortunately, the aircraft crashed during this sequence, leading to the tragic loss of the pilot.

Bell X-2 (Starbuster, 1952 - 56), a swept-wing, high-speed experimental aircraft.

Douglas X-3 (*Stiletto*, **1954 - 1956**) was developed to investigate high-speed aerodynamic phenomena, test titanium construction, and assess the capability of taking off and landing under its own power. Additionally, it conducted tests on lateral and directional stability, which led to a better understanding of inertia coupling. Distinguished as one of the sleekest early experimental aircraft featuring low-aspect ratio wings, the *X-3*'s research outcomes had a profound impact. Findings from the *X-3* experiments influenced modifications that enhanced the performance of notable aircraft like the *F-100A Super Sabre* and the *Lockheed F-104 Starfighter* which emerged as a versatile multi-role aircraft capable of sustained Mach number 2 flight. It played a significant role with the U.S. Air Force during the *Vietnam War*, showcasing the practical applications and importance of the research conducted by the *X-3* program.

Douglas X-3 (Stiletto, 1954 - 56), designed to test high-speed aerodynamic phenomena and explored the use of very short wings and titanium airframe construction.

Northrop X-4 (*Bantam*, 1950 - 53) was a research aircraft featuring swept wings and a semi-tailless configuration–lacking conventional horizontal tail surfaces–to address stability challenges. Its purpose was to explore the behavior of the configuration at transonic speeds. The *X-4* provided significant insights into the complex interactions of combined pitching, rolling, and yawing movements, crucial for the development of advanced military aircraft with high performance. However, the results indicated that tailless aircraft might not be well-suited for transonic flight conditions.

Bell X-5 (1952 - 55) was designed to explore the possibility of altering an aircraft's wing sweep angle while in flight. This innovative concept aimed to assess the feasibility of variable geometry wings. The *X-5*'s successful tests of its variable sweep-wing mechanism contributed significantly to the devel-

Northrop X-4 (*Bantam*, 1950 - 53), an experimental aircraft developed to test swept wings and a semi-tailless configuration.

opment of the versatile and adaptable *General Dynamics F-111 Aardvark* fighter/bomber, which utilized variable geometry wings to optimize performance across different flight conditions.

Bell X-5 (1952 - 55), built to test the feasibility of variable sweep angle.

Lockheed X-7 (Kingfisher, 1951 - 60): The *Lockheed Skunk Works* developed the *X-7* aircraft, aiming to explore the use of 'ramjet' engines for achieving supersonic and hypersonic speeds. Over time, its role evolved to encompass comprehensive testing of various power plants.

The *X-7* achieved an exceptional record by attaining a remarkable speed of Mach number 4.31 (4,636 km/h) and reaching a notable altitude of 32.3 km. Over a span of more than nine years, the research conducted on the X-7 significantly contributed to the practical advancement of ramjet technology. This advancement expanded the operational capabilities of atmospheric flight, paving the way for groundbreaking developments such as the *Bomarc* surfacr-to-air missile. These innovative technologies played pivotal roles as frontline *Cold War* missiles, fortifying America's defense capabilities.

Lockheed X-7 (*Kingfisher*, 1951 - 60), developed to test supersonic ramjet engines technology and missile guidance and control components.

Ryan X-13 (*Vertijet*, 1955 - 57) was developed to demonstrate the capability of a jet-powered aircraft to perform vertical take-off and landing (VTOL), hover, and smoothly transition to horizontal forward flight. It marked the initial achievement of successful VTOL flight solely through jet thrust. It was a *tailsitter* VTOL aircraft that takes off and lands on its tail, then tilts horizontally for forward flight.

Ryan X-13 (*Vertijet*, 1955 - 57), a tail-sitting VTOL flight demonstrator.

Bell X-14 (1957 - 81) was an aircraft designed to test Vertical Take-Off and Landing (VTOL) technology and its suitability as a trainer for the *Apollo 11* mission. However, it was not deemed suitable for this purpose, leading NASA to develop the *Lunar Landing Research Vehicle*. The *X-14* was the first VTOL aircraft to utilize a jet thrust diverter system for vertical lift. The insights gained from its service contributed to the development of the iconic *Harrier* fighter/attack aircraft, known for its combat VTOL capabilities.

Bell X-14 (1957 - 81), a vertical takeoff fighter demonstrator.

Bell X-18 (1959 - 61) and X-22A (1966 - 84): The *X-18* served as a Vertical and/or Short Take-Off and Landing (V/STOL) demonstrator with engines mounted on tilt-wings. These wings could be moved to enable the aircraft to shift between horizontal and vertical flight modes.

The insights got from the *X-18*'s operations played a pivotal role in the development of the *swing-wing cargo* aircraft *Vought XC-142*. This second-generation V/STOL transport aircraft was utilized by the US Air Force before being transferred to NASA in 1966. The *X-18*'s impact extended to NASA's own research efforts, which continued until 1970.

Bell X-18 (1957 - 81), built to test the tiltwing concept for VTOL flight

Vought XC-142, cargo transport aircraft based on *X-18* experimental aircraft

The *X-22A* was created to study dual-tandem-ducted propeller configuration and V/STOL handling through a variable stability system design. The flights of the *X-22A* effectively validated the practicality of the ducted fan concept. Furthermore, these flights showcased a wide array of technologies essential to the broader spectrum of V/STOL vehicles.

Bell X-22A, built to test the dual-tandem-ducted propeller configuration for V/STOL flight

Curtiss-Wright X-19 (1964 - 65) was designed to showcase VTOL capabilities employing radial lift for a four-seat executive-class aircraft. By utilizing short, broad propeller blades with substantial twist, and positioning nacelles at the wing tip, the aircraft leveraged the lift-increasing effect as the propellers transitioned from horizontal to vertical flight. The insights gathered from the *X-19*'s experiments contributed to the development of the *Bell-Boeing V-22 Osprey* aircraft. *V-22 Osprey* was an American multi-mission, tiltrotor military aircraft with both VTOL and STOL capabilities. It is designed to combine the functionality of a conventional helicopter with the long-range, high-speed cruise performance of a turboprop aircraft.

Curtiss-Wright X-19.

Bell-Boeing V-22 Osprey, an American multi-mission, tiltrotor military aircraft with both VTOL and STOL capabilities. It is designed to combine the functionality of a conventional helicopter with the long-range, high-speed cruise performance of a turboprop aircraft.

Northrop Corporation X-21A (1963 - 64) was developed to test full-scale laminar flow control (LFC) on larger aircraft to significantly enhance range and endurance by at least 50% and between two to four times, respectively, while simultaneously reducing drag by around 25%.

Utilizing the *Douglas B-66D* light bomber as a testbed, the *X-21A* underwent modifications that included re-mounting engines at the rear of the fuselage to reduce turbulence and noise, adding a prominent hump/fairing atop the fuselage to mitigate wing flow turbulence, and incorporating LFC slots and holes on the wings, along with underwing pumps to extract boundary layer air from the wing surfaces.

This technology path found success in later endeavors like the *F-16XL supersonic* LFC, *F-111* Transonic Aircraft Technology (TACT) demonstrator, and *F-14* refined TACT with an advanced glove, all of which achieved laminar flow and contributed to advancements in aviation technology.

Northrop Corporation X-21A (1963 - 64), designed to test wings with laminar flow control.

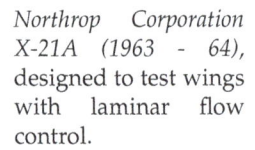

Bensen Aircraft X-25 (1968) was designed to test and refine the Discretionary Descent Vehicle (DDV), a solution developed to address the challenges of rescuing endangered crews, particularly American pilots downed over enemy territory during the Vietnam conflict. The DDV concept involved compact autogyros, featuring an aluminum tubular structure, crew seat, twin-bladed rotor, rudder, and piston engine, ingeniously packed into cockpits to facilitate the secure transportation of at-risk crew members. This innovative approach showcased the potential of the DDV to effectively succeed in combat scenarios.

Bensen Aircraft X-25 (1968), designed to test light autogyro for downed pilots.

Schweizer Aircraft, Lockheed Missiles X-26B (1967 - 88): The *X-26*, the longest-lived of the X-plane programs, included the *X-26A Frigate* sailplane and the motorized *X-26B Quiet Thruster* versions, aimed to develop an ultra-quiet surveillance aircraft, spurred by the demands arising from the context of the Vietnam War. The *X-26B* (*Lockheed YO-3 Quiet Star*), embarked on a crucial mission to Vietnam in mid-1968 conducting extensive nighttime missions over the course of two months. These missions provided invaluable real-time data on enemy movements, highlighting the aircraft's battlefield effectiveness. The insights gained from these operations also paved the way for enhancements to their design, setting the stage for future platform improvements.

George Pereira, Osprey Aircraft X-28 (Air Skimmer, 1971): The goal of the X-28 was to assess the potential of a compact, single-engine seaplane for civil police patrols in Southeast Asia. This aircraft was designed as a short take-off-and-landing (STOL) flying boat, aiming to enhance operational capabilities in maritime and coastal environments.

Schweizer Aircraft, Lockheed Missiles X-26B (1967 - 88), an ultra-quiet surveillance experimental aircraft

George Pereira, Osprey Aircraft X-28 (Air Skimmer, 1971), a low-cost aerial policing seaplane.

Grumman Aerospace X-29 (1984 - 90) project aimed to test and validate the forward-swept wing design, incorporating canard control surfaces, advanced composites, and other aerodynamic advancements. This experimental aircraft marked a significant innovation by being the first forward-swept wing aircraft to achieve supersonic flight in level conditions. It featured an unconventional configuration where its forward-swept wings were positioned farther back on the fuselage, and its canards (horizontal stabilizers for pitch control) were situated ahead of the wings, in contrast to the traditional tail placement. This unique combination of wing and canard geometries resulted in remarkable maneuverability, supersonic capabilities, and a lightweight structure. The *X-29* project highlighted the potential benefits of novel wing designs in enhancing aircraft performance and agility.

Grumman Aerospace X-29 (1984 - 90), the most aerodynamically unstable aircraft ever built–demonstrated forward-swept wing technology for supersonic fighter aircraft.

Rockwell International, Deutsche Aerospace, X-31A (1990 - 95) aimed to demonstrate *Enhanced Fighter Maneuverability* by gathering critical data on control in the post-stall flight regime. This experimental aircraft showcased the effectiveness of thrust vectoring, a technique involving the redirection of engine exhaust flow, in conjunction with advanced flight control systems. This combination enabled controlled flight at exceptionally high angles of attack. Notably, the *X-31A* marked the first international cooperation in X-plane development, involving Germany. The project achieved in-flight thrust vectoring and expanded the post-stall flight envelope to a remarkable 40° angle of attack. To collect precise data during low tolerance for side-slip conditions, the aircraft incorporated a distinctive *Kiel probe*, a nose boom bent downward by 10 degrees. The *X-31A* pushed boundaries further, successfully achieving controlled flight at an astonishing 70° angle of attack. Furthermore, it executed a post-stall maneuver, performing a minimum radius, 180-degree turn, showcasing its ability to operate well beyond the conventional aerodynamic limits of standard aircraft.

Rockwell International, Deutsche Aerospace, X-31A (1990 - 95), built to demonstrate thrust vectoring supermaneuverability.

Boeing X-32 and the Lockheed Martin X-35 (1996 - 2006): Emerging as the victor in the *Joint Strike Fighter (JSF)* program, the X-35 triumphed in the most extensive and ambitious fighter development program. Following an eight-year competition with the *Boeing X-32*, the *X-35* was chosen in 2001, which paved the way for production commencing in 2006 under the designation of *F-35 Lightning II*. The aim of the JSF program was to create a cutting-edge fifth-generation supersonic multirole fighter jet that could seamlessly handle air-to-air combat, striking, and ground attack missions. The *F-35* boasts supersonic speed, extended range, versatile weapons capacity, highly integrated sensors, advanced electronic warfare capabilities, and classified state-of-the-art features to excel in air superiority missions. The aircraft's combat debut took place in 2018 during Israel–Iran clashes, when it was deployed by the Israeli Air Force. Although X-32 was not selected in JFS program but yielded many advances in stealth technology and design and manufacturing methods and helped in Boeing's other programs including *F/A-18E/F Super Hornet, X-45A UCAV* etc.

Lockheed Martin X-35, a fifth-generation supersonic multirole fighter jet emerging as the winner of the *Joint Strike Fighter (JSF) program* in 2001.

Boeing X-32, a concept demonstrator aircraft that was designed for the JSF competition.

NASA/Boeing, X-36 (1997): The *NASA/Boeing X-36*, a Tailless Fighter Agility Research Aircraft (TFAR), was designed to showcase and prove the potential of tailless fighter aircraft utilizing advanced technologies to enhance maneuverability and survivability in future fighter plane designs. Making its inaugural flight on May 17, 1997, the *X-36* successfully completed a series of 31 research flights, demonstrating its exceptional agility at both low speeds and high angles of attack, as well as at high speeds and low angles of attack. These flights were conducted at altitudes of up to 6 km and achieved maximum angles of attack of 40 degrees. The Air Force Research Lab played a significant role by developing the 'Reconfigurable Control for Tailless Fighter Aircraft' software, which showcased the adaptability of a neural-net algorithm to compensate for in-flight damage or malfunctions affecting control surfaces like flaps, ailerons, and rudders.

X-36, a remotely piloted research aircraft.

NASA X-43 (1996 - 2004): The *X-43, Hypersonic Experimental Vehicle (Hyper-X)*, program embarked on a high-risk, high-payoff research journey, starting in 1996, to address unprecedented challenges and achieve groundbreaking results. This initiative led to the creation of three uncrewed *X-43A* research aircraft: the initial two vehicles were designed for Mach number 7 flights, while the third aimed for Mach number 10 speeds. In March 2004, the *Hyper-X* project achieved a remarkable milestone by conducting successful flights of a scramjet-powered aircraft at hypersonic velocities, reaching Mach number 6.8. This achievement was followed by a second flight in November 2004, during which the aircraft attained Mach 9.6. The program's crowning achievement occurred during the third and final flight, in which the vehicle set a world speed record for an air-breathing aircraft, flying at an astonishing speed of around 11,250 km/h at an altitude of 33.5 km. The *X-43A* was carried by a *B-52B aircraft*, dropped from 12 km altitude, and its booster propelled it to its designated test speed and altitude. To withstand the higher temperatures generated during the Mach number 10 flight, the research vehicle's vertical fins were equipped with carbon-carbon composite material for enhanced thermal protection.

NASA X-43 (1996 - 2004), a hypersonic experimental vehicle for testing scramjet technology.

Boeing X-45 and Northrop Grumman X-47 (2002 - 2006): *The Joint Unmanned Combat Air System (J-UCAS)* was a collaborative effort between the United States Air Force and the United States Navy aimed to create an Uninhabited Combat Air Vehicle (UCAV) demonstrator, showcasing a networked system of highly autonomous aircraft capable of efficiently executing modern combat missions. The program involved the development and testing of various prototype aircraft, including the *Northrop Grumman X-47* and the *Boeing X-45*. These aircraft showcased innovative features such as stealth capabilities, autonomous flight systems, and the integration of advanced technologies for communication and coordination. The *X-45* featured retractable landing gear and a composite, fiber-reinforced epoxy skin, accommodating two internal weapons bays within its fuselage. Notably, the *X-45A* introduced yaw axis thrust vectoring, enabling advanced maneuverability. This program achieved several significant milestones, including the first autonomous flight of a high-performance, combat-capable UAV, the inaugural weapons release from an autonomous UAV, and the successful execution of autonomous multi-vehicle coordinated flights. The X-45's impact extended to the development of the *Northrop Grumman X-47*, a variant of the *X-45*. The *X-47* stood as a jet-powered, tailless, blended-wing-body UAV with the unique capability of aerial refueling, showcasing the evolution and potential of unmanned aerial systems.

Boeing Phantom Works/NASA, X-48B (2007) research project was undertaken to explore the benefits of the *Blended Wing Body (BWB)* concept, which combines aspects of both traditional aircraft and

Boeing X-45, the first stealthy, swept-wing modern UAV jet.

X-47B, a naval carrier-based variant of the X-45.

flying wing designs, offering structural, aerodynamic, and operational advantages. Constructed under *Boeing*'s guidance by *Cranfield Aerospace* in the United Kingdom, the *X-48B* embarked on flight trials in early 2007. Its objective was to showcase remarkable traits like high fuel efficiency, minimal noise, and a spacious payload capacity relative to its size. The *X-48B* boasts an approximate maximum airspeed of 224 km/h and can reach altitudes of around 3 km.

Boeing Phantom Works/NASA, X-48B, a blended wing body demonstrator.

Piasecki, X-49 (*SpeedHawk*, 2007) served as a testbed for experimenting the *Vectored Thrust Ducted Propeller (VTDP)* technology on a compound helicopter. This innovative approach enhances the speed of conventional helicopters through the integration of additional thrust and small fixed wings. The *X-49A* is built upon the framework of a *Sikorsky YSH-60F Seahawk*, incorporating the VTDP design along with supplementary lifting wings. Its inaugural flight on June 29, 2007 showcased remarkable capabilities such as hovering, pedal turns, and controlled slow flight in various directions, utilizing VTDP for anti-torque, directional, and trim control functions. This compound helicopter platform signifies a crucial step in advancing helicopter speed and maneuverability.

Piasecki, X-49 (*SpeedHawk*, 2007), four-bladed, twin-engined experimental high-speed compound helicopter build to test Vectored Thrust Ducted Propeller (VTDP).

Boeing, X-50A (*Dragonfly*, 2002 - 2007): The *X50, Canard Rotor/Wing Demonstrator*, was developed as a VTOL rotor wing UAV to showcase a groundbreaking concept: a helicopter's rotor could be stopped in flight and act as a fixed wing, enabling seamless transitions between rotary-wing and fixed-wing flight. Rooted in the 1980s *Sikorsky S-72 X-Wing program*, the *X-50A* was conceived with this innovation in mind. This aircraft employed a wide-chord, two-blade rotor for vertical take-off and landing capabilities. The control surfaces featured fully-movable canards and a tailplane with twin endplates, complemented by a fixed tricycle landing gear. The ingenious design redirected engine exhaust to rotor tips for helicopter mode, eliminating the need for anti-torque mechanisms. As the UAV gained forward speed and generated lift through tail and canards, the engine thrust transitioned to the tailpipe, eventually stopping the rotor entirely, enabling conventional fixed-wing flight.

The *X50* Canard Rotor/Wing Demonstrator stands as a testament to pioneering advancements in aerial technology.

Boeing, X-50A (Dragonfly, 2002 - 2007), a VTOL rotor wing UAV vehicle built to demonstrate Canard Rotor/Wing (CRW) configuration.

Boeing Phantom Works, X-53

The X-53 (*F/A-18 Active Aeroelastic Wing (AAW)*) project aimed to advance the concept of wing-warping, a method of aerodynamically induced wing twist inspired by the *Wright Brothers'* early approach, to enhance aircraft roll control across full-scale aircraft at transonic and supersonic velocities. For this advanced AAW research, the U.S. Navy chose a modified *F/A-18A* as the test aircraft. By replacing certain wing skin panels with thinner and more flexible alternatives near the trailing-edge flaps and ailerons, the aircraft's roll control capabilities were enhanced. The flight data included aerodynamic, structural and flight control characteristics that demonstrated and measured the AAW concept in a comparatively low-cost, effective manner.

Boeing Phantom Works, X-53, an 'Active Aeroelastic Wing' (AAW) demonstrator.

Lockheed Martin, X-55 (2007 - 2009): The X-55, an *Advanced Composite Cargo Aircraft (ACCA)* served as a platform to showcase cutting-edge manufacturing techniques that harnessed advanced composites and innovative structural concepts. These methods aimed to streamline design processes, diminish aircraft size and weight, expedite manufacturing, and trim production costs.

In the ACCA program, the *Skunk Works* team adopted a unique strategy of dissecting a functional *Dornier 328JET* commuter jetliner to maintain cost efficiency. This approach was projected to reduce development expenses by 25%, with added operational savings stemming from reduced weight leading to lower fuel consumption. In comparison to traditional metallic components, the composite structure remarkably reduced the number of parts required, utilizing about 300 structural elements in contrast to 3,000 metallic parts. Additionally, mechanical fasteners were optimized, with approximately 4,000 used compared to the 40,000 employed in a traditional counterpart. The *X-55* initiative marked a significant leap forward in aviation manufacturing.

Lockheed Martin, X-55 (2007-2009), Advanced Composite Cargo Aircraft (ACCA) demonstrator.

Dornier 328JET, a commuter airliner, was used as a base model to design *X-55*.

Lockheed Martin-Skunk Works X-56 (2013): The X-56A, a 'Multi-Utility Technology Testbed', was created as an unmanned aircraft to explore active aeroelastic control methods aimed at reducing flutter and alleviating gust loads on flexible wing structures. Continuing its mission, NASA extended testing with the *X-56A* to further investigations into lightweight structures and advanced control technologies. This research has implications for the development of environmentally conscious, efficient transport aircraft in the future. The design emphasis lies in elongated, slender, high-aspect-ratio wings, which, although susceptible to uncontrollable vibrations, are vital components in the blueprint of upcoming long-range aircraft such as fuel-efficient commercial planes and cargo carriers.

Lockheed Martin-Skunk Works X-56 (2013), an advance aeroservoelastic technology demonstrator.

NASA X-57 Maxwell: The *X-57*, low emission plane powered entirely by electric motors, the project had the goal of advancing and validating more efficient, quieter, and environmentally friendly aviation technologies through the development of distributed electric propulsion system. In collaboration with *Tecnam*, an Italian aircraft manufacturer, NASA integrated their *Scalable Convergent Electric Propulsion Technology and Operations Research (SCEPTOR)* project into an electric propulsion system, resulting in the creation of the *X-57* aircraft. Designed by NASA, the experimental high-aspect ratio wing features twelve small electric propellers which are primarily utilized for takeoff and landing lift generation. During the cruise, the two outboard motors take over. Powering the aircraft is a battery system.

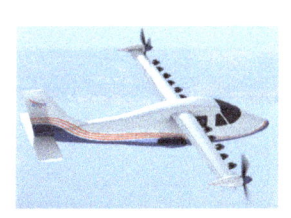

NASA X-57 Maxwell (Artist's concept is shown), an electric propulsion demonstrator

Lockheed Martin and NASA X-59: The X-59 *Quiet SuperSonic Technology (QueSST)* project aimed to build a technology that mitigates the loudness of a sonic boom for the people on the ground. The *X-59* is projected to maintain a cruising speed of Mach 1.42 (1,510 km/h) at an altitude of 16.8 km, producing a low perceived sound level of 75 decibels to assess the acceptability of supersonic transport.

Lockheed Martin and NASA X-59, a Quiet SuperSonic Technology (QueSST) demonstrator.

Dynetics, X-61 (Gremlins): The X-61 (*Air-launched and air-recoverable reconnaissance unmanned air vehicle*) project aimed to showcase a cost-effective unmanned aerial vehicle (UAV) with digital flight controls and navigation systems, designed to be recoverable mid-air by a modified transport aircraft. The *X-61A* is capable of accommodating various payloads, including electro-optical sensors, infrared imaging equipment, electronic warfare systems, and weaponry. Its inaugural free flight occurred on January 17, 2020; however, an incident transpired as the main parachute failed to deploy during recovery, leading to the loss of the aircraft. In a positive turn, in October 2021, an airborne *X-61 Gremlins* was successfully captured for the first time using a mechanical arm and towed docking mechanism deployed from the cargo ramp of a *Lockheed Martin C-130* aircraft.

Dynetics, X-61 (Gremlins).

Airborne *X-61 Gremlins* was caught by *Lockheed Martin C-130*.

General Dynamics, X-62 (2021): The X-62 (*Variable Stability In-flight Simulator Test Aircraft*) is a modified version of the *F-16D Fighting Falcon*, featuring a multi-axis thrust vectoring (MATV) engine nozzle for enhanced control in post-stall scenarios. With its MATV engine nozzle, the X-62 possesses exceptional maneuverability, maintaining pitch and yaw control even at angles of attack that conventional control surfaces cannot effectively manage.

General Dynamics, X-62, a Variable Stability In-flight Simulator Test Aircraft.

In this section, we looked at some of the experimental aerospace programs pursued by USA which brought about significant changes in the way we see aviation today. These programs will have implications in the design of future aerospace vehicles. In the next section, we look at some of the major aviation accidents which brought out major changes in the design and/or operation of aerospace vehicles. They show the utmost attention paid by the aerospace designers and certifying agencies in the relentless pursuit for providing a safe and comfortable experience for the passengers.

1.9 Aviation Accidents

Aviation accidents, while relatively rare compared to the vast number of flights taking place daily, are sobering reminders of the inherent risks in air travel. These incidents encompass a range of factors, from human error and technical malfunctions to adverse weather conditions and external threats. The aviation industry places a strong emphasis on safety, with rigorous regulations, advanced training, and technology continually evolving to minimize the likelihood of accidents. In the aftermath of such events, comprehensive investigations aim to uncover the root causes and improve safety protocols, contributing to the industry's ongoing commitment to enhancing air travel's safety and reliability.

The black box: The flight data recorder (FDR) and cockpit voice recorder (CVR) jointly referred to as "the black box," are vital components of aviation safety and accident investigation. Typically, these devices are located in the aircraft's tail section to enhance their chances of survival in the event of an accident. The FDR is responsible for storing a plethora of data related to an aircraft's performance, including altitude, airspeed, heading, and various system parameters. It acts as an invaluable source of information for investigators, helping them reconstruct the sequence of events leading up to an accident. On the other hand, the CVR records audio from the flight deck, capturing conversations between the flight crew, alarms, and ambient sounds. This data is essential for understanding the factors that may have contributed to an accident. Together, these devices provide a comprehensive record of what transpired during a flight, aiding investigators in determining the causes of accidents and improving safety standards.

David Warren, an Australian scientist, is credited with the invention of the flight data recorder and cockpit voice recorder in the late 1950s. His innovative work was driven by a personal tragedy – the mysterious disappearance of an aircraft carrying his father in the early 1930s. *Warren* recognized the need for a device that could record critical flight information and cockpit conversations to aid in accident investigations. Although *Warren*'s invention initially garnered limited attention, it later gained momentum and was eventually adopted for development in both the United States and the United Kingdom several years afterward. Australia led the way in making this technology mandatory, with the United States subsequently following suit in 1967, making installation mandatory on all commercial aircraft. In November 2008, *Qantas* honored *Warren*'s contributions to aviation by naming one of their *Airbus A380*s after him.

Major aviation accidents

Statistics of the aviation accidents will tell that flying in a jetliner is extraordinarily safe. But air travel only became so safe because previous accidents ushered in crucial safety improvements. From midair collisions, to onboard fires, these infamous airplane accidents triggered major technological advances in flight safety that keep air travel routine today. Here, we list out some of these accidents.

1956 Grand Canyon, mid-air collision: A *United Airlines Douglas DC-7* headed to Chicago from Los Angeles and a *Trans World Airlines Lockheed L-1049 Super Constellation* on its way to Kansas City from Los Angeles collided mid-air on June 30, 1956 killing all 128 passengers and crew onboard.
Technology/Policy upgrade: Federal Aviation Administration (FAA) was created in 1958 in USA to oversee air safety. Airliners are equipped with 'Traffic Collision Avoidance Systems', which detect potential collisions.

1977 Tenerife Airport (Spain) Disaster: On March 27, 1977, two *Boeing 747* passenger jets (*KLM Flight 4805* and *Pan Am Flight 1736*) collided on the runway of the airport. The impact and resulting fire resulted in 583 fatalities with only 61 survivors.
Technology/Policy upgrade: The crash illustrated the need for better communication protocol between pilots and ATC.

1978 Portland, United Airlines Flight 173 (DC-8): On December 28, 1978, the *DC-8* approaching Portland, Oregon, with 181 passengers, circled near the airport for an hour troubleshooting a landing gear problem before running out of fuel and crashing resulting in 10 fatalities. The captain waited too long to begin his final approach ignoring the flight engineer's warning about the rapidly diminishing fuel supply.
Technology/Policy upgrade: Cockpit teamwork: 'Cockpit Resource Management (CRM)' was introduced abandoning the traditional *Captain is God* in airline hierarchy, CRM emphasized teamwork and communication among the crew, and it has since become a standard.

1983 Cincinnati, Air Canada Flight 797: On June 2, 1983, the *DC-9* from Dallas to Toronto developed an in-flight fire starting from the rear lavatory filling the plane with smoke. The fire also burned through electrical cables that disabled most of the instrumentation in the cockpit, forcing the plane to divert to Cincinnati. After the doors and emergency exits were opened, the fresh oxygen engulfed the cabin in fire killing 23 of the 46 people aboard.
Technology/Policy upgrade: The FAA mandated that aircraft lavatories be equipped with *smoke detectors* and automatic fire extinguishers. All jetliners were retrofitted with fire-blocking layers on seat cushions and floor lighting leading to exit doors.

1985 Dallas/Fort Worth, Delta Air Lines Flight 191: When the *Lockheed L-1011* aircraft was approaching to land at DFW airport on August 2, 1985, it encountered a microburst wind shear causing it to lose 100 km/h of airspeed in a few seconds. The aircraft impacted ground 1.6 km short of the runway and bounced across a highway, crushing a car and killing the driver. The plane then crashed into two huge water tanks killing 136 of 163 people onboard.
Technology/Policy upgrade: Downdraft detection The crash triggered a NASA/FAA research effort, which led to installing the onboard forward-looking radar wind-shear detectors that became standard equipment on airliners in the mid-1990s.

1985 Manchester England, British Airtours Flight 28M (Boeing 737-236): On August 22, 1985, the aircraft caught fire before take-off at Manchester Airport. Though the pilot followed protocol, the seats were placed too close to each other, causing delay while deplaning the passengers.
Technology/Policy upgrade: After the incident, plane manufacturers changed the internal layout to make evacuation easier.

1988 Maui, Aloha Airlines Flight 243 (Boeing 737): On April 28, 1988, a *Boeing 737-297* serving the flight between Hilo and Honolulu in Hawaii suffered extensive damage after an explosive decompression in flight, caused a portion of its fuselage to blow off. The plane was able to land safely at Kahului Airport on Maui. A flight attendant, who was swept out of the plane, was killed. The NTSB blamed a combination of corrosion and fatigue damage, the result of repeated pressurization cycles during the plane's 89,000-plus flights.
Technology/Policy upgrade: Airworthiness of aging air-

craft. The FAA began the National Aging Aircraft Research Program in 1991, which tightened inspection and maintenance requirements for high-use and high-cycle aircraft.

1989 Sioux City Iowa, United Airlines Flight 232: The *DC-10* flying from Denver to Chicago crash-landed on July 19, 1989, after suffering a failure of its tail-mounted engine due to an unnoticed manufacturing defect in the engine's fan disk, which resulted in the loss of many flight controls. Of the 296 passengers and crew on board, 112 died in the accident.

Technology/Policy upgrade: Engine safety improvements: The accident led the FAA to order modification of the *DC-10*'s hydraulic system and to require redundant safety systems in all future aircraft, and it changed the way engine inspections are performed.

1994 Pittsburgh, US Airways Flight 427 (Boeing 737) On September 8, 1994, the *Boeing 737* flying from Chicago to Florida crashed while landing in Pittsburgh. The aircraft suddenly rolled to the left and plunged to the ground, killing all 132 people onboard. The aircraft's rudder malfunctioned and went hard over in a direction opposite to that commanded by the pilot. The NTSB attributed the accident to a jammed valve in the rudder-control system that had caused the rudder to reverse.

Technology/Policy upgrade: In response to conflicts between the airline and the victims' families, Congress passed the Aviation Disaster Family Assistance Act, which transferred survivor services to the NTSB. As a result, *Boeing* spent $500 million to retrofit all 2,800 of the world's most popular jetliners.

1996 Miami, ValuJet Flight 592 (DC-9): On May 11, 1996, the *ValuJet Airlines McDonnell Douglas DC-9* operating between Miami and Atlanta crashed into the Everglades after taking off from Miami as a result of a fire in the cargo compartment caused by mislabeled and improperly stored hazardous cargo. All 110 people on board died.

Technology/Policy upgrade: FAA responded by mandating smoke detectors and automatic fire extinguishers in the cargo holds of commercial airliners. Rules regarding hazardous cargo on aircraft became stringent.

2000, Air France Flight 4590 (Concorde): On July 25, 2000, after taking off from Paris, the aircraft crashed into a hotel killing all 109 people on board and 4 on the ground. Investigations revealed a series of events that led to the crash. A piece of metal lying on the runway caused a tyre to rupture and tyre fragments caused the fuel tanks to rupture and catch fire. The engulfing fire, loss of thrust and control eventually led to the crash.

Technology/Policy upgrade: The accident became the trigger for grounding the supersonic airliner forever in 2003.

2014, Malaysia Airlines 370 (Boeing 777-200): The disappearance of *Malaysia Airlines Flight 370 (MH370)* on March 8, 2014, remains one of the most perplexing mysteries in the history of aviation. The aircraft, a *Boeing 777-200ER*, was en route from Kuala Lumpur to Beijing with 239 people on board when it vanished from radar screens. Despite extensive search efforts and investigations, the main wreckage of the plane was not discovered till date. This unprecedented incident raised questions about aviation safety, the capabilities of tracking systems, and the need for enhanced communication protocols between aircraft and air traffic control.

Technology/Policy upgrade: Real-time flight tracking: ICAO has ordered all airlines to install equipment that will keep closer tabs on planes and manufacturers are also developing black boxes that can eject and float.

The Boeing 737 MAX accidents and groundings: In October 2018, *Lion Air Flight 610*, operating a *Boeing 737 MAX 8*, crashed into the Java Sea shortly after takeoff, resulting in the tragic loss of 189 lives. Just five months later, in March 2019, *Ethiopian Airlines Flight 302*, another *Boeing 737 MAX 8*, crashed shortly after departure, claiming the lives of all 157 individuals on board. These accidents prompted a comprehensive reevaluation of aircraft certification and safety procedures within the aviation industry. *Technology/Policy upgrade:* In response to these accidents, aviation authorities around the world, including the FAA grounded all the *Boeing 737 MAX* series till thorough investigations into the causes of the crashes. These investigations primarily focused on the aircraft's Maneuvering Characteristics Augmentation System (MCAS), a system designed to prevent stalls but which was implicated in both accidents. Subsequent scrutiny revealed design and training deficiencies, prompting *Boeing* to work on significant modifications and enhancements to the *737 MAX*'s systems and pilot training programs.

The grounding of the *737 MAX* had far-reaching implications, disrupting airline schedules and supply chains while impacting *Boeing*'s revenue. The events led to a broader reevaluation of aircraft certification processes and the relationship between aircraft manufacturers and regulatory bodies. Following extensive changes to the aircraft's systems and enhanced training for pilots, regulatory agencies worldwide gradually began lifting the grounding orders, allowing the *737 MAX* to return to service in 2020.

Aviation incidents involving terrorism

Some of the major terrorist related incidents in aviation are listed below.

September 11, 2001 (9/11 Attacks): The most infamous terrorist act in aviation history, 9/11 involved the hijacking of four commercial airliners by terrorists. Two of the planes were flown into the Twin Towers of the World Trade Center in New York City, leading to their collapse, while another plane struck the Pentagon in Washington, D.C. The fourth plane, *United Airlines Flight 93*, crashed in Pennsylvania after passengers bravely attempted to retake control. The attacks resulted in nearly 3,000 deaths, profound changes in aviation security, and the establishment of the Transportation Security Administration (TSA) in the United States.

Lockerbie Bombing (1988): *Pan Am Flight 103*, a transatlantic flight from London to New York, was targeted by a terrorist bombing over Lockerbie, Scotland. The explosion resulted in the deaths of all 270 people on board. Libyan intelligence officer Abdelbaset al-Megrahi was eventually convicted for the attack, which remains one of the deadliest acts of aviation terrorism.

Air India Flight 182 (1985): This tragedy unfolded when *Air India Flight 182 Boeing 747*, flying from Montreal to Delhi, was destroyed by a bomb over the Atlantic Ocean, killing all 329 people on board. The bombing was attributed to Sikh separatists in Canada.

MetroJet Flight 9268 (2015): A Russian MetroJet flight departing from Sharm El-Sheikh, Egypt, was downed by a bomb, killing all 224 people on board. The Islamic State (ISIS) claimed responsibility for the attack.

Indian Airlines Flight 814 (IC 814) hijack in December 1999 remains a stark reminder of the challenges posed by aviation terrorism. The aircraft, an *Airbus A300* en route from Kathmandu to Delhi, was hijacked by a group of terrorists. After a tense stand-off and negotiations, the aircraft was eventually allowed to leave Indian airspace, leading to a 7-day ordeal involving stops in multiple countries. The passengers and crew endured immense physical and psychological stress during this period. Ultimately, the aircraft was released in exchange for the release of three jailed terrorists.

1.10 Future of Aviation

Since the advent of the *jet age*, the volume of air transportation has exhibited a doubling trend approximately every fifteen to twenty years, solidifying its status as one of the fastest-growing sectors in transportation. Across various domains of aerospace technology, there is a consistent drive for advancement. Among the promising avenues for future aircraft propulsion, we find concepts like open rotors, boundary layer ingestion, and electric/hybrid-electric propulsion gaining prominence. Equally captivating are innovative airframe configurations like the strut-braced wing, blended wing body, double-bubble fuselage, and box-wing aircraft, which have recently showcased notable advancements. The trajectory demonstrates the ongoing dynamism and transformative potential of the aviation industry.

Evolutionary Aircraft Technologies

The forthcoming generation of aircraft is expected to maintain a similar overall configuration to its predecessors, yet stands out due to the integration of retrofits, serial upgrades, and newly designed components and systems. These innovations collectively enable these aircraft to achieve an impressive 15% to 20% increase in fuel efficiency compared to their forerunners, a phenomenon often referred to as Evolutionary Aircraft Technologies. Aiding in this efficiency are cutting-edge techniques like Natural Laminar Flow (NLF) control and Hybrid Laminar Flow (HLF) Control, both of which will significantly decrease drag by preventing airflow turbulence over the aircraft's surface. The LADE (Breakthrough Laminar Aircraft Demonstrator in Europe) project has delved into NLF control since 2017, exemplified by *A340* flight tests that replaced outer wing sections with laminar profiles spanning about 10 meters in width. On the other hand, NASA's Environmentally Responsible Aviation (ERA) research program has pursued HLF control, conducting test flights on a *B-757* to maintain laminar flow through boundary-layer suction. Other modern aircraft enhancements include variable camber with new control surfaces, Spiroid Wingtips, riblets, winglets, or raked wingtips, all contributing to superior aerodynamics.

Furthermore, modern engines are pivotal in achieving enhanced fuel efficiency, powering airliners like the *MRJ*, *Embraer E2* family, *A220*, *A320neo*, and *737MAX*. These engines, featuring higher bypass ratios (BPRs) ranging from 9 to 12, lead to fuel burn reductions of approximately 15% compared to their predecessors with BPRs typically between 5 to 6. Innovations such as the *Rolls-Royce Advance Engine* (three-shaft architecture with a novel high-pressure core) and the *UltraFan* engine (a two-shaft configuration coupled to a geared turbofan) under the *Clean Sky 2* program, along with *Safran*'s Ultra-High-Bypass Ratio (UHBR) turbofan engines featuring a minimum bypass ratio of 15, have redefined engine architectures. In the United States, under the *CLEEN II* national research program, the development of advanced engine designs like the *GE9X* engine powering *Boeing*'s new *777X* aircraft is expected to enhance fuel efficiency by 10%. The use of sustainable aviation fuel(SAF), produced from sustainable feedstocks, results in significant reduction in carbon emissions compared to the conventional jet fuel. The use of SAF is expected to replace a major portion of the jet aviation fuel in the next few decades and thereby reducing the carbon footprint of aviation.

Revolutionary Aircraft Technologies

The industry's ambitious long-term climate objective of halving global net aviation CO_2 emissions by 2050, relative to 2005, relies not only on innovative aircraft concepts but also on a diverse range of novel configurations. In contrast to the conventional tube-and-wing design of current commercial aircraft, future fuel-efficient airframes explore cutting-edge configurations such as the *strut-braced wing design (SBW)*, the *blended wing body (BWB) or hybrid wing body (HWB)*, the *Flying-V*, the *Double-bubble fuselage*, and the *box-/joined-wing*. Boeing has developed a 154-seat high aspect-ratio, low induced-drag *SBW* aircraft as part of the *Subsonic Ultra Green Aircraft Research (SUGAR)* initiative.

The *Blended Wing Body (BWB)* configuration primarily features a large flying wing design where the central body's shape smoothly merges with the outer wings. This airfoil-shaped body enables the entire aircraft to generate lift. By eliminating the empennage and incorporating top-mounted engines, *BWB*s also mitigate aircraft noise and offer enhanced survivability with reduced radar and infrared signatures. This design accommodates cylindrical hydrogen fuel tanks thereby reducing the carbon foot print compared to the fossil fuels. In recent years, *Airbus* introduced a 200-passenger BWB design powered by hybrid-hydrogen turbofan engines as part of its *ZEROe initiative*, while *Bombardier* revealed the *EcoJet*, focused on reducing emissions through aerodynamic and propulsion enhancements.

Another noteworthy design is the V-shaped, highly-swept double-wing configuration that lowers aerodynamic drag, with two wings accommodating passengers, cargo, and fuel tanks. NASA's *D8* experimental aircraft employs a double-bubble fuselage for additional lift generation. The box-wing design connects offset horizontal wings, offering minimum induced drag and fuel savings compared to traditional aircraft.

As weight reduction using composite materials and light alloys continues, efforts are underway to explore revolutionary materials that enhance weight efficiency and aircraft performance. NASA's *Span-wise Adaptive Wing* project explores shape memory alloy that regains its original shape upon heating. The morphing wing, featuring overlapping pieces resembling scales or feathers, can increase flight efficiency and reduce fuel consumption.

In terms of propulsion, groundbreaking technologies such as open-rotor design, *boundary layer injection (BLI)*, and electric aircraft are expected to yield significant fuel savings in the future. Open rotor engines, a hybrid between propellers and turbofans, feature two counter-rotating, unshrouded fans. *BLI* technology positions the engines in the rear, allowing air to flow over the fuselage before entering the engine, resulting in backward acceleration. These innovations are poised to reshape the aviation industry, ensuring more fuel-efficient and environmentally friendly air travel in the years to come. Electric propulsion plays a pivotal role in achieving the aviation industry's ambitious 2050 environmental objective, resulting in a significant reduction in emissions. All-electric systems exclusively utilize batteries as the primary power source for aircraft propulsion. In contrast, hybrid-electric systems employ gas turbine engines both for generating propulsion power and charging the onboard batteries. These batteries supply the necessary energy for propulsion during various flight phases.

Pipistrel Velis Electro, the first type-certified electric airplane. It has a maximum cruise speed of 181 km/h (stall speed 83 km/h), can reach altitudes of up to 3.7 km, and has an endurance of 50 minutes. Power plant: *Pipistrel E-811-268MVLC* electric motor, 57.6 kW.

Jean-Baptiste Loiselet's Innovavis Sol.Ex.2, solar-powered motorglider fly uses converted energy from its solar panels to fly. The panels' photovoltaic cells are infused between thin pieces of fiberglass.

Bibliography

R. G. Grant (2010) *Flight: the complete history of aviation*, Dorling Kindersley, London.

M. H. Gorn and G. De Chiara (2021) *X-Planes from the X-1 to the X-60*, Springer-Praxis books in space exploration, Gewerbestrasse, Switzerland

J. D. Anderson Jr. (1997) *A history of aerodynamics: and its impact on flying machines*, Cambridge University Press.

E. Torenbeek and H. Wittenberg (2009) *Flight physics: essentials of aeronautical disciplines and technology, with historical notes*, Springer, Dordrecht.

T. Kessner (2010) *The flight of the century: Charles Lindbergh and the rise of American aviation*, Oxford University Press.

R. Van Der Linden, A. M. Spencer, and T. J. Paone (2016) *Milestones of flight: the epic of aviation with the National Air and Space Museum*, Zenith Press.

A. M. Millbrooke (2006) *Aviation history*, Jeppesen.

R. Curley (2011) *The complete history of aviation: From ballooning to supersonic flight*, Rosen Educational Services : Britannica Educational Pub., New York.

J. D. Anderson Jr. (2002) *The airplane: a history of its technology*, American Institute of Aeronautics and Astronautics, Reston, VA.

J. R. Hansen (1989) *Aviation history in the wider view*, The Johns Hopkins University Press.

C. Chant (2002) *A century of triumph: the history of aviation*, Free Press, New York.

J. D. Anderson Jr. (1989) *Hypersonic and high temperature gas dynamics*, McGraw-Hill, New York.

Bravo-Mosquera, P. D., Catalano, F. M., and Zingg, D. W. (2022). *Unconventional aircraft for civil aviation: A review of concepts and design methodologies*, Progress in Aerospace Sciences, https://doi.org/10.1016/j.paerosci.2022.100813.

BIBLIOGRAPHY   121

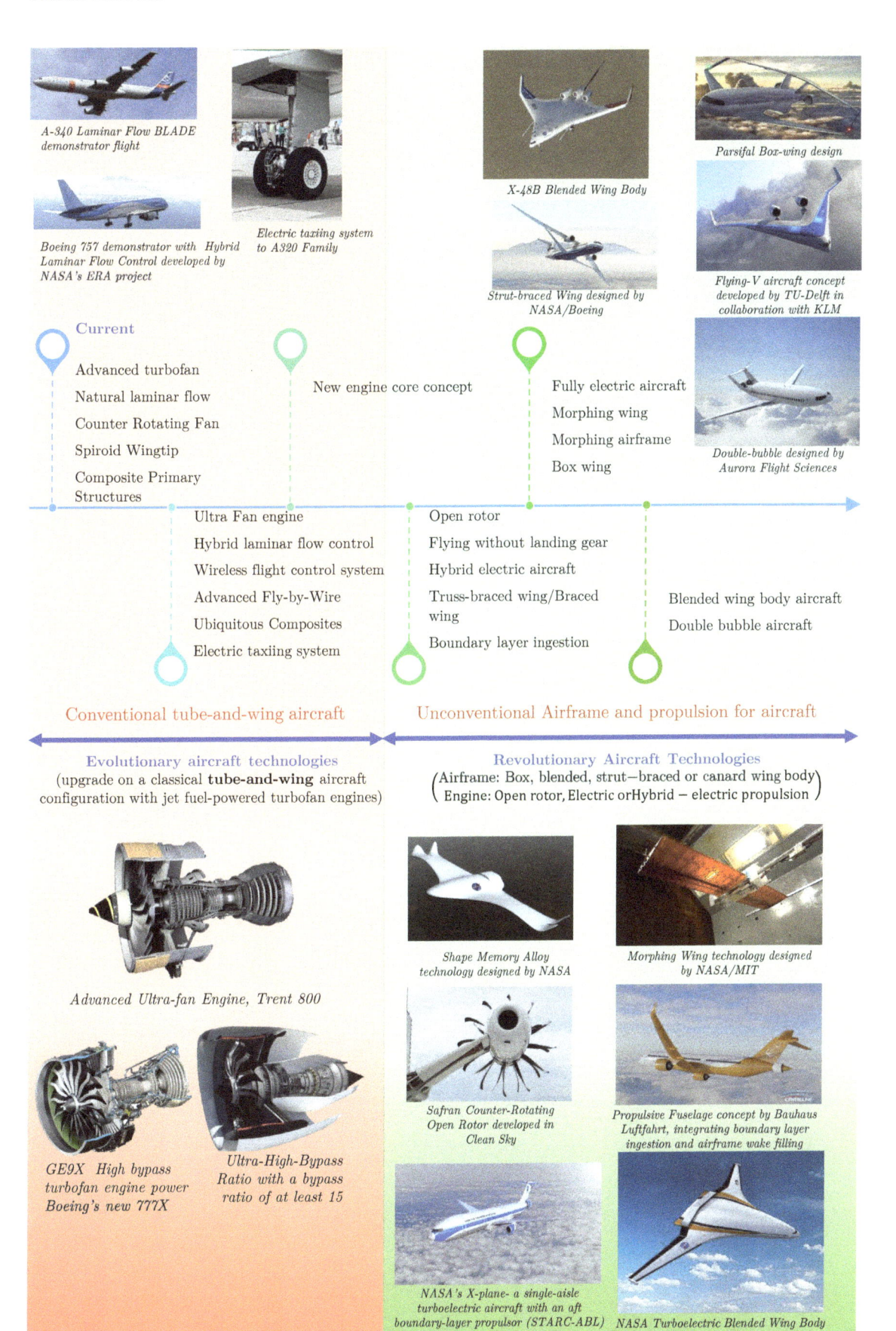

A-340 Laminar Flow BLADE demonstrator flight

Boeing 757 demonstrator with Hybrid Laminar Flow Control developed by NASA's ERA project

Electric taxiing system to A320 Family

X-48B Blended Wing Body

Strut-braced Wing designed by NASA/Boeing

Parsifal Box-wing design

Flying-V aircraft concept developed by TU-Delft in collaboration with KLM

Double-bubble designed by Aurora Flight Sciences

Current

- Advanced turbofan
- Natural laminar flow
- Counter Rotating Fan
- Spiroid Wingtip
- Composite Primary Structures

New engine core concept

- Fully electric aircraft
- Morphing wing
- Morphing airframe
- Box wing

- Ultra Fan engine
- Hybrid laminar flow control
- Wireless flight control system
- Advanced Fly-by-Wire
- Ubiquitous Composites
- Electric taxiing system

- Open rotor
- Flying without landing gear
- Hybrid electric aircraft
- Truss-braced wing/Braced wing
- Boundary layer ingestion

- Blended wing body aircraft
- Double bubble aircraft

Conventional tube-and-wing aircraft ← → **Unconventional Airframe and propulsion for aircraft**

Evolutionary aircraft technologies
(upgrade on a classical **tube-and-wing** aircraft configuration with jet fuel-powered turbofan engines)

Revolutionary Aircraft Technologies
(Airframe: Box, blended, strut—braced or canard wing body)
(Engine: Open rotor, Electric or Hybrid – electric propulsion)

Advanced Ultra-fan Engine, Trent 800

GE9X High bypass turbofan engine power Boeing's new 777X

Ultra-High-Bypass Ratio with a bypass ratio of at least 15

Shape Memory Alloy technology designed by NASA

Safran Counter-Rotating Open Rotor developed in Clean Sky

NASA's X-plane- a single-aisle turboelectric aircraft with an aft boundary-layer propulsor (STARC-ABL)

Morphing Wing technology designed by NASA/MIT

Propulsive Fuselage concept by Bauhaus Luftfahrt, integrating boundary layer ingestion and airframe wake filling

NASA Turboelectric Blended Wing Body

Evolutionary and revolutionary aircraft technologies

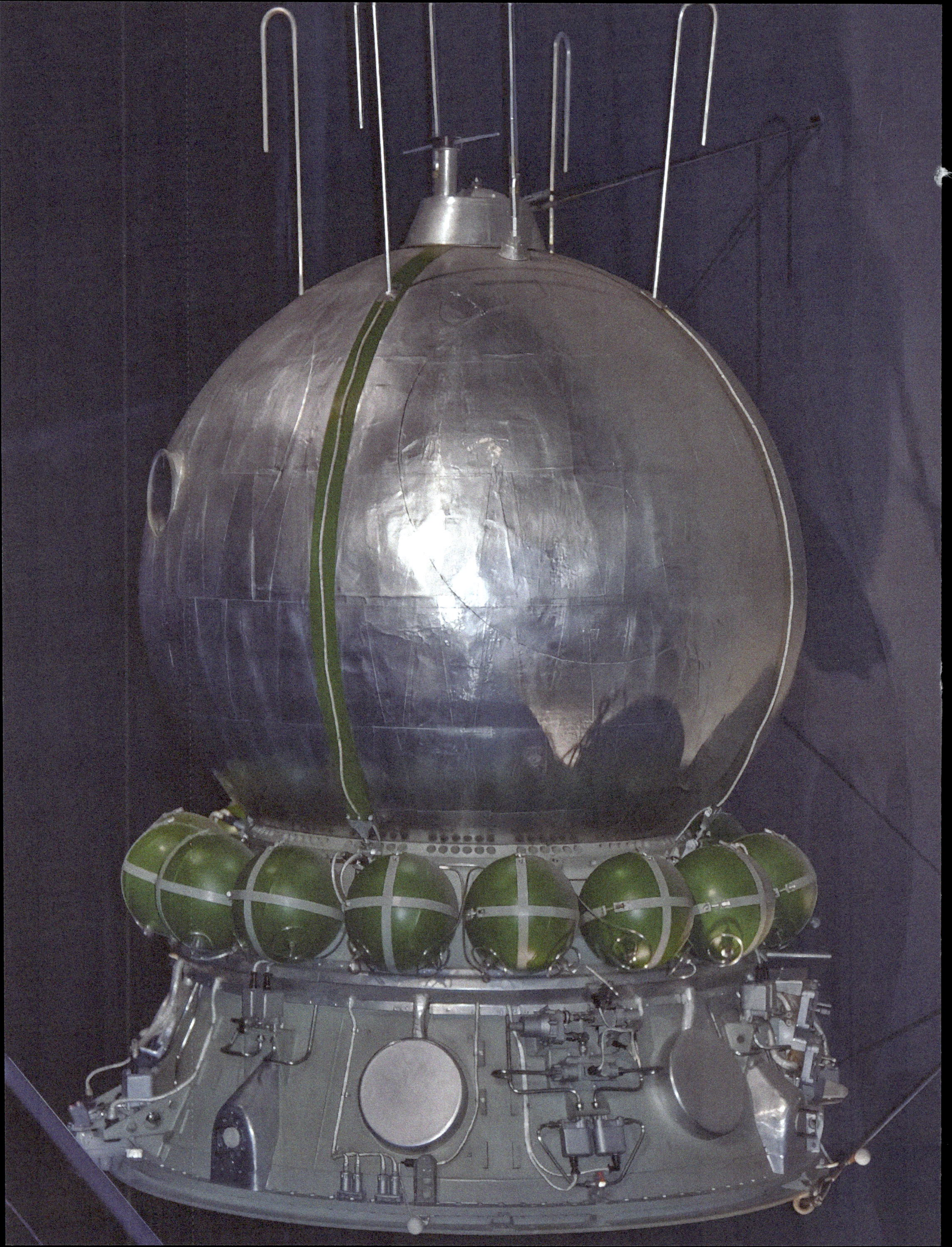

Chapter 2

History of Rocketry and Space Flight

Contents

2.1	Introduction		124
2.2	Visionaries and Pioneers		124
2.3	Near Earth Space Explorations		130
	2.3.1	Technological Advancements	130
	2.3.2	First Artificial Earth Satellite	132
	2.3.3	The First Human in Space	135
	2.3.4	Space Stations	136
	2.3.5	Space Shuttle	139
	2.3.6	Hubble and James Webb Space Telescopes	144
2.4	Missions to Celestial Bodies		145
	2.4.1	Luna Mission to Moon	145
	2.4.2	Apollo Space Program	145
	2.4.3	Space Missions to Mars	152
	2.4.4	Space Missions to Venus	153
	2.4.5	Voyager Mission to Jupiter, Saturn, Uranus and Neptune	155
	2.4.6	Pioneer Mission to Jupiter and Saturn	155
	2.4.7	NASA's Juno Space Probe to Jupiter	156
	2.4.8	Cassini–Huygens Mission to Saturn	156
	2.4.9	New Horizons Mission to Pluto	156
	2.4.10	NEAR-Shoemaker Mission to Asteroid	156
	2.4.11	JAXA's Hayabusa Missions	157
2.5	USA's Experimental Programs		157
2.6	Current and Future Space Explorations		162
	2.6.1	Lunar Missions	162
	2.6.2	Planetary Explorations	166
	2.6.3	Missions to Sun	167

2.1 Introduction

The life of human beings in modern society is significantly different from that of their predecessors, essentially because of the gadgets that are being used. In the current age, the use of a parabolic antenna to receive signals from geostationary satellites enables users to view TV channels. The use of a global navigation system to maneuver through the roads and reach the correct destination is possible through the use of satellite technology. Weather prediction has become more accurate with the use of remote-sensing satellites. Through the images sent by the *Hubble* telescope and other interplanetary probes, we can have a view of the other planets and outer space itself. The use of satellite technology has helped the armed forces to be better equipped for any combat situations. With all these and in many other ways, the life of human beings today is significantly different from that of their predecessors.

The evolution of flight, from its practical inception with the *Wright* brothers' inaugural flight of a heavier-than-air, powered aircraft on December 17, 1903, has been propelled by a single driving force: the relentless pursuit of greater speed and altitude. It is evident from the history of flight that throughout the 20th century flight vehicles have witnessed a dramatic increase in both velocity and altitude. It all began with the *Wright Flyer*, reaching ~56 km/h at sea level in 1903, advancing to World War II fighters flying at speeds around 650 km/h at altitudes of 10 km. This progression continued with supersonic aircraft in the 1960s and 1970s, capable of achieving speeds of 2000 km/h at altitudes of 20 km.

A remarkable highlight in this journey was the USA's experimental *X-15* hypersonic aircraft, which achieved a remarkable Mach number 7 and soared to an altitude of 107 km on August 22, 1963. This incredible saga culminated with the *Space Shuttle*, the pinnacle of manned aircraft, returning to Earth's atmosphere at a breathtaking speed of around Mach number 25, following its journey from a low Earth orbit of 320 km above the planet's surface. The era of high-speed missiles and spacecraft witnessed remarkable advancements, including the creation of the intercontinental ballistic missiles during the 1950s which fly around Mach number 25. The 1960s brought about the development of spacecraft such as the *Mercury*, *Gemini*, and *Vostok*, designed for manned orbital missions. Undoubtedly, a monumental milestone was reached with the historic *Apollo spacecraft*, which commenced the historic journey of bringing humans back from the Moon, commencing in 1969.

This chapter looks at the contributions by some of the great scientists and engineers who made it possible for human being's exploration of the outer space. Further, some of the noteworthy near-Earth and outer space explorations by human beings are discussed.

2.2 Visionaries and Pioneers

Galileo Galilei (1564 - 1642)

Galileo Galilei, a pioneering Italian astronomer, physicist, and mathematician of the 17th century, made groundbreaking contributions to science, which had a profound impact on space engineering and our understanding of the cosmos. *Galileo* was among the first to use telescopes for astronomical observations. His meticulous observations of celestial bodies, including the Moon, Jupiter's moons, sunspots, and the phases of Venus, revolutionized our understanding. *Galileo* laid the groundwork for the modern understanding of motion and mechanics. He formulated principles related to inertia, acceleration, and the laws of falling bodies, which later formed the basis for *Isaac Newton*'s laws of motion and universal gravitation. *Galileo* has been called the father of observational astronomy, modern-era classical physics, the scientific method, and modern science. *Galileo* delved into the realms of speed, velocity, gravity, free fall, inertia, and projectile motion. Additionally, he made significant contributions to applied science and technology. His investigations extended to describing the characteristics of the pendulum and "hydrostatic balances," shedding light on their properties and functionalities. He proposed both his astronomical solution and an accurate sea clock—the first clock ever to have a pendulum. Galileo died before making the clock, but his son built a model in 1649.

Galileo, an Italian polymath renowned for his expertise in astronomy, physics, and engineering, is often hailed as a pioneer in various fields. His observations, laws of motion, and empirical approach continue to be integral in designing spacecraft, planning missions, and exploring the depths of space.

Johannes Kepler (1571 - 1630)

Johannes Kepler, a renowned German mathematician and astronomer of the 17th century, made significant contributions to the field of astronomy and celestial mechanics that had a profound impact on space engineering and exploration. His discoveries and laws describing planetary motion laid the groundwork for our understanding of orbits and the mechanics of celestial bodies, forming the basis for much of modern space engineering. His groundbreaking work resulted in three laws that describe the motion of planets around the Sun. *Kepler*'s laws revolutionized the understanding of planetary motion and celestial mechanics. His profound insights not only influenced luminaries like *Isaac Newton* but also laid a cornerstone for *Newton*'s theory of gravitation. *Kepler*'s extensive and influential body of work established him as a foundational figure in the realms of modern astronomy, the scientific method, and the development of natural and modern science.

Johannes Kepler (1571 - 1630), a renowned German mathematician and astronomer of the 17th century, played a pivotal role in the scientific revolution. He is most celebrated for formulating the laws governing planetary motion.

Sir Isaac Newton (1642 - 1727)

Sir Isaac Newton, an English mathematician, physicist, and astronomer of the 17th and 18th centuries, made monumental contributions to science that significantly influenced space engineering and our comprehension of celestial mechanics. His groundbreaking discoveries and laws laid the groundwork for space exploration and engineering. *Newton* formulated three fundamental laws of motion that revolutionized our understanding of how objects move. *Newton*'s law of universal gravitation postulated that every particle in the universe attracts every other particle with a force that is directly proportional to the product of their masses and inversely proportional to the square of the distance between their centers. *Newton*'s development of calculus provided a powerful mathematical tool that enabled scientists and engineers to solve complex problems in dynamics, orbits, and the motion of celestial bodies.

Sir Isaac Newton (1642 - 1727), was a versatile English figure renowned for his expertise across various disciplines. His laws continue to be instrumental in the planning, execution, and success of space missions, propelling humanity's quest to understand and navigate the vast reaches of space.

Konstantin Tsiolkovsky (1857 - 1935)

Konstantin Tsiolkovsky is renowned as the Russian space scientist who conclusively established the scientific principles that form the basis of space exploration. Inspired at an early age by *Jules Verne*'s science fiction stories of space travel, he began to write science fiction stories himself. Remarkably, he infused his stories with scientific and technological elements, grappling with problems like the intricate control of rockets navigating gravitational fields. Gradually, *Tsiolkovsky* transitioned from the realm of science fiction to the realm of academia, writing seminal theoretical papers on many topics that form the basis for space exploration. His profound contributions extended to the pioneering use of liquid propellant rockets, as well as innovative designs for rockets equipped with steering thrusters, multistage boosters, space stations and descent trajectories for space vehicles.

Notably, *Tsiolkovsky*'s dedication to advancing the field of aerodynamics was evident in the establishment of Russia's inaugural aerodynamics laboratory. In 1897, he achieved a significant milestone by constructing Russia's inaugural wind tunnel, characterized by an open test section. Supported by a grant from the *Russian Academy of Sciences*, he measured the drag coefficients of basic shapes like spheres, flat plates, cylinders and cones.

Konstantin Tsiolkovsky (1857 - 1935), a Soviet rocket scientist, overcame early obstacles, including scarlet fever that almost left him deaf. Yet, he became a founding figure in modern rocketry and astronautics, inspiring leading Soviet engineers like Sergei Korolev, crucial to the Soviet space program's success.

Commencing in 1896, he systematically delved into the theory of rocketry. As early as 1883, he pondered the use of the rocket principle in space, culminating in the development of rocket propulsion theory in 1896. *Tsiolkovsky* formulated what he termed the *formula of aviation*, linking the change in a rocket's speed (δV), the engine's exhaust velocity (v_e), and the initial (M_0) and final (M_1) mass of the rocket as $\delta V = v_e \ln \left(\dfrac{M_0}{M_1} \right)$. Currently, the equation is known as the *Tsiolkovsky rocket equation*. *Tsiolkovsky*'s work titled *Exploration of Outer Space Employing Rocket Devices*, published in May 1903, calculated the minimum velocity required for an orbit around the Earth. He proposed achieving this through a multistage rocket fueled by liquid oxygen and liquid hydrogen, guided by the *Tsiolkovsky* equation. In 1911, *Tsiolkovsky* released the second part of his work, *Exploration of Outer Space by Means of Rocket Devices*, where he calculated the energy needed to overcome gravity, determined the required speed for interplanetary travel (*escape velocity*), and examined flight time calculations. This publication made a profound impact in the scientific community, garnering Tsiolkovsky numerous fellow scientist admirers.

Robert Goddard (1882 - 1945)

Robert Goddard, often hailed as *the pioneer of modern rocket propulsion*, found his lifelong passion for space exploration ignited by early encounters with science fiction literature. His journey into rocketry commenced during his academic years and continued as a professor at Clark University. In 1914, *Goddard* earned two U.S. patents-one for a liquid-fueled rocket and another for a two- or three-stage rocket utilizing solid fuel.

In 1919, he garnered media attention by asserting that solid-fuel rocket engines could propel a rocket to the Moon. However, *Goddard*'s most significant breakthrough came on March 16, 1926, when he successfully launched the world's first liquid-fueled rocket in Auburn, Massachusetts. Notably, his exploratory work attracted the interest of aviator *Charles Lindbergh*, who facilitated funding from the *Guggenheim Foundation*. This financial support enabled *Goddard* to conduct ambitious rocket experiments in the 1930s, based in New Mexico. By 1935, one of his rockets reached an impressive altitude of 2,300 meters.

Dr. Robert Goddard (1882 - 1945) is renowned for creating the world's first liquid-fueled rocket and substantial contributions to the development of sophisticated control systems for rockets.

Hermann Oberth (1894 - 1989)

Hermann Oberth, originally from Romania, found himself in Italy to recover from scarlet fever. During this time, he immersed himself in *Jules Verne*'s novel *From the Earth to the Moon*, which sparked his fascination with space exploration. By the tender age of 14, *Oberth* had a vision of a recoil rocket capable of traversing space by expelling exhaust gases from its base. Upon his return to the university after World War I, he shifted his focus to physics and began to grasp the significance of employing multiple stages in rocket propulsion. In 1922, *Oberth* submitted his doctoral thesis on rocketry, but it was met with rejection. His ideas were considered utopian at that time! Undeterred, he published an extended and more comprehensive version of his work, titled *The Rocket into Planetary Space*, a year later. This publication garnered widespread recognition for *Oberth* as it explained the principles of escaping Earth's gravitational pull using rockets. Notably, *Wernher von Braun* a prominent rocket scientist, was deeply influenced by this book, sparking his own interest in rockets and space travel.

In 1929, *Oberth* conducted a static firing experiment involving his first liquid-fueled rocket motor, which he named the *Kegeldüse*. This engine, constructed by *Klaus Riedel* in a workshop space provided by the *Reich Institution of Chemical Technology*, briefly operated, albeit without a cooling system. During this experiment, a young student named *Wernher von Braun* lent his assistance. Following the patenting of his rocket design, *Oberth*'s first rocket was successfully launched on May 7, 1931, near Berlin.

Interestingly, while Russia's *Tsiolkovsky* and America's *Goddard* conducted parallel research and reached similar conclusions, there is no evidence to suggest that they were aware of each other's work. Therefore, all three of these pioneering scientists are rightfully acknowledged as the *Fathers of Rocketry*.

Hermann Oberth (1894 - 1989), a key figure in rocketry and astronautics, proposed liquid-fueled long-range missiles for the German army. He was inducted into the International Air & Space Hall of Fame in 1980.

Sergei Korolëv (1907 - 1966)

Sergei Korolëv was a prominent Soviet rocket engineer and spacecraft designer who played a pivotal role during the *Space Race* between the United States and the Soviet Union in the 1950s and 1960s. Many consider him the father of practical astronautics. While the world recognized his achievements, such as the development of the *R-7* (the world's first ICBM used to propel *Sputniks* into Earth orbit and *Luna* spacecraft to the Moon), *Sputnik 1* (the first satellite placed in orbit), *Sputnik 2* (the first satellite to carry dogs into Earth's orbit), *Sputnik 5* (the first spaceflight to send two dogs into orbit and safely return them), *Luna 3* (a robotic spacecraft that captured the first images of the Moon's far side), and his contributions to the Soviet human spaceflight programs *Vostok*, *Voskhod*, and *Soyuz*, *Korolëv*'s true identity remained shrouded in secrecy until his death, as it was a closely guarded state secret. Despite facing the hardships of *Stalinism*, including imprisonment and forced labor, he emerged as an icon of Russian rocketry, and his rocket and spacecraft designs continue to be used to this day.

Korolëv's journey in aeronautical engineering began when he joined the Moscow GIRD, a rocket club, in 1931. By 1933, he had risen to the position of deputy chief at a newly established government rocket institute, where he supervised early Soviet experiments with liquid-propellant rockets. However, his career took a dark turn when he was arrested during *Stalin*'s purges in 1938. He endured the harsh conditions of a Siberian concentration camp and subsequent imprisonment in various facilities until 1944. Following his release, *Korolëv* went on to manage the Soviet Union's ballistic missile program and assumed leadership of its space program from the historic launch of *Sputnik* in 1957 until his passing in 1966. The trajectory of the Soviet space program would have taken a different path if he had lived longer.

Sergei Korolëv (1907 - 1966), the enigmatic figure behind the historic milestones of human spaceflight, is credited with groundbreaking achievements including the *R-7* rocket, the launch of Sputnik, and the development of spacecraft like *Vostok* and *Soyuz*.

Wernher von Braun (1912 – 1977)

Wernher von Braun, a towering figure in both German and American rocket engineering, is renowned for his pivotal contributions to rocketry. Inspired by *Hermann*'s 1923 book *The Rocket into Planetary Space*, von Braun delved into the mathematics and physics of rocketry. His passion for spaceflight led him to join the *German Society for Space Travel (VfR)* in 1928. Subsequently, he was recruited by the German army for its weapons program. In 1932, *von Braun* obtained a Bachelor's Degree in Aeronautical Engineering from the Berlin Institute of Technology, followed by a Ph.D. in Physics, funded by army research, from the University of Berlin in 1934.

Although *von Braun* maintained that he was not interested in weapons research, he used military funding to advance rocket technology with the ultimate goal of space travel. In 1936, at the age of 24, he was appointed as the head of a top-secret Nazi research and development facility that produced the pioneering *V-2 rocket*, the world's first long-range guided ballistic missile powered by a liquid-propellant rocket. In the U.S., he and his team adapted the *V-2*'s design to create America's inaugural ballistic missile, the *Redstone*, in 1953. Under *von Braun*'s guidance, a modified *Redstone* rocket (*the Jupiter* C) launched the inaugural American satellite, *Explorer 1*, into space in 1958, leading to the discovery of the *Van Allen radiation belts*. In 1961, *von Braun*'s *Redstone* rockets carried out the historic Project *Mercury missions*, launching the first two American astronauts into space.

von Braun continued his groundbreaking work as the director of NASA's Marshall Space Center, overseeing the development of increasingly massive rockets for the *Apollo* program. Beginning in 1961, his team crafted the *Saturn I*, *Saturn IB*, and the colossal *Saturn V*, which successfully transported *Apollo 11* to the Moon in 1969. In 1970, he was called upon to lead strategic planning for NASA in Washington, D.C. Apart from his technical achievements, *von Braun* authored two widely read books, *Conquest of the Moon*(1953) and *Space Frontier* (1967).

Arthur Clarke (1917 – 2008)

While serving in the Royal Air Force, Arthur Clarke proposed the concept of global communication satellites in a 1945 letter to the British magazine *Wireless World*. This letter, titled *Extra-terrestrial Relays–Can Rocket Stations Give World-wide Radio*

Wernher von Braun (1912 – 1977) was the rocket engineer responsible for designing the feared *V-2* missile used by the Nazis during World War II. He later became the visionary architect behind the *Apollo* program, which successfully landed man on the Moon.

Coverage? presented the first public vision of a technology that would revolutionize global communications. *Clarke* envisioned a network of three geostationary satellites evenly spaced around the equator, enabling worldwide communication. In addition to his pioneering work in telecommunications, Clarke was a renowned science-fiction writer and co-authored the screenplay for the iconic 1968 film *2001: A Space Odyssey*, considered one of the most influential movies ever made.

Arthur Clarke (1917 – 2008), an English science writer, conceived the idea of utilizing rocket technology to launch a spacecraft into orbit and enable communication to and from Earth.

Neil Armstrong (1930 - 2012)

Neil Armstrong, an iconic figure in space exploration history, made an indelible contribution to humanity's journey beyond Earth. His most renowned achievement was becoming the first human to set foot on the Moon during NASA's *Apollo 11* mission on July 20, 1969. Armstrong's famous words, *That's one small step for man, one giant leap for mankind*, echoed worldwide, signifying a monumental step forward in space exploration.

Armstrong's role in this historic mission extended beyond his momentous lunar landing. As the *mission commander*, he displayed remarkable piloting skills and composure during a challenging descent to the lunar surface, narrowly avoiding boulders and craters. His actions ensured the safe landing of the lunar module, *the Eagle*. *Armstrong* and his colleague, *Buzz Aldrin*, conducted experiments, collected lunar samples, and set up scientific instruments on the Moon, contributing valuable data to our understanding of the Moon.

Neil Armstrong embarked on his inaugural space journey as the command pilot of *Gemini 8* in March 1966. This historic mission marked Armstrong as NASA's first civilian astronaut to venture into space. Alongside pilot *David Scott*, *Armstrong* achieved a significant milestone by executing the first-ever docking of two spacecraft in orbit.

However, the mission encountered a perilous situation when a thruster malfunction caused a severe and uncontrollable roll of the spacecraft. In response to this critical issue, *Armstrong*'s quick thinking and exceptional piloting skills came to the forefront. He used a portion of the re-entry control fuel to stabilize the spacecraft, averting a potentially disastrous outcome. This decisive action ensured the safety of the mission, despite its ultimate termination.

Renowned for his exceptional engineering skills and piloting expertise, *Armstrong* achieved a remarkable feat by becoming one of the select 12 pilots to operate the cutting-edge North American *X-15* aircraft. As an accomplished test pilot, he made crucial contributions to the development of the *X-15 rocket* plane and various other aircraft. His experiences as an aviator and astronaut provided insights into aviation and space technology, influencing future missions and the design of spacecraft.

Yuri Gagarin (1934 – 1968)

Yuri Gagarin, a Soviet pilot and cosmonaut, achieved the monumental feat of becoming the first human in space. He embarked on a remarkable 108-minute orbital journey around Earth aboard his *Vostok 1 spacecraft*. *Gagarin*'s fascination with space exploration ignited during his youth when he joined

Neil Armstrong (1930 - 2012), a NASA astronaut, made history as the first human to set foot on the Moon during the *Apollo 11* mission on July 20, 1969. *Armstrong*'s stellar career also included participation in NASA's *Gemini 8* mission in 1966.

the *AeroClub* in Saratov, where he honed his skills in flying light aircraft.

Subsequently, Gagarin pursued formal flight training at the Orenburg Military Pilot's School. His dedication and talent led to his commission as a Lieutenant in the Soviet Air Force. This significant milestone set him on the path to realizing his dream of venturing into the cosmos.

In 1960, when the Soviet Union decided to send a human into space, a secret nationwide selection process was initiated. *Yuri Gagarin* emerged as the choice from a group of 19 pilots. The selection was based on a rigorous evaluation of various criteria, including performance in training, physical attributes, and tests assessing both physical and psychological endurance. Gagarin tragically lost his

Yuri Gagarin (1934 – 1968) became the first human in space, orbiting the Earth aboard *Vostok 1*. His historic journey catapulted him to global celebrity status.

life in a plane crash on March 27, 1968, at the young age of 34. The aircraft he was piloting, a *MiG-15* jet trainer, crashed near the town of Kirzhach in the Vladimir region of the former Soviet Union.

In the next two sections, we will scan some of the space explorations that have captured the attention of the world. Some of these missions were for establishing technological capabilities whereas others were for scientific explorations that revealed intriguing information about outer space and the celestial bodies.

2.3 Near Earth Space Explorations

Space flight, a breathtaking and transformative endeavor, represents humanity's quest to explore the vast and mysterious expanse beyond our planet. It is a remarkable fusion of science, engineering, and human determination and aspirations that has expanded our understanding of the universe, pushed the boundaries of technology, and inspired generations to reach for the stars. From the legendary launch of the first artificial satellite, *Sputnik*, in 1957, to the breathtaking Moon landings of the *Apollo* program in 1969 and the recent *Chandrayan 3* mission to the Moon in 2023, space flight has captured the collective imagination and propelled us into a new era of discovery.

2.3.1 Technological Advancements

V-2 rocket

The initial development of liquid-propellant rockets in Germany during the 1930s can be largely attributed to the efforts of the *Verein für Raumschiffahrt* (Society for Space Travel), a group of German rocket enthusiasts. They successfully launched several rockets in Berlin, achieving altitudes of approximately 1.5 kilometers. Notably, this work was closely observed by the German army, which was bound by the restrictions of the Versailles Peace Treaty, prohibiting the development of certain weapons.

However, the German army recognized that the treaty did not specifically include rockets in its list of prohibited weaponry. Consequently, the military decided to enlist the expertise of these German rocket enthusiasts, among whom was the young *Wernher von Braun*. In 1936, *von Braun* assumed leadership of a top-secret research and development facility located in *Peenemünde*, a remote area along the Baltic Sea. By 1943, the operation had expanded significantly, employing around 6,000 scientists, technicians, and engineers, as well as a significant number of forced laborers and prisoners of war. Thanks to *von Braun*'s pioneering work, the *V-2 (V for Vergeltungswaffe, or vengeance weapon)* was developed. When launched vertically, this rocket achieved an astounding altitude of 175 km.

The German *V-2* rocket, which marked the world's inaugural large-scale liquid-propellant rocket vehicle and served as the precursor to contemporary large rockets and launch systems, was officially disclosed to the public in November 1944. This announcement occurred just two months after its initial deployment as a weapon. The *V-2* possessed an impressive maximum range of approximately 320 kilometers and could carry a one-ton warhead.

The *V-2* missile introduced a groundbreaking technology to both the Western Allies and the Soviet Union. Acquiring this technology became a paramount objective for both parties in 1945. As World War II concluded, *Wernher von Braun* and some of his colleagues were apprehended and transported to the United States. In America, they continued their work, contributing to the development of the *Redstone*, the country's first nuclear ballistic missile. This missile underwent its inaugural test launch from Cape Canaveral, Florida, in 1953.

Conversely, the Soviets occupied *Peenemünde* as Germany and its allies faced defeat. They seized control of the *V-2* production facility in Nordhausen, along with several lower-level scientists and technicians involved in the German rocket program. Leveraging the fundamental *V-2* design, the Soviets created the Russian Intercontinental Ballistic Missile known as the *M-101*, which served as the precursor to numerous significant Russian missile initiatives.

However, prior to these developments, some space enthusiasts in the Soviet Union had formed rocket clubs, such as the Moscow Group for the Study of Reaction Motion (*MosGIRD*). In 1932, *MosGIRD* launched its first rockets, one of which reached an altitude of 400 meters. These independent rocket clubs eventually became integrated into the state apparatus. By 1937, most Soviet rocket experts, including *Sergei Korolëv*, were engaged in military research for the Soviet Union.

Bumper Program

The *Bumper program*, initiated by the U.S. Army, resulted in a series of rocket experiments in February 1949. This test firing of the *V-2* rocket with *WAC Corporal*, a slender, needle-like rocket, mounted on top is the first meaningful attempt to demonstrate the use of a multi-stage rocket for achieving high velocities and high altitudes. It marked a significant milestone in the early days of space exploration and served as a precursor to the more ambitious space endeavors that followed. The program's primary objective was to investigate the feasibility of launching rockets into the upper atmosphere and ultimately into space.

One of the most notable aspects of the *Bumper program* was the use of the *V-2* rocket, originally developed by Nazi Germany during World War II. These captured *V-2* rockets were modified and improved

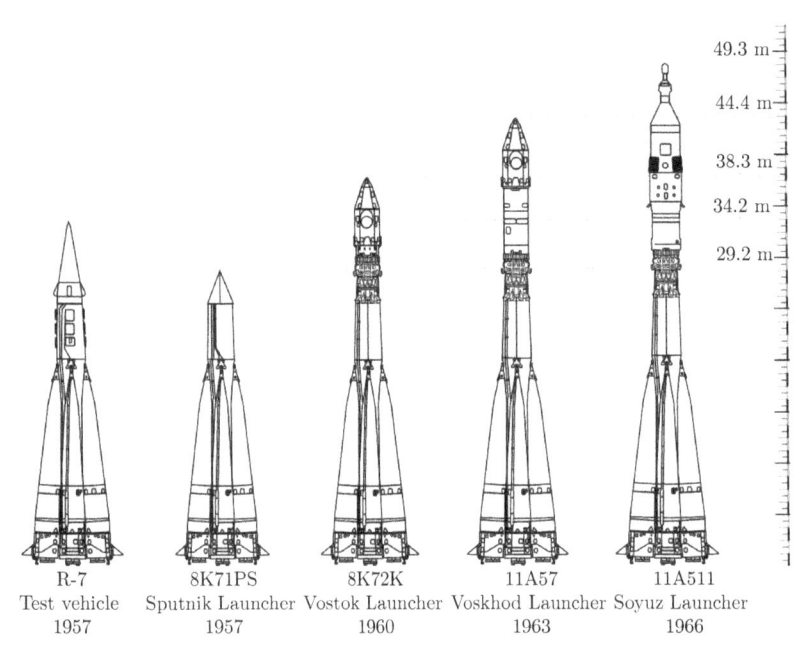

R-7	8K71PS	8K72K	11A57	11A511
Test vehicle	Sputnik Launcher	Vostok Launcher	Voskhod Launcher	Soyuz Launcher
1957	1957	1960	1963	1966

Heights marked: 49.3 m, 44.4 m, 38.3 m, 34.2 m, 29.2 m

R-7, the world's first ICBM, was developed by Soviet rocket designer *Sergei Korolëv* who headed development from 1954 to 1957 and successfully flight tested in 1957. The *R-7* missile was powerful enough to launch a nuclear warhead across vast distances including the United States.

Also, the *R-7* family of rockets has launched many space missions into orbit including *Sputnik* (the world's first artificial satellite) in 1957, *Luna 2* (a space probe launched to make an impact on the Moon's surface) in 1959, *Vostok* (the first manned spacecraft) in 1961, *Voskhod* (the first human spaceflight to carry more than one crewman into orbit without the use of spacesuits) in 1964.

upon by American scientists and engineers. The program saw a collaboration between key figures in the emerging field of rocketry, including *Wernher von Braun* and his team, who had been brought to the United States as part of *Operation Paperclip* after World War II. The *V-2* rocket reached an altitude of 160 km at a velocity of 5650 km/h, at which point it ignited the *WAC Corporal*. This slender upper stage accelerated to a maximum velocity of 8300 km/h and reached an altitude of 390 km, surpassing the previous record set by a *V-2* alone exceeding by 210 km. After reaching this peak, the *WAC Corporal* nosed over and returned to the atmosphere at speeds exceeding 8000 km/h. In accomplishing this feat, it became the first human-made object to achieve hypersonic flight. These events took place at the White Sands Proving Ground in New Mexico, a location chosen for its remote nature and extensive test facilities. The successful launches of the *Bumper program* laid the foundation for further advancements in rocket technology and paved the way for future space exploration endeavors by the United States.

2.3.2 First Artificial Earth Satellite

Following World War II, *Sergei Korolëv* and other Soviet rocket experts were released from the *Gulag*. They were granted the opportunity to contribute to the development of the Soviet rocket program. *Sergei Korolëv*, in particular, played the leadership role in the creation of the *R-7 Semyorka*, the world's first Intercontinental Ballistic Missile (ICBM), boasting an impressive range of 6,400 km and a payload capacity of one ton. The rocket had a remarkable thrust capability of 455 tons. In comparison, by the early 1950s, the Soviets had developed rocket motors capable of generating over 90 ton of thrust, far surpassing the *Redstone*'s motor capacity of 35 tons. This marked a significant milestone in the development of powerful rocket technology and marked the dawn of the space race. The late 1950s was a period marked by the Cold War, an intense geopolitical rivalry between the United States and the Soviet Union. The competition extended beyond politics and ideology into a race for technological supremacy. At the center of this race was the quest to conquer space, symbolizing national prestige and military advantage.

Sputnik 1: Amid this backdrop, the Soviet Union embarked on an audacious project: the creation of the world's first artificial Earth satellite, known as *Sputnik 1*. *Sergei Korolëv* was instrumental in designing the satellite and the rocket that would launch it. *Sputnik 1* was meticulously designed to fulfill specific objectives. It prioritized simplicity and reliability, characteristics adaptable for future projects. This spherical satellite, 58.5 cm in diameter, consisted of two hermetically sealed 2 mm thick hemispheres connected by 36 bolts. Its mass was 83.6 kg, and it featured a highly polished 1 mm thick heat shield made of an aluminum–magnesium–titanium alloy (AMG6T). The satellite incorporated two pairs of antennas, Each made up of two whip-like parts, 2.4 and 2.9 m in length, designed for efficient radiation pattern. Power was supplied by three silver-zinc batteries, with two powering the radio transmitter and one for temperature regulation.

On October 4, 1957, the world held its breath as the Soviet Union launched *Sputnik 1* from Baikonur Cosmodrome at Tyuratam in Kazakhstan (then part of the former Soviet Union) into an elliptical low Earth orbit using an *R-7 Semyorka* rocket. The orbit had a perigee altitude of 215 km, an apogee altitude of 939 km and an inclination of $\sim 65°$ to the equatorial plane. The successful launch marked a seismic shift in the balance of power, as the Soviet Union had achieved a remarkable technological feat. *Sputnik 1* circled the Earth once every hour and 36 minutes, traveling at an astonishing speed of 29000 km/h for three months. When it finally fell out of orbit in January 1958, Sputnik had traveled 70 million km around the planet. The only payload onboard Sputnik was a low-power radio transmitter, which broadcast a beeping noise at regular intervals. This beeping could be heard by radio listeners around the world. During the launch, *Sputnik 1* was safeguarded by an 80-centimeter cone-shaped payload fairing that separated simultaneously from the satellite and the spent *R-7* second stage. Rigorous testing was carried out at OKB-1 (Russian manufacturer of spacecraft and space station components) under the supervision of *Oleg Ivanovsky*, ensuring the satellite's readiness for its historic mission.

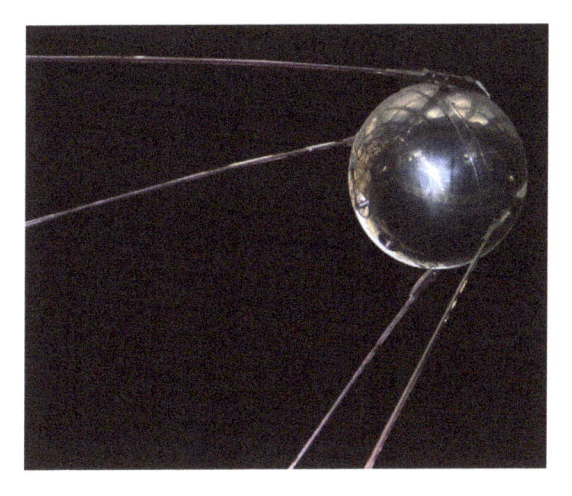

Sputnik 1, the world's first artificial Earth satellite, was launched into an elliptical low Earth orbit on October 4, 1957, was a significant milestone for the Soviet space program. It marked the dawn of the space age and the beginning of human space exploration.

Exploded view of *Sputnik 1*

The impact of *Sputnik 1* was profound and far-reaching. Firstly, it marked the dawn of the space age and the beginning of human space exploration. The successful launch of an artificial satellite demonstrated the Soviet Union's mastery of rocket technology and placed them ahead in the *space race*. Secondly, the launch had significant political ramifications. It fueled anxiety and fear in the United States, leading to increased investment in science, technology, and education, particularly in the fields of mathematics and science. This investment ultimately led to the creation of NASA and a commitment to space exploration. Thirdly, *Sputnik 1* triggered a rapid escalation in the development of satellite technology, fostering global communication, navigation systems, and weather forecasting. It laid the foundation for a multitude of applications, from telecommunications to remote sensing, with satellites becoming indispensable tools of modern life. Just a month later, the Soviets achieved an even more remarkable feat by launching a dog into orbit aboard *Sputnik-2*, further solidifying their space achievements on the global stage.

Sputnik 2: On November 3, 1957, the Russians launched the second spacecraft into low Earth orbit which carried a dog named *Laika*. The dog died on the fourth orbit due to overheating caused by an air conditioning malfunction. *Sputnik 2* was a 4 m high cone-shaped capsule with a base diameter of 2 m that weighed around 500 kg.

Korabl-Sputnik 2 or Sputnik 5: A third Soviet artificial satellite was launched on August 19, 1960. The third test flight and the first spaceflight to send animals (two dogs, *Belka* and *Strelka*) into orbit and return them safely back to Earth.

Explorer 1 (USA): Recognizing that the Soviet Union had forged ahead in the realm of technologies with potential implications for American security, the United States grew increasingly concerned. In response, the United States embarked on two unsuccessful attempts to launch a satellite into space before achieving success with a rocket carrying a satellite named *Explorer* on January 31, 1958. This accomplishment was largely attributed to the contributions of German rocket engineers who migrated to the USA.

The launch of *Explorer 1*, the United States' maiden successful orbiting satellite, marked a pivotal moment in the nation's history, particularly in the context of the escalating space race with the Soviet Union. *Explorer 1* was equipped with an array of scientific instruments designed for conducting experiments in space. Among these instruments was a Geiger counter, specifically employed for detecting cosmic rays. This experiment yielded groundbreaking results, confirming the existence of what is now known as the *Van Allen radiation belts* encircling the Earth.

Around the same period in the United States, a newly established government agency, the National Aeronautics and Space Administration (NASA), commenced its operations in October 1958. NASA's inception marked a transformative development, with much of its foundation originating from the National Advisory Committee for Aeronautics (NACA).

Explorer 1, the United States' first successful orbiting satellite launched on 31 January 1958, was 203 cm long and 15.2 cm in diameter and weighing 13.97 kg. Data from *Explorer 1* and two subsequent *Explorer* satellites led to the discovery of the *Van Allen radiation belt* (intense radiation surrounding the Earth).

2.3.3 The First Human in Space

The quest to send a human into space was a defining challenge of the Cold War era, driven by the rivalry between the United States and the Soviet Union. In this backdrop, the Soviet Union took the remarkable step by selecting *Yuri Gagarin* as the first human to venture into the cosmos. The mission marked a culmination of scientific innovation, political ambition, and human courage.

On April 12, 1961, *Yuri Gagarin*, a 27-year-old Soviet Air Force pilot, embarked on the *Vostok 1* mission, becoming the first human in space. *Gagarin*, encased in a *Vostok-1* spacecraft consisted a pressurized sphere measuring 2 m in diameter, accompanied by an equipment module housing electronics and thrusters. These thrusters were crucial for the vehicle's safe return to Earth. In the unlikely event of a thruster failure, *Gagarin* was equipped to enter Earth's atmosphere naturally after 10 days, carrying sufficient food and supplies to sustain him in such a scenario.

Gagarin and *Vostok 1* embarked on their journey around noon aboard a *Vostok-K* rocket from the Baikonur Cosmodrome in Kazakhstan. The pad where *Vostok 1* blasted off is now named Gagarin's Start and is still used for crewed *Soyuz* launches. Shortly after the launch, the cosmonaut exclaimed *Poyekhali!* (translated to *Off we go*) over the radio. A few minutes later, he entered orbit and gained a captivating view of Earth through a solitary window positioned near his feet.

Model of *Vostok* space capsule, which carried the first human into orbit.

Most of the flight was automated, with *Gagarin* retaining the capability to assume manual control in case of emergencies. The spacecraft completed one orbit around the Earth with a perigee altitude of 181 km and an apogee altitude of 327 km at an inclination of $\sim 65°$, taking approximately 108 minutes. During his historic flight, *Gagarin* experienced weightlessness and reported back to Earth on his observations of the cosmos. As *Vostok 1* prepared for entry into Earth's atmosphere, the planned jettisoning of its equipment module encountered an issue. The module did not fully separate, remaining tethered to Gagarin's capsule by a strand of wires.

This unexpected extra weight induced a spin, subjecting *Gagarin* to a force of 8 Gs–significantly higher than typical flight conditions. As a trained fighter jet pilot, he managed to stay conscious until the wires finally released, stabilizing his descent. At an altitude of approximately 7 km above Earth, Gagarin executed his planned ejection from *Vostok 1* and safely descended via parachute. Onlookers, including a local farmer and his daughter, witnessed the spherical metal vessel of *Vostok 1* making an impact with the ground, followed by Gagarin gracefully parachuting down in his distinctive orange flight suit. Gagarin later recounted reassuring the observers, saying, *Don't be afraid; I am a Soviet citizen like you, descending from space, and I need to find a telephone to call Moscow!*.

Gagarin's historic flight marked a resounding success for the Soviet space program, catapulting him into the status of a national hero within the Soviet Union. His fame transcended borders, transforming him into a global celebrity. His biography and the details of his remarkable journey were prominently featured in newspapers worldwide. In a grand procession, *Gagarin* was escorted through the streets of Moscow in a lengthy motorcade, flanked by high-ranking officials. This procession culminated in a lavish ceremony held at the Kremlin, where *Nikita Khrushchev*, then Premier of USSR, bestowed upon him the prestigious title of *Hero of the Soviet Union*. Across various cities in the Soviet Union, massive demonstrations were organized, rivaling the scale seen only during the Victory Parades of World War II.

Yuri Gagarin in Warsaw, 1961.

Three weeks later, NASA launched its astronaut *Alan Shepard* into space on a suborbital trajectory—a flight that goes into space but does not go all the way around Earth. *Shepard*'s flight lasted just over 15 minutes. Subsequently on May 25, President *John F. Kennedy* challenged the US to an ambitious goal, declaring: "I believe that this nation should commit itself to achieving the goal, before the decade is out, of landing a man on the moon and returning him safely to Earth."

Meanwhile, the Russians achieved several other space milestones ahead of the United States. In addition to launching the first artificial satellite, the first dog in space, and the first human in space, they launched *Luna 2* in 1959, which became the first human-made object to hit the Moon. Later, the Russians launched *Luna 3*, which photographed the far side of the Moon. Few months after *Gagarin*'s flight in 1961, a second Soviet human mission orbited a cosmonaut around Earth for a full day. The USSR also achieved the first spacewalk and launched the *Vostok 6* mission, which made *Valentina Tereshkova* the first woman to travel to space.

2.3.4 Space Stations

Space stations, a type of space habitat that is capable of supporting a human crew in orbit for an extended period of time, represented the next frontier in space exploration. In 1966, the Soviet Union initiated the development of a classified military space station project named *Almaz*, with the inaugural launch scheduled for the early 1970s. The *Almaz* concept comprised a 20-ton module periodically inhabited by crews of 2 to 3 cosmonauts who utilized *Soyuz* crew transport spacecraft for travel. The first-ever space station to orbit Earth was the Soviet *Salyut 1*, launched in 1971.

Illustration of the Soviet Union's *Salyut* space station, the world's first space station placed into orbit on April 19, 1971, with a *Soyuz* crew transport spacecraft approaching at upper left. The *Salyut*, designed for a 6-month on-orbit operational lifetime, hosted three crew (*Georgi Dobrovolski, Vladislav Volkov, and Viktor Patsayev*) for a then record-setting 24-day mission. The flight ended tragically when the crew died due to the sudden depressurization of their *Soyuz 11* spacecraft shortly before entering into the Earth's atmosphere.

Subsequently, NASA introduced the *Skylab* space station, serving as the inaugural orbital laboratory where astronauts and scientists conducted comprehensive research on Earth and the impact of space travel on human physiology.

Mir Space Station

The *Mir space station* functioned in low Earth orbit operated by the Soviet Union between 1986 and 2001. *Mir* stood as the pioneering modular space station, gradually assembled in orbit from 1986 to 1996. At the time it was the largest artificial satellite in orbit, succeeded by the *International Space Station (ISS)*. Launched by the Soviet Union in 1986, *Mir* was the first modular space station, serving as

America launched its first space station, *Skylab 1*, on May 14, 1973. The *Skylab* weighing around 76 tons included a pressurized workshop, an airlock, and a suite of telescopes, with an overall habitable volume of around 350 m^3. Its highly publicized crash back to Earth, during which huge chunks of hardware are dropped into the Indian Ocean and across Western Australia, took place on July 11, 1979.

a pivotal platform for scientific research, technological advancements, and human habitation in orbit for more than 15 years.

During its operational life, *Mir* hosted numerous scientific experiments that spanned various disciplines, including biology, astronomy, physics, and materials science. It served as a microgravity laboratory, allowing scientists to conduct long-duration studies on the effects of space travel on the human body and to explore the potential of space for research and technological development. Moreover, *Mir* provided a foundation for understanding prolonged space missions and laid the groundwork for future space stations, including the *International Space Station (ISS)*.

Mir's endurance in orbit fostered international cooperation, as it welcomed astronauts and cosmonauts from multiple countries, transcending geopolitical boundaries. Its collaborative missions paved the way for joint scientific endeavors and fostered friendships among space explorers from different nations. The knowledge gained from *Mir*'s operations significantly contributed to humanity's understanding of living and working in space, leaving an indelible mark on the history of space exploration. *Mir*'s legacy continues to inspire future generations of space scientists and engineers, serving as a testament to the power of international partnerships in advancing our understanding of the universe.

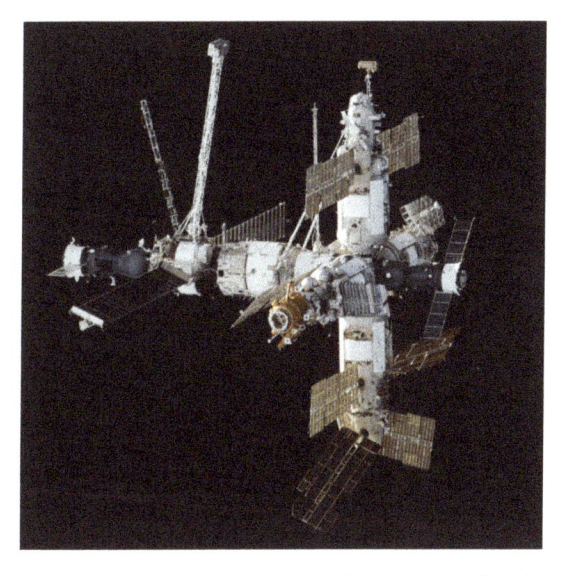

The *Mir* Space Station viewed from Space Shuttle *Endeavour* during the STS-89 (a Space Shuttle mission to the Mir space station) rendezvous on January 28, 1998.

Mir established a historic milestone by holding the record for the longest continuous human presence in space, totaling 3,644 days. However, this record was eventually eclipsed by the International Space Station (ISS) on October 23, 2010. Notably, *Mir* also boasted the distinction of hosting the longest single human spaceflight, as cosmonaut *Valeri Polyakov* spent an astounding 437 days and 18 hours aboard the station between 1994 and 1995. Throughout its fifteen-year lifespan, *Mir* remained occupied for approximately twelve and a half years, demonstrating its capability to accommodate a resident crew of three individuals, or larger crews for brief visits.

Tiangong Space Station

The *Tiangong* space station stands as a testament to China's burgeoning prowess in space exploration, embodying the nation's commitment to establishing a permanent human presence in space. Under the auspices of the China National Space Administration (CNSA), the *Tiangong* station represents a significant milestone in China's space program, positioning the country as a key player in crewed space missions and scientific research in low Earth orbit (LEO).

Comprising multiple modules, the *Tiangong* space station is designed to serve various functions, including scientific research, technological experiments, and supporting human habitation in space. It aims to facilitate an array of research endeavors spanning fields such as astronomy, materials science, biology, and Earth observation. As China progressively assembles and expands the station, the vision for *Tiangong* entails accommodating not only scientific missions but also fostering international collaborations through partnerships with other nations and space agencies.

The *Tiangong* space station is a testament to China's ambitions for long-term space exploration and its commitment to international cooperation in space endeavors. Assembling the station and conducting scientific research within its modules signifies a critical step forward in advancing China's space capabilities. *Tiangong* serves not only as a symbol of national achievement but also as a beacon for scientific discovery, fostering global collaborations in the quest for deeper insights into space exploration and human presence beyond Earth.

The International Space Station (ISS)

The *International Space Station* (*ISS*) is a remarkable and esteemed symbol of international cooperation and human achievement in space exploration. It stands as one of humanity's most ambitious and enduring achievements in space science and technology. In 1984, US President *Ronald Reagan* instructed *NASA* to create a permanent research laboratory in low Earth orbit, which would serve as an Earth-orbiting space station. He also advocated for international collaboration on this project. Following several revisions, the partnership involving the United States, Canada, Japan, and the European Space Agency extended an invitation to Russia to join the *ISS* Program in 1993. A significant Russian contribution to the *ISS* was the *Zvezda* Service Module, the third module launched to the station, that enabled the commencement of continuous human habitation in November 2000 by providing all of the station's life support systems.

The primary construction phase spanned from 1998 to 2011, but the station remains an ever-evolving platform, accommodating new missions and experiments. The assembly process began in 1998 with the launch of the Russian module *Zarya* and continued with numerous missions over the

The layout of the station, highlighting the *Tianhe* module (the first module to launch the *Tiangong* space station) positioned at the center. Adjacent to *Tianhe*, a *Tianzhou* automated cargo spacecraft is seen docked on its aft port, while the *Wentian* (laboratory cabin module) occupies its starboard port. The *Mengtian* module (a major module of the space station) is located on its portside port. Additionally, the illustration depicts two *Shenzhou* spacecraft, both utilizing the station's multi-docking hub for their docking configurations.

years. Each component of the ISS was built by a different space agency and delivered to the station using spacecraft like the *Space Shuttle*, *Russian Soyuz* and *Progress spacecraft*, and more recently, commercial vehicles like *SpaceX*'s *Dragon* and *Northrop Grumman*'s *Cygnus*. Astronauts from different countries worked together to install and connect these modules.

Since November 2, 2000, the *ISS* has been continuously inhabited. As of May 2022, the *International Space Station* has hosted 258 individuals from 20 countries, with the United States (158 individuals) and Russia (54 individuals) being the leading participating nations. The allocation of astronaut and research time aboard the station is determined by space agencies based on their respective financial or resource contributions, including modules and robotics.

The *ISS* is not just a space station; it's an orbiting laboratory. It houses state-of-the-art research facilities, living quarters, and life support systems. The station functions as a research laboratory for microgravity and space environment studies across various scientific domains, including astrobiology, astronomy, meteorology, and physics. Additionally, the *ISS* serves as a valuable platform for testing spacecraft systems and equipment necessary for potential long-duration missions to the Moon and Mars.

2.3.5 Space Shuttle

NASA's *Space Shuttle* program, which ran from 1981 to 2011, was a remarkable achievement in space exploration. This program introduced the Space Shuttle, a partially reusable spacecraft designed by NASA for low Earth orbit missions as part of the Space Shuttle program named *Space Transportation System (STS)*. The program had a dual purpose: firstly, to decrease the expenses associated with space travel by substituting the prevailing approach of launching capsules on disposable rockets with reusable spacecraft; and secondly, to provide backing for ambitious future initiatives, which included

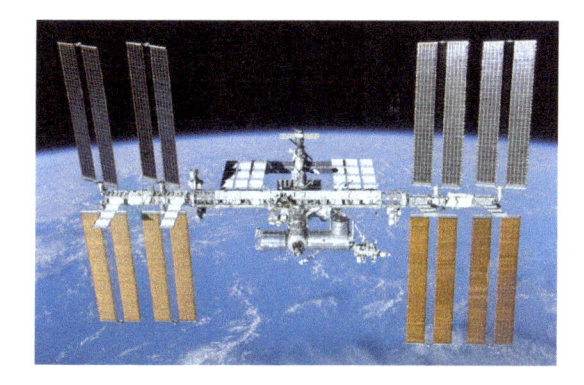

The *International Space Station* (*ISS*) — the largest structure humans ever put into space in low Earth orbit was developed by five space agencies: the United States' *NASA*, Russia's *Roscosmos*, Japan's *JAXA*, Europe's *ESA*, and Canada's *CSA*. The *International Space Station* is 109 meters in length with a mass of $\sim$ 420 tons without visiting vehicles. It has 13,696 cubic feet of habitable volume for crew members. The space station has seven sleeping quarters, with the ability to add more during crew handover periods, two bathrooms, a gym, and the cupola– a 360-degree-view bay window of the Earth. The space station orbits Earth at an altitude of approximately 400 km, with its orbital path taking over 90 percent of the Earth's population. It circles the globe every 90 minutes at a speed of about 28,000 km/h.

establishing enduring space stations in orbit around both Earth and the Moon, as well as embarking on a human mission to land on Mars. Over the course of its operation, from the inaugural launch on April 12, 1981, to the historic final landing on July 21, 2011, NASA's fleet of space shuttles–*Columbia*, *Challenger*, *Discovery*, *Atlantis*, and *Endeavour*–executed an astounding total of 135 missions. The *Space Shut-*

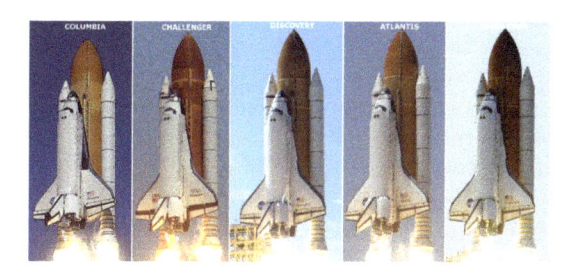

Space shuttles (From left): *Columbia*, *Challenger*, *Discovery*, *Atlantis*, and *Endeavour*

tle was a sophisticated spacecraft system consisting of an orbiter launched alongside two reusable solid rocket boosters and a disposable external fuel tank.

This advanced vehicle could transport up to eight astronauts and carry payloads weighing up to 23 tons to low Earth orbit (LEO). Upon accomplishing its mission objectives, the orbiter would execute a graceful entry into Earth's atmosphere, landing much like a glider at one of two designated locations: either the Kennedy Space Center or Edwards Air Force Base.

Three primary parts of space shuttle: the Orbiter, serving as the crew's habitat; a sizable external tank responsible for storing fuel for the main engines; and two solid rocket boosters, crucial for generating the majority of the Shuttle's lift during the initial two minutes of its journey. Remarkably, all components, except the external fuel tank, are reusable. After each launch, the external fuel tank undergoes destruction, burning up on returning to the atmosphere.

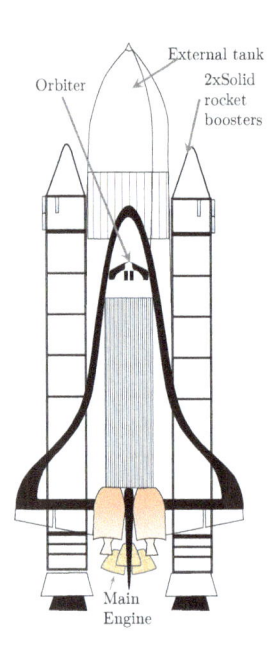

The *Space Shuttle*'s missions encompassed a range of tasks, including transporting substantial payloads to diverse orbital destinations such as the *International Space Station* (*ISS*). In addition, it played a crucial role in facilitating crew rotations for the space station and conducting essential servicing missions for the *Hubble Space Telescope*. Furthermore, the orbiter occasionally performed the intricate task of retrieving satellites and other payloads, including those from the *ISS*, from their orbits and safely returning them to Earth, although this particular function was relatively infrequent.

The space shuttle *Endeavour* docked to the *International Space Station* at an altitude of approximately 355 km. The photo was taken in 2011 by *Expedition 27* crew member *Paolo Nespoli* from Russian human spaceflight to the *ISS*, the *Soyuz TMA-20* spacecraft.

The launch of the first orbital spaceflight, STS-1 *Columbia*, on April 12, 1981, the first American crewed space flight. It returned on April 14, 1981, 54.5 hours later, having orbited the Earth 37 times.

On April 12, 1981, the *Space Shuttle Columbia*, carrying astronauts *John Young* and *Robert Crippen*, embarked on the program's maiden flight (*STS-1*) marking the beginning of the *Space Shuttle* era. Tragically, on 28 January 1986, the Space Shuttle *Challenger* disintegrated 73 seconds after liftoff, killing seven crew, including *Christa McAuliffe*, a teacher from New Hampshire who would have been the first civilian in space. This disaster, caused by a faulty O-ring in one of the solid rocket boosters, resulted in the suspension of shuttle flights and a thorough redesign of the boosters. After a hiatus of over two years following the *Challenger* disaster, the *Space Shuttle* program returned to flight with the successful launch of *Discovery* on *STS-26* in September 1988. This marked a significant milestone in the program's recovery. In April 1990, the *Space Shuttle (STS-31) Discovery* deployed the *Hubble Space Telescope* into orbit. The *Hubble* has since provided invaluable insight into the universe.

On April 24, 1990, the Space Shuttle *Discovery* deployed the *Hubble Space Telescope* into low Earth orbit on 35th mission of *NASA*'s *Space Shuttle* program.

Space Shuttle Endeavour's inaugural flight designated *STS-49* in May 1992 featured a daring spacewalk to retrieve the *Intelsat VI (Intelsat 603)*, a communications satellite operated by *Intelsat*, which had remained stranded in low Earth orbit for two years due to a malfunction. The plan involved securing the satellite to a new upper stage and then launching it once again, this time successfully, to its intended geosynchronous orbit. The most notable achievement of STS-49 was the daring and unprecedented spacewalk performed by astronauts *Pierre Thuot*, *Richard Hieb*, and *Thomas Akers*. They conducted a series of spacewalks to reach the satellite, capture it, and attach a new upper-stage rocket to it. Once the astronauts secured the satellite, they used the shuttle's robotic arm to attach a new motorized upper-stage rocket to *Intelsat 603*. This rocket successfully propelled the satellite to its intended geosynchronous orbit, saving it from becoming space debris.

Three crew members of mission *STS-49* hold onto the 4.5 ton *INTELSAT VI* satellite during an extravehicular activity (EVA). The satellite which was stranded in an unusable orbit since its launch aboard a *Titan* vehicle in March 1990 was captured and equipped with a new perigee motor. The satellite was subsequently put into a geosynchronous orbit using the new motor.

On February 1, 2003, *Space Shuttle Columbia (STS-107)* disintegrated upon entering into Earth's atmosphere, resulting in the tragic loss of all seven crew members including *Kalpana Chawla*, an Indian-born American astronaut. This incident, caused by damage to the shuttle's thermal protection system, led to the suspension of shuttle flights. Launched from the Kennedy Space Center on January 16, 2003, the *Columbia* mission spent 15 days, 22 hours, 20 minutes and 32 seconds in orbit. The crew conducted several scientific experiments before the disaster occurred during the return to Earth. The failure was attributed to a piece of foam that detached during launch, damaging the thermal protection system on the leading edge of the orbiter's left wing. As the damaged wing heated up during return, it gradually disintegrated, resulting in the loss of control and the vehicle's ultimate disintegration. All the remaining *Space Shuttles* were grounded till safety issues were addressed. It was not until July 2005 that flights resumed with the launch of *Discovery* on mission *STS-114*.

Kalpana Chawla (1962 – 2003), an Indian-born American astronaut, was the first woman of Indian origin in space. She flew on the Space Shuttle *Columbia* in 1997 and tragically lost her life during her second flight and the final mission of the shuttle *Columbia*, during the failed return to Earth.

As the space shuttle program suffered a couple of fatal accidents and the cost of the *ISS* program continued to escalate from an originally agreed $17 billion to $50-100 billion, the aging *Space Shuttle* fleet needed to be phased out. A shuttle program was renewed to a fully reusable launch system, flying under its own power into space and back. A new crop of experimental *X-planes* with the ambition of creating a *space liner* that would transport people back and forth between bases on Earth and the *ISS*– and transform the economics of journeys into space. Many observers felt, though, that NASA's human space flight program was lacking in direction and focus. The sense of drift ended in January 2004, when President *George Bush* announced his vision for space exploration. There was to be a return to the Moon to establish a permanent base and a future journey onward to Mars.

The Space Shuttle program's final mission, designated *STS-135*, concluded on July 21, 2011, as the *Atlantis* shuttle gracefully touched down at its home base, NASA's *Kennedy Space Center* in Florida. This moment marked the end of an era in space exploration, leaving an indelible legacy of achievements and inspiration for future endeavors beyond Earth's boundaries.

NASA'S HYPER-X: X-43A hypersonic technology demonstrator which flew in March 2004 was the first time a scramjet-powered aircraft had flown freely. The scramjet was created by *Micro Craft* for *NASA* to explore the possibility of hypersonic flight at speeds up to Mach number 10 using an air-breathing engine. A potential use for such an aircraft would be as a reusable space launch vehicle.

Atlantis after its final landing, marking the end of the *Space Shuttle* Program.

Soyuz Program

The *Soyuz* program, launched by the Soviet Union in the early 1960s, is one of the most enduring and successful space exploration initiatives in history. It was originally a part of a Moon landing project intended to put a Soviet cosmonaut on the Moon. It followed the earlier *Vostok* (1961 – 1963, placed the first human *Yuri Gagarin* into space on April 12, 1961) and *Voskhod* (1964 – 1965) programs, making it the third Soviet human spaceflight initiative.

The *Crew Dragon* (background) and four of the astronauts of its first two crewed missions(foreground).

Comprising the *Soyuz* spacecraft and the Soyuz rocket, this program is currently managed by the Russian space agency, *Roscosmos*. Notably, following the *Space Shuttle*'s retirement in 2011, *Soyuz* remained the sole means of transporting humans to the *International Space Station* until May 30, 2020, when the *Crew Dragon* spacecraft, reusable spacecraft developed by *Space Exploration Technologies Corporation*, commonly referred to as *SpaceX*, conducted its inaugural astronaut mission to the ISS.

The *Soyuz* program traces its roots back to the early days of the Space Race during the Cold War. The first crewed *Soyuz* mission, *Soyuz 1*, took place in 1967, and it ended tragically with the death of its cosmonaut, *Vladimir Komarov*, due to a parachute failure during reentry. However, over the years, *Soyuz* spacecraft have proven to be highly reliable and robust over the decades. They served as the workhorse for transporting astronauts and cosmonauts to and from the *International Space Station*.

Soyuz 11 was intended to be the world's first space station mission. The crew was to visit the *Salyut 1* space station, which was the Soviet Union's pioneering attempt at establishing a long-term orbital space station. *Soyuz 11* successfully docked with the *Salyut 1* space station, and the crew began conducting experiments and observations. Tragically, during their 23-day stay on the station, a cabin vent valve construction defect caused a cabin depressurization incident on June 30, 1971. The cabin depressurization led to the death of all three cosmonauts on board. The loss of the *Soyuz 11* crew led to a suspension of Soviet space station missions while investigations took place. Changes and improvements were made to *Soyuz* spacecraft to prevent similar accidents in the future.

Later *Soyuz 19/Apollo-Soyuz Test Project* (1975) marked the historic rendezvous and docking of a Soviet *Soyuz* spacecraft with an American *Apollo* spacecraft in Earth orbit. A highly critical mission, *Soyuz T-13* in 1985 involved the repair and reactivation of the orbiting space station *Salyut 7*, which had been uncrewed and malfunctioning for months.

Soyuz TM-32 mission carried American businessman *Dennis Tito* to the International Space Station (ISS) as the first space tourist. It marked a significant step in the commercialization of space travel. In the wake of the *Space Shuttle Columbia* disaster, *Soyuz TMA-3* (2003) became a lifeline for the *ISS* crew, providing them with a way back to Earth while *Space Shuttle* flights were suspended.

The launch of *Soyuz TMA-19* mission in 2010 marked the first time an American astronaut, *Tracy Caldwell Dyson*, flew on a *Soyuz* as a result of *NASA*'s partnership with *Roscosmos*. The mission *Soyuz MS-01* (2016) gained significance due to the spaceflight of American astronaut *Kate Rubins*, who became the first person to sequence DNA in space aboard the *ISS*. A launch abort during the mission *Soyuz MS-10* in 2018 demonstrated the effectiveness of the *Soyuz* spacecraft's launch escape system, saving the lives of the two crew members, American astronaut *Nick Hague* and Russian cosmonaut *Alexey Ovchinin*.

The *Soyuz* program holds several records in human spaceflight, including the longest continuous human presence in space and the most crewed launches as on 2023. Over the years, different variants of the *Soyuz* spacecraft have been developed for various purposes, including cargo missions (*Soyuz Progress*) and interplanetary exploration (*Soyuz TMA*). Despite being over half a century old, the *Soyuz* program continues to operate successfully and remains a critical component of human spaceflight, especially in ferrying astronauts to and from the *ISS*.

Soyuz TMA-13, a *Soyuz* mission to the *International Space Station* (*ISS*), lifting off from *Gagarin's Start* at Baikonur Cosmodrome in 2008.

The *Hubble Space Telescope* orbits around Earth with a time period of 95 minutes at an altitude of approximately 570 km. It has a 13-meter-long tube opened to space on one end. Light travels through the tube to a bowl-shaped, 2.4-meter-wide main mirror and bounces up to a secondary mirror. That mirror concentrates the light that travels back through a hole in the middle of the primary mirror which is directed through various scientific instruments for analysis.

In 2009, astronauts flew to the *Hubble* telescope on the *space shuttle*. This was the fifth time astronauts went to fix *Hubble*. They put new parts and cameras in the telescope. It still takes beautiful pictures of objects in space and is expected to be operational till 2040.

Recently, *NASA* built on the legacy of the *Hubble Space Telescope*, *James Webb Space Telescope*, a multi-purpose observatory launched in December 2021. The telescope, named after *James Webb*, NASA's second administrator from 1961 to 1968, will help to determine whether planets orbiting other stars could support life, and observe galaxies. *Webb* uses infrared light, which cannot be perceived by the human eye, to study different phases in cosmic history.

Next, NASA is building *Roman Space Telescope* (*RST*) and is expected to be launched in 2026 to understand dark matter and to image exoplanets. *RST*'s camera is just as sensitive as the *Hubble* telescope's, but with a field of view 100 times bigger. That means no matter what *RST* is looking at, it will be able to collect a lot more data at one time. The *RST* will be able to detect some of the light wavelengths coming from the exoplanets it directly im-

2.3.6 Hubble and James Webb Space Telescopes

NASA launched *Hubble Space Telescope*, named after American astronomer Edwin Hubble (1889 - 1953), in 1990 for a multipurpose observatory in Earth orbit and used for astrophysics and planetary science. The *Hubble* telescope has revolutionized the understanding of the cosmos. Earth's atmosphere distorts the light from celestial objects, causing them to twinkle. Therefore, a telescope above our planet's atmosphere is required to study outer space. The telescope has helped scientists learn about our solar system, the comets, and the planets. *Hubble* telescope even discovered moons around Pluto that had not been seen before. The telescope has helped scientists understand how planets and galaxies form.

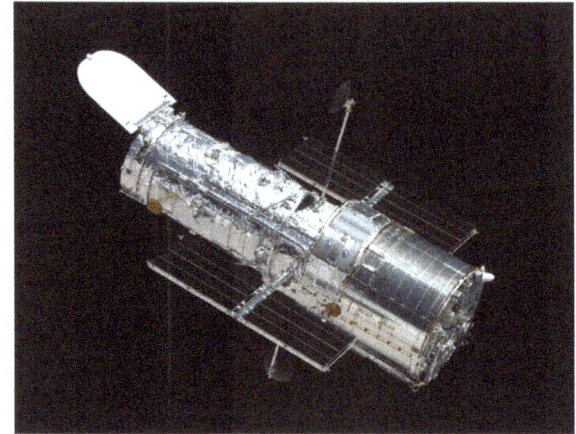

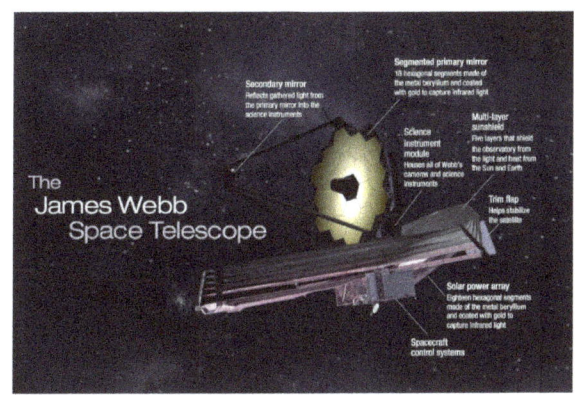

2.4 Missions to Celestial Bodies

2.4.1 Luna Mission to Moon

The Soviet Union, during the Space Race era, conducted a series of missions aimed at exploring the Moon, showcasing pioneering efforts in lunar exploration. The Soviet lunar program, known as the *Luna program*, encompassed several missions from the late 1950s through the 1970s, each contributing valuable data and insights into lunar exploration. Out of a total of 44 missions, 29 ended in failure while 15 achieved success, each fulfilling roles as either orbiters or landers. These missions notably marked numerous milestones in space exploration, pioneering groundbreaking achievements. Additionally, they conducted a multitude of experiments, delving into studies related to the Moon's chemical composition, gravitational forces, surface temperatures, and radiation levels.

In 1959, *Luna 1* became the first spacecraft to reach the vicinity of the Moon, marking a significant milestone in space exploration. Although it missed a lunar impact, it provided crucial data by flying past the Moon and studying the space environment.

Luna 2, launched later in 1959, achieved the first successful impact on the lunar surface, providing essential data about the Moon's composition. *Luna 9*, in 1966, made history as the first spacecraft to achieve a soft landing on the Moon, transmitting images back to Earth. *Luna 16, 20,* and *24* (launched in 1970 and 1972) were groundbreaking sample return missions. These missions successfully collected lunar soil from the Moon's surface and brought it back to Earth, offering scientists the opportunity to analyze and study material directly obtained from the lunar landscape.

2.4.2 Apollo Space Program

The Cold War rivalry between the United States and the Soviet Union was a driving force for moon mission. The Soviets had achieved several space

James Webb Space Telescope can see a much larger portion of the infrared spectrum than *Hubble* telescope and collects six times more light. It complements and extends *Hubble*'s observations, becoming the world's latest premier space observatory.

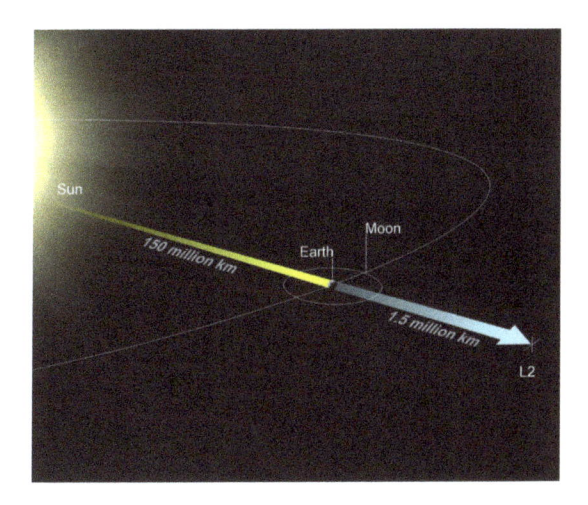

The *James Webb* telescope orbits the Sun at a position called *L2*, located 1.5 million kilometres from Earth.

ages which will help scientists understand the composition of the exoplanets' atmospheres.

Roman Space Telescope (RST).

Luna 1 impactor. An *impact probe* that undergoes a forceful landing, causing damage or destruction upon arrival, a lander executes a soft descent, ensuring a gentle touchdown that allows the probe to continue functioning following its arrival on the surface. A *lander* refers to a spacecraft that navigates downward and gently settles on the surface of a celestial body apart from Earth.

milestones, including launching the first artificial satellite, *Sputnik*, and sending the first human, *Yuri Gagarin*, into space. Americans aimed to reestablish their prestige and leadership in space exploration.

On May 25, 1961, President *John F. Kennedy* stood before a joint session of the United States Congress and said the now-famous words, "I believe that this nation should commit itself to achieving the goal, before this decade is out, of landing a man on the Moon and returning him safely to the Earth". This proclamation marked the beginning of an extraordinary journey filled with challenges, dedication, and an unwavering commitment to the mission.

NASA faced a daunting challenge with many problems to be solved before being able to land a human being on the Moon. They lacked the necessary rockets, portable computers, spacesuits, spacecraft, tracking stations, and even a clear plan for the journey. The level of unpreparedness was so profound that they didn't even possess a comprehensive list of what was needed. They also lacked the knowledge of how to navigate to the Moon or what they might encounter there, such as lunar conditions. Concerns arose from various experts, including physicians worried about cognitive abilities in microgravity and mathematicians unsure of how to precisely rendezvous and dock spacecraft in orbit.

The *Apollo Project* mobilized a workforce three times larger than that of the *Manhattan Project*, responsible for developing the atomic bomb. In 1961, when *President Kennedy* formally introduced the *Apollo* initiative, *NASA*'s annual expenditure was $ 1 million. Fast forward just five years, and NASA was allocating approximately $ 1 million every three hours, around the clock, solely for the *Apollo* program. At that time, the American space program incurred an annual cost of approximately $ 4 billion. *Apollo* was the third US human spaceflight program to fly, preceded by projects *Mercury* (1958 - 1963) and *Gemini* (1965 - 1966) conceived to extend spaceflight capability in support of *Apollo*. Project *Mercury*, was the first human spaceflight program by *NASA*, designed to put astronauts into orbit around the Earth. *Mercury*'s historic success included *Alan Shepard*'s suborbital flight in 1961 and *John Glenn* becoming the first American to orbit the Earth in 1962. Project *Gemini*, conducted from 1961 to 1966, bridged the gap between *Project Mercury* and the *Apollo Program*, focusing on crucial spaceflight techniques and orbital maneuvers. It achieved vital milestones such as the first American spacewalk, rendezvous and docking, and extended endurance in space, paving the way for lunar exploration.

The journey began with *Apollo 1*, intended as a simple Earth orbit test. Tragically, a cabin fire during a pre-launch test on January 27, 1967, claimed the lives of three astronauts: *Gus Grissom*, *Ed White*, and *Roger Chaffee*. It was a devastating setback, but it only fueled the determination to honor their sacrifice by continuing the mission.

2.4. MISSIONS TO CELESTIAL BODIES

Apollo 7, launched on October 11, 1968, marked the triumphant return to space. The first crewed flight in the Apollo program, Astronauts *Walter Schirra*, *Donn Eisele*, and *Walter Cunningham* orbited the Earth, testing the Command and Service Module (CSM) with crew performance and demonstrating *Apollo* rendezvous capability.

The prime crew of the first manned *Apollo 7* space mission from left to right are: Command Module pilot, *Donn Eisele*, Commander, *Walter Schirra* and lunar module pilot *Walter Cunningham*

Then came the pivotal *Apollo 8*, a mission that would forever change how humanity viewed its place in the cosmos. It was the first crewed spacecraft to leave low Earth orbit and the first human spaceflight to reach the Moon. On December 21, 1968, Commander *Frank Borman*, Command Module Pilot *James Lovell*, and Lunar Module Pilot *William Anders* embarked on a daring journey to orbit the Moon. For the first time, humans witnessed the far side of the Moon and the Earthrise, a breathtaking sight that ignited a new appreciation for our fragile planet. The crew orbited the Moon ten times without landing and then departed safely back to Earth on December 27, 1968.

With newfound inspiration, the world eagerly anticipated *Apollo 11* which had a clear and daring mission to land humans on the Moon and return them safely to Earth, the goal set by President *John F. Kennedy* on May 25, 1961. The objectives included scientific exploration, and collecting lunar samples.

Apollo 8 lifted off on December 21, 1968, utilizing the three stage *Saturn V* rocket to attain Earth orbit. The first two stages splashed down in the Atlantic Ocean and the third stage placed the spacecraft into Earth's orbit and remained connected to execute the Trans-Lunar Injection (TLI) burn, setting the spacecraft on its trajectory toward the Moon. After the third stage had placed the mission on course for the Moon, the Command and Service Modules (CSM), the remaining *Apollo 8* spacecraft, separated from it.

On July 16, 1969 at 9:32 AM (EDT), the mighty *Saturn V* rocket roared to life from *Kennedy Space Center* in Florida, carrying astronauts *Neil Armstrong* (Commander), *Buzz Aldrin* (Lunar Module Pilot), and *Michael Collins* (Command Module Pilot) toward their lunar destination.

The spacecraft consisted of three modules: the Command Module (CM), named *Columbia*, the Lunar Module (LM) named *Eagle* and the Service Module. The service module contained the main spacecraft propulsion system and consumables. The Lunar Module was the two-person craft used by *Armstrong* and *Aldrin* to descend to the Moon's surface. The Command Module is the only portion of the spacecraft to return to Earth. A *Saturn V* rocket, one of the most powerful rockets ever built, was used to propel the spacecraft into space. The spacecraft entered Earth orbit, where it underwent checks and maneuvers to ensure everything was functioning

Earthrise is a famous photograph captured during the *Apollo 8*, the first manned mission to the moon, on December 24, 1968. It shows the Earth rising above the lunar horizon as seen from the Moon's surface. The photograph was taken by astronaut *William Anders* while the spacecraft was in lunar orbit. It is a spectacular image that symbolizes the beauty and fragility of our planet and the human exploration of space. *Earthrise* is often celebrated as the most influential environmental photograph ever taken.

The *Apollo 11* lunar landing mission crew: Commander *Neil Armstrong* (left), Command Module (Columbia) Pilot *Michael Collins* (middle) and Lunar Module (Eagle) Pilot *Edwin E Aldrin* (right)

The *Apollo 11* Command Module (CM), *Columbia* a conical pressure vessel, was the living quarters for the three-person crew during most of the first crewed lunar landing. Primary Materials: Aluminum alloy, Stainless steel, Titanium. Dimensions: maximum diameter of 3.9 m at its base and a height of 3.65 m., weighing around 4141.3 kg)

properly. To travel from Earth to the Moon, *Apollo 11* executed a Translunar Injection burn. This critical maneuver occurred about 2.5 hours after launch. The burn took place in Earth orbit and put the spacecraft on a trajectory to intercept the Moon. After three days of coasting towards the Moon, *Apollo 11* entered lunar orbit on July 19, 1969. The CM *Columbia* orbited the Moon while the LM *Eagle* remained attached.

On July 20, the LM *Eagle* separated from the CSM Columbia and descended to the lunar surface. On July 20, 1969, *Neil Armstrong* and *Buzz Aldrin* piloted the *Eagle* to the lunar surface, while *Michael Collins* remained in lunar orbit aboard the *Columbia*. As the world watched, *Neil Armstrong* famously descended from the LM *Eagle*, to become the first human to set foot on the Moon, uttering the unforgettable words, "That's one small step for [a] man, one giant leap for mankind".

The descent to the lunar surface was challenging, with limited fuel and the need to avoid hazardous terrain. At 4:17 PM EDT, *Neil Armstrong* became the first human to set foot on the Moon, followed by *Buzz Aldrin*. They conducted experiments, collected lunar samples, and deployed scientific instruments. After spending about 2.5 hours on the lunar surface, they returned to the Lunar Module and prepared for ascent. On July 21, 1969, the Lunar Module, *Eagle* ascended from the Moon's surface to rendezvous (orbital maneuvers to bring two spacecraft on the same orbit closer to each other by matching orbital velocities and position vectors) with the CSM *Columbia* in lunar orbit.

2.4. MISSIONS TO CELESTIAL BODIES

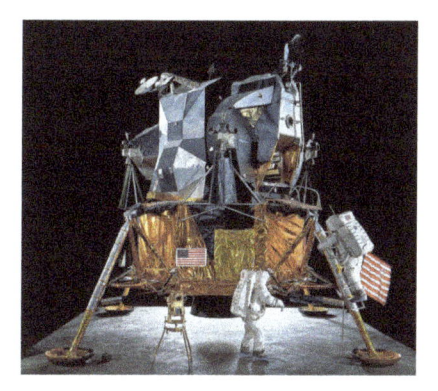

The *Apollo 11* Lunar Module (LM), *Eagle* was a two-stage space vehicle designed to transport two astronauts from lunar orbit to the lunar surface and back. The upper stage consisted of a pressurized crew compartment, equipment areas, and an ascent rocket engine. The lower descent stage had the landing gear and contained the descent rocket engine and lunar surface experiments. Materials used: Aluminum, titanium, aluminized Mylar and aluminized Kapton blankets. Dimensions: $654 \times 654\,cm^2$, weighing 3855.5 kg).

After a successful lunar orbit rendezvous, the CSM *Columbia* and the Lunar Module *Eagle* docked (orbital maneuvers to physically join two separate free-flying space vehicles). On July 22, 1969, the spacecraft left lunar orbit and began its journey back to Earth. On July 24, 1969, *Apollo 11* entered Earth's atmosphere, and the Command Module separated from the Service Module. The Command Module safely splashed down in the Pacific Ocean, approximately 21 km from the recovery ship *USS Hornet*. The astronauts were recovered by a helicopter and brought aboard the *USS Hornet*.

A historic celebration, ticker tape parade in New York City, that captivated the nation and the world ensued. On August 13, 1969, an estimated six million spectators gathered along the parade route to honor astronauts *Neil Armstrong*, *Buzz Aldrin*, and *Michael Collins*, who had recently achieved the remarkable feat of landing on the Moon and safely returning to Earth. As the astronauts rode in a motorcade through the city's streets, countless tons of ticker tape, confetti, and shredded paper rained down from the towering skyscrapers, creating a surreal and joyous spectacle. It was a moment of

Three astronauts – strapped into *Apollo 11* spacecraft on top of the vast *Saturn V* rocket – were propelled into Earth's orbit from Cape Kennedy in over 11 minutes on July 16, 1969. Four days later, on July 20, *Armstrong* first and 19 minutes later *Aldrin* became the first humans to set foot on the Moon's surface. They spent about 2 hours and 15 minutes exploring the site they landed and returned to Earth with 21.5 kg of lunar material on July 24.

Aldrin on the Moon, photographed by *Neil Armstrong*. An estimated 650 million people watched *Armstrong*'s televised image and heard his voice describe the event as he took *...one small step for a man, one giant leap for mankind* on July 20, 1969.

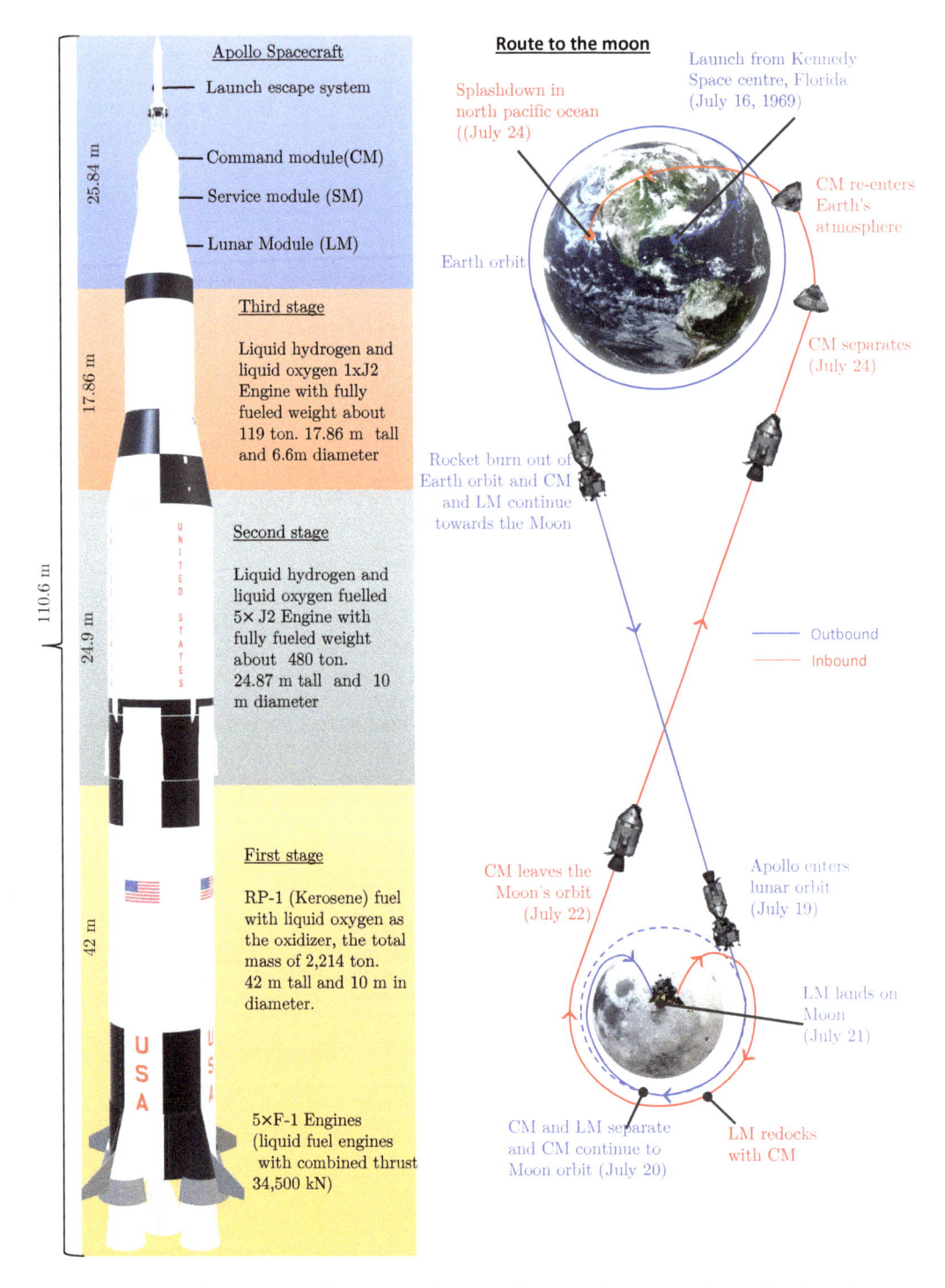

Apollo Spacecraft
- Launch escape system
- Command module (CM)
- Service module (SM)
- Lunar Module (LM)

25.84 m

Third stage
Liquid hydrogen and liquid oxygen 1xJ2 Engine with fully fueled weight about 119 ton. 17.86 m tall and 6.6m diameter

17.86 m

Second stage
Liquid hydrogen and liquid oxygen fuelled 5× J2 Engine with fully fueled weight about 480 ton. 24.87 m tall and 10 m diameter

24.9 m

First stage
RP-1 (Kerosene) fuel with liquid oxygen as the oxidizer, the total mass of 2,214 ton. 42 m tall and 10 m in diameter.

42 m

5×F-1 Engines (liquid fuel engines with combined thrust 34,500 kN)

110.6 m total

Route to the moon
- Launch from Kennedy Space centre, Florida (July 16, 1969)
- Splashdown in north pacific ocean ((July 24)
- CM re-enters Earth's atmosphere
- Earth orbit
- CM separates (July 24)
- Rocket burn out of Earth orbit and CM and LM continue towards the Moon
- Outbound
- Inbound
- CM leaves the Moon's orbit (July 22)
- Apollo enters lunar orbit (July 19)
- LM lands on Moon (July 21)
- CM and LM separate and CM continue to Moon orbit (July 20)
- LM redocks with CM

Astronauts were strapped into their *Apollo 11* spacecraft on top of the around 100 m long *Saturn V* rocket and were propelled by burning around 20 tons of fuel a second at launch (propellant accounted for 85% of its overall weight of 2,800 tons) into orbit in just over 11 minutes.

collective pride and awe as New Yorkers and people from all walks of life came together to express their admiration for the courage, dedication, and scientific achievement represented by the *Apollo 11* mission.

The *Apollo 11* mission was a historic achievement, as it marked the first successful human landing on the Moon and the safe return of astronauts to Earth.

2.4. MISSIONS TO CELESTIAL BODIES

A historic celebration, ticker tape parade in New York City, on August 13, 1969, an estimated six million spectators gathered along the parade route to honor astronauts that captivated the nation and the world.

It remains a symbol of human exploration and technological achievement, demonstrating the incredible capabilities of science and engineering.

Apollo 12–17 followed, each with its unique triumphs and challenges. Scientists and astronauts conducted experiments, explored the lunar surface, and brought back precious lunar samples. The missions solidified humanity's presence in space, expanded our understanding of the Moon and Earth, and paved the way for future exploration.

Apollo 12, launched on November 14, 1969, was the sixth crewed mission in the *Apollo* program and the second to land on the Moon. It aimed to explore the lunar surface, specifically the *Surveyor 3* landing site. The mission successfully landed astronauts *Charles Conrad* and *Alan Bean* on the Moon, where they conducted experiments and collected samples. Despite a lightning strike during launch, which temporarily endangered the mission, *Apollo 12* demonstrated NASA's ability to overcome challenges and furthered lunar exploration.

Launched on April 11, 1970, *Apollo 13* was intended to be the third lunar landing mission, but an oxygen tank explosion in the service module forced the mission to be aborted. The focus shifted to a safe return to Earth. *Apollo 13* is best remembered for the famous phrase "Houston, we've had a problem," and the heroic efforts of the crew and mission control to bring the astronauts safely home. The mission highlighted NASA's problem-solving capabilities and reinforced the risks and challenges of space exploration.

Apollo 14, launched on January 31, 1971, aimed to explore the lunar highlands and conduct scientific experiments. Astronauts *Alan Shepard* and *Edgar Mitchell* explored the *Fra Mauro* region. The mission was a success, with *Shepard* famously hitting golf balls on the lunar surface. *Apollo 14* helped advance lunar geology and furthered our understanding of the Moon's geological history.

Apollo 15, launched on July 26, 1971, was the first J-class mission, which included a lunar rover for extended surface exploration. The mission focused on lunar geological studies. The lunar rover allowed astronauts *David Scott* and *James Irwin* to cover more ground and conduct in-depth geological investigations. *Apollo 15*'s findings significantly expanded our knowledge of lunar geology and the Moon's volcanic history.

Launched on April 16, 1972, *Apollo 16* aimed to explore the *Descartes Highlands* region on the Moon's surface. It was the second J-class mission with a lunar rover. Astronauts *John Young* and *Charles Duke* conducted three moonwalks and collected a wealth of geological samples. The mission added to our understanding of lunar history, including evidence of volcanic activity and impact cratering.

Apollo 17, launched on December 7, 1972, was the final mission of the *Apollo Program*. It explored the *Taurus-Littrow Valley* and conducted experiments. The mission included the longest lunar stay (nearly three days) and the most extensive moonwalks. Astronauts *Eugene Cernan* and *Harrison Schmitt* collected valuable samples and conducted experiments. *Apollo 17* marked the end of the *Apollo Program* and represented the culmination of lunar exploration during that era.

As the final *Apollo* mission, *Apollo 17*, departed the Moon's surface on December 14, 1972, and the lunar module's ascent stage disappeared into the cosmos, it marked the end of an extraordinary chapter in history. The legacy of the *Apollo* missions lives on, inspiring generations to reach for the stars, to explore the unknown, and to dream of what lies beyond the horizon.

Interplanetary missions are ambitious space exploration endeavors designed to travel beyond Earth's orbit and investigate celestial bodies within our solar system. Interplanetary missions target a wide range of celestial bodies, including planets like Mars, Venus, Jupiter, Saturn, and the outer planets, their moons, asteroids, and comets. Each mission is tailored to the specific characteristics and scientific goals of its target. Notable interplanetary missions include the Mars rovers (e.g., *Spirit*, *Opportunity*, *Curiosity*, *Perseverance*), the *Voyager* spacecraft that have entered interstellar space, and the *Cassini-Huygens* mission to Saturn and its moon Titan.

These missions are conducted to closely investigate planets, asteroids, and comets. Notably, no crewed missions have been dispatched to any planet within the Solar System, except for NASA's Apollo program, which successfully delivered twelve astronauts to the Moon and safely brought them back to Earth.

Space probes have achieved orbit around all five planets recognized in ancient times. These missions included Venus (*Venera 7* in 1970), Mars (*Mariner 9* in 1971), Jupiter (*Galileo* in 1995), Saturn (*Cassini/Huygens* in 2004), and most recently Mercury (*MESSENGER* in March 2011). These probes have collected valuable data about these celestial bodies and their natural satellites.

Furthermore, the *New Horizons* probe conducted a flyby of Pluto, and the *Dawn* spacecraft is currently in orbit around the dwarf planet Ceres. The *Voyager 1* and *Voyager 2* spacecraft ventured beyond the boundaries of the Solar System. Additionally, *Pioneer 10*, *Pioneer 11*, and *New Horizons* are en route to eventually exit the Solar System.

2.4.3 Space Missions to Mars

Today, Mars hosts more operational spacecraft than any other planet, second only to Earth. These spacecraft include orbiters, landers, and rovers. An *orbiter* is a spacecraft designed to orbit a celestial body, such as a planet, moon, or asteroid. It revolves around the object without landing or making physical contact with its surface. Orbiters are equipped with instruments and sensors to observe, study, and collect data about the celestial body from a distance. They often provide comprehensive mapping, conduct remote sensing, and analyze the surface features, composition, atmosphere, and magnetic field of the celestial body. A *lander* is a spacecraft that is specifically designed to land safely on the surface of a celestial body, such as a planet or moon. Unlike an orbiter that remains in orbit, a lander descends and makes direct contact with the surface. It's designed to execute a controlled landing, often using propulsion systems, landing gear, and protective measures to ensure a soft touchdown. Landers typically carry scientific instruments and equipment to conduct experiments, collect samples, and analyze the surface characteristics of the celestial body where they land. A *rover* is a mobile robotic vehicle designed to move across the surface of a celestial body after landing. Rovers are often part of lander missions, deployed to explore the terrain, conduct scientific investigations, collect samples, and transmit data back to Earth. These vehicles are equipped with wheels or tracks, navigation systems, cameras, scientific instruments, and sometimes arms or drills for sample collection. Rovers play a vital role in conducting close-up studies, traversing varied landscapes, and investigating regions that are inaccessible from orbit.

One of the notable achievements in the early American space program was the *Mariner 4* mission. It embarked on a journey to Mars, reaching its closest point on July 15, 1965, and capturing the very first images of another planet from space. After a 228-day voyage, *Mariner 4* passed Mars at a distance of 9,846 kilometers. In the realm of Soviet space exploration, the uncrewed *Mars 2* space probe, part of the *Mars* program, was launched on May 19, 1971. Following the success of *Mars 2*, the *Mars 3* missions also achieved success. Both missions involved identical spacecraft, each equipped with an orbiter and an attached lander.

Mars Odyssey, developed by NASA, and contracted out to *Lockheed Martin*, was a robotic spacecraft orbiting the planet Mars. It was a crucial component of NASA's Mars Exploration Program and played a vital role in studying Mars from orbit. *Mars Odyssey* was launched on April 7, 2001 on a *Delta II* rocket from Cape Canaveral Air Force Station in Florida and reached Mars orbit on October 24, 2001. The mission of *Mars Odyssey* involved the utilization of spectrometers and a thermal imager to identify signs of historical or current water and ice on the planet, alongside an examination of its geological features and radiation surroundings. The overarching goal was to harness the data collected by *Odyssey* to address two critical inquiries: whether Mars once harbored life and to assess the potential radiation exposure future Mars-bound astronauts might encounter. Furthermore, *Mars Odyssey* serves as a vital intermediary for transmitting communications between Earth and various Martian missions, including the *Curiosity* rover, as well as previous missions such as the *Mars Exploration Rovers* and the *Phoenix Lander*.

More recently, ISRO performed India's first interplanetary mission, *Mars Orbiter Mission (MOM)* (also known as *Mangalyaan*) that enabled a space probe orbiting Mars since 2014. *Mangalyaan* represents a significant milestone in India's space exploration history. India's maiden venture to Mars began on November 5, 2013, when ISRO initiated its journey. The mission was designed for in-depth study of the Red Planet and to validate crucial technologies essential for future inner solar system exploration. A significant milestone was reached on September 23, 2014, as the Mangalyaan spacecraft successfully entered orbit around Mars. This achievement distinguished ISRO as only the fourth

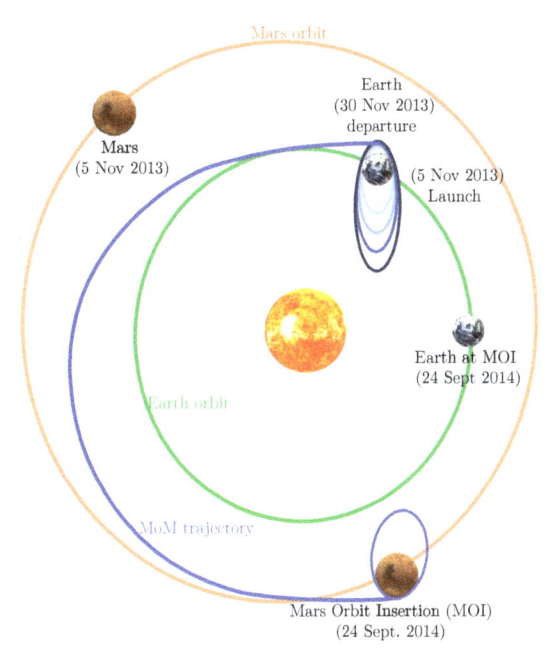

Orbit trajectory diagram of Mars Orbiter Mission (MOM)

space agency globally to accomplish this feat. Before India's success, Mars exploration had been achieved solely by the United States, the Soviet Union, and the European Space Agency (ESA).

2.4.4 Space Missions to Venus

Interplanetary missions to Venus have been conducted by various space agencies and organizations over the years to study and explore Earth's neighboring planet. These missions have included both flybys and orbiters, landers, and even balloon probes. As a part of NASA's *Mariner Program*, *Mariner 2* became the first successful mission to another planet when it flew as close as 34,773 km by *Venus* on December 14, 1962. The spacecraft made several discoveries about the planet and marked another first by measuring the solar wind, a constant stream of charged particles flying outward from the sun. *Mariner 2* indicated that Venus is very hot and has no measurable magnetic fields or radiation belts. Five years later, *Mariner 5* followed up on *Mariner 2*'s study of Venus with a closer look at the planet's atmosphere. The mission not only taught NASA about Venus but also how to operate a spacecraft far from Earth.

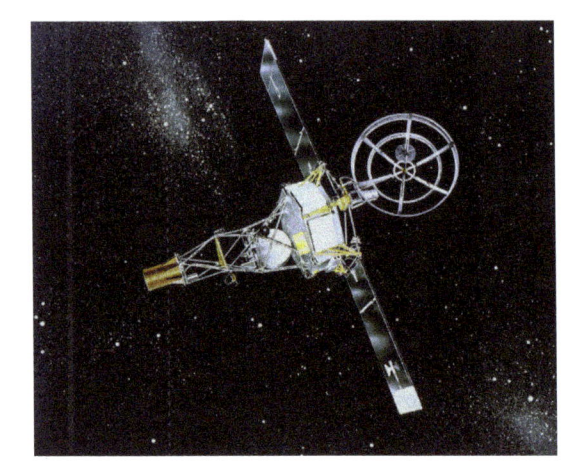

Depiction of *Mariner 2*, the world's first successful interplanetary spacecraft, in space. The *Mariner* probe was comprised of a hexagonal bus with a diameter of 1 m. This bus served as the central structure to which solar panels, instrument booms, and antennas were securely affixed. Among the scientific instruments carried on the *Mariner* spacecraft were the following: a pair of radiometers, each designated for the microwave and infrared segments of the electromagnetic spectrum, a micrometeorite sensor, a sensor for solar plasma, another for charged particles, and a magnetometer.

The *Venera Program*, conducted by the Soviet Union, was one of the most extensive missions to Venus. It included a series of spacecraft sent to Venus from the 1960s to the 1980s. Among these, around ten probes achieved the remarkable feat of safely reaching the planet's surface. However, due to the exceedingly harsh conditions on Venus, the probes had only a brief lifespan on the surface, ranging from as short as 23 minutes to a maximum of two hours. *Venera 3* (1966) was the first human-made device to enter the atmosphere of another planet. *Venera 7* (1970) was the first successful lander on Venus and it transmitted data for 23 minutes before succumbing to the harsh environment. *Venera 9* and *10* (1975) landers sent back the first images of the Venusian surface. *Venera 13* and *14* (1982) missions returned color images and extensive data including sound from the surface, providing valuable insights into Venus' environment.

NASA's *Magellan* spacecraft, also known as the *Venus Radar Mapper*, was the first interplanetary mission to be launched from the Space Shuttle. *Magellan* consisting of a robotic space probe launched on May 4, 1989, went into orbit around Venus in 1990. It spent four years mapping the planet's surface using synthetic aperture radar and provided detailed topographical data and contributed to the understanding of Venus' geology. It used radar to investigate what lay beneath the clouds that covered the planet. The images it sent back showed that the Venus landscape was covered with huge lava flows from hundreds of volcanoes. There were also spidery cracks, called arachnoids, on the planet's surface.

The *Magellan* spacecraft and its Inertial Upper Stage in the payload bay of *Space Shuttle Atlantis* during the *STS-30* mission. The core structure of the spacecraft, originally a spare component from the *Voyager* missions, took the form of a 10-sided aluminum bus. This bus housed critical components such as computers, data recorders, and various subsystems. The spacecraft had dimensions of 6.4 m in height and 4.6 m in diameter weighing around 3.4 tons.

Other notable missions to Venus include *Venus Express* mission (2005) launched by *European Space Agency* (ESA), *Akatsuki (Venus Climate Orbiter)* mission (2010)launched by the *Japan Aerospace Exploration Agency* (JAXA). These missions conducted a comprehensive study of Venus' atmosphere and climate, providing valuable data on the planet's atmospheric dynamics and composition and the planet's super-rotational winds and its cloud cover. The *JAXA* launched its first successful mission to another planet, *Akatsuki* or *Planet-C* or the *Venus Climate Orbiter*, on May 20, 2010, with the mission's main goal to understand the dynamics of Venus' atmosphere. Initially, it did not enter orbit around Venus in December 2010 as planned and instead remained in orbit around the sun in hibernation until 2015, when JAXA engineers were able to fire the thrusters up again and put the spacecraft in its intended orbit around Venus.

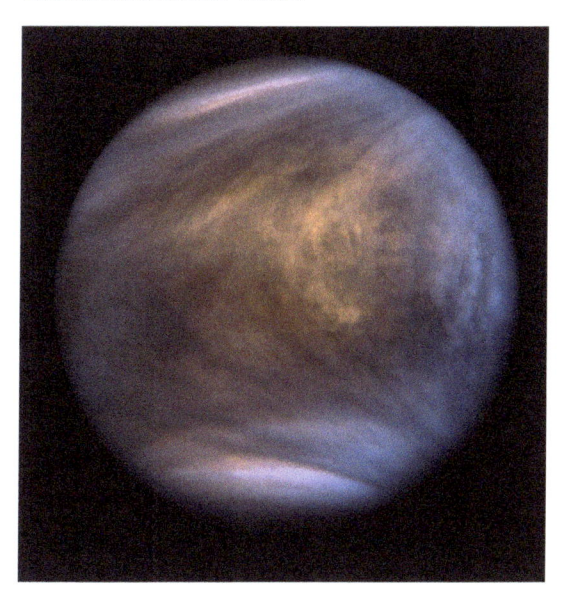

Global view of Venus from *Akatsuki*.

2.4.5 Voyager Mission to Jupiter, Saturn, Uranus and Neptune

NASA's *Voyager Interstellar Mission (VIM)* launched two robotic interstellar probes, *Voyager 1* and *Voyager 2* in 1977 to extend the NASA exploration to the outer limits of the Sun's sphere of influence, and possibly beyond. In August 2012, *Voyager 1* entered into interstellar space, the region between stars, filled with material ejected by the death of nearby stars millions of years ago. Initially, the program was launched to explore two gas giants *Jupiter* and *Saturn*. After making a series of discoveries there, the mission was extended and *Voyager 2* went on to explore Uranus and Neptune, and is still the only spacecraft to have visited those outer planets. The most distant spacecraft, *Voyager 1* and *Voyager 2* have left the Solar System as of December 2018. However, what sets the *Voyager* mission apart and has garnered significant attention is their continued journey into interstellar space, marking humanity's farthest reach into the cosmos. As of 2023, the *Voyagers* are still in operation past the outer boundary of the heliosphere in interstellar space. They collect and transmit useful data back to Earth.

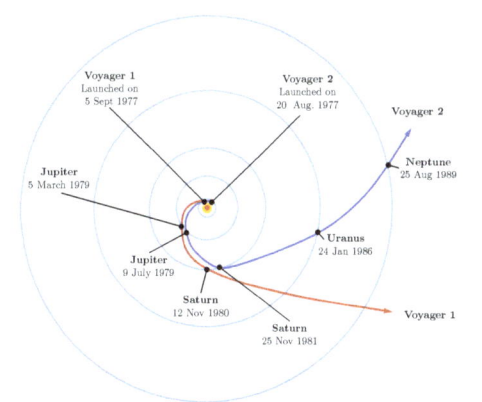

The twin *Voyager 1* and *2* space vehicles are exploring where nothing from Earth has flown before. Continuing on their journey since their 1977 launches, they each are much farther away from Earth and the sun than Pluto. As of 2023, the *Voyagers* are still in operation past the outer boundary of the heliosphere in interstellar space. *Voyager 1* is the farthest human-made object from Earth and the first space vehicle to reach interstellar space, 23.81 billion km away moving with a velocity of 60988 km/h.

2.4.6 Pioneer Mission to Jupiter and Saturn

Pioneer 10 and *11* are *NASA*'s space vehicles launched on March 2, 1972, and April 5, 1973, respectively, to study the asteroid belt, the environment around Jupiter and Saturn, solar winds, and cosmic rays. *Pioneer 10* is credited with being the first spacecraft to traverse the asteroid belt and fly by Jupiter. *Pioneer 11* was the first space vehicle to do a flyby of Saturn. *Pioneer 11*'s mission ended in 1995 and *Pioneer 10* sent its last signal to Earth in 2003.

2.4.7 NASA's Juno Space Probe to Jupiter

The *Juno* spacecraft mission, led by NASA, is a remarkable endeavor designed to unlock the mysteries of Jupiter, the largest planet in our solar system. Launched in 2011, *Juno*'s primary objective is to study Jupiter's composition, gravity field, magnetic field, and polar magnetosphere in unprecedented detail. What sets *Juno* apart is its unique polar orbit, which brings it much closer to Jupiter than previous missions, allowing for a more comprehensive and intimate exploration of the gas giant. Equipped with a suite of scientific instruments, including a microwave radiometer, magnetometers, and gravity science equipment, *Juno* has provided a wealth of data about Jupiter's internal structure, magnetic field dynamics, and the distribution of its key atmospheric constituents, such as water and ammonia. Additionally, *Juno* has been studying Jupiter's powerful auroras, offering insights into the planet's magnetosphere and its interactions with its moon *Io*. The mission's findings have challenged existing models of Jupiter's formation and evolution, shedding light on the planet's deep interior and atmospheric processes. *Juno*'s ongoing mission continues to unveil the secrets of Jupiter, enhancing our understanding of not only this gas giant but also the broader mechanisms at play in the solar system's formation and evolution.

2.4.8 Cassini–Huygens Mission to Saturn

Cassini–Huygens was a space-research mission by NASA, the European Space Agency (ESA), and the Italian Space Agency to send a space probe to study the planet *Saturn* and its system, including its rings and natural satellites. In 2004 the *Cassini* spacecraft reached the ringed planet Saturn after a seven-year journey from Earth. *Cassini* was programmed to release a probe called *Huygens* into the thick atmosphere of Saturn's largest moon, *Titan*, and to land it on the moon's surface. *Cassini* revealed in great detail the true wonders of Saturn, a giant planet governed by raging storms.

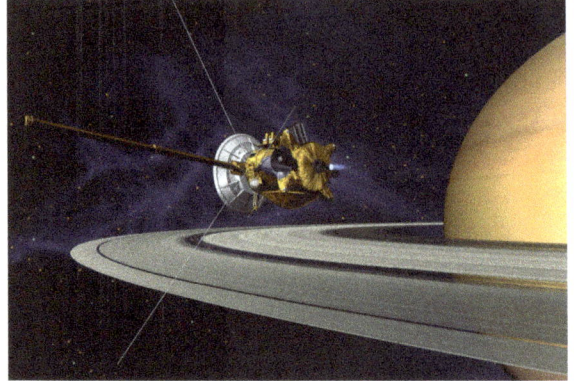

Artist's impression of the *Cassini-Huygens* spacecraft orbiting Saturn and its magnificent rings.

2.4.9 New Horizons Mission to Pluto

Remotely guided space probes have flown by all of the observed planets of the Solar System from Mercury to Neptune. The *New Horizons* spacecraft was launched in 2006 to explore Pluto up close, flying by the dwarf planet and its moons. The spacecraft flew by the solar system's largest planet, Jupiter, for a gravity assist maneuver in 2007 and passed the halfway point to Pluto in 2010. The spacecraft began its approach phase toward Pluto on January 15, 2015, and pictures of Pluto began to reveal distinct features by April 29, 2015, with detail increasing week by week into the approach. The flyby had been fully successful on July 14, 2015. Besides collecting data on Pluto and its moon Charon, *New Horizons* also observed Pluto's other moons, Nix, Hydra, Kerberos, and Styx.

2.4.10 NEAR-Shoemaker Mission to Asteroid

A spacecraft called *NEAR-Shoemaker* was launched in 1996 to study the near-Earth asteroid Eros (a near-Earth asteroid about 22 million km from Earth) from close orbit over a period of a year. Spending about a year in orbit of the Eros, it made an unexpected landing on the asteroid in February 2001. At the end of the mission, the scientists decided to let the craft get closer and closer to the surface, taking pictures as it went. To their surprise, the craft survived a hard landing on the asteroid, and one instrument continued to work for several days afterward.

2.4.11 JAXA's Hayabusa Missions

The *Hayabusa* missions represent a remarkable series of space exploration endeavors that have captured global attention. These missions were designed with the primary objective of collecting samples from the near-Earth asteroids named *Itokawa* and *Ryugu*, providing invaluable insights into the origins and composition of these ancient celestial bodies. *Hayabusa*, the first mission in the series, journeyed to asteroid *Itokawa* and, through a series of complex maneuvers, successfully collected a tiny sample of material from its surface in 2005. This marked a historic achievement as the first-ever retrieval of asteroid samples by a spacecraft. The samples returned to Earth in 2010, enabling scientists to study further about its composition.

Building on the success of *Hayabusa*, JAXA launched *Hayabusa2* in 2014, setting its sights on the asteroid *Ryugu*. This ambitious mission not only collected surface samples but also deployed small landers and an impactor to study *Ryugu*'s geology and composition. In December 2020, *Hayabusa2* returned to Earth with its precious cargo of asteroid samples, further enriching our understanding of the early solar system. The *Hayabusa* missions stand as a testament to Japan's space exploration prowess and its commitment to advancing our knowledge of the cosmos, particularly regarding the fascinating and scientifically significant realm of near-Earth asteroids.

2.5 USA's Experimental Programs

The United States of America has been a pioneer in designing and testing experimental spacecraft that challenge traditional flight principles. These vehicles not only contribute to scientific understanding but also drive progress in aerospace engineering, and our quest to explore the cosmos. In the realm of space exploration, the USA has been at the forefront with innovative spacecraft that have expanded our understanding of the cosmos and opened new frontiers. In this section, we will look at some of the experimental programs pursued by the USA that will push the limits of the performance of aerospace vehicles. The USA's programs are mentioned here because of the ease of availability of information in the open domain. It is not to discredit the efforts of many other countries in the space domain.

Aerojet General X-8 (1947 - 56): The *X-8*, an unguided, spin-stabilized sounding rocket designed to launch a 68 kg payload to 61.0 km altitude, played a key role in the expansion of the United States' influence in space and continues to be a cornerstone for ongoing research in ballistic missiles. Towards the end of World War II, the U.S. Army, in collaboration with the California Institute of Technology's Jet Propulsion Laboratory, had successfully developed a meteorological sounding rocket known as the *WAC Corporal*. Simultaneously, they had acquired a significant cache of components, potentially enough to construct approximately 100 German *V-2 missiles*. Recognizing the immense potential, the U.S. Army expanded its Project *Hermes*, aiming to assemble and launch a series of *V-2* missiles for military, technological, and scientific purposes. However, it became apparent that a substantial portion of the *V-2* components had suffered damage or rendered unusable. Consequently, the initial plan of the Army was to launch a limited quantity of only 20 missiles. This limitation in the number of available *V-2*s prompted a parallel exploration into the design and development of several competing sounding rockets, ensuring that the pursuit of scientific knowledge and technological advancements could continue despite the constraints imposed by the *V-2* supply. The development of experimental planes like X-8 arose as a response to the necessity for a sounding rocket to fill the void left by the diminishing availability of *V-2* rockets.

The *Aerojet General X-8* remains a testament to the spirit of innovation and exploration that defined the aerospace industry during its formative years, leaving a lasting legacy in the realm of aerospace research and technology development. Its significance is evident in various endeavors, including the NACA's *Wallops Island missile launches*, the renowned *X-15 program*, the Navy's *Viking rocket*, and the enduring missile projects undertaken by JPL.

Aerojet Engineering X-8 (1947 - 56), Upper Atmospheric Research Vehicle (UARV) and parachute recovery system

Bell X-9 (Shrike, 1949 - 53): The *Bell X-9 (Shrike)* was developed with the primary purpose of testing air-to-surface missile systems, including their guidance mechanisms. Throughout its operational period, the *X-9* conducted tests that offered valuable insights into various aspects of missile technology. By examining the data collected from the *X-9* experiments, significant knowledge was gained about effective bomber launch platforms and the intricate guidance and control systems required for precision strikes. Additionally, the *X-9* investigations delved into the integration of liquid rocket propulsion with nuclear warheads, a critical aspect of modern missile development. The *X-9*'s flight tests contributed to a better grasp of the fundamental aerodynamics involved in advanced air-to-surface weaponry. This enhanced understanding played a vital role in the advancement of the *GAM-63 Rascal* missile, showcasing the practical application of the *X-9*'s findings in the development of cutting-edge military technology.

Bell X-9 (Shrike), a prototype surface-to-air, liquid-fueled guided missile.

North American Aviation X-10 (1955 - 59): The *North American Aviation X-10* was designed as a dedicated platform for testing various components related to cruise missiles. It played an important role in establishing the technology foundation for remote control systems. Furthermore, the *X-10* marked a significant achievement as the first target drone capable of reaching Mach number 2 speeds. The *X-10*'s flight tests yielded a wealth of valuable data, contributing to the advancement of both aerodynamics and navigation systems. This information proved instrumental in the development of the *Navaho* missile. Beyond its immediate objectives, the *X-10* also served as a crucial testbed that laid the groundwork for future generations of remote-controlled, high-performance aircraft. The knowledge and insights gained from the *X-10*'s operations extended far beyond its initial scope, shaping the trajectory of military aviation technology in subsequent years

Turbojet-powered *X-10* tested flight characteristics and guidance, navigation and control systems.

Convair Astronautics Division X-11 (1956 - 58): *X-11*, a test bed for *Convair SM-65A Atlas* or *Atlas A*, stands as an important milestone in the history of American aerospace engineering and the development of intercontinental ballistic missiles (ICBMs). The *Atlas A* was the United States' first operational ICBM developed during the Cold War. Designed by a team of engineers led by *Karel Bossart* at *Convair*, a division of *General Dynamics*, the *Atlas A* represented a remarkable leap in technology and capability.

Karel Jan Bossart (1904 - 1975), a rocket designer renowned for his role in conceiving the Atlas ICBM.

The *Atlas A*'s successful development and deployment marked a critical turning point in the arms race between the United States and the Soviet Union during the Cold War. It not only bolstered the United States' strategic deterrence capabilities but also paved the way for the development of more advanced versions of the *Atlas* missile family. The legacy of the *Atlas A* extends beyond its military significance, as the technology and expertise gained from its development would later contribute to the success of the *Apollo* program, enabling human spaceflight to the moon. The *Convair SM-65A Atlas A* remains an enduring symbol of American innovation and determination in the realm of missile technology and space exploration.

Convair Astronautics Division X-11 (1956 - 58), The first ICBM prototypes and provided flight data for Atlas missile.

North American Aviation X-15: The *North American X-15*, a rocket-powered research aircraft, bridged the divide between conventional manned flight within Earth's atmosphere and the extraordinary realm of manned spaceflight. Its initial test flights in 1959 marked a historic turning point, as the *X-15* became the first winged aircraft to attain astounding velocities, reaching Mach numbers up to 6. Notably, on October 3, 1967, under the skilled piloting of *William Knight*, the *X-15* reached its pinnacle speed of 7,274 km/h and soared to an altitude of 31.12 km. To endure these high-speed endeavors, the *X-15* was meticulously engineered to withstand aerodynamic temperatures soaring to around 1,200 °F, and it was expertly crafted using a specialized, high-strength nickel alloy known as *Inconel X*.

This groundbreaking aircraft, launched from a modified *Boeing B-52 Stratofortress*, demanded both conventional aerodynamic control surfaces for atmospheric flight and unique thruster reaction control rockets strategically placed in its nose and wings, enabling pilots to maintain control on the cusp of space. Such was the *X-15*'s space-like design that during the early days of *Project Mercury*, the USA's inaugural attempt to send a human into orbit, North American and NASA engineers contemplated employing a modified version of the *X-15* for orbital missions, although this notion was eventually set aside in favor of a blunt-body vehicle. To safeguard pilots operating within the *X-15*'s near-space environment, they donned specially developed full-pressure protection 'spacesuits' during their daring flights. In total, three *X-15* research aircraft were constructed and completed an impressive 199 research missions.

North American Aviation X-15 (1958), rocket-powered experimental aircraft developed to explore problems of space and atmospheric flight at very high speeds and altitudes (Hypersonic high-altitude flight).

Lockheed Missiles X-17: The *X-17*, a three-stage solid-fuel research rocket conducted experiments focused on the high Mach number atmospheric entry environment. This program furnished crucial flight data necessary for refining the aerodynamic profiles of ballistic missile warheads and shaping the design principles for forthcoming space capsules. Notably, on April 24, 1957, an X-17 rocket achieved an impressive speed of 14,000 km/h during a test launch at *Patrick Air Force Base*, and it demonstrated its capabilities by simulating atmospheric entry conditions at velocities reaching up to an astonishing Mach number 14.5 as it hurtled back towards Earth.

Martin Marietta X-24A: The *X-24*, an innova-

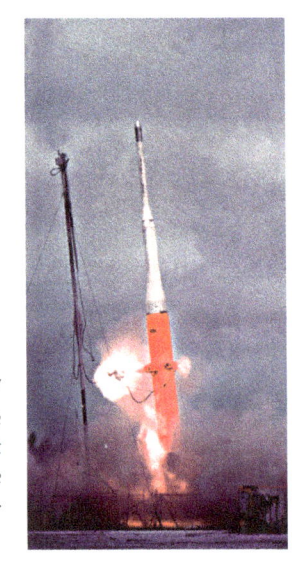

Lockheed Missiles X-17 (1958), a three-stage solid-fuel research rocket developed to explore atmospheric entry characteristics.

Martin Marietta X-24A (1958), research of aerodynamics, flight characteristics of manned vehicle with *FDL-7* configuration.

Martin Marietta X-24B (1958), an evolution of the X-24A, contributed valuable data on lifting body aerodynamics, ultimately influencing the design of future spaceplanes and the *Space Shuttle* program.

tive experimental aircraft, emerged from a collaborative effort between the United States Air Force and NASA, under the banner of the *PILOT* program. This groundbreaking project aimed to explore the theoretical benefits of a lifting body configuration for hypersonic, trans-atmospheric aircraft. The term *lifting body* refers to a design that relies on the shape of the fuselage itself to generate lift, rather than conventional wings.

The *X-24B*'s design originated from a family of entry shapes with improved lift-to-drag ratios, proposed by the Air Force Flight Dynamics Laboratory. The *X-24A* underwent modifications, transforming its shape into a "flying flatiron," featuring a rounded top, flat bottom, and a double delta planform ending in a pointed nose. *John Manke* piloted the *X-24B*, conducting its inaugural glide flight on August 1, 1973, and its first powered mission on November 15, 1973. The X-24B was particularly notable for its contributions to the development of spaceplanes and future space shuttle designs. Its unique shape and flight testing provided valuable data that played a role in the design and operational principles of vehicles like the *Space Shuttle*, which needed to enter Earth's atmosphere safely at orbital speeds. The *X-24* program thus stands as a testament to the successful collaboration between the military and NASA in advancing aerospace tech-

nology and expanding the understanding of hypersonic flight and spaceplane design.

Skunk Works X-33 (1999 - 2001): The *X-33*, an advanced technology demonstrator for a Reusable Launch Vehicle (RLV), was designed as an unpiloted craft to launch vertically like a rocket and land horizontally like an airplane, boasting capabilities to reach altitudes of around 80 km and speeds exceeding Mach number 11. In 1996, NASA awarded *Lockheed Martin* the contract to construct and operate the *X-33* test vehicle, aimed at showcasing advanced technologies for a prospective reusable launch vehicle to replace the *Space Shuttle*. *Lockheed Martin* referred *X-33* as a technology demonstrator for the *VentureStar*, a single-stage-to-orbit reusable launch system. Designed by the *Lockheed Skunk Works*, the vehicle featured a lifting body shape, aerospike rocket engines, and a metallic thermal protection system, representing a single-stage-to-orbit solution. Unfortunately, in 2001, NASA terminated the project due to insurmountable technical challenges, rendering this model a poignant relic of an ambitious yet unsuccessful attempt to validate a groundbreaking spaceplane concept.

Orbital Sciences Corp., X-34 (1996 - 2000): In 1996, NASA enlisted *Orbital Sciences Corp.* to develop and assess the *X-34*, a compact, reusable technology demonstrator for a launch vehicle. This

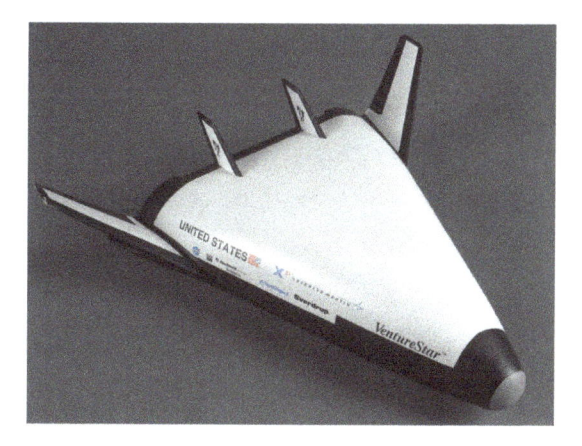

Lockheed Martin X-33 (Advanced Technology Demonstrator test vehicle), a proposed uncrewed, sub-scale technology demonstrator suborbital spaceplane. The X-33 aimed to flight-test a spectrum of technologies considered essential by NASA for single-stage-to-orbit reusable launch vehicles (SSTO RLVs) including metallic thermal protection systems, composite cryogenic liquid hydrogen fuel tanks, the aerospike engine, autonomous uncrewed flight control, expedited flight turnaround procedures achieved through streamlined operations, and advanced lifting body aerodynamics.

Orbital Sciences Corp., X-34 (1996 - 2000), a small, pilotless, reusable technology demonstrator for a launch vehicle.

Boeing X-37 (2010), Orbital Test Vehicle (OTV), a reusable spacecraft boosted into space by a launch vehicle, then enters Earth's atmosphere and lands as a spaceplane.

single-engine rocket was designed for air launch, with an *Orbital Sciences L-1011* aircraft as its launch platform. The *X-34* prototype conducted its maiden captive-carry flight attached to the modified *L-1011* on June 29, 1999, followed by two more successful captive-carry flights on September 3 and September 14 of the same year. The *X-34* was originally planned to achieve Mach number 8 speeds and reach an altitude of 76.2 km. Regrettably, NASA discontinued funding for the program in 2001, marking an early end to this promising venture.

Boeing Phantom Works X-37 (2010): The *X-37*, an orbital research vehicle intended for launch aboard the Space Shuttle or a reusable launch vehicle, designed to return to Earth autonomously. *Boeing Phantom Works* was tasked with its construction through a cooperative agreement signed in July 1999. The *X-37* was anticipated to measure 8.92 m in length, approximately half the size of the *Shuttle* payload bay, with an estimated weight of around 6 tons and a wingspan of 4.55 m. The initial *X-37* test took place in 2006, while its maiden orbital mission was successfully launched in April 2010, utilizing an *Atlas V* rocket. This inaugural mission concluded in December 2010 upon the spacecraft's return to Earth. Subsequent flights gradually extended mission durations, culminating in the fifth mission's impressive achievement of spending 780 days in orbit. Notably, this mission marked the first launch using a *Falcon 9* rocket.

The sixth mission in the series embarked on its journey aboard an *Atlas V* rocket on May 17, 2020. This mission wrapped up on November 12, 2022, setting a record of 908 days spent in orbit. The seventh mission was launched on December 28, 2023, utilizing a *Falcon Heavy* rocket. **NASA Boeing X-40A (1998)**: The *X-40A*, an 80% scale model of the X-

37 autonomous spaceplane technology demonstrator, did not include propulsion or thermal protection systems. This vehicle was carried into the sky by an Army *CH-47D Chinook* helicopter, reaching an altitude of 4.5 km, and then released to execute a fully autonomous 75-second descent, culminating in a successful landing on the primary runway at Edwards Air Force Base. Boeing originally constructed the *X-40A* for the Air Force as a component of the *Space Maneuver Vehicle* program.

Boeing X-40 Space Maneuver Vehicle (1998).

Generation Orbit Launch Services, X-60 (2014): The *X-60* was a single-stage, air-launched liquid rocket vehicle with the goal of conducting hypersonic flight research through suppressed trajectories. This technology enables the rocket to follow unique paths, enhancing the flexibility of experimental payloads in high-speed, high-pressure test conditions. On July 20, 2014, the tubular rocket, named *GO1* (*GOLauncher 1*), took to the skies beneath the wings of the *Learjet 35* aircraft for its inaugural flight test. The *X-60A* has wide-ranging applications, including facilitating access to elevated altitudes for various purposes such as microgravity experiments, astrophysics research, hypersonics testing, and advancements in avionics studies.

Generation Orbit Launch Services, X-60 ((GOLauncher 1 or GO1), 2014), an air-launched rocket for hypersonic flight research vehicle.

2.6 Current and Future Space Explorations

There are various space missions planned by different countries and space agencies with varying objectives. These missions are planned to take a payload to outer space or to any of the celestial bodies. These celestial bodies include the planets, moons of these planets, Sun and asteroids within the solar system. Some of these missions aim to go even beyond the solar system. The missions to the celestial bodies aim for fly-by or a landing on these celestial bodies. For a soft landing on the celestial body, a decelerating mechanism will be required to ensure that the payload is functional after landing. To explore a larger segment of the celestial body, sometimes a rover is employed. The payload carried by these missions will differ depending on the objectives of the mission. The payload can include scientific instruments, rover, and astronauts. Few of these missions aim to return sample materials from these celestial bodies. Returning to the Earth poses additional technological challenges and propulsion requirements. We will look at some of these exciting missions in more detail in this section.

2.6.1 Lunar Missions

Here, we look at the recent *Chanrayaan-3* mission to Moon by ISRO in detail.

Chandrayaan-3

On August 23, 2023, at approximately 6 pm Indian time, India achieved a momentous milestone in its space exploration narrative, marking a historic moment for the country. This remarkable feat of engineering prowess propelled India into an exclusive league, becoming only the fourth nation, following the Soviet Union, the United States, and China, to master a precisely controlled lunar landing. Notably, India distinguished itself by achieving this feat at the lunar south polar region, a remarkable touchdown occurring approximately 600 kilo-

2.6. CURRENT AND FUTURE SPACE EXPLORATIONS

Chandrayaan-3 was launched aboard an *LVM3-M4* rocket on July 14, 2023, from *Satish Dhawan Space Centre*, Sriharikota, entering an Earth parking orbit with a perigee of 170 km and an apogee of 36,500 km

meters from the pole.

Chandrayaan-3 embarked on its celestial journey on *GSLV Mark 3 (LVM 3)* heavy lift launch vehicle from the *Satish Dhawan Space Centre*, Sriharikota on 14th July, 2023. The spacecraft transitioned into lunar orbit by the 5th August and touched down near the lunar south pole on the 23rd August. The lander executed a precise hop and repositioning maneuver on the 3rd of September, subtly shifting itself by 30 to 40 centimeters from its original landing site.

The mission consisted of two key components: the autonomous propulsion module and the integrated module consisting of a specialized lander and advanced rover. The propulsion module transported the integrated lander and rover module from the injection orbit to a lunar orbit situated approximately 100 kilometers above the lunar surface where the lander module separates. In addition to this primary task, it accommodated a Spectropolarimetry of Habitable Planetary Earth (SHAPE) payload, which was designed for the examination of Earth's spectral and polarimetric characteristics from its lunar orbit vantage point. The propulsion module adopts a box-like configuration with one side of this structure featuring a solar panel, while the top showcased a cylinder referred as the Intermodule Adapter Cone, serving as the primary mounting platform for the lander. Positioned on

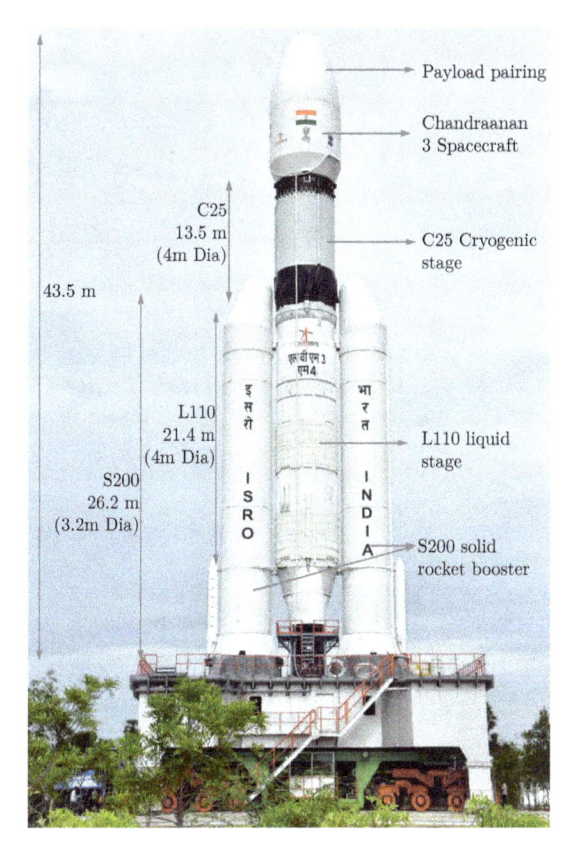

Launch Vehicle Mark-3 (LVM3)-M4, India's largest and heaviest launch vehicle.

Height	43.5 m
Lift-off Mass	642 tons
Propulsion	Stages
Strap-on Motors	2 x S200 – solid propellant (HTPB), 204.5 tons each
Core Stage	L110 – liquid propellant(UH25 + N_2O_4), 115.8 tons
Upper Stage	C25 – cryogenic, LH_2 & LOX, 28.6 tons

Event	Flight time (s)	Altitude (km)	Inertial velocity (km/s)
2xS200 Ignition	0.00	0.024	0.45
L110 Ignition	108.10	44.66	1.78
2xS200 Separation	127.00	2.171	61.96
Payload Pairing (PLF) Separation	194.96	114.80	2.56
L110 Separation	305.56	175.35	4.62
C25 Ignition	307.96	176.57	4.62
C25 Shut-off	954.42	174.69	10.24
payload Separation	969.42	179.19	10.26

LVM3-M4/Chandrayaan-3 mission flight sequence.

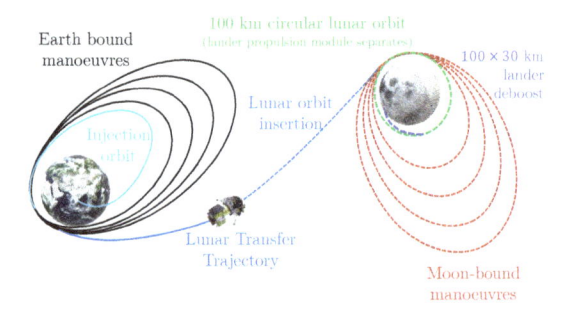

Chandrayaan-3 – Mission Profile. On July 14, 2023, at 2:35 pm IST, *Chandrayaan 3* was launched on a GSLV Mark 3 launch vehicle from the *Satish Dhawan Space Center* in Sriharikota, India and placed into an elliptical Earth parking orbit of $170 \times 36,500$ km. Subsequently, a sequence of maneuvers spanning roughly 40 days was executed to propel the spacecraft toward its lunar destination.

On August 5, the spacecraft was placed into a $164 \times 18,074$ km lunar orbit by a 30-minute engine firing. Subsequent thrusts by the propulsion module meticulously maneuvered the lander/rover into a circular polar lunar orbit of 100 km above the lunar surface by August 17.

The *Vikram* lander autonomously detached and initiated its powered descent towards the Moon's surface on August 23 and landed in the lunar south polar region.

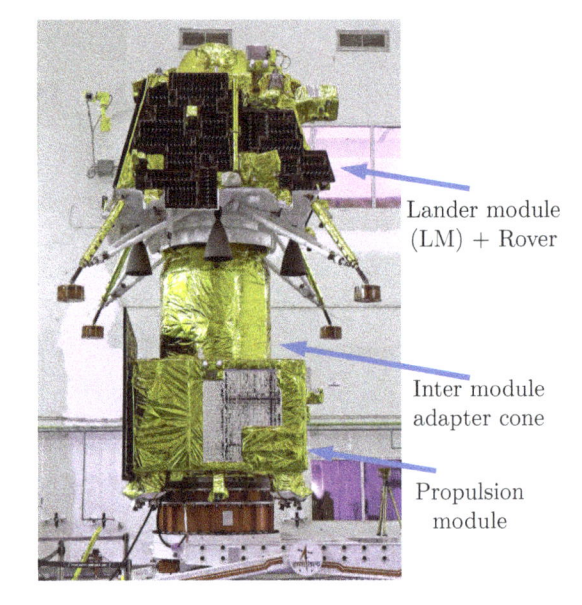

Chandrayaan-3 – Integrated module consists of lander and rover configuration connected with propulsion module using inter-module adapter cone. The propulsion module has a mass of 2145 kg, while the lander module, which includes the rover weighing 26 kg, totals 1752 kg.

the module's underside was the primary thruster nozzle. In terms of mass, it weighed 2145.01 kg, with a significant portion, amounting to 1696.39 kg, designated for the bi-propellant propulsion system, MMH (mono-methyl hydrazine) + MON3 (Mixed oxides of nitrogen). Communications were via S-Band and attitude sensors include a star sensor, Sun sensor, and inertial reference unit and accelerometer package. The Chandrayaan-3's integrated module is composed of a lunar lander bearing the name *Vikram* and a lunar rover named as *Pragyan*. The *Vikram* lander (named after Indian space program pioneer Vikram Sarabhai) is responsible for the soft landing on the Moon's surface. Resembling a box-shaped structure, it featured four landing legs and four potent landing thrusters, each capable of generating an impressive 800 N of thrust. This versatile lander served as the rover's transport and was equipped with an array of scientific instruments to facilitate on-site analysis. *Chandrayaan-3*'s lander boasted four variable-thrust engines with the capability to adjust thrust levels dynamically throughout its descent. This design refinement was instrumental in addressing one of the key factors contributing to *Chandrayaan-2*'s landing setback, namely attitude variation during the camera coasting phase. To enhance precision, the *Chandrayaan-3* lander was equipped with a Laser Doppler Velocimeter (LDV) to measure attitude in three dimensions. Strengthening impact legs, enhancing instrumentation redundancy, and refining landing region targeting based on high-resolution imagery from the Orbiter High-Resolution Camera (OHRC) onboard *Chandrayaan-2*'s orbiter were among the strategic improvements implemented by ISRO. Additionally, structural robustness was enhanced, instrument polling frequency and data transmission were increased, and supplementary contingency systems were integrated to bolster the lander's resilience in the face of potential challenges during descent and landing.

The lander carry three important instruments

- *Radio Anatomy of Moon Bound Hypersensitive ionosphere and Atmosphere (RAMBHA) Langmuir Probe* to measure the near-surface plasma (ions and electrons) density and its changes with time

- *Chandra's Surface Thermo-physical Experiment (ChaSTE)* to carry out the measurements of thermal properties of the lunar surface near-polar region

- Instrument for Lunar Seismic Activity (ILSA) to measure seismicity around the landing site and delineating the structure of the lunar crust and mantle.

The *Pragyan* rover is a six-wheeled vehicle having a rectangular chassis weighing 26 kg and measuring $917 \times 750 \times 397 mm^3$ in dimensions. This rover is equipped to conduct a multitude of measurements to facilitate in-depth research on various aspects, including the composition of the lunar surface, the potential presence of water ice within lunar soil, the historical record of lunar impacts, and the transformation of the Moon's atmosphere over time. The *Pragyan* rover carried two instruments to study the local surface elemental composition

- Alpha Particle X-Ray Spectrometer (APXS) to derive the chemical composition and infer the mineralogical composition.

- Laser Induced Breakdown Spectroscope (LIBS) to determine the elemental composition of lunar soil and rocks.

Other Moon Missions

The *Artemis program*, led by *NASA* in collaboration with six major partner agencies, including the *European Space Agency (ESA)*, the *German Aerospace Center (DLR)*, the *Japan Aerospace Exploration Agency (JAXA)*, the *Canadian Space Agency (CSA)*, the *Israel Space Agency (ISA)*, and the *Italian Space Agency (ASI)*, aims to reestablish human presence on the Moon after nearly five decades since the *Apollo*

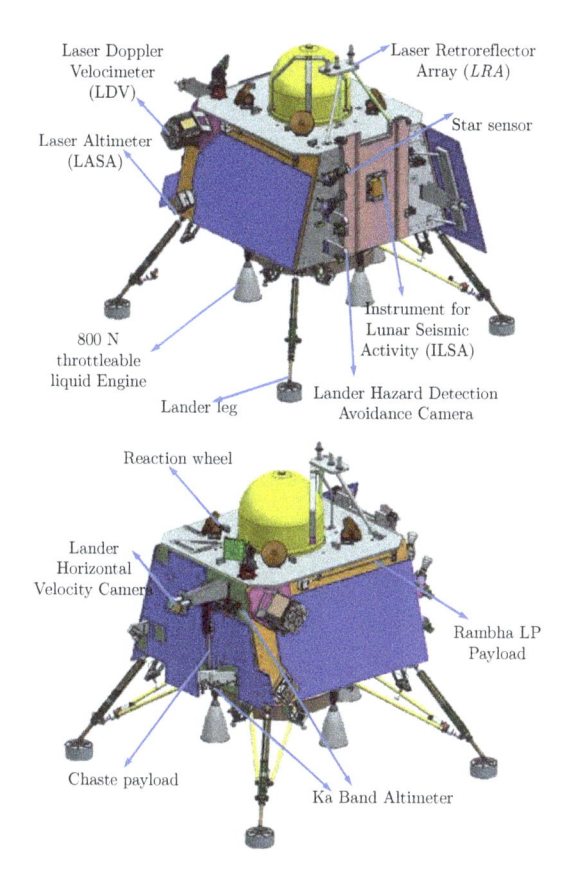

The Chandrayana mission's lander is called *Vikram*. The *Pragyan* rover is accommodated inside the Vikram lander. Major specifications of the lander are

Dimensions	$2000 \times 2000 \times 1166$ mm^3
Mass	1749.86 kg including Rover
Mission life	14 Earth days
Power	738 W (Winter solstice)
Communication	ISDN, Ch-2 Orbiter, Rover
Payloads	RAMBHA-LP Langmuir Probe, ChaSTE and ILSA

mission. Key components of the program include the *Space Launch System (SLS)*, the *Orion* spacecraft, the *Lunar Gateway* space station, and commercial *Human Landing Systems*. The ultimate objective of *Artemis* is to establish a permanent lunar base, paving the way for future human missions to Mars.

Formally established in 2017, the *Artemis program* incorporates components of *Orion* spacecraft, developed during the earlier *Constellation Program* (2005 – 2010) and subsequent efforts following its cancellation. *Orion*'s first launch, and the first use of the *Space Launch System*, was originally set in

(July 2022) and *Lunar Reconnaissance Orbiter* (2009), China's *Chang'e-5* returned lunar samples to Earth in 2020 and *Chang'e-4* performed the first landing on the Moon's far side in 2018. *Chandrayaan-1* was the first Indian lunar probe under the Chandrayaan program. It was launched by the Indian Space Research Organisation (ISRO) in October 2008, and operated until August 2009.

NASA is planning to launch the *Lunar Trailblazer orbiter* to map the Moon's water, returning humans to the Moon later scheduled to launch in 2024 through *Artemis* mission and launching the *Volatiles Investigating Polar Exploration Rover* (*VIPER*) in 2024 to map water on the Moon's south pole. After *Beresheet1*, a private mission to the Moon launched in 2019, Israel is scheduled to launch *Beresheet 2* in 2025 carrying the mothership (orbiter) and two landers that will be released for landing at different locations on the Moon.

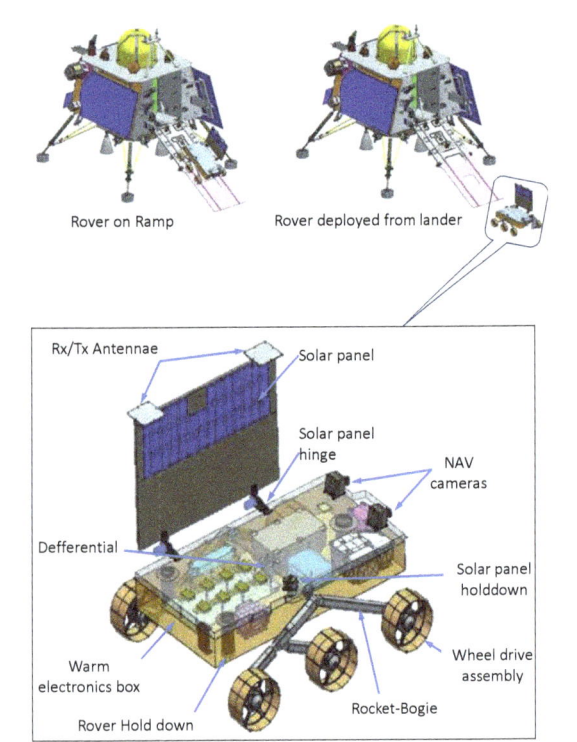

Chandrayaan-3 rover called *Pragyan*. Major specifications of the rover are

Dimensions	$917 \times 750 \times 397$ mm^3
Mass	26 kg including Rover
Mission life	14 Earth days
Power	50 W
Communication	Lander
Payloads	APXS and LIBS

2016, but was rescheduled and launched in November 2022 as the *Artemis 1* uncrewed Moon-orbiting mission, with robots and mannequins aboard. As per the plan, the crewed *Artemis 2* launch is scheduled for 2024, followed by the *Artemis 3* crewed lunar landing in 2025, *Artemis 4*'s docking with the *Lunar Gateway* in 2028, and annual lunar landings in the years that follow. However, some observers express concerns about potential cost overruns and delays in the program's timeline.

South Korea joined the club by launching *Danuri*, the *Korean Pathfinder Lunar Orbiter*, in 2022 to study the Moon's surface. With the success of *Danuri*, South Korea aims for a robotic landing mission on Moon and more. Other current space missions to Moon are *NASA*'s *CAPSTONE* spacecraft

2.6.2 Planetary Explorations

BepiColombo mission to Mercury

A joint mission of the *ESA* and the *JAXA* to Mercury was launched in October 2018. It is expected to arrive near Mercury in late 2025 the smallest and least explored terrestrial planet in our Solar System. The mission — comprising the Mercury Planetary Orbiter (MPO) spacecraft and the Mercury Magnetospheric Orbiter (MIO) spacecraft — will endure temperatures over 350 °C and gather data during its one-year nominal mission, with a possible one-year extension. *ESA*'s *MPO* will study the planet's surface and interior and *JAXA*'s *MIO* will study the planet's magnetic field. Although primarily a mission to Mercury, it had a Venus flyby in 2020 on its way to Mercury. During the flyby, it collected data on Venus to assist in future missions. Only two other spacecraft that have visited Mercury are *NASA*'s *Mariner 10* (1973 - 1975) and *MESSENGER* (2004 - 2015).

The Venus Climate Orbiter

There are a few upcoming space missions to Venus viz., *NASA*'s *DaVinci* (to provide detailed measurements of composition and structure of the Venusian atmosphere and expected to launch after 2029), *Veritas* (to map out Venus' topography and expected to launch after 2027), *ESA*'s *EnVision* (to investigate the atmosphere and overall structure of Venus and will launch after 2031), and *ISRO*'s Venus orbiter *Shukrayaan*.

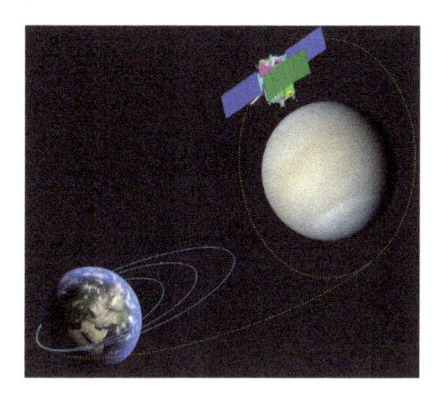

ISRO's Venus orbiter *Shukrayaan*

Mars Missions

Currently, there are many Mars missions from various space research agencies. NASA's *Perseverance* rover landed in February, 2021 and is searching for past life and collecting samples for return to Earth, *MAVEN* orbiter studies what happened to Mars' atmosphere, and the *Curiosity* rover explores an ancient lake bed that once had conditions that could have supported life. *Mars Reconnaissance Orbiter* studies the planet with a high-powered camera and relays communications between the surface and Earth and long-lived *Odyssey* monitors surface changes. ESA's *ExoMars* Trace Gas Orbiter searches for atmospheric gases linked to life and *Mars Express* surveys the planet and searches for subsurface water. China's *Tianwen-1* is an orbiter and rover mission that arrived in February 2021 to study the planet. The UAE's *Hope orbiter* arrived in February 2021 and will build a complete picture of the Martian atmosphere.

Japan is planning to launch *Martian Moons eXploration mission* in 2024 to collect samples of Phobos for return to Earth. ESA is expected to launch *ExoMars* rover after 2028 to find signs of life on Mars. The NASA-funded *ESCAPADE* twin orbiters are to be launched in 2024 to explore how the solar wind strips away Mars' atmosphere. *NASA* and *ESA* are planning a series of missions, *Mars Sample Return*, to return samples from Mars to Earth in the early 2030s.

Space Missions to Jupiter

Apart from *NASA*'s *Juno* which was launched in 2011 to learn more about Jupiter's origins and how the planet has changed, *ESA* launched the *Juice* spacecraft in April 2023 to study Jupiter and its three largest icy moons — Europa, Ganymede, and Callisto. NASA is scheduled to launch *Europa Clipper* in 2024 to determine whether Jupiter's moon Europa could support life.

Space Missions to Saturn, Uranus and Neptune

NASA is expected to launch *Dragonfly* mission in 2027 to explore Saturn's moon Titan. The spacecraft is an eight-bladed drone-like craft called a quadcopter that will make short flights around the surface. NASA's *Voyager 2* is the only spacecraft to have visited Uranus and Neptune. It flew past Uranus and Neptune in 1986 and 1989, respectively, and is now exploring interstellar space.

2.6.3 Missions to Sun

Throughout the decades, a variety of instruments have been deployed to delve deeper into the mysteries of the Sun, probing its outer layers and investigating its impact on the surrounding space. These include optical telescopes, capable of peering into the Sun's photosphere, which is its lowest atmospheric layer.

In an effort to explore the extended solar atmosphere and monitor solar activity, NASA initiated the *Pioneer 6* through 9 spacecraft missions between

1965 and 1969. These spacecraft formed a network of stations strategically positioned along Earth's orbit to provide early warnings about approaching solar storms. Additionally, in 1973, NASA established a human-operated solar observatory on its inaugural space station, *Skylab*. Japan's *Yokoh* satellite, launched in 1991, dedicated itself to a comprehensive study of the Sun during an entire solar cycle, focusing on X-ray observations, where solar activity is more pronounced compared to visible light.

In a collaborative effort, the *European Space Agency* (*ESA*) and *NASA* executed the *Ulysses* mission, which utilized a gravity-assist maneuver from Jupiter to shift its solar orbit and gather pioneering data about the Sun's polar regions. After years of traversing the Solar System, *NASA*'s twin *Voyager* spacecraft successfully crossed the Sun's magnetic field boundary. Further advancing solar observation, the *ESA-NASA* Solar and Heliospheric Observatory (SOHO) mission has been continuously monitoring the Sun since 1995, utilizing a coronagraph to obscure sunlight and provide clearer views of the solar corona while also tracking solar eruptions. Notably, SOHO has the distinction of discovering over 4,000 comets during its mission.

Since 2006, *NASA*'s STEREO-A (Solar TErrestrial RElations Observatory) spacecraft has been instrumental in enhancing the understanding of solar phenomena known as Coronal Mass Ejections (CMEs). These explosive events, originating from the Sun, can expel over 10 billion tons of the Sun's magnetized atmosphere into the vast expanse of interplanetary space, reaching staggering speeds exceeding 5 million kilometers per hour. When these powerful CMEs collide with Earth's magnetosphere, they have the potential to generate severe geomagnetic storms and other forms of space weather, which can disrupt and even incapacitate satellite operations, communication systems, and terrestrial power grids, posing a significant risk to both technology and astronaut safety.

NASA's Solar Dynamics Observatory (SDO) mission launched in February 2010, the first mission launched Living With a Star (LWS) Program, is designed to understand the causes of solar variability and its impacts on Earth.

NASA launched the historic *Parker Solar Probe* in 2018 by *Delta IV-Heavy* is arguably the most dangerous robotic mission ever to study the Sun's atmosphere by being in it. In 2025, the *Parker* probe will be just 6 million km (151.31 million km is the average distance between Earth and Sun) above the Sun's surface—nine times closer than scorching Mercury. *Parker* probe aims to study the corona from within and intends to solve one of the biggest fundamental mysteries, why the Sun's corona is much hotter than its surface.

The *ESA*-led *Solar Orbiter* mission — the most complex scientific laboratory ever to have been sent to the Sun launched in February 2020 — is imaging the Sun and huge swaths of its corona up close for the first time. Solar Orbiter will take images of the Sun from closer than any spacecraft before and for the first time look at its uncharted polar regions.

Aditya-L1, a coronagraph spacecraft designed and developed by *ISRO* in collaboration with various Indian research institutes, is poised to embark on a groundbreaking mission. Positioned approximately 1.5 million kilometers from Earth, it will gracefully orbit in a halo orbit encircling the L1 Lagrange point, strategically positioned between Earth and the Sun. This pioneering spacecraft is dedicated to an intricate study of the solar atmosphere, diligently scrutinizing the solar magnetic storms and their profound effects on the Earth's surrounding environment.

On September 2, 2023, *Aditya-L1* embarked on its celestial journey, courtesy of the *PSLV*, an expendable medium-lift launch vehicle. In the upcoming phase of its mission, *Aditya-L1* is poised to gradually approach its intended orbit at the L1 point, a remarkable feat expected to transpire approximately 127 days following its launch.

ESA's *Proba-3*, short for "Project for Onboard Autonomy 3", is an innovative space mission that aims to demonstrate precise formation flying of multiple spacecraft in close proximity. Comprising two separate spacecraft, one acting as a corona-graph and the other as a star-tracker, *Proba-3* is designed to maintain a controlled separation distance between them with extreme accuracy. The corona-graph will obscure the Sun, allowing for the study of the solar corona, while the star-tracker will precisely measure the relative positions of stars and the corona-graph's edge. This mission is a pioneering endeavor in the realm of autonomous formation flying and holds immense potential for advancing our understanding of the Sun's outer atmosphere and for the development of future space-based observatories.

Bibliography

R. G. Grant (2010) *Flight: the complete history of aviation*, Dorling Kindersley, London.

J. D. Anderson Jr. (1989) *Hypersonic and high temperature gas dynamics*, McGraw-Hill, New York.

D. Edberg and W. Costa (2022) *Design of Rockets and Space Launch Vehicles*, AIAA, Incorporated.

M. H. Gorn and G. De Chiara (2021) *X-Planes from the X-1 to the X-60*, Springer-Praxis books in space exploration, Gewerbestrasse, Switzerland

Part II

Fundamentals of Flight

Chapter 3

Flight Environment

Contents

3.1	**Introduction**		**174**
3.2	**Gravitational Force**		**175**
3.3	**Solar System**		**175**
	3.3.1	Sun	176
	3.3.2	Planets and Natural Satellites	176
	3.3.3	Dwarf Planets, Comets, Asteroids and Meteors	178
3.4	**Earth's Atmosphere**		**179**
	3.4.1	Altitude, Longitude and Latitude	179
	3.4.2	Pressure, Density and Temperature Variation	180
	3.4.3	Real Atmosphere	182
	3.4.4	Aviation Meteorology	183

3.1 Introduction

Imagine that you plan to go for a picnic. Once you have decided where you want to go, you will have to decide on the route you will take towards the picnic spot. The location and the route will be the deciding factors for the type of vehicle you will choose for the travel. The clothing, food and other accessories that you will take along with you depends on where you are going and what you plan to do over there. The amusement during the trip is affected by the local weather. The weather conditions may play a spoilsport and can even completely ruin your plan. All the other arrangements for the picnic can be in vain if the weather conditions are not suitable. In a similar fashion, it is important to know about the environment in which the flight vehicles operate in order to have an efficient and safe flight. The design of these vehicles are determined by the characteristics of the environment in which the vehicle is meant to be used and the operational requirements of the vehicle. The designers' choice of propulsion system for a flight vehicle and the materials that can be used for manufacturing the vehicle are strongly affected by the ambient conditions in which it operates. The forces acting on flight vehicles while in flight or even when grounded are also functions of its environment. So, the ambient environment has a paramount effect on the performance characteristics of flight vehicles. Extreme weather conditions can even affect the safe operation of aerospace vehicles. The environment in which these vehicles operate plays a crucial role in the life expectancy of these vehicles. This chapter gives a broad overview of the pertinent aspects of the environment which affect the design and flight of aerospace vehicles.

Though most of the flight vehicles operate within the Earth's atmosphere, space vehicles operate outside the Earth's environment and some like the *Voyager 2* and *Pioneer 10* are moving towards the outer bounds of the Solar System. It is the gravitational force exerted on these vehicles by the celestial bodies which determines the trajectory of these space vehicles. This chapter gives a basic understanding of this fundamental force that dictates the motion of the planets, satellites, and the space vehicles. For a successful space mission to a celestial body, an understanding of the gravitational force exerted by the body on the space vehicle and the environmental conditions of the celestial body is of utmost importance. In this context, the nature of some of the prominent members of the Solar System are glanced at.

Most of the flight vehicles operate within the Earth's atmosphere. The forces and moments experienced by these vehicles during flight are functions of the density and pressure of atmospheric air. For high speed vehicles like the *Space Shuttle*, the severe heating experienced is determined by its shape, speed, trajectory and the environmental conditions. The external conditions not only affect the safety and comfort experienced by the payload and the flight vehicles, it is the deciding factor for the arrangements to be made for safely carrying the payload. The thermodynamic properties, viz., pressure, temperature, and density, of the air do not remain constant within the Earth's atmosphere. They vary with altitude, longitude and latitude. At a particular location, there can be variations with respect to time. However, it is important to have a standard model for the atmosphere while designing flight vehicles and calculating performance parameters of these vehicles. Standard Atmosphere tabulates the thermodynamic variables with altitude. These tables may also contain variations of sound speed, coefficient of viscosity, mean free path and atmospheric composition with altitude. In a Standard Atmosphere, the temporal, longitudinal and latitudinal variations of these variables are averaged over to give a mean profile of these quantities with height. Tables for Standard Atmosphere are published by *International Civil Aviation Organization (ICAO)*, *International Organization for Standardization (ISO)* and various Government Agencies. Weather balloons and sounding rockets are typically used for collecting the relevant data required to prepare these tables. It is common to assume that the temperature varies in a piece-wise linear manner with altitude. This combined with the assumption of perfect gas equation and the hydrostatic equation, which relates pressure variation with altitude as $dp = -\rho g dh$ (ρ is density of air and g is acceleration due to gravity), is used to get variation of pressure and density with altitude. There can be differences across these tables based on the locations chosen for collecting the data. A broad understanding of the Standard Atmosphere is also presented in this chapter. Some of the important weather phenomena which affects aviation are also briefly discussed.

3.2 Gravitational Force

Before the notion of gravitational force was established, it was common to come across stories like *Atlas* holding the celestial bodies on his shoulders. The Law of Universal Gravitation by *Isaac Newton* (1642 - 1727) helped to explain how the celestial bodies are able to move in their orbits.

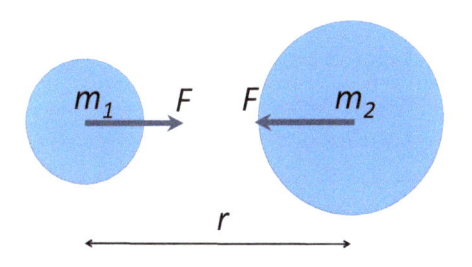

Figure 3.1: Gravitational force.

According to the law, every particle attracts every other particle in the universe with a force directly proportional to the product of the masses of the particles and inversely proportional to the square of the distance between them. i.e.,

$$F \propto \frac{m_1 m_2}{r^2}$$

$$F = \frac{G m_1 m_2}{r^2}$$

where, G is called the universal gravitational constant. Later, the *Cavendish* experiment was used to determine the value of G. With subsequent improvements in the estimation, the value of G is taken as $\sim 6.67 \times 10^{-11} Nm^2/kg^2$. The gravitational force is a weak force and for small masses the magnitude of the force is rather minuscule. For example, the gravitational force of attraction felt by two bodies of 1 kg each separated by a distance of 1 m is $6.67 \times 10^{-11} N$, which is hardly perceptible! But, when one of the masses is large like the Sun or the planets, the magnitude of the force is significant. This is perceived as the acceleration due to gravity near those bodies. For example, when one of the bodies is the Earth and the other body (m) is on the surface of the Earth, the force is $F = \frac{GM_E m}{R_E^2}$, where M_E is the mass of the Earth and R_E is the radius of the Earth. By applying the Newton's second law of motion, the acceleration experienced on the surface of the Earth is $\frac{GM_E}{R_E^2}$, which is $\sim 9.8\,m/s^2$. When a planet is moving around the Sun or a satellite is moving around a planet, it is the gravitational pull of the parent body which acts as the centripetal force to keep the body in the orbit.

3.3 Solar System

The Solar System refers to the assembly of Sun and the celestial bodies within the region of the universe which is bound by the gravitational pull of the prominent star in it, the Sun. The Sun is the biggest and the most massive object in the Solar System. The other bodies in the Solar System include the planets, natural satellites of the planets, asteroids, comets and meteors which are in continuous motion subjected to the gravitational force exerted on them by the other members of the system. The Oort Cloud, a spherical shell consisting of many icy debris orbiting at around 1.6 light years away from the Sun, is typically considered as the outer boundary of the Solar System.

3.3.1 Sun

Sun is the closest star to Earth and it accounts for more than 99 % of the mass of the Solar System. It is the largest object in the Solar System and its diameter is nearly 109 times that of the Earth. Sun is so massive that the acceleration due to gravitation on the Sun's surface is almost 28 times that on the Earth's surface. The average distance of the Earth from the Sun is about 150 million (15 crore) kilometers, typically referred to as one Astronomical Unit. Sun contains significant quantity of hydrogen which undergoes nuclear fusion to generate energy which it radiates as heat and light. The photosphere is the visible surface of the Sun from which is emitted most of the light. The Sun's average surface temperature is estimated to be around 5,700 K. To get a relative understanding how high the temperature is, tungsten is the metal with the highest melting point and it melts around 3,700 K. Surprisingly, the temperature in the outermost layer of Sun's atmosphere called the Corona reaches millions of Kelvin! Due to this high temperature, the Sun radiates heat and light which plays a key role in the atmospheric conditions experienced in the Solar System. It is also a major source of energy for the space vehicles. As expected, the intensity of the solar heat flux decreases with distance from the Sun. The solar heat flux is expressed in terms of solar irradiance which is the power per unit area received from the Sun in the form of electromagnetic radiation. It's average value outside the Earth's atmosphere is around 1360 W/m^2 which drops to nearly 1000 W/m^2 at sea level conditions due to atmospheric attenuation. The Sun also ejects a stream of charged particles which travels at hundreds of kilometers per second which is called the solar wind and it's interaction with the magnetic fields of the planets plays a key role in the space weather. This can affect the satellites and space vehicles which are exposed to it. Phenomena like coronal mass ejection from the Sun can also significantly alter the space weather temporarily. These can be detrimental to satellites and astronauts in space.

3.3.2 Planets and Natural Satellites

The Solar System consists of eight planets, viz., Mercury, Venus, Earth, Mars, Jupiter, Saturn, Uranus and Neptune. Previously, Pluto was also considered as a planet. In 2006, its status was downgraded to dwarf planet. Planets are approximately spherical in shape and are typically classified as terrestrial planets and gas giants. Mercury, Venus, Earth and Mars are considered terrestrial planets, whereas the other four planets are considered as gas giants. The four innermost planets in the Solar System are primarily made of rocks. The gas giants are composed of frozen hydrogen and helium. The table 3.1[1] gives a relative comparison of the size and acceleration due to gravity of these planets.

These planets move around the Sun in elliptical orbits with the Sun at one of the foci. The closest and furthest points from the Sun are called perihelion and aphelion, respectively. The plane in which the Earth orbits around the Sun is called the ecliptic plane. The orbital planes of the other planets are inclined at small angles to the ecliptic plane. All these planets rotate about their own axis, causing day and night. These orbital parameters are summarized in table 3.2. It should be noted that a precise understanding of these orbital characteristics is necessary for planning interplanetary missions.

The pressure on Mercury and Mars is almost negligible, whereas that on Venus is almost 92 times that on Earth. The pressure on the gas giants is expected to be more than a thousand times that on the Earth. The atmospheres of Venus and Mars contain predominantly CO_2 whereas that of the gas giants contain H_2 and He. Venus is the warmest planet with a mean temperature of around 460° C and Neptune is freezing with a mean temperature of $-200°$C. The gas giants have a ring system around them, the most noticeable one being the rings around Saturn. Mercury and Venus do not have a moon, Earth has one moon, and the other planets have multiple moons orbiting around them.

---
[1]Source : https://nssdc.gsfc.nasa.gov/

3.3. SOLAR SYSTEM

Table 3.1: Planets: size and acceleration due to gravity

Planet	Radius of planet/ Radius of Earth	Acceleration due to gravity/ Acceleration due to gravity on Earth
Mercury	0.382	0.378
Venus	0.949	0.908
Earth	1	1
Mars	0.532	0.378
Jupiter	11.209	2.357
Saturn	9.449	0.918
Uranus	4.007	0.888
Neptune	3.883	1.122

Table 3.2: Planets: orbit parameters

Planet	Perihelion (10^6 km)	Aphelion (10^6 km)	Orbital period (Days)	Orbital inclination (Degrees)	Sidereal rotation period (Hours)	Obliquity to orbit (Degrees)
Mercury	46	69.8	88.0	7.0	1407.6	0.034
Venus	107.5	108.9	224.7	3.4	-5832.5	177.4
Earth	147.1	152.1	365.2	0.0	23.9	23.4
Mars	206.7	249.3	687.0	1.8	24.6	25.2
Jupiter	740.6	816.4	4331	1.3	9.9	3.1
Saturn	1357.6	1506.5	10,747	2.5	10.7	26.7
Uranus	2732.7	3001.4	30,589	0.8	-17.2	97.8
Neptune	4471.1	4558.9	59,800	1.8	16.1	28.3

Moon, the natural satellite of the Earth, has a mean radius of 1737 km and the acceleration due to gravity on its surface is 1.62 m/s^2. It is orbiting around Earth in an elliptical obit. The closest point on the orbit called the perigee is at a distance of 0.363×10^6 km and the furthest point called the apogee is at a distance of 0.406×10^6 km from the Earth. Its orbit is inclined to the ecliptic plane at 5.1°. There is hardly any perceptible atmosphere on the Moon and the temperature on its Equator varies from 95 K to 390 K.

3.3.3 Dwarf Planets, Comets, Asteroids and Meteors

A dwarf planet is a relatively small object that is in direct orbit of the Sun, smaller than any of the eight classical planets. The differentiating criterion between planets and dwarf planets is that planets gravitationally dominate their neighbourhood whereas dwarf planets do not. They have a nearly spherical shape. Astronomers are of the broad consensus that there are at least nine dwarf planets including Pluto in the Kuiper belt and Ceres in the asteroid belt.

Comets, which are much smaller than dwarf planets, are giant pieces of ice, dust and rock that travel through space. Most comets remain perpetually frozen in the outer solar system. Some of them which enter the inner solar system develops the tail which results from the ice melting. Comets usually have highly eccentric elliptical orbits. Short-period comets originate in the Kuiper belt, which lies beyond the orbit of Neptune. Long-period comets are thought to originate in the Oort cloud. Halley's comet which was visible to the naked eye from Earth in 1986 has an orbital period of around 75 years and is considered a short period comet. Comet Hale-Bopp discovered in 1995 has an orbital period of more than 2000 years and is considered a long period comet. It is estimated that there are thousands of comets in the Solar System with most of them remaining in the outer regions of the Solar System.

Asteroids are rocky objects that orbit within the inner Solar System. They are generally irregular in shape and do not contain any atmosphere. Their dimensions are smaller than 1000 kilometers. Most of them are located between the orbits of Mars and Jupiter known as the asteroid belt. There are a large number of asteroids in the Solar System with the total number close to a million. The smaller chunks of rock that orbit the Sun are called meteoroids. Some of them enter the Earth's atmosphere as meteors leaving a visible tail, popularly known as 'shooting stars'. Most of them vaporize in the atmosphere and those that reach the ground are known as meteorites.

The vast expanse of the Solar System is mind-boggling and there are mysteries to be unravelled. The future will see human expeditions to many of these celestial bodies in the Solar System. A detailed and precise understanding of the gravitational force exerted by the celestial bodies is necessary while planning to send a space vehicle to these bodies. It is also important to know the trajectory of these bodies to execute a successful space mission. For space missions that enter the atmosphere of these bodies, an accurate model of the atmospheric characteristics on the celestial bodies is required for designing the space vehicles. As can be noted, the temperature, pressure, heat flux, and atmospheric constituents on these celestial bodies are significantly different from that of the Earth. The absence of significant quantities of oxygen, water and vegetation on these bodies make them markedly different from Earth. Addressing these differences are some of the formidable challenges that human beings will face if they have to make habitable colonies on these celestial bodies. Also, the distances of these bodies from Earth are measured in millions of kilometers which takes days or even months to traverse with the currently available technology. Addressing some of these challenges will pave the way for greater exploration of the Solar System.

Having looked at the intriguing space of the So-

3.4 Earth's Atmosphere

The atmosphere of Earth is the gaseous envelope surrounding the planet, the dominant components being nitrogen and oxygen. Like other planetary atmospheres, the earth's atmosphere figures centrally in transfers of energy between the sun and the planet's surface and from one region of the globe to another; these transfers maintain thermal equilibrium and determine the planet's climate.

Knowledge of the vertical distribution of quantities such as pressure, temperature, density, and speed of sound is required for purposes of the pressure altimeter calibrations and performance and design of flight vehicles. Also, space vehicles such as satellites, the *Apollo* lunar vehicle, and deep-space probes, which operate outside the sensible atmosphere, do encounter the earth's atmosphere during their blastoffs from the earth's surface and again during their return and recovery after completion of their missions.

Any point in the Earth's atmosphere is uniquely identified by its altitude, longitude and latitude. Below, we look at how these quantities are defined. They help not only in tabulating the properties of the atmosphere, but also in the navigation of aerospace vehicles.

3.4.1 Altitude, Longitude and Latitude

According to the Geodetic Glossary (National Geodetic Survey 1986), height is defined as, "The distance, measured along a perpendicular, between a point and a reference surface". The difficulty comes in determining the reference surface because of which the definition of height falls broadly into two categories: those that employ a reference ellipsoid as their datum and those that employ the Earth's gravitational field as their datum. There are certain features of the Earth that makes the definition of this datum a difficult task. The most prominent one being that the Earth is not a perfect sphere. The Earth has an equatorial diameter of ∼12,756 km, and a polar diameter of ∼12,713 km. The diameter of the Earth at the equator is about 43 kilometers more than that at the poles. If this difference is not taken into account, there can be grossly wrong estimates of height especially close to the Earth's surface. Traditionally, the mean sea surface was taken as the vertical datum. The "Mean Sea Level" is the average location of the interface between ocean and atmosphere, over a period of time sufficiently long so that all random and periodic variations of short duration average to zero. Due to various physical features of the Earth, not all "Mean Sea Level" stations are at the same distance from the center of mass of the Earth. For the above reasons, reference ellipsoids are used as models of the Earth's shape and fall into two categories: local and equipotential. Local reference ellipsoids are geometric in nature and are determined by geometrical means. Equipotential ellipsoids include the geometric considerations of local reference ellipsoids, but they also include information about the Earth's mass and rotation. A more meaningful datum is the geoid which The National Geodetic Survey (1986) defines as "The equipotential surface of the Earth's gravity field which best fits, in a least squares sense, global mean sea level". The orthometric height is defined as, "The distance between the geoid and a point measured along the plumb line and taken positive upward from the geoid". Due to its basis in the gravitational potential, the orthometric height is a physically and geometrically significant measure of height. However, it is not straightforward to measure it directly. Ellipsoid heights are the straightline distances normal to a reference ellipsoid produced away from the ellipsoid to the point of interest. Ellipsoid height can be readily determined using GPS receivers. Orthometric heights can be obtained from ellipsoid heights by having a mathematical model for the geoid and the three quantities are related as $H = h - N$, where, H = orthometric height, N = geoid height and h = ellipsoid height (Refer Figure 3.2).

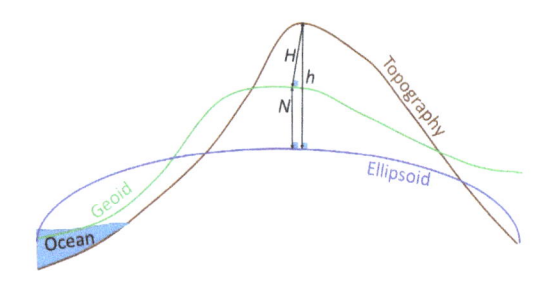

Figure 3.2: Geopotential height and ellipsoid height.

On a map or globe, the lines running east to west are known as lines of latitude and the lines running north to south are known as lines of longitude. The latitude indicates the north-south position of a point on Earth. Points on the equator have 0° latitude and those at the North and South Poles are at 90° latitude. Geodetic latitude is the angle between the normal to the reference ellipsoid and the equatorial plane. Lines of longitude are imaginary lines that divide the Earth. They run from pole to pole of the Earth and they measure the distance east or west. They intersect the equator at right angles. The prime meridian, which runs through Greenwich, England, has a longitude of 0°. It divides the Earth into the eastern and western hemispheres. There are 360 degrees of longitude, +180° eastward and −180° westward. Both latitude and longitude are measured in degrees, minutes, and seconds.

3.4.2 Pressure, Density and Temperature Variation

The *standard atmosphere* is a simplified model for the Earth's atmosphere whose properties are changing dynamically across altitude, longitude and latitude. Typical standard atmosphere tabulates the variation of thermodynamic variables like pressure, temperature and density of the air with altitude. There can be information about the variation of sound speed, coefficient of viscosity, thermal conductivity, species concentration, mean free path, collision frequency, molecular weight etc. with altitude. The variation with longitude and latitude are averaged over in these tables. So, there can be significant deviations in the actual physical properties at any point in the atmosphere when compared to the standard atmosphere. These errors become more prominent especially at higher altitudes. Nevertheless, the standard atmosphere helps to relate flight tests, wind tunnel results, and general airplane design and performance to a common reference. The air in the model is assumed to obey the perfect gas law and the hydrostatic equation, which taken together, relate temperature, pressure, and density variations in the vertical.

The first standard atmospheric models were developed in the 1920s in both Europe and the United States. The slight differences between the models were reconciled and an internationally accepted model was introduced in 1952 by the International Civil Aviation Organization (ICAO). The current standard atmosphere is that adopted in 1976 and is a slight modification of the one adopted in 1952 by the (ICAO). The International Standard Atmosphere is defined assuming the mean sea level (MSL) conditions as pressure $p_0 = 101325$ N/m^2, density $\rho_0 = 1.225$ kg/m^3, Temperature $T_0 = 288.15$ K, speed of sound $a_0 = 340.294$ m/s, acceleration due to gravity $g = 9.806$ m/s^2. In the lower region of the atmosphere called the troposphere, the temperature decreases with altitude. In the next layer called the stratosphere, the temperature remains constant with altitude in the lower portion and subsequently starts increasing with altitude. Tropopause marks the boundary between troposphere and stratosphere. The altitude of tropopause varies with location on the Earth and with seasons. Fig. 3.3 illustrates the temperature variations in the standard atmosphere. One of the important parameters is altitude. The direct measurement of the vertical distance from the ground which is at sea level is called *the geometric altitude* h_G. If the vertical distance is measured from the center of the earth, then the vertical distance is called *the absolute altitude* h_a,

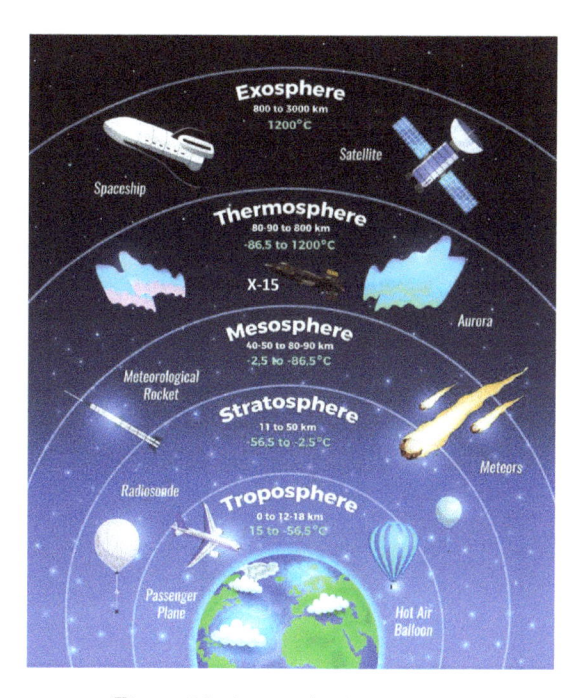

Figure 3.3: Atmospheric structure.

which means $h_a = h_G + R_E$, where, R_E is the radius of the earth. The absolute altitude is important, especially for space flight as the local acceleration of gravity g varies with h_a which is given by Newton's law of gravitation

$$g = g_0 \left(\frac{R_E}{R_E + h_G} \right)^2 \qquad (3.1)$$

Hydrostatic equation, $dp = -\rho g dh_G$, obtained by a force balance on an element of fluid at rest. It relates an infinitesimally small change in pressure dp to a corresponding infinitesimally small change in altitude dh_G, where, dp and dh_G are differentials. The assumption that g is constant throughout the atmosphere and equal to its value at sea level g_0, leads to a new hydrostatic equation, $dp = -\rho g_0 dh$. Here, h is the *geo-potential altitude* which differs from the geometric altitude (h_G). Geo-potential altitude is related to the geometric altitude by the following relation $dh = (g/g_0)dh_G$. Using the Eq. (3.1), it can be expressed as

$$dh = \left(\frac{R_E}{R_E + h_G} \right)^2 dh_G \quad \text{on integration}$$
$$h = \frac{R_E h_G}{R_E + h_G} \qquad (3.2)$$

It shows that there is little difference between geopotential and geometric altitude for low altitudes.

Based on the US standard atmosphere of 1976, Fig. 3.4 to 3.11 plot temperature, pressure, density, mean free path, and molecular weight with altitude. In the troposphere (from sea level to ~ 11 km in the standard atmosphere), it is seen that the temperature decreases linearly with altitude. A generally accepted value for the temperature lapse rate in the troposphere is 6.5 K/km. In the stratosphere, it first remains constant at about 217 K before increasing again. Both the density and pressure are seen to decrease rapidly with altitude. The density curve is of particular importance since the lift and drag on a flight vehicle is directly dependent on the density. The thrust produced by air breathing engines is also dependent on density.

The variation of the temperature $T = T(h)$ can be established using measurements and then pressure $p = p(h)$ and density $\rho = \rho(h)$ follow from the laws of physics. The hydrostatic equation, $dp = -\rho g_0 dh$ is divided by the equation of state yields

$$\frac{dp}{p} = -\frac{\rho g_0 dh}{\rho R T} = -\frac{g_0}{RT} dh \qquad (3.3)$$

In the gradient layers of the standard atmosphere (shown in fig. 3.4), it is found that the temperature variation is linear and is given as

$$\frac{T - T_1}{h - h_1} = \frac{dT}{dh} \equiv a \qquad (3.4)$$

where, a is a specified constant, sometimes called the lapse rate, for each of the gradient layers obtained from the defined temperature variation in Fig. 3.4. Now, $dh = dT/a$ is substituted into the hydrostatic equation, Eq. (3.3) gives

$$\frac{dp}{p} = -\frac{g_0}{aR} \frac{dT}{T} \qquad (3.5)$$

For the isothermal (constant-temperature) layers, equation (3.3) can be integrated to give the variation of pressure and density with altitude as

$$\frac{p(h)}{p_1} = \frac{\rho(h)}{\rho_1} = e^{-\frac{g_0}{RT}(h - h_1)}, \qquad (3.6)$$

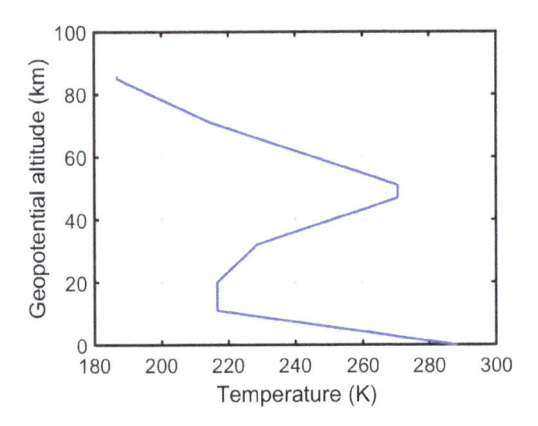

Figure 3.4: Variation of temperature with altitude in the standard atmosphere.

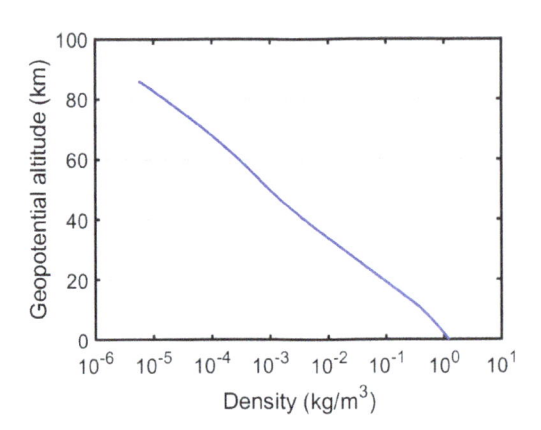

Figure 3.6: Variation of density with altitude in the standard atmosphere.

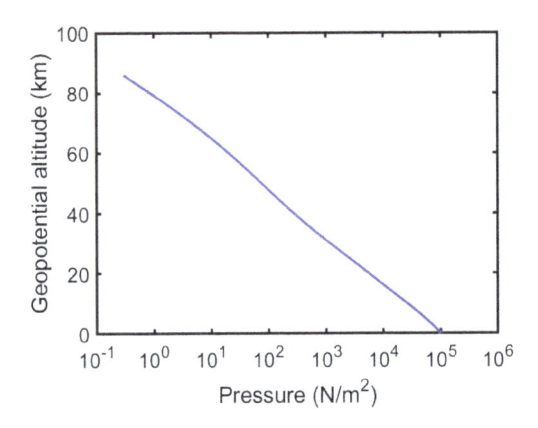

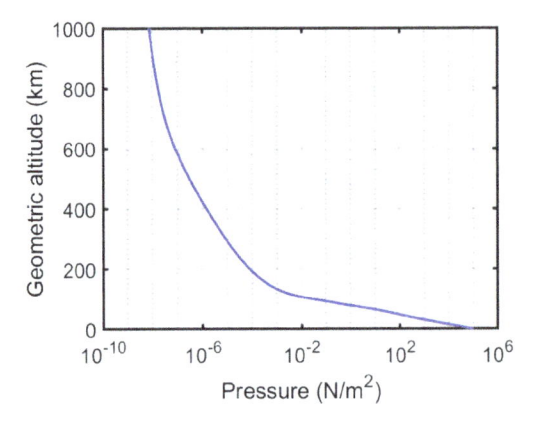

Figure 3.5: Variation of pressure with altitude in the standard atmosphere.

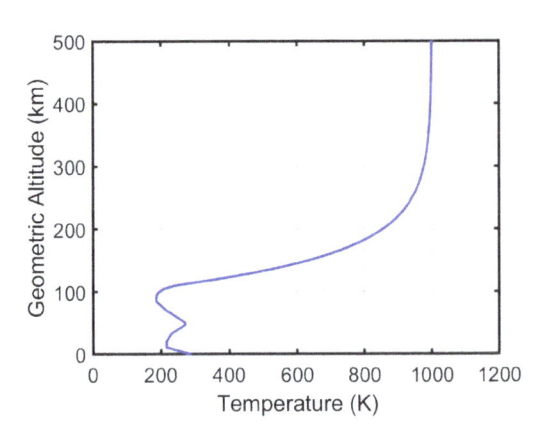

Figure 3.7: Variation of temperature with altitude in the standard atmosphere.

where, p_1 and ρ_1 are the pressure and density at the base of the isothermal layer(h_1), respectively. For gradient layers, Eq. (3.5) relates base location of the layer h_1 with values T_1, p_1 and ρ_1 to any point on the layer h with values $T(h)$, $p(h)$ and $\rho(h)$ as

$$\frac{p(h)}{p_1} = \left(\frac{T(h)}{T_1}\right)^{-\frac{g_0}{aR}}, \quad \text{using equation of state,}$$

$$\frac{\rho(h)}{\rho_1} = \left(\frac{T(h)}{T_1}\right)^{-\frac{g_0}{aR}-1}, \quad (3.7)$$

where, $T(h) = T_1 + a(h - h_1)$.

3.4.3 Real Atmosphere

Due to the combined effects of solar radiation, the presence of landmasses and oceans, the Earth's

Figure 3.8: Variation of pressure with altitude in the standard atmosphere.

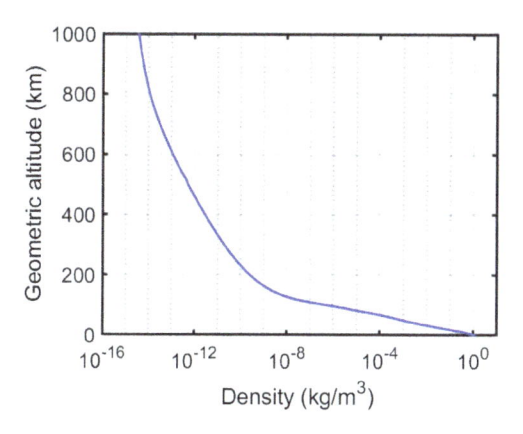

Figure 3.9: Variation of density with altitude in the standard atmosphere.

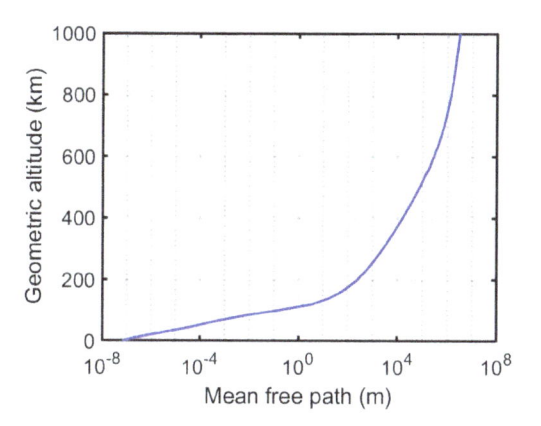

Figure 3.10: Variation of mean free path with altitude in the standard atmosphere.

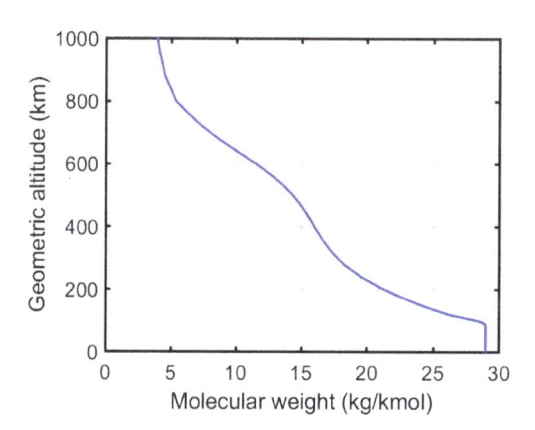

Figure 3.11: Variation of molecular weight with altitude in the standard atmosphere.

spinning motion about its own axis and the rotation about the Sun effect, the Earth's atmosphere is in a dynamical state of motion. Although a standard atmosphere provides the preliminary data necessary for the design of aerospace vehicles, it is essential to anticipate nonstandard performance in the real atmosphere. This nonstandard performance shows up in numerous ways: winds, turbulence, precipitation, icing, hailstorm, dust storm, fog etc.

The air mass through which a flight vehicle operates is in a state of constant motion with respect to the surface of the Earth. The motion of the atmosphere may be large-scale motion known as wind or small-scale motion known as turbulence. Wind affects both the navigation and the performance of a flight vehicle. In general, air turbulence may cause discomfort and in extreme cases injuries to the passengers. Also, excessive shaking or vibration of the aircraft may render the pilot unable to read instruments. During precision operations such as air-to-air refueling, turbulence-induced motions of the aircraft are a nuisance. Turbulence-induced unsteady loads over a long period can cause fatigue in the airframe and in extreme cases heavy turbulence may cause the loss of control of an aircraft or even structural failure.

Water in the air, in either its liquid or vapor form, is not accounted for in the pure dry standard atmosphere and will affect an aircraft in varying degrees. The forms of precipitation adversely affect aircraft performance such as icing on the wings, zero visibility in fog or snow, and physical damage caused by hail. Water vapor is less dense than dry air and consequently humid air (air containing water vapor) will be less dense than dry air. Because of this, an aircraft requires a longer take-off distance in humid air than in more dense dry air.

3.4.4 Aviation Meteorology

Meteorology is the branch of science which deals with the study of the Earth' atmosphere and the physical processes happening within the Earth's atmosphere. Aviation meteorology focuses on those aspects of the weather which affects aviation. The performance of flight vehicles while operating within the Earth's atmosphere are dictated by the atmospheric conditions, especially density, pressure, temperature and humidity. There are a plethora of weather conditions which have significant impact on aviation, the notable ones being poor visibility, severe wind conditions and turbulence, icing, rain, thunderstorm and lightning. Severe weather is considered as one of the significant causes for many aviation accidents.

There are different types of pressure systems that are formed on the Earth. A "cyclone" is a region of low pressure surrounded by regions of high pressure. These are associated with strong winds which blow around in an anticlockwise direction in the northern hemisphere and clockwise direction in southern hemisphere. An "anticyclone" is a region of relatively higher pressure surrounded by region of low pressure. It is characterized by regions of descending and diverging air.

From an aviation perspective, turbulence is associated with disturbed air. Air turbulence is caused by up and down currents which interfere with the horizontal movement of air. It can be produced due to the convection currents or due to presence of obstructions to the winds or due to the currents within the clouds. It can occur in clear air particularly near the tropopause and is associated with the jet streams, clouds and the pressure systems at those altitudes. A gust is a rapid increase in wind strength with a time duration less than a minute. A squall is a sudden increase of wind speed of at least 16 knots (8.2 m/s) which lasts more than a minute. A gale is a condition where the wind speed exceeds 33 knots and for hurricane it exceeds 63 knots. Jet streams are relatively strong winds concentrated as narrow currents between 9 to 13 kms altitude. The polar-front jet stream encircles the globe in segments thousands of kilometers long, hundreds of kilometers wide, and several kilometers thick. It flows generally from west to east in great curving arcs and the wind speeds can reach 400 kilometers per hour. Westbound flights plan to avoid the jet-stream head winds. Eastbound flights take advantage of the time-saving tail winds. However, they can produces strong wind shears in some regions because of large changes in wind speeds over short vertical and horizontal distances.

Clouds are indicators in the sky to the pilots for possible weather problems like turbulence, poor visibility, precipitation and icing. Clouds are classified into three basic forms: stratiform, cumuliform and cirriform. Stratiform clouds show a horizontal development, cumuliform ones show a vertical development and the cirriform are feathery clouds. There are a variety of clouds with different compositions and which form at different altitudes. Some of them can affect the visibility whereas few are known to create atmospheric turbulence. Tiny ice crystals which are detrimental to the life of jet engines are found in some type of clouds.

The objective of this section is to get some preliminary understanding of the atmosphere's role in aviation. A comprehensive review of all the weather phenomena that affects the flight of aerospace vehicles is beyond the scope of this book.

Bibliography

J. D. Anderson Jr. (2005) *Introduction to Flight*, McGraw-Hill Higher Education.

U.S. Standard Atmosphere, 1976, National Oceanic and Atmospheric Administration, National Aeronautics and Space Administration, and United States Air Force.

T. A. Talay (1975) *Introduction to the Aerodynamics of Flight*, NASA-SP-367.

T. H. Meyer, D. R. Roman, and D. B. Zilkoski (2004) *What Does Height Really Mean? Part I: Introduction*, Thomas H. Meyer's Peer-reviewed Articles.

Joint Aviation Authorities Airline Transport Pilots License Theoretical knowledge Manual: Meteorology (2001), Jeppessen GmbH, Frankfurt, Germany.

Chapter 4

Basics of Aerospace Engineering

Contents

4.1	**Introduction**		**186**
4.2	**Aerodynamics**		**186**
	4.2.1	Basics Definitions	186
	4.2.2	Airfoil	189
	4.2.3	Aerodynamic Forces	190
	4.2.4	Wings	193
4.3	**Propulsion**		**194**
	4.3.1	Aspects of Thermodynamics	195
	4.3.2	Reciprocating Engines	197
	4.3.3	Gas Turbine Engines	199
	4.3.4	Ramjet and Scramjet Engines	207
	4.3.5	Pulsejet Engines	210
	4.3.6	Rocket Propulsion	210
4.4	**Flight Mechanics**		**215**
	4.4.1	Performance of Flight Vehicles	215
	4.4.2	Principles of Stability	222
4.5	**Flight Structures**		**224**
	4.5.1	Material Types	228
	4.5.2	Aircraft Structures	230
	4.5.3	Manufacturing	231
4.6	**Orbital Mechanics**		**234**
	4.6.1	Basic Laws	234
	4.6.2	Space Vehicle Trajectories	235

4.1 Introduction

In the Part I, we saw the evolution of aerospace vehicles from a historical perspective. In the current chapter, the fundamentals of aerospace engineering which are required to understand the working principles of aerospace vehicles will be introduced. The fundamentals of flight are discussed in five sections. They are (i) Aerodynamics, (ii) Propulsion, (iii) Flight Mechanics, (iv) Flight Structures and (v) Orbital Mechanics.

4.2 Aerodynamics

The term "Aerodynamics" is generally used for studying the problems related to the flight of aerospace vehicles in the air. It mainly deals with the forces and moments experienced by the vehicles during their flight. The principles can be used for studying the more general problem of motion of bodies in fluids, for example, the study of flapping wings of insects and the study of forces experienced by a racing car. The concepts help in understanding the working of compressor and turbine blades of jet engines and propellers. It can also be used for the study of the motion of gases. For example, the working principles of shock tubes, which are experimental set-ups used to study a variety of problems from diverse domains, can be understood based on the fundamental principles of aerodynamics. The rules of aerodynamics explain how an airplane is able to fly. A rocket blasting off the launch pad and a kite in the sky react to aerodynamics. During cruise flight, the airplane flies in equilibrium under the balance of four forces: lift, weight, thrust and drag. Any change in these forces makes it move up/down or turn around or fly faster/slower. Hence, aerodynamics is very important to understand how the motion of aerospace vehicles are controlled.

By definition, lift is the component of aerodynamic force experienced by a body in flight which acts perpendicular to the motion of the body. Lift for an airplane comes predominantly from its wings. The fuselage and other components on the airplane can contribute to the lift generated. Depending on the shape of the vehicle and orientation, the lift experienced on a vehicle in motion can be zero.

Drag is the component of aerodynamic force experienced by the body in flight which acts in a direction opposite to the motion of the body. For a body moving within the atmosphere with a finite value of velocity, drag is always positive. As a result, a non-propelled flight vehicle in motion in the air will lose its energy. If drag were to be zero or negative for a flight vehicle, we could have created perpetual motion vehicles which is an impossibility.

Below, we will look at some of the basic variables and concepts that are used in the study of aerodynamics.

4.2.1 Basics Definitions

Density is a measure of mass per unit of volume (ρ) and expressed in kg/m^3. For example, when we say the density of a gas is 1.2 kg/m^3, it means the gas filled in a cube of volume 1 m^3 weighs 1.2 kg. In fact, density is a point property that can vary from point to point in the gas. The density of the air depends on other properties like temperature and pressure.

Temperature is a property that quantitatively expresses the degree of hotness or coldness of the medium. It indicates the thermal energy possessed by the air. The temperature also is a measure of the average kinetic energy of the air particles. Its SI unit is Kelvin. However, there are a multitude of units like Celsius, Fahrenheit and Réaumur which are in common use.

Pressure: In order to understand pressure, let us consider heating a pressure cooker that is partially filled with water. The pressure cooker works by raising the temperature of boiling water and then forming vapors. As more and more vapors add up, the pressure inside the cooker increases. The force due to pressure acts normally on the wall of the cooker uniformly. When this

4.2. AERODYNAMICS

force is sufficient to lift the dead weight, we hear the whistling sound of the cooker. From this example, we can see that fluid pressure can act on solid surfaces to produce a net force. It is well known that sailboats use the pressure exerted on the sail for navigation.

By definition, pressure is the normal force acting per unit area and has the unit of N/m^2 (Pascal). Pressure can also be expressed in a multitude of units like bar, atmosphere, pounds per square inch, kg/cm^2 and torr.

Velocity: Before defining the velocity, take an example of the flow of air over an airfoil, which is a cross-section of the aircraft's wing. Imagine yourself traveling with a particle of air or maybe a small mass of air, generally called a fluid element. You experience that the velocity of the fluid element is changing at every point on the surface of the airfoil. Somewhere on the upper surface, you experience the highest velocity. Now, imagine you are sitting outside and gazing from a point on the airfoil. Different fluid elements pass through that point with the same velocity. With this example, we can say the magnitude of velocity of air at a point as the distance traveled by the fluid element per unit time. Also, velocity has a direction associated with it which is given by the tangent to the path of the fluid element at the point of interest. It is also clear that velocity is a point property and may vary from point to point in the flow.

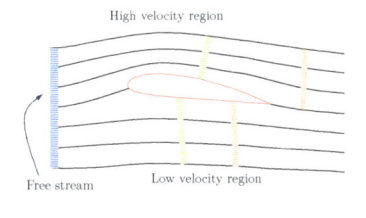

Sound: Imagine you create a disturbance in stationary water by dropping a stone. You see that disturbances start moving in the form of a pressure wave with a velocity in all directions. Similarly, pressure waves may be created when the air is disturbed. The disturbance can be a bird chirping or a loud speaker or a jet engine or any other source of sound. The velocity at which the pressure wave travels isentropically through a medium is the speed of sound. The speed of sound is a function of the elasticity of the medium. For ideal gases, the speed of sound is given as $\sqrt{\gamma RT}$, where γ is the ratio of specific heats, $R = R_U/MW$ is the gas constant, R_U is the universal gas constant $(8.314\ J/(K.mol))$, MW is the molecular weight of the gas and T is the temperature of the medium expressed in Kelvin. In the standard atmosphere, the speed of sound in the air at sea level conditions is about 343 m/s.

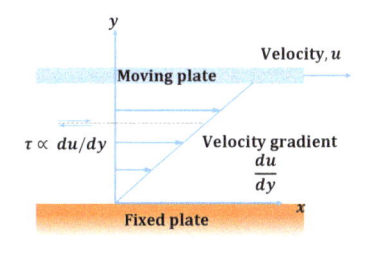

Shear Stress: When you want to move a study table on the floor, you will apply a force on the table in the horizontal direction. Now you see that the table moves only when the applied force is sufficient to overcome the friction force that exists between the table and the floor. Similar to the friction that exists when two solid bodies in contact move relative to each other, fluid also experience a force when it slides relative to the solid surface or fluid itself. It can be felt as the mild resistance offered when you try to push a flat swimming float over water. This force that acts tangential to the surface of interest is called shear force. This shear force can be interpreted as the measure of the resistance of one layer of the fluid sliding past another layer in the same fluid or a solid surface. This shear force (F) per unit area is called as shear stress, $\tau = F/A$, where A is the contact area between the solid surface and the fluid. However, under the influence of shear stress, the fluid layer deforms continuously and the rate of this fluid deformation is the velocity gradient, du/dy. Consider an example of the flow of fluid between the plates as shown in the figure. For Newtonian fluids like air, the rate of deformation is linearly proportional to the shear stress i.e., $\tau \propto du/dy$.

Viscosity: The shear force between layers of a moving fluid is different for different fluids. The property of a fluid that represents this shear force is the coefficient of viscosity (μ). For the situation shown above, we can write the shear stress as $\tau = \mu du/dy$. The viscosity of air which is a function of temperature is expressed in Ns/m^2 or Pa.s. The value of μ of air at $15°C$ is 1.81×10^{-5} Ns/m^2.

Boundary Layer: We see from the figure that due to viscous force, the fluid layer close to the plate adheres to the plate. Such a condition where the flow velocity is zero is termed the no-slip condition. As we move in the y-direction, the velocity of each layer is more than a layer adjacent to it. At some particular distance, the fluid layer will have a velocity close to the free-flow velocity of the air. That means that there is a thin region of retarded flow in the neighborhood of the plate that exists because of the viscosity. This region of fluid close to the solid boundary where viscosity plays an important role is called a boundary layer.

Boundary layer over a flat plate

Let us see some history of the boundary layer. Hydrodynamicists in the 18th century attempted to calculate the resistance experienced by the body moving through the fluid. Theoretical calculations based on the inviscid (fluid which has no viscosity) and incompressible (fluid with constant density) flow over various two-dimensional bodies by *d'Alembert* predicted no resistance to motion, which contradicts the real-life observation. *Ludwig Prandtl* (1875 - 1953) introduced the boundary layer theory in 1903 which showed that there is a thin region near the solid boundaries where the viscous effects are dominating that caused the resistance to the body. The theory helped to debunk the centuries-old paradox regarding the resistance to motion of bodies in fluids.

Reynolds Number is an important dimensionless number in aerodynamics named after *Osborne Reynolds*. To define Reynolds number, let us consider flow over a flat plate as shown in the above figure. Here, U^∞ is the free stream air velocity and x is the distance measured from the leading edge along the flow at which the Reynolds number is determined. Now the Reynolds number is defined as $R_e = \rho U^\infty x/\mu$, where, ρ and μ are the density and viscosity of the air, respectively. Non-dimensionalization of the Navier-Stokes equation will show that the Reynolds number is the ratio of inertial forces to viscous forces within a fluid. At a lower Reynolds number, the viscous forces are dominant and flow is laminar. As the Reynolds number gets higher, the inertial forces begin to dominate and the flow turns turbulent.

Mach Number (M) is an important non-dimensional parameter named after the Austrian physicist *Ernst Mach* (1838 - 1916). It is the ratio of the actual speed of the air (u) (or an object in the air) to the speed of sound in the air (c) at the same condition, $M = u/c$. Mach number depends on the speed of sound, which is a function of temperature. Therefore, aircraft flying at the same velocity can have different Mach numbers at different locations based on the state of the air. Aerodynamic flows can be divided into different regimes based on the value of the Mach number as

- Sonic when $M = 1$
- Subsonic when $M < 1$
- Supersonic when $M > 1$
- Transonic when M is near one that is when the flow around an object generates regions of both subsonic and supersonic airflow around that object
- Hypersonic when $M >> 1$ (Usually the flight speeds above which aerodynamic heating becomes important in flight vehicle design. Generally, $M > 5$ is considered to be hypersonic, though it is not a strict demarcation).

Compressibility: We learned that the density of air changes with a change in its temperature or pressure. It can be observed that air experiences contraction under the application of pressure and expands when the pressure is released. Therefore, air acts like elastic solids with respect to pressure. The property analogous to Young's modulus of solids, is a property of fluid called compressibility. It is the ratio of the change in pressure to the fractional change in the density of the air while the temperature remains constant. Mach number can also be interpreted as the ratio of inertial forces to elastic forces. If M is less than about 0.3, the flow may be approximated as incompressible.

One of the key questions in the minds of students of aerodynamics is *"How do an airplane's wings produce lift?"*. Before understanding the complex aerodynamics of a wing, it makes sense to understand the flow around an airfoil which is the two-dimensional cross-section of the wing.

4.2.2 Airfoil

The airfoil can be interpreted as an infinitely long wing with no tips. Now, let's look at the airfoil to understand its terminologies. An imaginary line passing through the points that are the midpoints of the upper and lower surfaces is called the mean camber line. This is measured normal to the *mean camber line* itself. The maximum distance between the upper and lower surface measured normal to the mean camber line is the thickness. The most forward and aft points of the mean camber line are called the leading edge and trailing edge, respectively. The straight line connecting the leading and trailing edges is the chord line and the distance from the leading edge to the trailing edge measured along the chord line is a chord. With these definitions, it is clear that if the airfoil is symmetric then the mean camber line and the chord line coincide. However, for an asymmetric airfoil, the maximum distance between the mean camber line and the chord line is called a camber. Note that this distance is measured normal to the chord line. Different distributions of thickness and shapes of the mean camber line produce different shapes of the airfoil. An infinite number of airfoil shapes are therefore possible. The aerodynamic characteristics of each of these airfoils can be different. The aerodynamic characteristics are obtained by wind tunnel testing or Computational Fluid Dynamics (CFD) simulations. The aerodynamic force experienced by a moving airfoil into still air is the same as that experienced by a stationary airfoil with the air moving in the opposite direction with the same speed. In other words, what matters for the calculation of aerodynamic forces is the relative velocity between the body and the air. This principle forms the basis of wind tunnel testing and CFD simulations. The velocity of the air far upstream of the airfoil is called the freestream air velocity. The angle between the chord line of the airfoil with the direction of the freestream air velocity is called the Angle of Attack (α). It is to be noted that α is not defined with reference to the local horizontal line.

(a) Aeroplane and wing

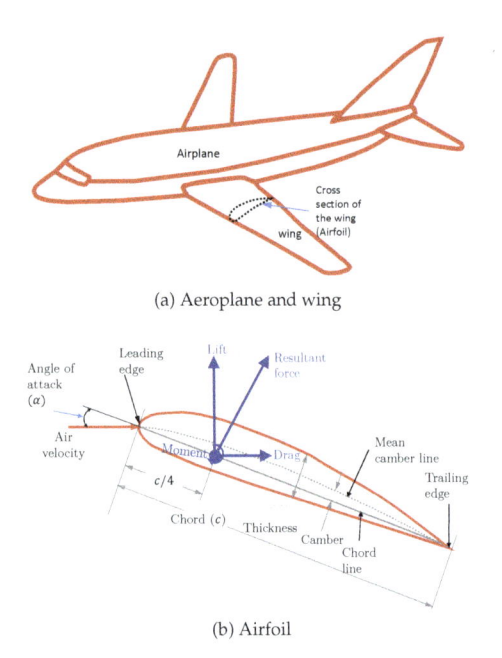

(b) Airfoil

4.2.3 Aerodynamic Forces

The airfoil shape of the wing is the primary factor affecting the aerodynamic characteristics of the wing. At small angles of attack, the mean pressure on the bottom surface of the airfoil is higher than that at the top which produces the lift on the airfoil. As air flows over the airfoil, every point on its surface experiences a normal pressure and tangential shear stress due to the viscosity of the air. If we do vector addition of the pressure and shear stress by integrating these two quantities over the surface of the airfoil, the resultant force can be obtained which acts at some point on the chord line of the airfoil. The resultant force can be resolved into two components parallel and perpendicular to the direction of the free-stream air velocity. The component of the aerodynamic force parallel to the freestream direction is called drag and the component of the aerodynamic force perpendicular to the direction of the freestream is called lift. In addition, to lift and drag, the surface pressure and shear stress distributions create a moment that tends to rotate the wing. Let us integrate the forces on the upper and lower surfaces of the airfoil separately. The total forces on the upper and lower surfaces are not equal. They do not act at the same point and are opposite to each other. This situation creates the aerodynamic moment and the value of which depends on the point about which we choose to take moments.

For force analysis, it is easier to represent the pressure and shear stress distribution as forces and moments acting at a point. Now let's define two points on the airfoil, the *aerodynamic center* and the *center of pressure*. We learned that the aerodynamic moment depends on the angle of attack and the point at which the moment is taken. However, there exists a certain point on the airfoil about which the aerodynamic moment does not change with the angle of attack for small values of it. This point at which the aerodynamic moment is independent of the angle of attack is called *Aerodynamic Center* ($dM/d\alpha = 0$ at aerodynamic center). The point of application of the net aerodynamic force is the *Centre of Pressure* (C_P) (analogous to the centroid of an area, C_P is the centroid of the pressure distribution). Note that the aerodynamic moment about the center of pressure is zero. The aerodynamic center is a better reference point because it doesn't change with the angle of attack whereas the center of pressure shifts with a change in angle of attack. Thin airfoil theory predicts the location of the aerodynamic center for an airfoil to be at the quarter-chord point.

Lift and Drag Coefficients

These aerodynamic forces and the moment on the airfoil depend on various parameters such as free stream air velocity (V), air properties like density, ρ, speed of sound of the medium, the coefficient of viscosity μ, shape of the airfoil, size (chord, c) of the airfoil and angle of attack (α). Using dimensional analysis, the expressions for lift, drag and moment per unit length of the airfoil are given by $L = c_l \frac{1}{2}\rho V^2 c$, $D = c_d \frac{1}{2}\rho V^2 c$, $M = c_m \frac{1}{2}\rho V^2 c^2$. Here, the c_l, c_d and c_m are the lift coefficient, drag coefficient and moment coefficient of the airfoil, respectively. The values of c_l, c_d and c_m are functions of the Reynolds number, Mach number and the angle of attack. Now let's consider the dependency

4.2. AERODYNAMICS

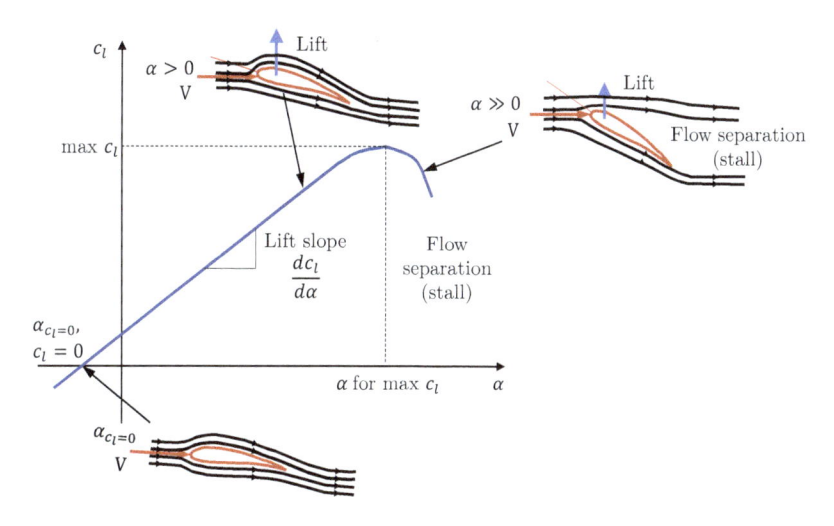

The variation of coefficient of lift with the angle of attack for a typical cambered aerofoil

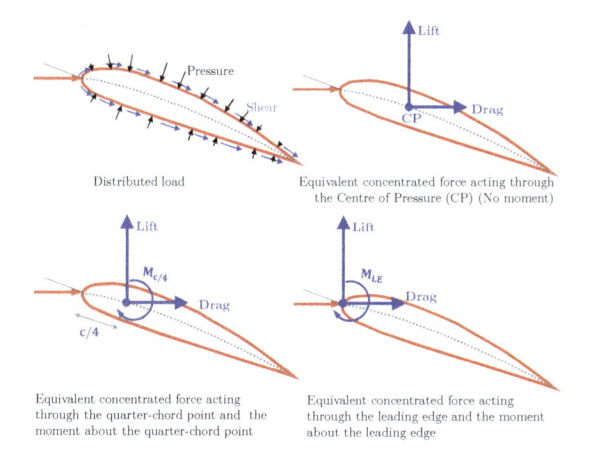

Ways of representing the distributed load exerted by pressure and shear stress on the surface of the airfoil by a concentrated force at a point and the moment at that point.

of these coefficients on the angle of attack which is important information for aircraft designers.

The typical variation of the lift coefficient with the angle of attack is plotted in the figure. If the airfoil is asymmetric and if it kept at zero angle of attack, we see that some lift is still produced i.e., $c_l > 0$ for $\alpha = 0$. As we reduce the angle of attack, at some value of α the c_l is found to be zero. This angle of attack at which lift is zero is called the zero-lift angle of attack $(\alpha_{(L=0)})$. However, $(\alpha_{L=0} = 0)$ for the case of a symmetric airfoil. Experiment results and thin airfoil theory by *Prandtl* predict that the lift coefficient varies linearly with the angle of attack for small values of α. The slope of the linear portion of the lift curve is called the lift slope and the thin airfoil theory predicts the slope to be $2\pi/\text{rad}$. However, for large values of angle of attack, the linearity of the lift curve breaks down. As the angle of attack is increased beyond a certain value, c_l peaks at some maximum value $c_{l,max}$ and then drops as α is further increased. In this situation, where the lift is rapidly decreasing at higher α, the airfoil is said to be stalled. The stall is caused by flow separation on the upper surface of the airfoil. When flow separates, the lift decreases drastically and the drag increases suddenly. Typical airfoil stalls around 12-15 degrees.

For an airfoil in subsonic flows, there are two general types of drag: *Pressure drag* (due to a net imbalance of surface pressure acting in the drag direction) and *Friction drag* (due to the net effect of shear stress acting in the drag direction). The component due to pressure drag becomes prominent when there is flow separation. The total drag coefficient for an airfoil in low-speed flows is given by $c_d = c_{d,f} + c_{d,p}$ where, $c_{d,f}$, and $c_{d,p}$ are *skin friction drag* (drag caused by the friction of a fluid against the surface of an object that is moving through it) and *pressure drag* (drag coefficients due to flow separation), respectively. For an airfoil in supersonic flow, due to the presence of shock waves, there is an additional component of drag called the wave drag and the coefficient of drag becomes $c_d = c_{d,f} + c_{d,p} + c_{d,w}$, where $c_{d,w}$ is the *wave drag* coefficient (drag inherently due to the pressure increase across the shock wave in supersonic flow). *Shocks* are flow phenomena observed in transonic and supersonic flows. They are thin regions in the flowfield across which there is an abrupt jump in the thermodynamic variables and the velocity. The shape, location and strength of the shock depend on the shape and orientation of the body placed in the flow and Mach number. Shock waves in transonic flow cause the prominent transonic drag rise near Mach number 1.

Drag Divergence

The typical variation of coefficient of drag of an aerodynamic surface with Mach number for subsonic and supersonic speeds is shown in fig. 4.1. In the subsonic flow regime c_d stays relatively constant with M up to, and slightly beyond, the *critical Mach number*, the free-stream Mach number at which sonic flow is first encountered at some location near the airfoil. For unseparated flows, the drag in the subsonic region is mainly due to friction, and the compressibility effect on friction in the subsonic regime is relatively small. The shock wave-boundary layer interaction in transonic flows greatly increases the pressure drag. As a result, the drag coefficient increases significantly in the transonic regime. The Mach number at which the drag coefficient starts increasing suddenly and significantly is called *drag divergence* Mach number.

Transonic airplane designers deal with the dramatic increase in drag near Mach number 1(refer fig. 4.1) by angling the wings back from the fuselage (swept wings), making wings thinner and using other features designed to reduce drag. Thinner wings reduces the structural strength and the capacity to store fuel, which are disadvantageous . The other approach is to adopt a specially shaped airfoil called a supercritical airfoil in order to push the drag divergence Mach number to a speed closer to Mach number 1.

Supercritical airfoils are special type of airfoils designed to delay and reduce the drag rise experienced at transonic Mach numbers. The significant increase in drag is due to presence of strong normal shocks and shock-induced boundary layer separation. For supercritical airfoils, the shape is designed with a relatively flat top surface. When the free-stream Mach number exceeds the critical Mach number, a pocket of supersonic flow is present over the top surface as usual; but because it is relatively flatter, the local supersonic Mach number is of lower magnitude than would exist in the case of a conventional airfoil. As a result, the pressure rise across the shock wave is less and consequently the drag is also less compared to a conventional airfoil. In turn, the supercritical airfoil can penetrate closer to sonic speed before drag divergence occurs. As compared with a conventional airfoil, a supercritical airfoil has a reduced camber, an increased leading edge radius, relatively flat top surface, and a concavity in the rear part of the bottom side. The first supercritical airfoils were designed and analyzed by *Richard Whitcomb* at NASA Langley in the 1950s. A standard *General Dynamics F-111* was modified with a supercritical wing. The supercritical airfoil allows to attain a higher cruise velocity for a given airfoil thickness or allows to increase airfoil thickness (straightforward structural design and increased fuel capacity) for a given lower cruise velocity. The *C-17*, heavy-lift, air-refuelable cargo transport, has a supercritical wing that gives it excellent performance.

In the supersonic flow regime, c_d gradually decreases, following approximately the variation $c_d \propto 1/\sqrt{M^2-1}$. A significant portion of the total drag at supersonic speeds is due to wave drag. Hence, the design of supersonic vehicles concentrates on reducing wave drag associated with supersonic flow. The shock wave strength is reduced by having a sharp nose, slender fuselage, and sharp wing and tail leading edges. These differences can easily be noticed when comparing the shape of a subsonic aircraft with that of a supersonic one.

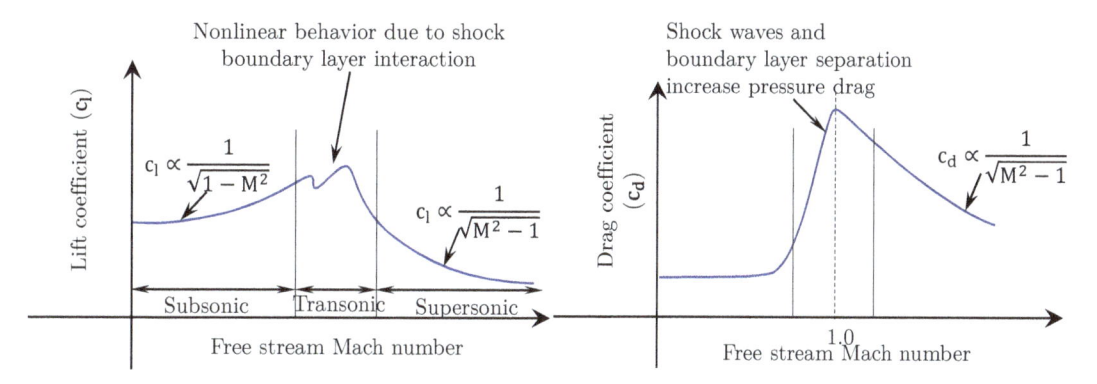

Figure 4.1: The generic variations of c_l and c_d with free-stream Mach number, M for a conventional airfoil

4.2.4 Wings

In the previous section, we considered an airfoil, which is essentially a wing with a reduction in a span-wise (along the length) dimension. Let's define some terminologies associated with the wing of the aircraft. The wingspan, b, is the distance from the left wing tip to the right wing tip measured perpendicular to the aircraft's longitudinal axis. The two-dimensional area of the wing as seen by looking down on the aircraft (planform view) is called wing area, S. An important dimensionless geometric parameter called *Aspect Ratio (AR)* is defined as $AR = b^2/S$. We have defined the chord in the previous section for an airfoil. The chord for the wing varies along the wingspan. The chord at the tip of the wing is the tip chord and the chord at the root (where wings are attached to the fuselage) is called the root chord. The ratio of the tip chord to the root chord is called the taper ratio.

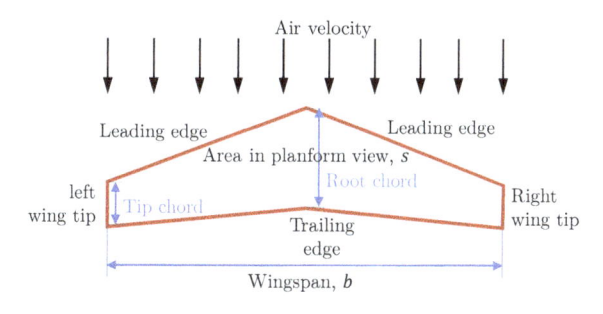

Till now, we looked at the aerodynamic characteristics of the cross-section of the wing called the airfoil. The obvious question is whether the aerodynamic coefficients of a uniform wing are the same as those of the airfoil. Unfortunately, the answer is no. However, the aerodynamic characteristics of an airfoil can be used to predict those of the wing. The difference between the two comes from the nature of the flowfield in both cases. In an airfoil, the flow is restricted to be two-dimensional in the plane of the airfoil. Whereas for a wing, the flow is three-dimensional, especially near the tips where the flow "leaks" from the high-pressure region below the wing to the low-pressure region above the wing. This flow along the span of the wing creates the wingtip vortices. Though magnificent to visualize, they create an additional component of drag called the "induced drag". They also reduce the effective slope of the lift versus the angle of the attack curve. The lifting line theory by *Prandtl* predicts that the coefficient of induced drag varies as the square of the lift coefficient and it varies inversely with the aspect ratio of the wing. The induced drag coefficient, $C_{D,i} = C_L^2/(\pi e AR)$ is defined, where, C_L is the coefficient of lift of the wing and $e \leq 1$ is the span efficiency factor (depends on the shape of the wing and equal to unity for an elliptical planform). Hence, subsonic airplanes designed to minimize induced drag have high-aspect-ratio wings. For instance, the narrow wings of the *Lockheed U-2* high-altitude reconnaissance aircraft have an aspect ratio of 14.3. Supersonic aircraft have a lower AR to reduce the wave drag. It can be understood that the

Airplane	Aspect Ratio
F-16	3.2
Grumman X-29	3.9
Vought F4U Corsair	5.4
Wright Flyer	6.4
Lockheed Vega	6.1
Boeing 747	7.0
Boeing 727	7.1
Boeing B-17	7.6
Douglas DC-3	9.1
Lockheed U-2	14.3
Schweizer SGS 1-35	23.3

induced drag is the penalty that is paid for the production of lift.

Another important type of configuration of a wing is a swept wing. The swept angle is the angle between the 25% chord line (quarter chord line) and the lateral axis. The idea of swept wings was proposed by *Adolf Busemann* in 1935 to reduce drag at high speeds. Sweeping the wing has its own advantages. The effective chord of the swept wing is longer compared to the wing with no sweep as a result there is an increase in the critical Mach number of the wing. However, sweeping the wing reduces the lift. Though a swept wing reduces the transonic and supersonic wave drag (an advantage for high-speed airplanes), it causes a reduction in lift coefficient at low speed. To compensate for the reduction in the lift coefficient during low-speed operations (landing and take-off), high-lift devices such as trailing edge flaps, leading edge flaps and slats, etc. are installed. The swept wings are also used to provide longitudinal stability for tailless aircraft, e.g. *Messerschmitt Me 163 Komet*. Most commercial aircraft have swept wings. For example, *Boeing 747* has a sweep angle of $37.5°$ and Airbus A380 has wings swept at an angle of $33.5°$. Some aircraft have forward-swept wings which provide the same benefits reducing the critical Mach number and eliminating the wingtip stall problem of backward-swept wings. Forward-swept wings make an aircraft harder to fly, but the advantages are mainly due to maneuverability. The examples of aircraft with forward-swept wings are *Grumman X-29* or *Sukhoi Su-47*.

Swept wings with triangle-shaped planforms as shown in Figure 4.2 are called delta wings. Delta wings are used on many airplanes designed for supersonic flight, for example, the *HAL Tejas* and the *Concorde*. The aerodynamic flow over a low-aspect-ratio delta wing at low speeds is significantly different from that of a straight wing or a high-aspect-ratio swept wing. A delta wing has the advantages of a large sweep angle and also greater wing area than a simple swept wing to compensate for the loss of lift usually experienced with swept-back wings.

Wing Dihedral is the upward angle of an aircraft's wing, from the wing root to the wing tip. The amount of dihedral determines the amount of inherent stability along the roll axis. Although an increase in dihedral will increase roll stability, it will also decrease lift, increase drag, and decrease the axial roll rate. Generally adapted in civilian aircraft, for example, the dihedral angle of Boeing 737 is $6°$. Highly maneuverable fighter planes have no dihedral and some fighter aircraft have wing tips lower than the roots, giving the aircraft a high roll rate. A negative dihedral angle is called an anhedral as seen in *AV-8B Harrier II*.

Area Rule was discovered by NASA aerodynamicist *Richard Whitcomb* in the 1950s. It states that, in order to produce the least amount of drag when approaching supersonic flight, the cross-sectional area of an aircraft body should vary smoothly along the aircraft's length. To account for this at the location on an aircraft where the wings are attached to the fuselage, the fuselage needs to be made narrower. This is why aircraft that are designed to fly in the transonic regime have a pinched fuselage where the wings are attached to the body.

4.3 Propulsion

For an airplane in steady level flight, the thrust from the propulsion system must balance the drag. The power in excess of the drag can be used for accelerating the airplane or to gain altitude. For mis-

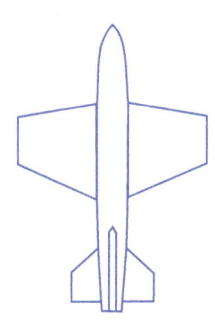

4.3. PROPULSION

Low AR wings - lower lift to drag ratio and better control, commonly used in

- Fighter aircraft for better manoeuvrability, (Ex. *Lockheed F-104 Starfighter with AR =2.45*) and
- High-speed aircraft to reduce the drag (Ex. *North American X-15 with AR=2.44 and* Concorde AR=1.55)

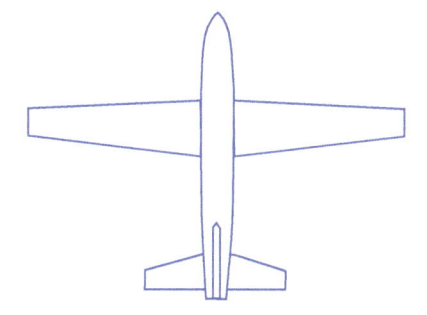

High AR wings : High lift to drag ratio, commonly used in

- Low-speed aircraft for high lift, (ex. Bombardier Dash 8 Q400, AR=12.8),
- Civilian aircraft (Ex. Boeing 777X with AR 10.74),
- Glider (Ex. Glaser-Dirks DG-500 AR 19.52 and ASH 31 glider AR=33.5)

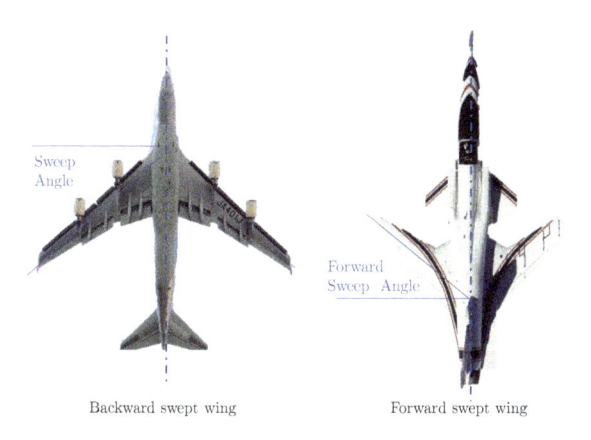

Backward swept wing   Forward swept wing

Swept wing.

siles, satellites and their launch vehicles, a significant portion of their overall mass is the propulsion system and the propellants. The propulsion system plays a key role in the performance characteristics of aerospace vehicles. In this section, we will look at some of the propulsion systems used in aerospace vehicles. To begin with, we will look at the basics of thermodynamics which forms the underlying working principle of propulsion systems.

4.3.1 Aspects of Thermodynamics

Thermodynamics deals with the transfer and transformation of energy from one form to another. Internal energy is stored energy in matter and can be viewed as a static form of energy. Another form of energy is macroscopic which is not stored but the energy gained or lost by a system during a process which can be viewed as dynamic energy. There are two forms of dynamic energy: heat and work. Heat is the form of energy that is transferred between two systems or a system and its surroundings by virtue of the temperature difference that exists between them. Whereas, work is associated with the force acting through a distance. In other words, if the energy transfer is due to temperature difference, the energy is heat otherwise it must be work.

An application of thermodynamics is the conversion of heat into useful work. You can also see that organized energy (work) can be converted to disorganized energy (heat) completely, but only a fraction of disorganized energy can be converted to organized energy by specially built devices called heat engines (like jet engines and internal combustion engines).

There are three thermodynamic laws governing energy transfer and transformation. The first law of thermodynamics states that energy can be neither created nor destroyed during a process; it can only change forms or energy is a conserved property.

The first law is concerned with the quantity of energy and the transformations of energy from one form to another. The second law introduces the concept of entropy as a physical property of a thermo-

(a) Taper ratio of wing

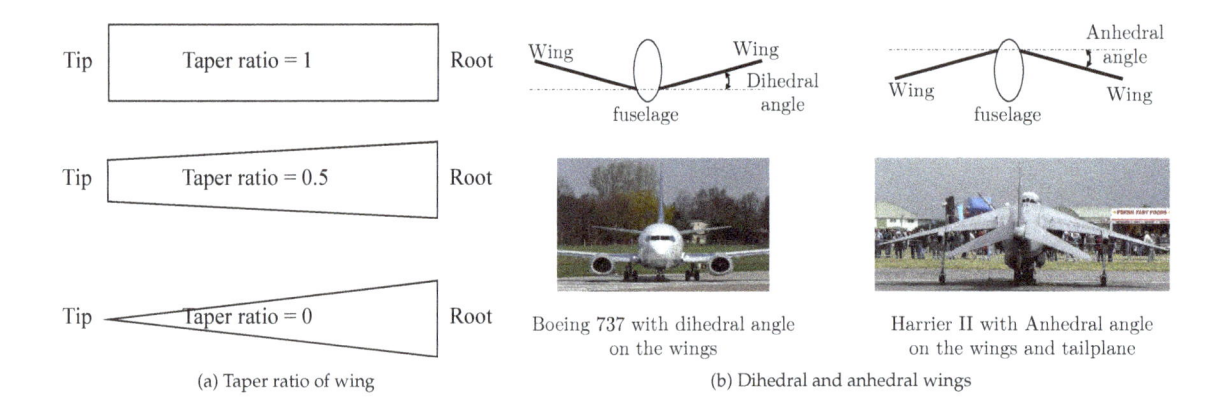

(b) Dihedral and anhedral wings

dynamic system. It predicts whether processes can actually happen despite obeying the first law.

Mechanism of Thrust Generation

Airplanes and rockets are propelled by accelerating a fluid flow in the opposite direction to motion. This is attained by either slightly accelerating a large mass of fluid as in a propeller-driven engine or greatly accelerating a small mass of fluid as in a turbojet engine or a combination of both like in the turboprop engine. The mechanism of thrust generated by the propeller and the jet need to be understood before getting into further details.

Wright brothers postulated propellers as twisted wings. The earlier propeller was driven by the reciprocating piston engines, but later gas turbine engines became common as in turboprop engines, propfan or advanced ducted propulsion. The propeller, like wings, are made up of airfoil sections designed to generate an aerodynamic force, provides thrust to push the airplane through the air. Unlike wings, a propeller is twisted such that the chord line of the airfoil changes from being almost parallel to airspeed at the root to almost perpendicular at the tip. The airflow seen by a propeller section is a combination of the airplane's forward motion and the rotation of the propeller itself as shown in the figure. The angle between the chord line and the propeller's plane of rotation is the pitch angle, β. The distance from the root to a given section is r. Due to the pitch angle and the rotation, the relative wind seen by the propeller section is the vector sum of airspeed (V) and the speed of the section due to rotation of the propeller, $r\omega$. At each airfoil section, a component of the aerodynamic force points in the forward direction. When this thrust is summed over the entire length of the propeller blades, the net thrust available is obtained.

The principle of thrust generation in the jet engine is based on the Newton's laws of motion. By accelerating the incoming air into the inlet of the engine as a high speed jet through the nozzle, the engine provides forward thrust to the aircraft. When a balloon is released with high-pressure air filled in, the air released at high speed, according to Newton's third law, produces a force opposite to the air velocity. The jet engine takes in air at the freestream velocity heats it by combustion of fuel inside the duct, and then ejects the hot mixture of air and combustion products out at the back end at a much higher velocity. The jet engine creates a change in the momentum of the gas by taking a small mass of air and giving it a large increase in velocity, thereby producing thrust. In the following section, the working principles of different propulsion systems such as reciprocating piston engines, gas turbine engines, and ramjet engines are covered. The thermodynamic cycles of each type of engines used in aerospace propulsion are presented.

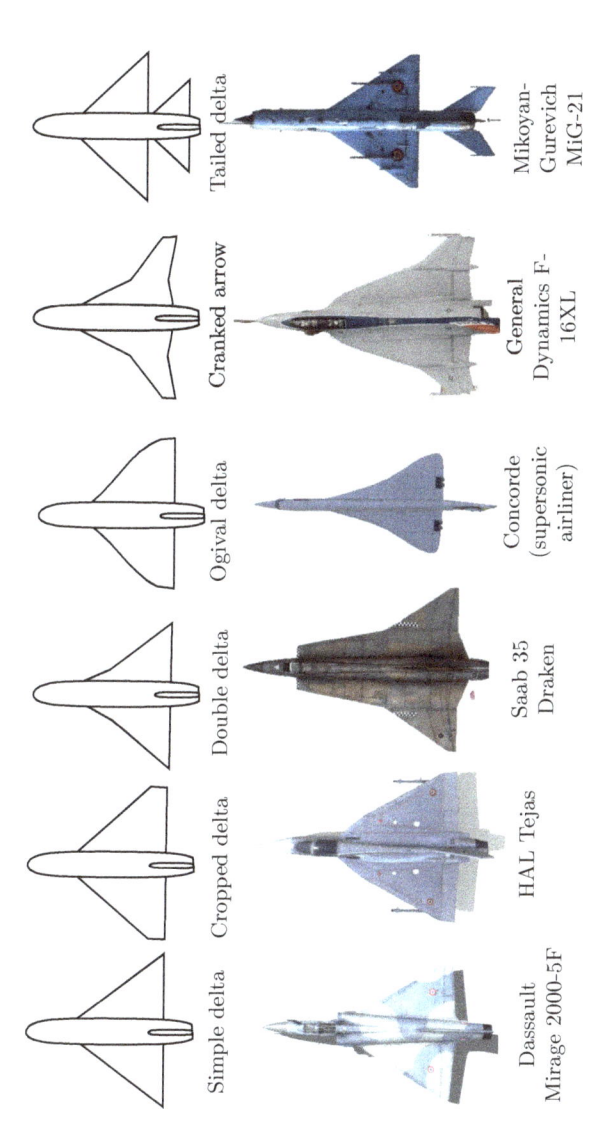

Figure 4.2: Delta wing configurations (Drawn NOT to scale).

4.3.2 Reciprocating Engines

The first airplane *Wright Flyer I* built by the *Wright brothers* and then nearly the first 50 years of manned flight were all propeller-driven airplanes, with propellers powered by engines essentially identical to automobile engines which are internal combustion, reciprocating, and gasoline-burning engines. These engines are still used today in airplanes designed to fly at speeds less than about 500 km/h, the range for the vast majority of light, private, general aviation aircraft. In the year 2000, the number of aircraft powered by jet engines (17200) equaled 10% of those powered by piston engines. Among the companies manufacturing piston-driven airplanes are the *Cirrus Aircraft Corporation*, *Cessna Aircraft Company*, *Beechcraft Corporation*, *Diamond Aircraft Industries*, and the *Mooney International Corporation*.

The reciprocating engine consists of a *piston-cylinder* arrangement as shown in figure 4.3. The piston moves back and forth in the cylinder between two fixed positions, *top dead center (TDC)* and the *bottom dead center (BDC)*. The charge (air or air–fuel mixture) is drawn into the cylinder through the *intake valve*, and the combustion products are expelled from the cylinder through the *exhaust valve*. The distance between the TDC and the BDC is the largest distance that the piston can travel in one direction, and it is called the *stroke* of the engine. The diameter of the piston is called the *bore*. The minimum volume formed in the cylinder when the piston is at TDC is called the *clearance volume*. The volume displaced by the piston as it moves between TDC and BDC is called the *displacement volume*. The ratio of the maximum volume formed in the cylinder to the minimum (clearance) volume is called the *compression ratio*.

Depending on how the combustion process in the cylinder is initiated, reciprocating engines can be classified as spark-ignition (SI) engines or compression-ignition (CI) engines. In the SI engines, the combustion of the air–fuel mixture is triggered by a spark plug. Whereas in CI engines, the air–fuel mixture is self-ignited due to the compression of the air-fuel mixture above its self-ignition temperature.

Thermodynamic Cycles

Heat engines such as pistons or gas turbines are designed for the purpose of converting thermal energy to work, and their performance is expressed in terms of the thermal efficiency η, which is the ratio of the net work produced by the engine to the total heat input. The engine that operates on a totally reversible cycle, such as the *Carnot cycle*, has the highest thermal efficiency among all other en-

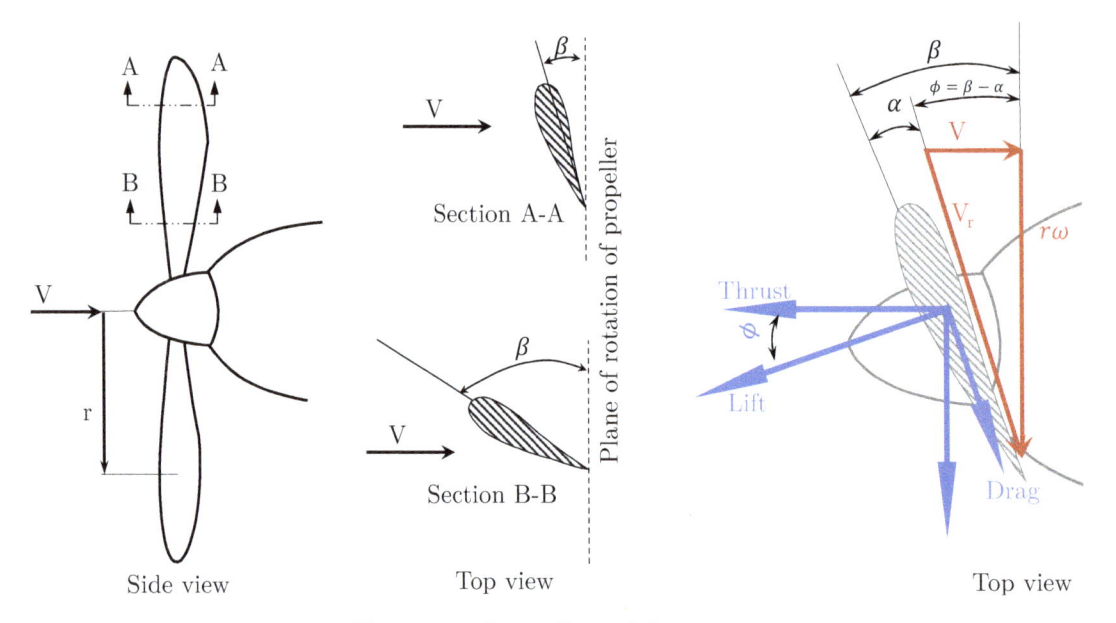

Illustration of propeller and thrust generation.

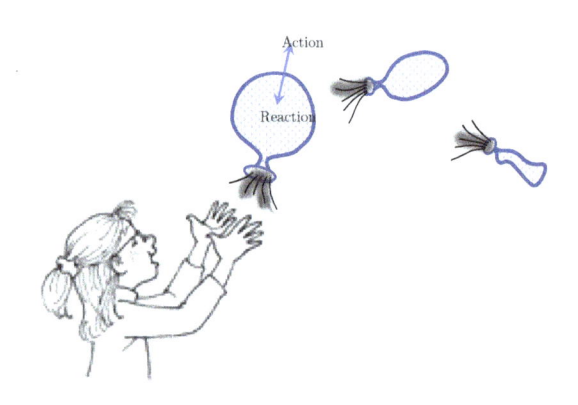

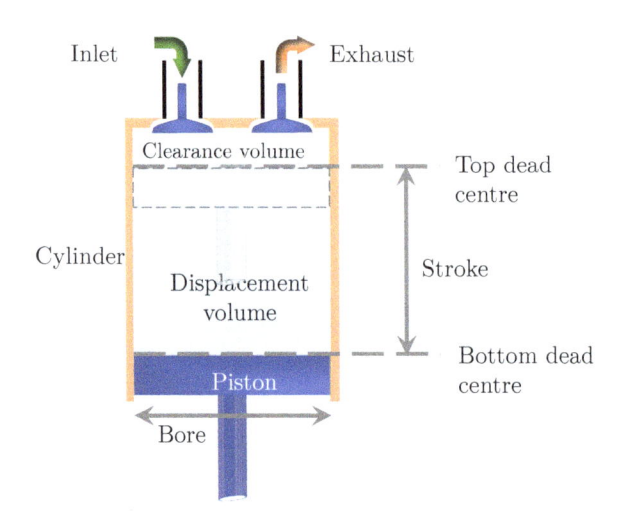

Figure 4.3: Elements of reciprocating engine.

gines operating between the same temperature levels. However, it is practically difficult to build an engine which operates on the Carnot cycle. Reversible isothermal heat transfer is an ideal concept which is practically difficult to achieve because it would require very large heat exchangers and it would take a very long time. The thermal efficiency of the ideal Carnot cycle is a function of the sink and source temperatures only. The thermal efficiency becomes 100 % when the source temperature is infinity or the sink temperature is 0 K. The highest temperature in the cycle is limited by the temperature that the components of the heat engine can withstand. The lowest temperature is limited by the temperature of the cooling medium us ed in the cycle such as the atmospheric air. The *Otto cycle* is the ideal cycle for spark-ignition type of engines. Its thermal efficiency depends on the compression ratio of the engine and the ratio of specific heat capacities of the working fluid. For a given working fluid, the thermal efficiency increases with the compression ratio. But, when high compression ratios are used, the temperature in some regions of the air–fuel mixture rises above the *autoignition* temperature of the fuel during the combustion process, causing an early and rapid burn of the fuel at some points ahead of the flame front, followed by almost instantaneous inflammation of the end gas. This premature ignition of the fuel, called autoignition, produces an audible noise, which is called *engine knock*. Autoignition in SI engines affects performance and can cause mechanical damage. This places an upper limit on the compression ratios that can be used in SI internal combustion engines. The *Diesel cycle* is the ideal cycle for CI reciprocating engines which operate at much higher compression ratios and thus are usually more efficient than the spark-ignition engines for similar conditions.

4.3.3 Gas Turbine Engines

Before World War II, turbine engines existed only as laboratory items for testing. By the end of the War, it was clear that the future of aviation lay with gas turbine engines. These turbine engines have great power and thrust in comparison to the reciprocating engines. The advantage of the gas turbine lies in the absence of reciprocating and rubbing members means that balancing problems are few, that the lubricating oil consumption is low, and that reliability can be high. The first flight of an airplane powered by a turbojet engine was the *He-178* German aircraft powered by the *Heinkel HeS-3* which was designed by *Hans von Ohain* while working at *Heinkel* in 1939. Later, the German engine designer *Anslem Franz* developed his turbojet engine that powered the fighter aircraft *Messerschmitt Me 262* in 1942. *Frank Whittle* in England, having no knowledge of *Ohain*'s engine, built his *W.1 (or Whittle W.1)* turbojet engine, which powered the *Gloster E28/39* aircraft in 1941.

The jet engine produces the thrust by blasting the gas out from the back end faster than it comes in through the front end. This is achieved by inducting a mass of air through the *intake* where the flow is reduced to a low subsonic Mach number in a diffuser. For a subsonic flow, the diffuser is a divergent duct and if the incoming flow is supersonic, the diffuser must be a convergent-divergent duct. The decrease in flow velocity is accomplished

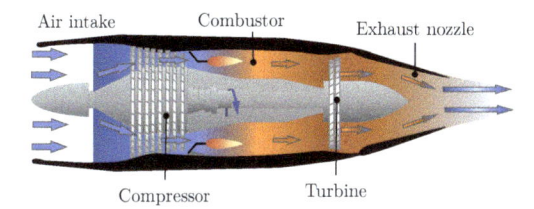

Figure 4.4: Components of jet engine

partly through shock waves. In the diffusion process, the static pressure is increased. Therefore, the intake section delivers a smooth and uniform stream of air at a reduced velocity which is suitable for the *compressor* where the air gets compressed to raise the pressure, temperature and density. By increasing the pressure of the air, the volume of the air is reduced, which means that the combustion of the fuel/air mixture will occur in a smaller volume. The pressurized air from the compressor enters the *combustion chamber*, where fuel is injected and burned, thus adding hugely more energy to the airflow which leads to a steep increase in temperature, whereas pressure remains almost constant. From here the airflow is processed to produce mechanical work in *turbine* efficiently. A turbine is a complementary part of a compressor, to which it is rigidly linked by a hollow shaft or spool. The task of the turbine is to convert gas energy into mechanical work to drive the compressor. In a *turbojet*, a major part of the heat and pressure energy of the gas is exhausted in the *exhaust nozzle* which is of characteristic tube-like shape to accomplish energy conversion from heat and pressure to velocity. High exhaust velocity is a prerequisite to the generation of thrust. The compressor can be centrifugal or axial based on the direction of the airflow. In the centrifugal compressor, the compression occurs by first increasing the velocity of the air (through rotation) and then diffusing the air where the velocity decreases and the pressure increases. The air in an axial compressor flows in an axial direction through a series of rotating rotor blades and stationary stator vanes that are concentric with the axis of rotation. Each set of rotor blades and stator vanes is

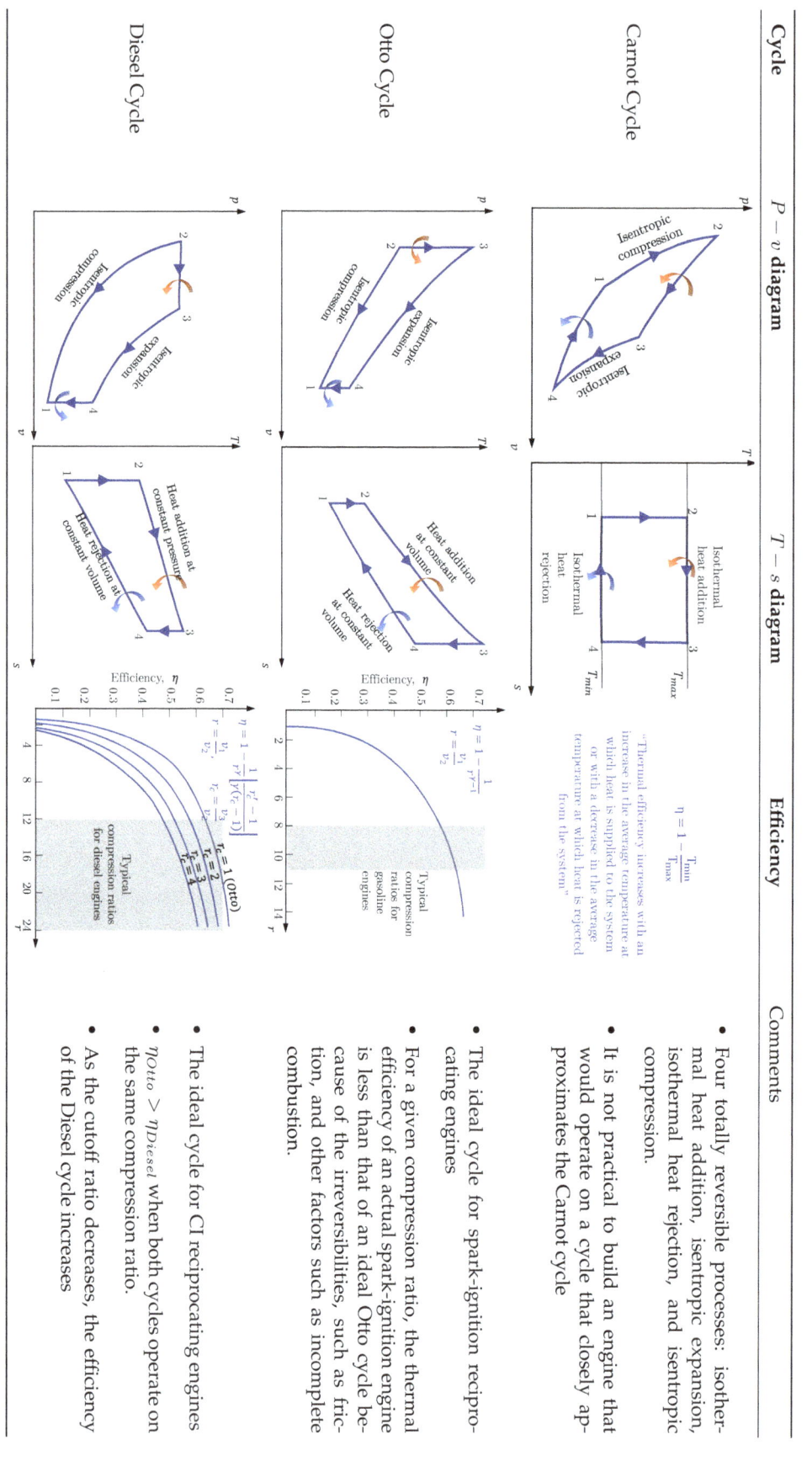

Table 4.1: Carnot, Otto and Diesel cycles

Cycle	$P-v$ diagram	$T-s$ diagram	Efficiency	Comments
Carnot Cycle			$\eta = 1 - \dfrac{T_{min}}{T_{max}}$ "Thermal efficiency increases with an increase in the average temperature at which heat is supplied to the system or with a decrease in the average temperature at which heat is rejected from the system"	• Four totally reversible processes: isothermal heat addition, isentropic expansion, isothermal heat rejection, and isentropic compression. • It is not practical to build an engine that would operate on a cycle that closely approximates the Carnot cycle
Otto Cycle			$\eta = 1 - \dfrac{1}{r^{\gamma-1}}$, $r = \dfrac{v_1}{v_2}$	• The ideal cycle for spark-ignition reciprocating engines • For a given compression ratio, the thermal efficiency of an actual spark-ignition engine is less than that of an ideal Otto cycle because of the irreversibilities, such as friction, and other factors such as incomplete combustion.
Diesel Cycle			$\eta = 1 - \dfrac{1}{r^{\gamma-1}}\left[\dfrac{r_c^{\gamma}-1}{\gamma(r_c-1)}\right]$, $r_c = \dfrac{v_3}{v_2}$, $r = \dfrac{v_1}{v_2}$, $r_c = 1$ (Otto)	• The ideal cycle for CI reciprocating engines • $\eta_{Otto} > \eta_{Diesel}$ when both cycles operate on the same compression ratio. • As the cutoff ratio decreases, the efficiency of the Diesel cycle increases

known as a stage. The flow path in an axial compressor decreases in the cross-sectional area in the direction of flow. The decrease in area is in proportion to the increased density of the air as the compression progresses from stage to stage. The axial compressor is more popular due to the following reasons. An average, single-stage, centrifugal compressor can increase the pressure by a factor of ~ 4. A similar average, single-stage axial compressor increases the pressure by only a factor of ~ 1.2. But it is relatively easy to link together several stages and produce a multistage axial compressor. In the multistage compressor, the pressure is multiplied from row to row (8 stages at 1.2 per stage gives a factor of $= 1.2^8 = 4.3$). It is much more difficult to produce an efficient multistage centrifugal compressor because the flow has to be ducted back to the axis at each stage. Because the flow is turned perpendicular to the axis, an engine with a centrifugal compressor tends to be wider, having a greater cross-sectional area than a corresponding axial compressor. This creates additional undesirable aircraft drag. For these reasons, most high-performance, high-compression turbine engines use multi-staged axial compressors. But, if only a moderate amount of compression is required, a centrifugal compressor is much simpler to use.

The performance of the ideal gas turbine cycles known as *Brayton cycle* is reviewed. The ideal gas turbine cycle was first proposed by *George Brayton*.[1] Gas turbines operate on the *Brayton* cycle where fresh air at ambient conditions is drawn into the compressor, where its temperature and pressure are raised (Isentropic compression in a compressor). The high-pressure air proceeds into the combustion chamber, where the fuel is burned at constant pressure (Constant-pressure heat addition). The resulting high-temperature gases then enter the turbine, where they expand to the atmospheric pressure while producing power (Isentropic expansion in a turbine). The exhaust gases leaving the turbine are thrown out to the ambient air (Constant-pressure heat rejection). The $T-s$ and $P-v$ diagrams of an ideal *Brayton* cycle are shown in Fig. 4.5 The steady flow energy equation neglecting changes in kinetic and potential energies is given by

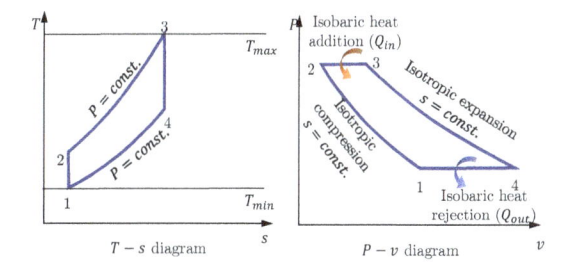

Figure 4.5: $T-s$ and $P-v$ diagrams for the ideal Brayton cycle.

$$(Q_{in} - Q_{out}) + (w_{in} - w_{out}) = h_{exit} - h_{inlet}$$
$$\text{where,} \quad Q_{in} = h_3 - h_2 = c_p (T_3 - T_2)$$
$$w_{in} = -(h_2 - h_1) = -c_p (T_2 - T_1)$$
$$w_{out} = (h_3 - h_4) = c_p (T_3 - T_4) \quad \text{and}$$
$$Q_{out} = h_4 - h_1 = c_p (T_4 - T_1).$$
(4.1)

h is the enthalpy, c_p is the specific heat capacity at constant pressure, T is the temperature. The subscripts refer to the states in the thermodynamic cycle. Then the thermal efficiency of the ideal Brayton cycle which is the ratio of net work out $(w_{out} - w_{in})$ and the heat input (Q_{in}) under the standard assumptions becomes

$$\eta = \frac{c_p (T_3 - T_4) - c_p (T_2 - T_1)}{c_p (T_3 - T_2)} = 1 - \frac{(T_4 - T_1)}{(T_3 - T_2)}$$

Using $p_1 = p_4$, $p_2 = p_3$ and an isentropic pressure-temperature relations

$$\frac{T_2}{T_1} = r^{(\gamma-1)/\gamma} = \frac{T_3}{T_4}, \quad \Rightarrow \quad \frac{T_4}{T_1} = \frac{T_3}{T_2}$$

where, $r = p_2/p_1 = p_3/p_4$ is the pressure ratio and

---

[1] American inventor *George Brayton* (1830 - 1892) is recognized as the most important of the early American inventors of vehicles utilizing internal combustion. *Brayton* patented a two-stroke kerosene stationary engine known as *Brayton's Ready Motor*, which had one cylinder for compression, a combustion chamber, and a separate cylinder in which the products expanded for the power stroke. It bore a marked resemblance to a steam engine with its rocking beam and flywheel. His engine needed no spark plug - it had a continuously burning flame to ignite each cycle of the engine. He demonstrated that prolonging the combustion phase of the cycle, by injecting fuel at a controlled rate, produced more power per unit of fuel consumed. However, much of the efficiency gained by this method was lost due to the lack of an adequate method of compressing the fuel mixture before ignition. *Brayton's* engine was considered the first safe and practical oil engine and also served as an inspiration to *George B. Selden* (1846 - 1922) to develop a combustion engine. For a few years, *Brayton's* engine was well regarded, but within a short time, the *Otto* engine became more popular.

γ is the specific heat ratio of gas, the cycle efficiency is then readily shown to be given by

$$\eta = 1 - \frac{T_1}{T_2}\frac{(T_4/T_1 - 1)}{(T_3/T_2 - 1)} = 1 - \frac{T_1}{T_2} = 1 - \left(\frac{1}{r}\right)^{(\gamma-1)/\gamma}$$

The thermal efficiency of an ideal Brayton cycle depends on the pressure ratio of the gas turbine and the specific heat ratio of the working fluid (the nature of the gas). The thermal efficiency increases with both of these parameters,

The specific work output is another important parameter upon which the size of the gas turbine depends. The specific work is given by $W = c_p(T_3 - T_4) - c_p(T_2 - T_1)$, which can be expressed in dimensionless form as

$$\frac{W}{c_p T_1} = t\left(1 - \frac{1}{r^{(\gamma-1)/\gamma}}\right) - \left(r^{(\gamma-1)/\gamma} - 1\right) \quad (4.2)$$

where $t = T_3/T_1$ is the ratio of the maximum temperature in the cycle that occurs at the end of the combustion process to the minimum temperature of the cycle. Therefore, the specific work output is a function of pressure ratio and maximum temperature (T_3) which is limited by the maximum temperature that the turbine blades can withstand. Early gas turbines used values of t between 3-5, but the introduction of air-cooled turbine blades allowed t to be raised up to 6. This also limits the pressure ratios that can be used in the cycle. For a fixed *Turbine Inlet Temperature* (T_3), the specific work output increases with the pressure ratio, reaches a maximum value and then starts to decrease. Therefore, there should be a compromise between the pressure ratio (thus the thermal efficiency) and the net work output. The specific work output is zero at $r = 1$ and $t^{\left(\frac{\gamma}{\gamma-1}\right)}$. The optimum value of the pressure ratio at which the specific work output is maximum is given by $r_{opt} = t^{\left(\frac{\gamma}{2(\gamma-1)}\right)}$. This condition implies that $T_2 = T_4$. Thus the specific work output is a maximum when the pressure ratio is such that the compressor and turbine outlet temperatures are equal. For all values of pressure ratio between 1 and $t^{\left(\frac{\gamma}{2(\gamma-1)}\right)}$, where specific work output increases with pressure ratio, T_4 will be greater than T_2. In such a scenario, the heat energy from the turbine exhaust can be used to preheat the air at the compressor outlet before entering into the combustor using the heat exchanger which effectively reduces the heat addition by fuel burning and consequently increases the efficiency. The expression for the efficiency of *heat-exchange cycle or regeneration cycle* is given by $\eta = 1 - \frac{r^{(\gamma-1)/\gamma}}{t}$. Thus it is evident that the efficiency of the *heat-exchange cycle* is not independent of the maximum cycle temperature, but increases as t is increased. Furthermore, for a given value of t, the efficiency increases with a decrease in pressure ratio and not with an increase in pressure ratio as for the simple cycle(refer 4.1). It can be noted that the heat-exchange cycle is effective only for the values of pressure ratio between 1 and $t^{\left(\frac{\gamma}{2(\gamma-1)}\right)}$ i.e., $T_4 > T_2$. For higher values of r for which $T_4 < T_2$, a heat exchanger would cool the air leaving the compressor and so reduce the efficiency. It can be concluded that to obtain an appreciable improvement in efficiency by heat-exchange cycle, a value of r should be appreciably less than r_{opt}.

The net work of a gas-turbine cycle is the difference between the turbine work output and the compressor work input, and it can be increased by either decreasing the compressor work or increasing the turbine work, or both. A substantial increase in work output can be obtained by splitting the expansion and reheating the gas between the high-pressure and low-pressure turbines utilizing *multistage expansion with reheating*. The work required to compress air between two specified pressures can be decreased by carrying out the compression process in stages and cooling the gas in between, that is, using *multi-stage compression with intercooling* the gas between the low pressure (LP) and high pressure(HP) compressors. It can be shown that the specific work output is maximum when the pressure ratios of the LP and HP compressors are equal. These arguments are based on the principle: The steady-flow compression or expansion work is proportional to the specific volume of the fluid. Therefore, the specific volume of the working fluid should be as low as possible during a compression process and as high as possible during an expansion process. It can be observed from the $T-s$ diagram (refer 4.6) that the vertical distance between any pair of constant pressure lines increases as the entropy increases, thus the turbine work is increased or compressor work is decreased is obvious. Thus for reheating, $(T_5 - T_6) + (T_7 - T_8) > (T_5 - T_{6'})$ and for intercooler $(T_2 - T_1) + (T_4 - T_3) < (T_{2'} - T_1)$. This is precisely what intercooling and reheating accomplish. It is noted from table 4.2 that the reheat markedly increases the specific output which is achieved at the expense of efficiency. This is to be expected because one is adding a less efficient cycle ($6'678$ in Fig. 4.6) to the simple cycle which is less efficient as it operates over a smaller temperature range. Note that the reduction in efficiency becomes less severe as the maximum cycle temperature is increased. The reduction in efficiency due to reheating can be overcome by adding a heat exchanger. The heat in the exhaust gas can be utilized. In fact, when a heat exchanger is employed, the efficiency is higher with reheat than without.

The specific work output of a gas-turbine cycle improves as a result of intercooling and reheating, but with a loss of thermal efficiency. This is because intercooling decreases the average temperature at which heat is added, and reheating increases the average temperature at which heat is rejected. However, if intercooling and reheating are accompanied by a heat exchanger (regeneration), the efficiency can also be improved. Therefore, in gas-turbine power plants, intercooling and reheating are always used in conjunction with regeneration. If the number of compression and expansion stages is increased, the ideal gas-turbine cycle with intercooling, reheating, and regeneration approaches the Ericsson cycle and the thermal efficiency approaches the theoretical limit of the Carnot efficiency. However, the contribution of each additional stage to thermal efficiency is less and less, and the use of more stages cannot be justified economically.

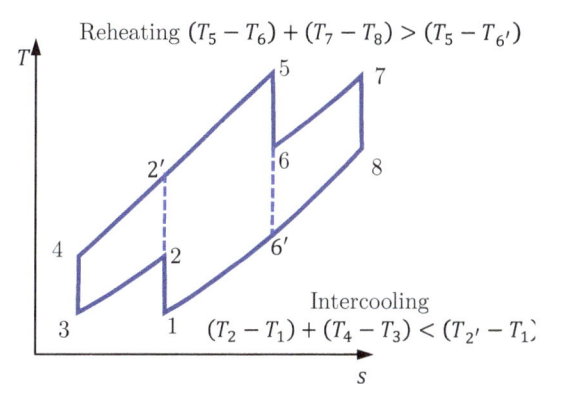

Figure 4.6: $T-s$ diagram of an ideal gas-turbine cycle with intercooling and reheating

The performance of real cycles differs from that of ideal cycles due to the irreversibilities associated with the diffuser, the compressor, the turbine and the nozzle. Assumptions of an ideal cycle and methods of accounting for losses in the actual cycle is presented in table 4.3. The deviation of the actual cycle from the ideal Brayton cycle is shown in Fig. 4.7. The effect of the irreversibilities is to reduce the thrust that can be obtained from a turbojet engine.

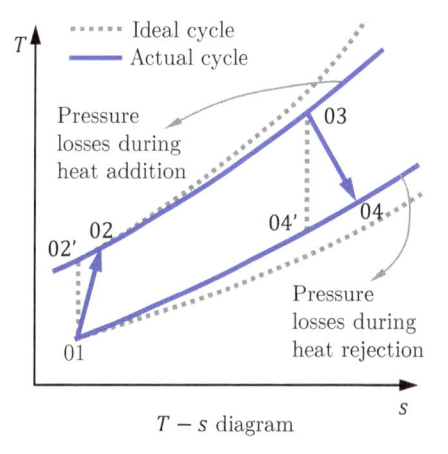

Figure 4.7: The deviation of an actual gas-turbine cycle from the ideal Brayton cycle as a result of irreversibilities. Primes attached to the symbol and the prefix 0 to the state denote the ideal and stagnation state, respectively.

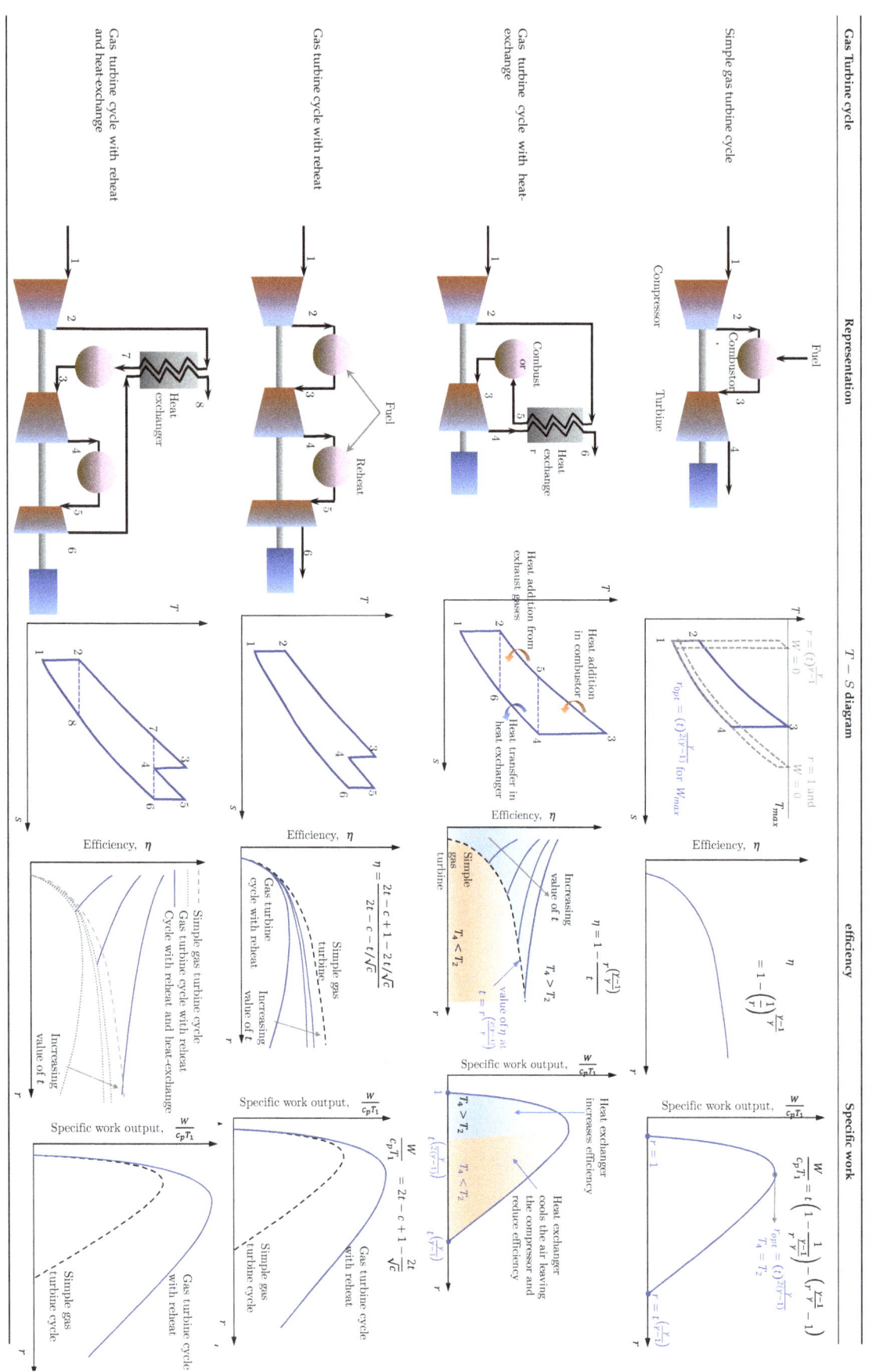

Table 4.2: Performance of different gas turbine ideal cycles

4.3. PROPULSION

Table 4.3: Assumptions of an ideal cycle and methods of accounting for losses (actual cycle)

Assumption in ideal cycle	Actual cycle
The change of kinetic energy of the working fluid between the inlet and outlet of each component is negligible	Because fluid velocities are high in turbomachinery the change in kinetic energy between the inlet and outlet of each component cannot necessarily be ignored. The kinetic energy terms can be accounted for using the concept of stagnation (or total) enthalpy (h_0) is the sum of gas enthalpy (h) and velocity (V), $h_0 = h + V^2/2$, $T_0 = TVC^2/2c_p$, $p_0 = p\left(\frac{T_0}{T}\right)^{\frac{\gamma}{\gamma-1}} = p\left(1 + \frac{\rho V^2}{2p}\frac{\gamma-1}{\gamma}\right)^{\frac{\gamma}{\gamma-1}}$.
Compression and expansion processes are reversible and adiabatic, i.e., isentropic.	The compression and expansion processes are irreversible and therefore involve an increase in entropy. To account for this, the compressor and turbine efficiencies (η_t and η_c) are defined as the ratio of ideal to the actual work output and actual to ideal work input, respectively. As a consequences $T_{02} - T_{01} = \frac{1}{\eta_c}\left(T'_{02} - T_{01}\right)$ and $T_{03} - T_{04} = \eta_t\left(T_{03} - T'_{04}\right)$ where, prime indicate ideal states,
There are no pressure losses in the inlet ducting, combustion chambers, heat-exchangers, intercoolers, exhaust ducting, and ducts connecting the components	Fluid friction results in pressure losses in combustion chambers and heat-exchangers, and also in the inlet and exhaust ducts. $p_{03} = p_{02} - \Delta p_b - \Delta p_{ha}$ and $p_{04} = p_a + \Delta p_{hg}$ where Δp_b is a loss in stagnation pressure that occurs in the combustion chamber due to the aerodynamic resistance of flame-stabilizing and mixing devices, and also due to momentum changes produced by the exothermic reaction. Δp_{ha} and Δp_{ha} are friction pressure losses in the passages of a heat-exchanger on the air-side and gas side, respectively.
Constant mass flow and same composition of a perfect gas with constant specific heats throughout the cycle	The values of c_p and γ of the working fluid vary throughout the cycle due to changes of temperature and, with internal combustion, due to changes in chemical composition. *Bleed flows*: Many modern engines make use of air-cooled turbine blades to permit operation at elevated temperatures which are achieved by extracting cool air from both stator and rotor blades. The required bleeds may amount to 15 percent or more of the compressor delivery flow in an advanced engine and must be properly accounted for accurate calculations.
Heat transfer in a heat-exchanger is complete	For an economic size heat exchanger, the compressed air cannot be heated to the temperature of the gas leaving the turbine. This is accounted by considering the heat-exchanger effectiveness
	Mechanical losses, combustion efficiency and cycle efficiency are to be considered in real cycle

Jet Propulsion

Unlike the Brayton cycle, aircraft gas turbine engines operate on an open jet-propulsion cycle where gases are not expanded to the ambient pressure in the turbine. The gases are expanded to a pressure such that the power produced by the turbine is just sufficient to drive the compressor and auxiliary equipment. The gases that exit the turbine at relatively high pressure are subsequently accelerated using a nozzle to provide the thrust to propel the aircraft. Also, aircraft gas turbines operate at higher pressure ratios (typically between 10 and 25), and the air passes through a diffuser first, where it is decelerated and its pressure is increased before it enters the compressor.

The thrust developed using a turbojet engine is the result of the difference in the momentum of the low-velocity air entering the engine and the high-velocity exhaust gases leaving through the nozzle which is a result of the Newton's second law. The pressure at the inlet and the exit of a turbojet engine can be equal to the ambient pressure for complete expansion in the nozzle.

The air enters the intake with a velocity V_i (relative to the engine) equal and opposite to the forward speed of the aircraft. The engine accelerates the air so that it leaves with the jet velocity V_j. The mass flow $\dot{m}$ is constant (i.e., the fuel flow rate is negligible relative to the airflow rate), and thus *the net thrust F due to the rate of change of momentum* (assuming the jet exit pressure to be equal to the ambient pressure, i.e., $p_j = p_a$) is $F = \dot{m}(V_j - V_i)$.

The thrust produced by the engines decreases with altitude as $\dot{m}$ is directly proportional to air density. Commercial airplanes save fuel by flying at higher altitudes since air at higher altitudes is thinner and exerts a smaller drag force on aircraft. However, the engine performance also gets affected with altitude. The *propulsive power* developed from the thrust of the engine is the propulsive force times the distance this force acts on the aircraft per unit time, i.e., the thrust times the aircraft velocity, $W = FV_i$. Unlike stationary gas-turbine engines, the efficiency of a turbojet engine (net work developed is zero) is defined in terms of *propulsive efficiency* (η_p) which is the ratio of the desired output (thrust power) to the sum of the thrust power and the unused kinetic energy of the jet. The propulsive efficiency can be expressed as

$$\eta_p = \frac{\dot{m} V_i (V_j - V_i)}{\dot{m} \left[V_i (V_j - V_i) + (V_j - V_i)^2 / 2 \right]} = \frac{2}{1 + (V_j / V_i)}$$

From the above equations, it is clear that η_p is a maximum when $V_j/V_i = 1$ but then the thrust is zero. This is the reason for the development of the family of propulsion systems based on the magnitude of the mass flow and the jet velocity.

Thrust Augmentation

The thrust of an engine is increased either by increasing the turbine-inlet temperature (in turn increasing the specific thrust and hence the thrust for a given engine size) or by increasing the mass flow rate through the engine without altering the cycle parameter. Any of the two methods imply some redesign of the engine, for example, the inlet area may be increased for large airflow, but this would be obtained at the expense of increased drag and weight. However, in certain circumstances, a temporary increase in thrust for a short time is required, for example, during takeoff, climb, and acceleration from subsonic to supersonic speed, or during combat maneuvers. In these circumstances, the thrust must be augmented for the fixed engine size. *Reheat or afterburning* is one of the methods to augmenting the basic thrust of an engine. Afterburning requires another combustion chamber located between the low-pressure turbine and the jet pipe propelling nozzle. Fuel is burnt in this second combustion chamber utilizing the unburned oxygen in the exhaust gas. This increases the temperature and velocity of the exhaust gases leaving the propelling nozzle and therefore increases the engine thrust. The turbojet with the addition of the afterburner requires a large fuel flow to rise the temperature and the penalty in increased fuel consumption is heavy. Another scheme for thrust augmentation for a given engine is *liquid injection* which is primarily useful for increasing take-off thrust. The thrust force depends to a large extent on the mass of the airflow passing through the engine. Spraying water into the compressor inlet causes evaporation of the water droplets, resulting in the extraction of heat from the air; the effect of this is equivalent to a drop in compressor inlet temperature. The technique is to inject a finely atomized spray of water or W/M (a mixture of water and methanol) into either the inlet of the compressor or into the combustion chamber inlet. Water injection was first used over 45 years ago on commercial transport aircraft *Boeing's 707-120* with *Pratt & Whitney JT3C-6* engines, and later on the *Boeing 747-100* and *-200* aircraft with *Pratt & Whitney JT9D-3AW* and *-7AW* engines for thrust augmentation.

Turbofan Engines

The propulsive efficiency of a simple turbojet engine can be improved by extracting a portion of the energy from the turbine to drive a ducted propeller called a fan. A larger diameter fan forces a considerable amount of air through the duct surrounding the engine. The fan exhaust leaves the duct at a higher velocity, contributing significantly to the total thrust of the engine. The fan increases the gas mass flow rate with an accompanying decrease in the required jet exit velocity for a given thrust, particularly for operation at high subsonic speeds. Another major advantage is that a lower jet velocity results in less jet noise. Thus the turbofan engine was conceived where a portion of the total incoming flow of air by-passes the compressor, combustion chamber, turbine and nozzle before being ejected through a separate nozzle. Thus the thrust has two components: the cold stream (or fan) thrust and the hot stream (which undergoes combustion) thrust. Sometimes, it is desirable to mix the two streams and eject them as a single jet of reduced velocity. An important parameter of turbofan engines is the *bypass ratio,* defined as the ratio of the mass flow through the cold stream to that of the hot stream. When both streams are expanded to atmospheric pressure in the propelling nozzles, the net thrust of the turbofan engine is given by

$$F = (\dot{m}_c V_{jc} + \dot{m}_h V_{jh}) - \dot{m}V_i,$$

$\dot{m} = \dot{m}_c + \dot{m}_h, \dot{m}_c = \dot{m}B/(B+1) \quad \dot{m}_h = \dot{m}/(B+1),$

where, $B = \dot{m}_c/\dot{m}_h$ is the bypass ratio and subscript c and h indicate the cold and hot jets, respectively. There are several advantages to the turbofan engine over both turboprop and turbojet engines. The fan is not as large as the propeller, so the increase of speeds along the blade is less. Thus a turbofan engine can power civil transport flying at transonic speeds as high as Mach number 0.9. Also, by enclosing the fan inside a duct or cowling, the aerodynamics are better controlled. There is less flow separation at higher speeds and less trouble with shock developing. Due to the higher mass flow rate, the turbofan can generate more thrust. Like the turboprop engine, the turbofan has low fuel consumption compared with the turbojet. Hence, turbofan engine is the best choice for high-speed, subsonic commercial airplanes.

Turboprop Engines

A turboprop engine is a hybrid engine that provides jet thrust and also drives a propeller. It is basically similar to a turbojet except that the turbine works through a shaft and speed-reducing gears to turn a propeller at the front of the engine. A turboprop engine has a concept similar to that of the turbofan, though having a higher bypass ratio. Instead of the turbine driving a ducted fan in a turbofan, it drives a completely external propeller. The turboprop uses a gas turbine core to turn a propeller. The propeller is driven either by the gas generator turbine or by another turbine classified as a free or power turbine. In this case, the free turbine is independent of the compressor-drive turbine and is free to rotate by itself in the engine exhaust gas stream.

As a general rule, propellers are more efficient than jet engines, but they are limited to low-speed and low-altitude operation since their efficiency decreases at high speeds and altitudes.

4.3.4 Ramjet and Scramjet Engines

Ramjet uses the forward motion of the engine to compress the incoming air, without a rotary compressor. Therefore, the compressor is eliminated, hence, the turbine is no longer required. Although a ramjet can operate at subsonic flight speed, the increasing pressure rise accompanying higher flight speeds renders the ramjet most suitable for supersonic flight. Ramjet is open at both ends like stovepipe and has only fuel nozzles in the middle. It has a properly shaped inlet-diffusion section to produce low-velocity and high-pressure air at the combustion section, also a properly shaped exhaust nozzle. High-speed air enters the inlet, is compressed as it is slowed down, is mixed with fuel and burned in the combustion chamber, and is finally expanded and ejected through the nozzle to produce the thrust. The air at the inlet of a high-speed engine (supersonic) is partially compressed across a shock system that precedes the inlet. The effect is called *ram effect*.

The operation of a ramjet depends on the flight speed and it cannot develop static thrust and therefore cannot accelerate a vehicle from a standing start. The vehicle must first be accelerated by other means to a sufficiently high speed prior to using a ramjet as a propulsive device. It may be dropped from a plane or launched by rocket assist.

When the flight speed is high enough, this compression can be sufficient to operate an engine without a compressor. The higher the speed, efficient is the combustion process as the air gets compressed sufficiently. Hence the ramjet is efficient for high-speed applications (Mach number >3). Ramjets become practical for military and missile applications which require flying at very high or supersonic speeds. Ramjets can operate at speeds as low Mach number as 0.2, but, fuel consumption is horrendous at these low velocities. At a lower Mach number, the turbojet engine performs better compared to the ramjet engine. To operate at higher Mach numbers, the turbojet must increase its combustion temperature. Therefore, the high-temperature material

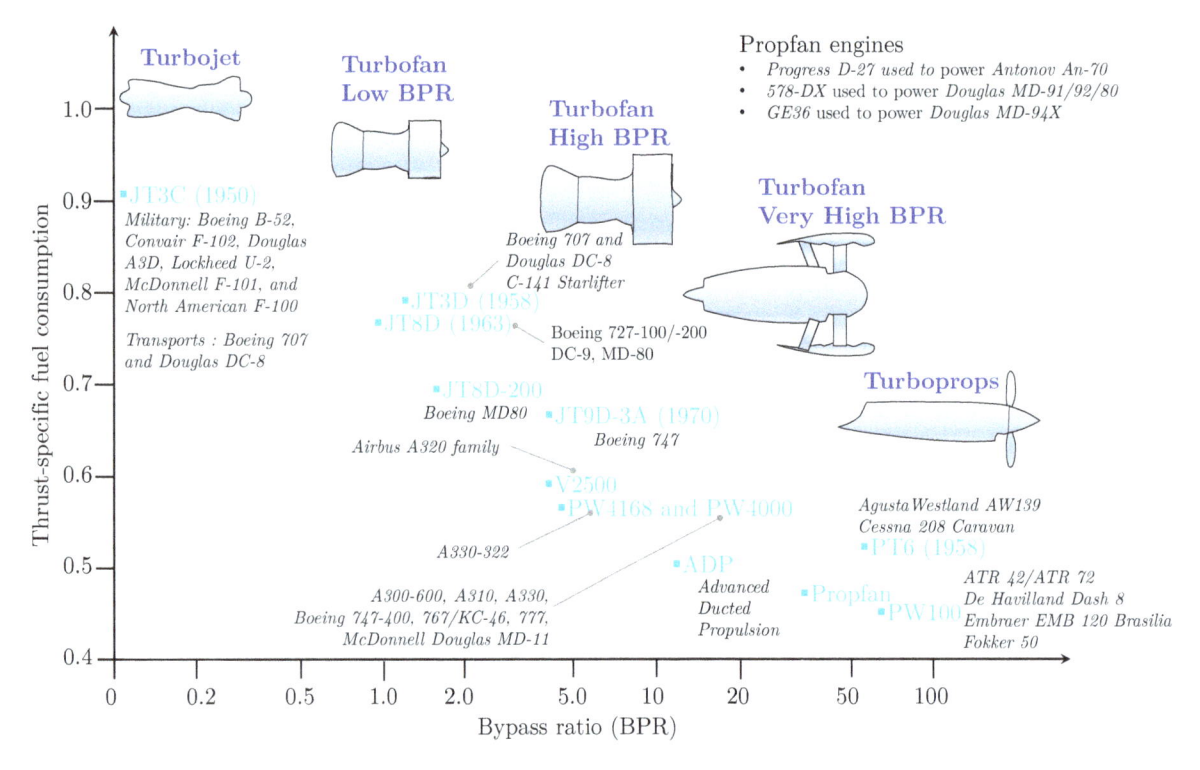

Figure 4.8: Trends in engine bypass ratio.

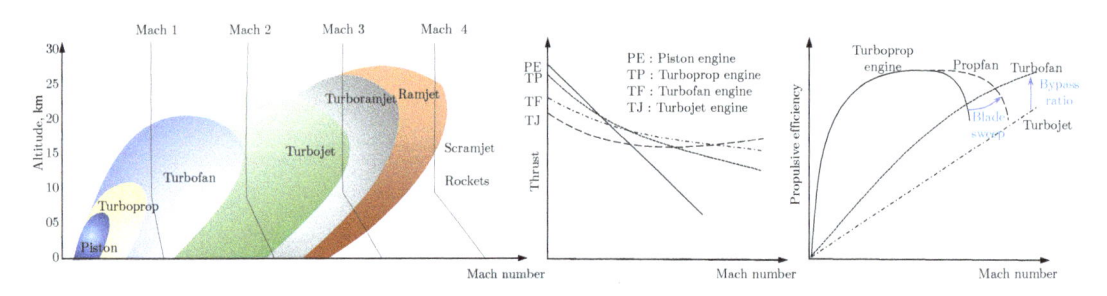

Figure 4.9: Comparison of the different propulsion systems in terms of thrust and efficiency.

properties of the turbine blades limit the conventional turbojet operation at a higher Mach number. Since ramjet has no turbine, its combustion temperatures can be much higher. Therefore, for sustained and efficient atmospheric flight at Mach numbers above 3, a ramjet is virtually the only choice, given our present technology. The fuel efficiency gets better as the Mach number increases, but the thrust produced per unit mass flow rate is maximum for a certain Mach number. Further, increase in the Mach number, the thrust decreases as shown in Fig. 4.11.

A hybrid engine, part turbojet at lower speeds, part ramjet at higher speeds (known as a turbo-ram

engine), was also used on the *SR-71* (the world's fastest and highest-flying air-breathing operational manned aircraft) and is a topic of current research interest for several possible future aerospace vehicles.

Scramjet (Supersonic Combustion Ramjet)

While ramjets operate by subsonic combustion of fuel in a stream of air compressed by the forward speed of the aircraft itself, in scramjet engines the airflow through the whole engine remains supersonic. An inlet airstream compression is efficient for the flight Mach number lower than 5 to decelerate the flow to subsonic velocity before it enters

4.3. PROPULSION

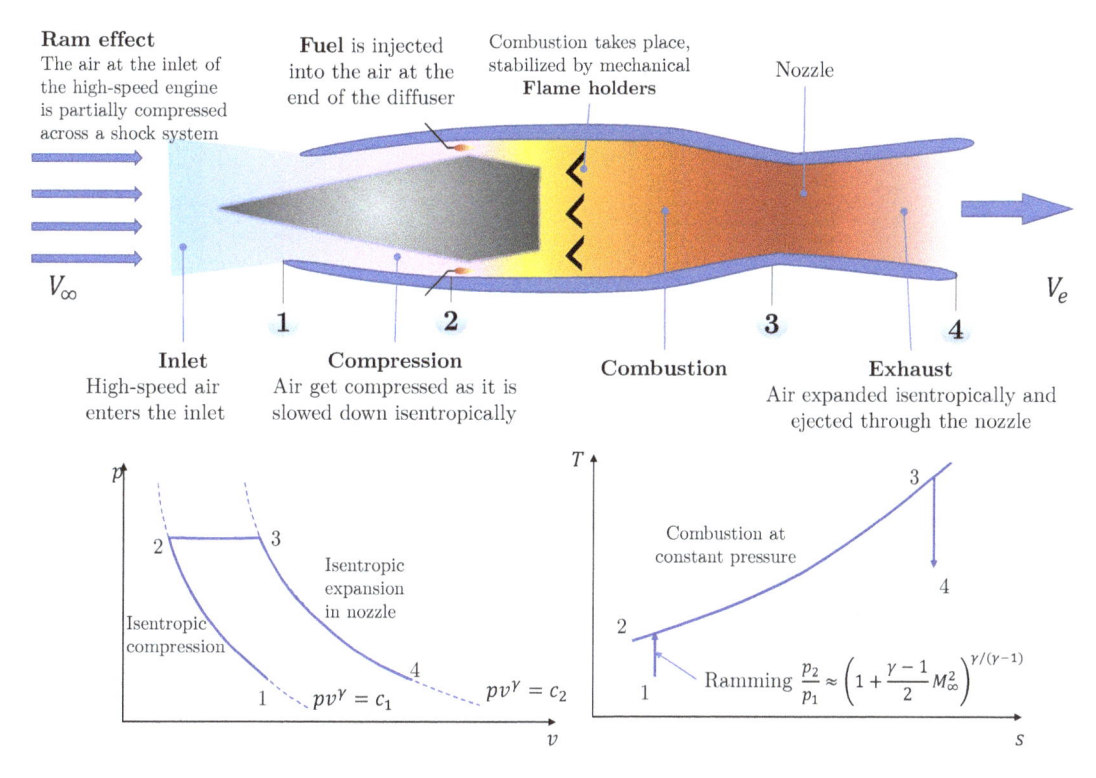

Figure 4.10: Working principle of ramjet engine

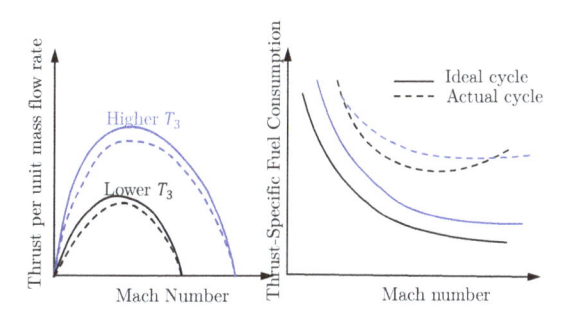

Figure 4.11: Specific thrust and TSFC for ideal and real ramjet engines

Table 4.4: Advantages and disadvantages of ramjet engine

Advantages	Disadvantages
Able to attain high speeds (around Mach number 5) and ram effect is efficient at higher speeds, so is the engine	The vehicle needs a booster to accelerate sufficiently high speed prior to using a ramjet as a propulsive device.
No moving parts so less wear and tear as well as minimum losses	It performs badly at lower speeds due to inefficient compression, hence higher fuel consumption (subsonic ramjet have a very small overall efficiency and too high TSFC (0.05 − 0.2 kg/kN.s)
Supersonic ramjet have very low Thrust-specific fuel consumption (TSFC)(less than 0.05 kg/kN.s)	
Lighter than turbine-based engines.	Maximum altitude is limited.
Higher temperatures can be employed	Since, it operates at a higher temperature, adequate material is required.

the combustion system. However, deceleration is more difficult and costly in terms of pressure losses, and it is necessary to make provision for the combustion chamber to burn its fuel in the supersonic airstream. Therefore, Scramjet operates at very high speeds and is projected to be fueled by a cryogenically liquified gas (e.g., hydrogen or methane). The advantage of using cold cryogenic fuel is to exploit greater heat values and it may be used as a heat sink for cooling a very high-speed and very hot engine and structure. Scramjet-powered vehicles are envisioned to operate at speeds up to Mach number 15. Long-duration, full-scale testing requires flight test speeds above Mach number 8. NASA's *Hyper-X*, an experimental unmanned hypersonic aircraft, will build knowledge, confidence and a technology bridge to very high Mach number flight. The *X-43* is the fastest jet-powered aircraft on record at approximately Mach number 9.6.

X-43A with scramjet attached to the underside

4.3.5 Pulsejet Engines

The pulsejet engine, like ramjet, is a very simple jet engine without a rotating turbine and compressor. The difference is in the intake where the air is supplied intermittently using a valve; thus, combustion occurs in pulses, resulting in a pulsating thrust rather than a steady or continuous one. A pulsejet is mechanically very simple and consists of a short inlet diffuser leading to a set of flow check valves, followed by a combustion chamber and a shaped tube. A fuel injection system is located downstream of the valves. The air flowing into the engine through the valves is mixed with a fuel spray. This fuel-rich mixture undergoes combustion using a spark plug in the combustion chamber, as a result, the pressure and temperature rise up. The valves close due to the momentary pressure rise and the gases expand to high velocity in the nozzle and the tail-shaped pipe. The short burst of expansion of gases creates a vacuum in the combustion chamber. Therefore the high-pressure fuel-air mixture upstream forces the valves to open starting a new cycle of recharging the combustion chamber, combustion, expansion and suction. The frequency of explosions can be as high as 300 per second. Its performance does not improve at high speeds. On account of low air density its performance deteriorates with altitude.

4.3.6 Rocket Propulsion

The atmosphere becomes thinner as we go up in the atmosphere. At very high altitudes, the atmosphere becomes so thin that the reciprocating engines and the jet engines which depend on the atmospheric air to produce thrust become inoperative. For such scenarios, rocket propulsion may be the only viable option. Even for some of the missiles which operate within the sensible atmosphere, rocket propulsion maybe a better design choice because of their reliability, cost effectiveness, ease of operation and storability. Due to its early development for military applications, rocket propulsion has been in use for a considerable more period than reciprocating engines and jet engines.

Rocket propulsion, a form of jet propulsion, generates thrust by accelerating the mass of gas. This mechanism is similar to a pressurized balloon, where thrust is produced by ejecting air through a hole, causing movement of the balloon in the opposite direction. In a chemical rocket, combustion products at high pressure and temperature are created in a chamber by the combustion of the chemical propellants. These gases are expanded through a nozzle to create a high speed jet which imparts the thrust based on Newton's laws of motion. The chemical propellants can be further classified based on the physical state of the propellants. Liquid propellant rocket engines use propellants in the liquid phase which are fed by pressure or pump into the thrust chamber. In the thrust chamber, they undergo combustion to produce hot gases which are exhausted through the nozzle. In solid propellant rocket motors, the propellant in the solid form is contained in the motor case and all the chemicals required for the combustion are found in the propellant. Although not very common, gaseous propellant is possible. It is also possible to have hybrid propellant rocket systems which use a combination of liquid and solid propellants. Chemical rockets are the main propulsion mechanisms for most missiles and satellite launch vehicles.

Converging-Diverging (CD) Nozzle: Nozzles are used to accelerate gases to produce thrust, the magnitude of which depends on the mass flow rate, the exit velocity of the flow, and the exit pressure.

These essential flow variables are dictated by the design of the nozzle. CD nozzle is a passage of varying area consisting of a *converging section* leading to a minimum area known as *throat*, followed by a *diverging section*. The subsonic flow is accelerated in the converging section and allow it attain the sonic speed (Mach number equal to one which is know as *choked flow*) by maintaining the appropriate pressure difference between inlet and exit pressure and then flow will isentropically expand to a supersonic Mach number. The exit Mach number depends on the area ratio of the exit to the throat. Note that if the sonic speed is not reached at the throat, the flow in divergence section actually decelerates. In addition to rocket engines and jet engine after-burners, CD nozzle also finds application in supersonic wind tunnels. The Mach number at any location in the CD nozzle is a function of the ratio of the local nozzle area (A) to the sonic throat area (A^*).

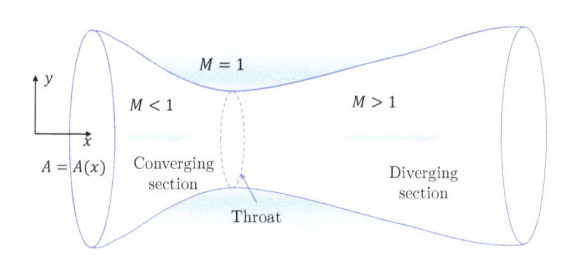

Converging-Diverging (CD) Nozzle

In electric rocket propulsion, electric energy is used in the generation of thrust. There are various categories of electric rocket propulsion. In electrothermal rocket propulsion, electric energy is used to heat the propellant and the hot gas is expanded to very high velocities using nozzles. Hydrazine, ammonia and hydrogen have been used as propellants in this type of propulsion. In ion propulsion, the propellant, typically xenon, is ionized and using strong electric field these ions are accelerated to very high velocities and thereby generating thrust. In the case of a magnetoplasma rocket, an electrical plasma is accelerated by means of combined electric and magnetic fields. The magnitude of thrust produced by the electric rocket propulsion is typically less than a few Newtons. So, they are used for more specialized applications like satellite orbit control.

Scaled down model of the *RD-171* liquid rocket engine developed by *Roscosmos* for use in *Soyuz-5* launch vehicle.

The performance of rocket propulsion systems is evaluated using following parameters.

Specific Impulse is the ratio of thrust (F) produced to the mass flow rate of propellant ($\dot{m}$) times the acceleration due to gravity at sea level conditions ($g_0 = 9.8 m/s^2$). i.e., $I_{sp} = \frac{F}{\dot{m} g_0}$ It has the units of seconds. The specific impulse multiplied by g_0 is called the **effective exhaust velocity** and it represents the average equivalent velocity with which propellant is ejected from the nozzle.

The **characteristic velocity** is defined as $c = \frac{p_1 A_t}{\dot{m}}$, p_1 is the chamber pressure and A_t is the throat area of the nozzle. It is used for evaluating the relative performance of different chemical propulsion system designs and is independent of the characteristics of the nozzle.

Table 4.5: Types of aerospace propulsion systems

Type	Applications	Characteristics	Working principle	Example
Rocket propulsion	Space launchers, Missiles	– Produce thrust by accelerating exhaust gases produced by combustion of propellant (fuel and oxidizer) – Jet velocity range, solid propellant – 1500-2600 (m/s) and liquid propellant – 2000-3500 (m/s) – Can operate outside Earth's atmosphere – Capable of very high thrusts – Can operate at high supersonic speeds – Heavy, non-air-breather (must carry oxidizer propellant) – Most are expendable (one-time use)		*S200* solid rocket motor using HTPB propellant, *L110* liquid engine using UDMH/N2O4 propellant and *C'25* cryogenic engine using LOX/LH2 in the *LVM3* space vehicle.
Piston engines	General aviation aircraft	– Relatively low cost – Relatively simple maintenance – Only capable of low thrusts – Limited to low subsonic speeds – Limited to low altitudes		*Lycoming O-360* piston engine used in *Cessna 172* aircraft
Turbojet engines	Fighter aircraft, long-range missiles	– Capable of high thrusts – Capable of supersonic speeds (normally an afterburner must be used) – Jet velocity range 350-1200 (m/s) – Less fuel efficiency relative to turbofan engines		*Tumansky R-25* engine used in *MiG-21* aircraft
Turbofan engines	Commercial aircraft, Business jets and Most combat aircraft	– Widely used today – Capable of medium to high thrusts – Capable of supersonic speeds (normally an afterburner must be used) – Jet velocity range 300-600 (m/s) – Better fuel efficiency than turbojet engines		*CFM56* engine used in *Boeing 737* aircraft

continued on following page

4.3. PROPULSION

Table 4.5, continued from previous page

Type	Applications	Working principle	Characteristics	Example
Turboprop engines	Short-range commercial aircraft, Cargo transports and Military troop cargo transports		– Fuel efficient – Short take-off and landing distances – Jet velocity range 30-200 (m/s) – Low to medium altitude limit – Subsonic speed limitation – Noisy, high vibration	*Pratt & Whitney Canada PW100* engine used in *ATR 72* aircraft
Turboshaft engines	Helicopter and Auxiliary power Unit		– Optimized to produce shaft power – Short in length – Jet velocity range 30 (m/s) – Generally not suitable for fixed-wing aircraft	*Honeywell T55* engine used in *Boeing Chinook* helicopter
Ramjet engines	Long range supersonic missiles, specialized aircraft		– Lighter and mechanically very simple as no moving parts so less wear and tear as well as minimum losses – Can operate efficiently at high supersonic Mach numbers (2.5 – 5.0) – Generally cannot operate at subsonic speeds, so requires a booster rocket – Jet velocity range 900-2400 (m/s)	Ramjet engine used in *BrahMos* missile
Scramjet engines	Experimental hypersonic vehicles		– Many difficult technical challenges, no operational models yet – Can operate at hypersonic Mach numbers (5.0 – 15.0) – Cannot operate at subsonic or low supersonic speeds, so requires a booster rocket	Scramjet engine used in NASA *X-43* program

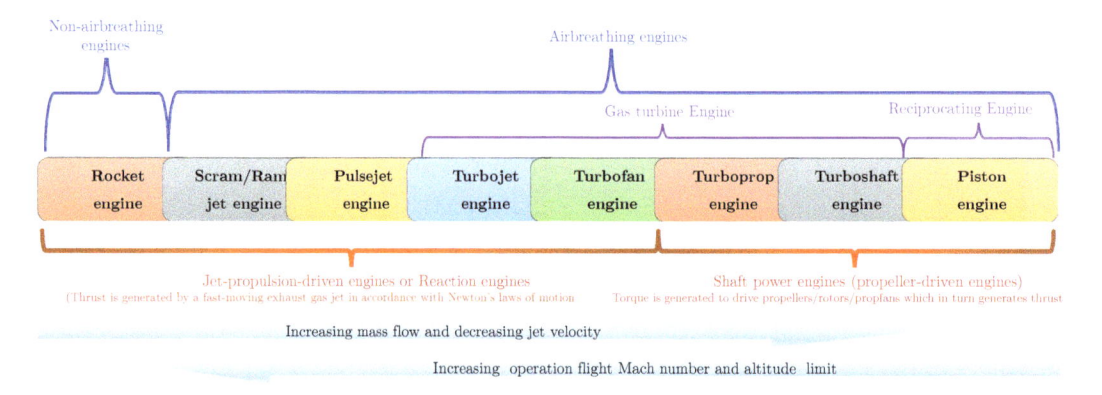

Figure 4.12: Overview of different propulsion systems used in aerospace vehicles.

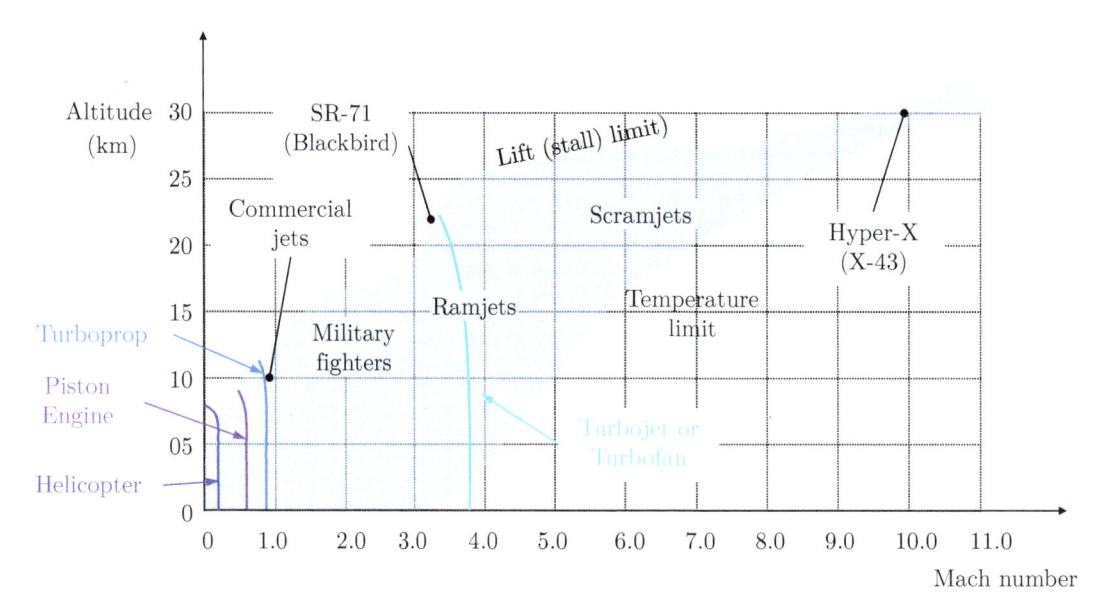

Figure 4.13: Flight envelope of aerospace vehicles with different propulsion system

4.4 Flight Mechanics

Imagine an Air Force is planning to decommission some of the old aircraft in its hangars and plans to get newer ones for combat roles. They will have to compare and contrast the performance characteristics of various aircraft that are available. A detailed understanding of these characteristics is crucial in such a situation. This section helps to understand the performance parameters of aircraft in general.

In the aerodynamics section, we have seen how aerodynamic forces are generated when an aircraft is flying. Flight mechanics deals with the movement of the airplane as it responds to the forces exerted on it: aerodynamic force, propulsive thrust and weight. The flight mechanics studies include performance, stability and control of the aircraft. Some of the details such as the runway length required to take off, at what rate the aircraft climbs to the safe altitude in the air, amount of fuel required to reach the destination etc. are part of the study. For these calculations, it is necessary to know how the drag varies with lift. Drag polar gives the variation of drag with lift for an aircraft.

Drag Polar for an aircraft is expressed as $C_D = C_{D,0} + KC_L^2$, where, C_D is the total drag coefficient, $C_{D,0}$ is the zero-lift drag coefficient or parasite drag coefficient at zero lift, which contains not only the profile drag of all the components of the aircraft that is exposed to the airflow. At transonic and supersonic speeds, $C_{D,0}$ also contains wave drag, an additional component of pressure drag due to the presence of shock waves. KC_L^2 is the drag due to lift, including both the contributions due to induced drag and the increment in parasite drag due to angle of attack different from zero angle of attack. Here, K is the proportionality constant for a given configuration. C_L is the total lift coefficient, including the contributions from the horizontal tail and fuselage. The drag polar depends on the configuration, Mach number and Reynolds number.

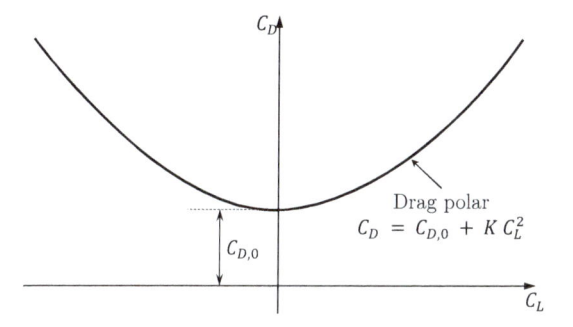

Figure 4.14: Schematic of a drag polar.

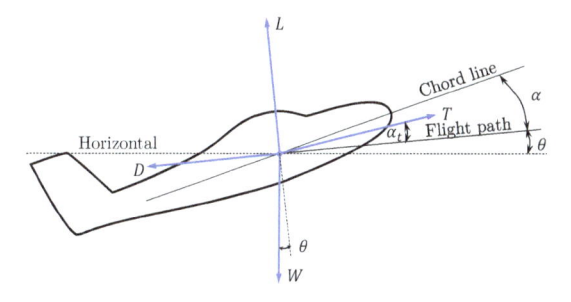

Figure 4.15: Force diagram for an airplane in flight. The flight path (same as relative wind direction) is the direction of motion of the airplane which is inclined at an angle θ, called the flight path angle, with respect to the horizontal. The mean chord line is at a geometric angle of attack α with respect to the flight path direction. Lift, L, is perpendicular to the flight path direction. Drag, D, is parallel to the flight path direction. Weight W, acts vertically downwards toward the center of the Earth and thrust T, in general, is inclined at the angle α_t with respect to the flight path direction.

4.4.1 Performance of Flight Vehicles

The fundamental equations of motion of flight for an airplane moving in a 2D plane are established using the force balance diagram as shown in Fig. 4.15. Equating the forces parallel and perpendicular to the path line yields the equations of motion for an airplane in translational flight.

$$\text{Parallel:} \quad T\cos\alpha_t - D - W\sin\theta = m\frac{dV}{dt}$$

$$\text{Perpendicular:} \quad L + T\sin\alpha_t - W\cos\theta = m\frac{V^2}{r_c} \quad (4.3)$$

where, $m\frac{dV}{dt}$ is the force as per Newton's law (Mass (m) multiplied by acceleration ($\frac{dV}{dt}$) along the flight path, V is the instantaneous value of the airplane's flight velocity. $\frac{V^2}{r_c}$ is the acceleration normal to a curved path with a radius of curvature r_c. Hence, $m\frac{V^2}{r_c}$ is the centrifugal force. The equations of motion given in Eq. (4.3) reduced to

$$T = D \quad \text{and} \quad L = W$$

for the following assumptions:

Static performance: Unaccelerated flight (acceleration, $m\frac{dV}{dt} = 0$ used in calculations of parameters such as maximum velocity, maximum rate of climb, maximum range, etc. which are vital interest in airplane design and operation

Level flight: The flight path is along the horizontal that is $\theta = 0$ and $r_c \to \infty$.

α_t *is very small*: For most conventional airplanes, the angle of attack is very small, hence, $\alpha_t \sim 0$ and $\sin \alpha_t \sim 0$

The aerodynamic forces acting on an airplane in steady, level (horizontal) flight is illustrated in Fig. 4.16. For the level flight, the weight W acts vertically downward. The lift L acts vertically upward, perpendicular to the relative wind velocity such that $L = W$ and the thrust T from the propulsive mechanism and the drag D are both parallel to wind velocity such that $T = D$. For most conventional flight situations, $L/D \approx 15 - 20$.

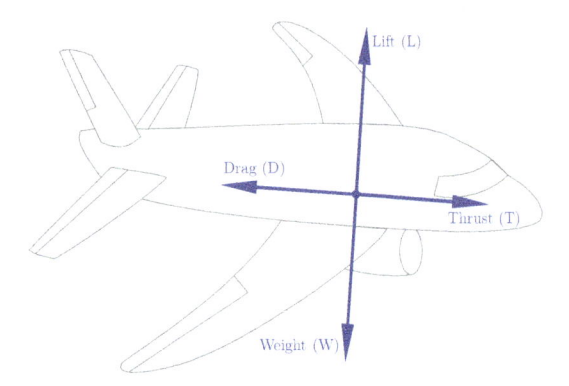

Figure 4.16: The four forces acting on an airplane in steady, level flight.

Thrust Required for Level, Unaccelerated Flight

For a level, unaccelerated flight, the airplane's power plant must produce a net thrust equal to the drag. The thrust required to maintain a steady velocity is given by $T = D = C_D(1/2)\rho V^2 S$ which can be written as

$$T = \frac{W}{C_L/C_D} = \frac{W}{L/D}. \tag{4.4}$$

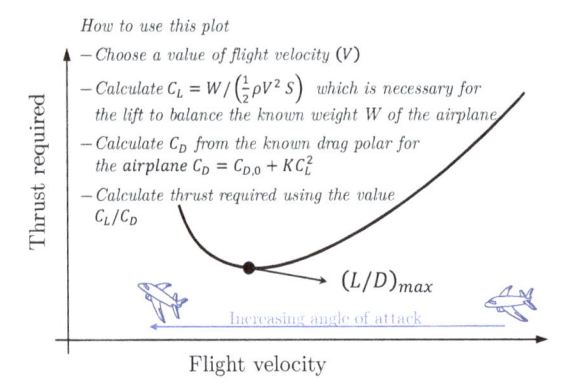

Figure 4.17: Thrust-required curve.

To obtain the expression for thrust, $L = W = C_L(1/2)\rho V^2 S$ for level, unaccelerated flight is used. The thrust required is dictated by the aerodynamics and weight of the airplane itself. The thrust-required curve for steady, level flight is plotted in Fig. 4.17 which is the locus of the thrust required to fly at the specific velocity in the flight range of the airplane. The thrust required varies inversely as L/D. Hence, the minimum thrust required will be obtained when the airplane is flying at a velocity where L/D is the maximum. The lift-to-drag ratio, L/D, which is a function of the angle of attack, measures the aerodynamic efficiency of an airplane. When an airplane is flying at the velocity for minimum thrust required, it is simultaneously flying at the angle of attack for maximum L/D, as given in the Equation 4.4. It is evident that different points on the thrust-required curve correspond to different angles of attack. As we move from right to left on the thrust-required curve, the airplane angle of attack increases.

Lift and drag coefficients play a strong role in the preliminary design and performance analysis of airplanes. The dependency of the lift and drag coefficients on the angle of attack and wind velocity is presented below. For an airplane of a given shape

4.4. FLIGHT MECHANICS

at a given Mach number and Reynolds number, C_L and C_D (capital letters for an aircraft) are only functions of the angle of attack, α of the airplane. C_L increases linearly with α until an angle of attack is reached when the wing stalls, the lift coefficient reaches a peak value, and then drops off as α is further increased. The lowest possible velocity at

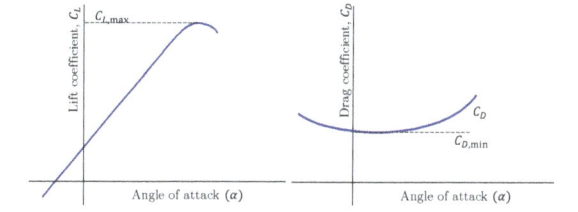

Figure 4.18: Schematic of lift and drag coefficients versus angle of attack, and maximum lift coefficient and minimum drag coefficient.

which the airplane can maintain steady level flight is the stalling velocity, V_{stall} which is depended on the value of $C_{L_{max}}$. The expression for V_{stall} can be obtained for level flight using the following relation $L = W = (1/2)\rho V^2 S C_L \Longrightarrow V = \sqrt{\frac{2W}{\rho S C_L}}$. For a given airplane flying at a given altitude, weight (W), free stream air density ρ and planform area (S) are fixed values and each value of velocity corresponds to a specific value of C_L. The stalling velocity, $V_{stall} = \sqrt{\frac{2W}{\rho_S C_{L_{max}}}}$, for a given airplane is determined by $C_{L_{max}}$ which is determined purely by laws of the aerodynamic flow over the airplane. There are certain mechanical devices called *high-lift devices* such as flaps, slats, and slots on the wing that when deployed by the pilot, serve to increase $C_{L_{max}}$ artificially beyond that of the basic airplane shape, and hence decrease the stalling speed. High-lift devices are usually deployed during landing and take-off.

The maximum velocity in steady, level flight for a given airplane configuration with a given thrust from the engine is determined by equating the thrust required to the thrust available from the engine. The thrust required curve as shown in Figure 4.17 equals the thrust available from the engine at two different velocities. The high speed intersection gives the maximum velocity that can be achieved in

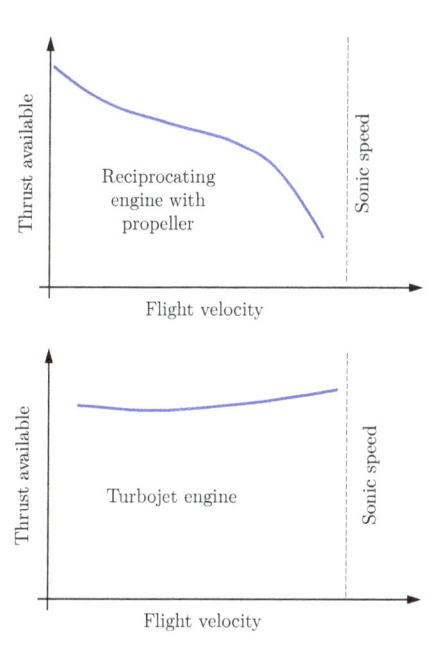

Figure 4.19: Thrust-available curves for piston engine-propeller combination and a turbojet engine.

steady, level flight. The low speed intersection is the minimum velocity with which the aircraft will be able to fly in steady, level flight. Typically, this velocity is less than the stall velocity and in such cases stall velocity as discussed before is the minimum velocity with which the aircraft will be able to fly in steady, level flight.

Thrust Available and Maximum Velocity

The thrust available (T_A) is the propulsive thrust provided by any power plant such as an engine-propeller combination, a turbojet, a rocket, or any other, which means T_A is strictly associated with the power plant. For reciprocating piston engines with propellers, the thrust at zero velocity (static thrust) is a maximum and decreases with forward velocity as shown in Fig. 4.19. At near-sonic flight speeds, the tips of the propeller blades encounter compressibility problems and the thrust available rapidly deteriorates. In contrast, the thrust of a turbojet engine is relatively constant with velocity. The intersection of the thrust required curve (dependent on the airframe, aerodynamics and weight) and the maximum thrust available curve (dependent on the engine) defines the maximum velocity of the airplane at the given altitude. Calculating the maximum velocity is an important part of the airplane design process.

Since piston engines are rated in terms of power (usually horsepower), the power required and the power available are the more relevant quantities for aircraft driven by the piston-engine-propeller combination. Since the power available from the engines depend on atmospheric density, the power available from both jet engines and piston engines decreases with altitude. The power required for an airplane is obtained by the product of thrust required and velocity, i.e., $P_R = T_R V$. The power available and power required for both propeller-driven and jet-propelled airplanes are given in Fig. 4.20.

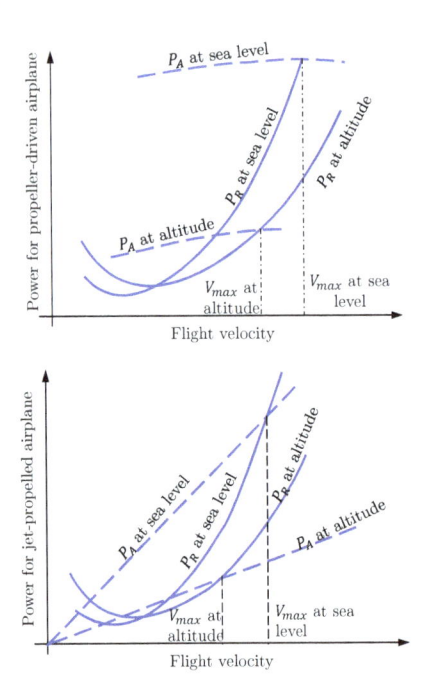

Figure 4.20: Power available and required. Effect of altitude on maximum velocity.

Maximum Velocity

In order to understand the design parameters that dictate the maximum velocity of an airplane, let us derive an expression for a maximum velocity for a steady, level flight, $T = D$ and $W = L$. We have

$$T = D = qSC_D = qS\left(C_{D,0} + KC_L^2\right) \quad (4.5)$$

where, $q = \frac{1}{2}\rho V^2$ is the dynamic pressure. Eliminating the C_L from the expression using the weight expression, $W = qSC_L$, one can get

$$T = qS\left[C_{D,0} + K\left(\frac{W}{qS}\right)^2\right],$$

solving the equation for q and taking the high speed solution,

$$q = \frac{\left(\frac{T}{W}\right)\left(\frac{W}{S}\right) + \left(\frac{W}{S}\right)\sqrt{\left(\frac{T}{W}\right)^2 - 4KC_{D,0}}}{2C_{D,0}} \quad (4.6)$$

Assuming the thrust in Eq. (4.6) equal to the maximum thrust available (full-throttle thrust), $(T_A)_{max}$, we obtain an expression for the maximum velocity, V_{max}, (obtained from $q_{max} = \frac{1}{2}\rho V_{max}^2$) as

$$V_{max} = \sqrt{\frac{\left(\frac{T_A}{W}\right)_{max}\left(\frac{W}{S}\right) + \left(\frac{W}{S}\right)\sqrt{\left(\frac{T_A}{W}\right)_{max}^2 - 4KC_{D,0}}}{\rho C_{D,0}}} \quad (4.7)$$

Examining Eq. (4.7) carefully, V_{max} depends not on thrust alone, or weight alone, or wing area alone, but rather only on certain ratios of these quantities. Two of the most important airplane design parameters are defined

- $\left(\frac{T_A}{W}\right)_{max}$: maximum thrust-to-weight ratio or power loading

- $\frac{W}{S}$: wing loading

Also, we can see from eq. (4.7) that V_{max} depends on ρ_∞ (altitude), the zero-lift drag coefficient $C_{D,0}$, and K which is related to $(L/D)_{max}$ (the maximum value of the lift-to-drag ratio for the airplane, an important design parameter). Therefore, it can be concluded that V_{max} can be increased by

1. Increasing the maximum thrust-to-weight ratio $\left(\frac{T_A}{W}\right)_{max}$

2. Increasing the wing loading $\frac{W}{S}$

3. Decreasing the zero-lift drag coefficient $C_{D,0}$

4.4. FLIGHT MECHANICS

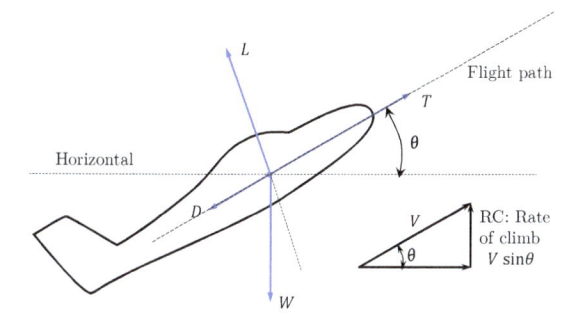

Figure 4.21: Airplane in climbing flight

Rate of Climb

Consider an airplane in steady, unaccelerated, climbing flight, as shown in Fig. 4.21. Thrust, T and the velocity, V, are along the flight path which is inclined to the horizontal at angle θ. The lift and drag are perpendicular and parallel to V, respectively and the weight is perpendicular to the horizontal. While climbing, the thrust not only works to overcome the drag but for climbing flight it is also supporting a component of weight. Hence, $T = D + W \sin \theta$. Now, the lift is smaller that the weight, $L = W \cos \theta$. These two equations represent the equations of motion for steady, climbing flight and are analogous. The vertical velocity of the airplane, *the rate of climb*, $R/C = V \sin \theta$, is obtained by multiplying the thrust equation by V and rearranging as $R/C = V \sin \theta = \frac{TV - DV}{W}$. Terms TV and DV represent the power available and the power required and the difference between them is called the *excess power*. Therefore, the rate of climb is the ratio of excess power to weight. A related concept is the rate of descent during gliding flight or power-off condition. The lift, drag and the weight are in equilibrium and the glide angle is given by $\tan \theta = \frac{1}{L/D}$ as shown in the Figure 4.23.

Absolute and Service Ceilings

It is clear from fig. 4.20 that the maximum velocity decreases as altitude increases. Also, the difference between power available and power required (the maximum excess power decreases), hence the rate of climb decreases for higher altitudes as shown in fig. 4.22. At some high altitude, the power available curve becomes tangent to the power required curve and the maximum excess power and rate of climb is zero at such altitude. The velocity at this point is the only value at which steady, level flight is possible. The altitude at which the aircraft has zero maximum rate of climb is defined as the *absolute ceiling* of the airplane. Further, the absolute ceiling does not represent the practical upper limit of steady, level flight. The term *service ceiling* is introduced which is defined as the altitude above which the maximum rate of climb falls below a prescribed value. The exact value depends on the type of aircraft and the configuration. The absolute and service ceilings can be determined using fig. 4.24.

Range and Endurance

The range is defined as the total distance, measured with respect to the ground, traversed by an airplane on a tank of fuel. Endurance is defined as the total time that an airplane stays in the air on a tank of fuel. One critical factor influencing range and endurance is the specific fuel consumption (SFC) which is a characteristic of the engine. It can be calculated as the ratio of mass flow rate ($\dot{m}$ in kg/hr) of the fuel to the thrust (N, for jet propulsion) or power (bhp, for propeller-driven) generated by the engine. The term thrust-specific fuel consumption (TSFC) is adopted for jet-propelled airplanes for which the thrust is used in the definition. Mathematically SFC can be expressed as

$$\text{SFC} = \frac{\text{kg/hr}}{\text{(bhp)}} \quad \text{or} \quad \text{TSFC} = \frac{\text{kg/hr}}{\text{N}}.$$

The amount of thrust an engine generates is important. But the amount of fuel used to generate that thrust is sometimes more important because the airplane has to lift and carry the fuel throughout the flight. Engineers use SFC as an efficiency factor to characterize an engine's fuel efficiency.

For maximum range, the engine is required to consume the minimum amount of fuel per kilometer distance traveled, which means kg/km has to be

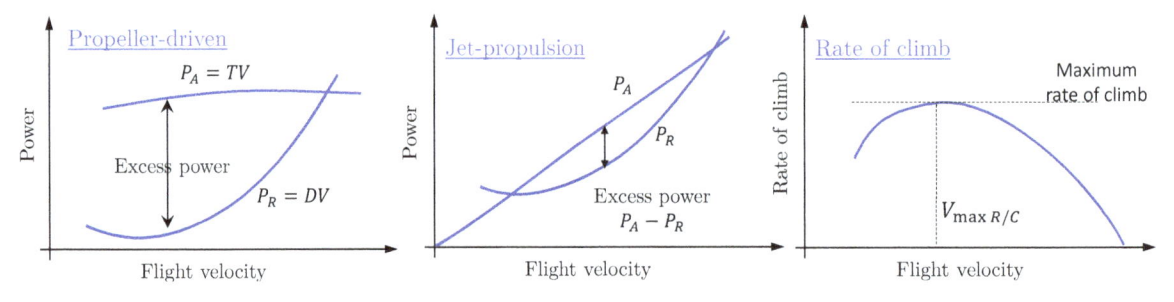

Figure 4.22: Airplane in climbing flight.

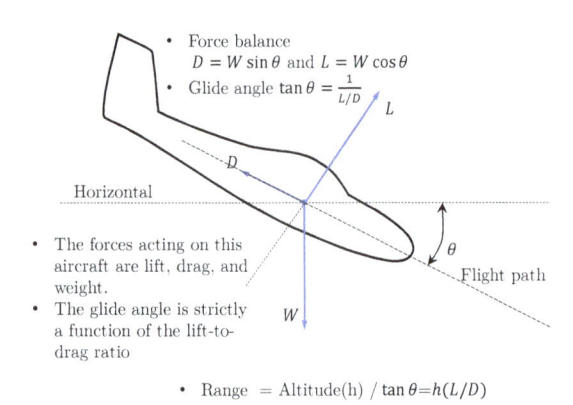

Figure 4.23: Airplane in power-off gliding flight.

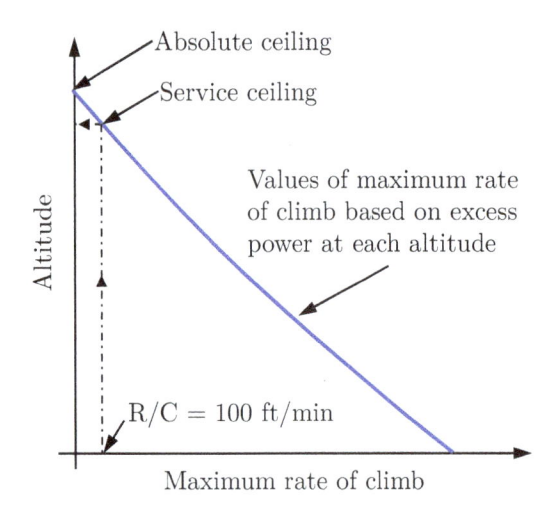

- Force balance
  $D = W \sin\theta$ and $L = W \cos\theta$
- Glide angle $\tan\theta = \dfrac{1}{L/D}$
- The forces acting on this aircraft are lift, drag, and weight.
- The glide angle is strictly a function of the lift-to-drag ratio
- Range = Altitude(h) / $\tan\theta = h(L/D)$

Figure 4.24: Determination of absolute and service ceilings

maximum. From the definition of SFC it can be concluded that
$$\frac{\text{kg}}{\text{km}} \propto \frac{(\text{SFC})\text{bhp}}{V},$$
here velocity V is in km/hr. Therefore, a minimum kg/km is obtained with a minimum bhp/V. The maximum range for a propeller-driven airplane occurs when the airplane is flying at a velocity such that C_L/C_D is at its maximum. The maximum range for a jet airplane occurs when the airplane is flying at a velocity such that $C_L^{1/2}/C_D$ is at its maximum.

For maximum endurance, the aircraft has to stay in the air for the longest time, which means using the minimum amount of fuel per hour ($\dot{m}$). Endurance is proportional to SFC and the thrust or power of the engine. Therefore, maximum endurance for a propeller-driven airplane occurs when the airplane is flying at minimum power required which is at a velocity such that $C_L^{3/2}/C_D$ is at its maximum. Maximum endurance for a jet airplane occurs when the airplane is flying at a velocity such that C_L/C_D is at its maximum. The point of maximum range and endurance for both propeller-driven and jet-propulsion airplanes are indicated in fig. 4.25

Takeoff and Landing

Schematics of a typical variation of forces acting on an airplane during take-off or lift-off are shown in fig. 4.26. To ensure a margin of safety during takeoff, the liftoff velocity is typically 20 percent higher than the stalling velocity i.e., $V_{LO} = 1.2\sqrt{2W/(\rho S C_{L,max})}$. By assuming that thrust is

4.4. FLIGHT MECHANICS

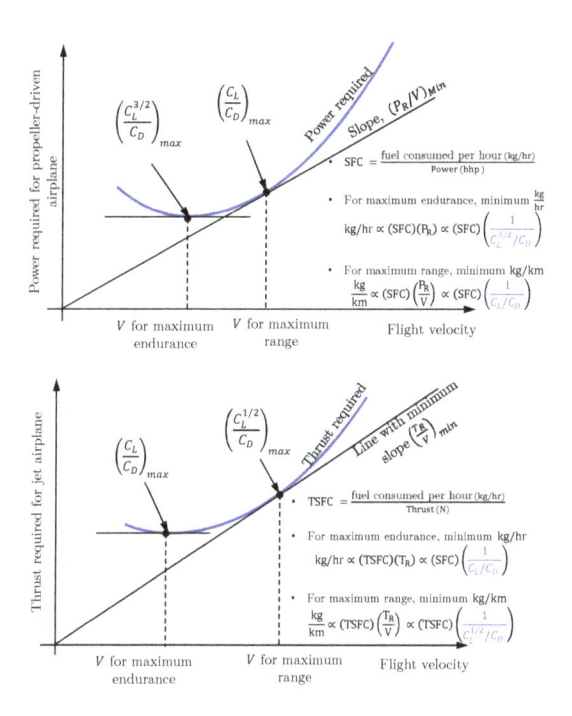

Figure 4.25: Maximum range and endurance on the power-required curve for a propeller-driven airplane

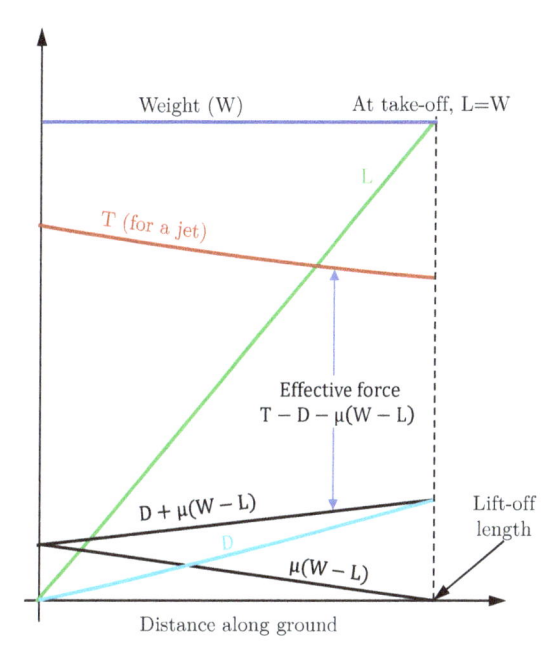

Figure 4.26: Schematic of a typical variation of forces acting on an airplane during takeoff. $\mu(W - L)$ is rolling friction between the tires and the ground, μ is the coefficient of rolling friction.

much larger than either drag or rolling friction between the tires and the ground during takeoff, the lift-off distance can be approximated as $S_{LO} = 1.44W^2/(g\rho S C_{L,max} T)$ which indicate that liftoff distance is very sensitive to the weight of the airplane, varying directly as W^2. Also, liftoff distance is dependent on the ambient density ρ and thrust is directly proportional to ρ, which means that liftoff distance is also varying directly as ρ^2. This explains why on hot summer days, when the air density is lower than that on cooler days, a given airplane requires a longer ground roll to get off the ground. Also, longer liftoff distances are required at airports that are located at higher altitudes compared to sea level. The liftoff distance can be decreased by increasing the wing area, increasing $C_{L,max}$ and increasing the thrust.

While landing, after the airplane has touched the ground, the force diagram during the ground roll is given in fig. 4.27. To minimize the distance required to come to a complete stop, the thrust is reduced at touchdown. To ensure a margin of safety during landing, the touch-down velocity is typically 30 percent higher than the stalling velocity i.e., $V_{TD} = 1.3\sqrt{2W/(\rho S C_{L,max})}$. The landing distance can be approximated as $S_{TD} = 1.69W^2/(g\rho S C_{L,max}(D + \mu(W_L)))$. Some aircraft may use features like spoilers, thrust reversal or parachutes to reduce the landing distance. In case of carrier landing, mechanisms like arresting wires and tailhook are used. The total takeoff distance, as defined in the Federal Aviation Requirements (FAR), is the sum of the ground roll distance S_{LO} and the distance (measured along the ground) to clear a 35 feet height (for jet-powered civilian transports) or a 50 feet height (for all other airplanes). Similarly, the total landing distance is the sum of the ground roll distance and the distance (measured along the ground) to achieve a touchdown in a glide from a 50 feet height.

There are some more performance parameters like the maximum turn rate, minimum turn radius and capability to perform various manoeuvres. Some features of advanced military aircraft

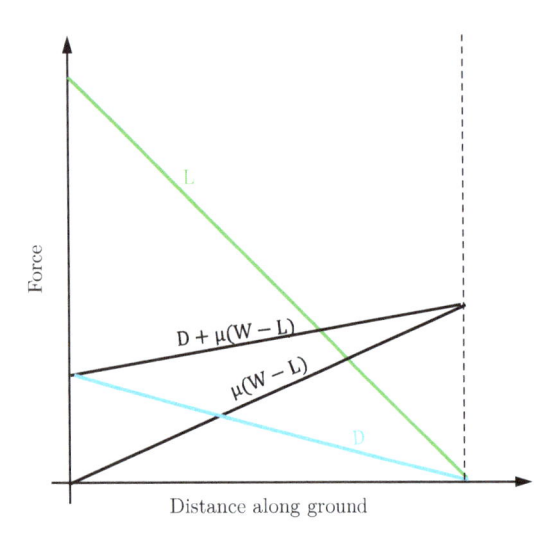

Figure 4.27: Schematic of a typical variation of forces acting on an airplane during landing. $\mu(W - L)$ is rolling friction between the tires and the ground, μ is the coefficient of rolling friction.

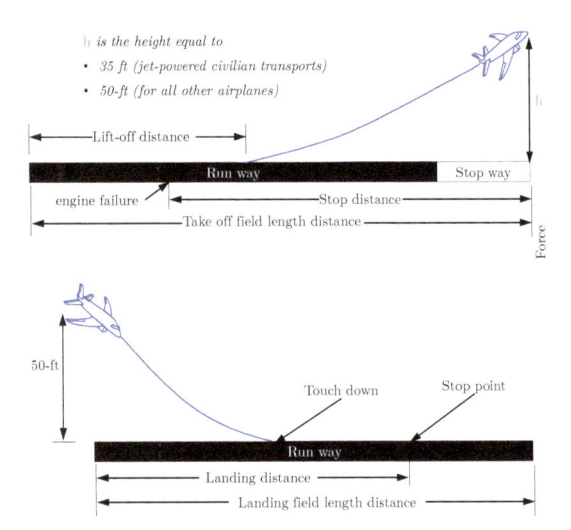

Figure 4.28: The total takeoff and landing distance

4.4.2 Principles of Stability

Airplane Components

The main parts of a conventional configuration airplane are shown in fig. 4.29. The fuselage (or body), which tends to link all of the other parts, holds crew, passengers, or cargo and other payloads. Tail assembly usually consists of a fixed vertical stabilizer, a moveable rudder, horizontal stabilizers (tailplane) and moveable elevators. The wings whose main function is to produce the lift required for flight usually have moveable ailerons towards the outboard (i.e., furthest from the fuselage) trailing edge to provide roll control. It may also have flaps on the inboard trailing edge to help the aircraft fly more slowly during landing and take-off. When aircraft is flying straight and level at a constant speed the lift produced must equal the weight of the aircraft. Therefore, it is important to maintain a correct angle of the wing to the airflow so that the correct amount of lift is generated to maintain a constant speed. In a conventional configuration aircraft, the stability about the pitch axis is achieved by the action of the horizontal stabilizer. If the wing angle increases slightly the body of the aircraft is rotated in a *tail down-nose up* direction and the horizontal stabilizer is also given an increased angle to the airflow. This increased angle produces a lift on the tail which levers the body and with it the wing returns to its original angles. A nose-down attitude is similarly corrected by the horizontal stabilizer working the opposite way and pushing the rear of the aircraft down.

When the aircraft is flying, it needs to be controlled by the pilot and it is controlled about the three axes as mentioned below. In that context, it

include the capability of mid-air refuelling, capability to carry extra fuel tanks, missiles and armaments, vertical take-off and landing, capability to engage in network-centric and electronic warfare, and stealth technology. The details on these aspects can be found in advanced textbooks.

4.4. FLIGHT MECHANICS

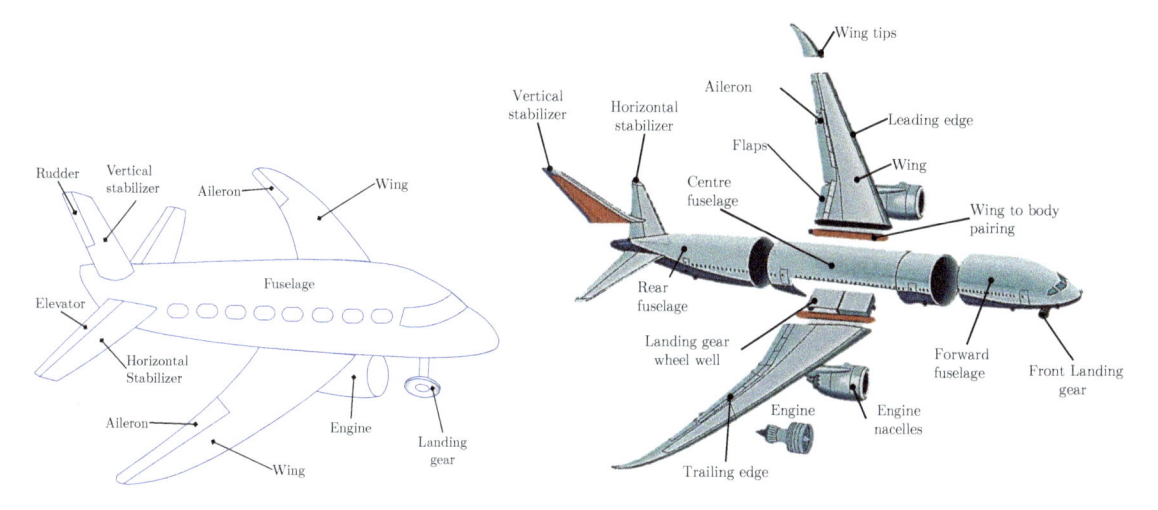

Figure 4.29: Parts of an airplane.

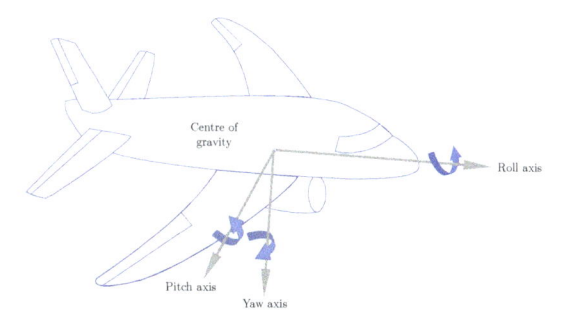

Figure 4.30: Aircraft axis

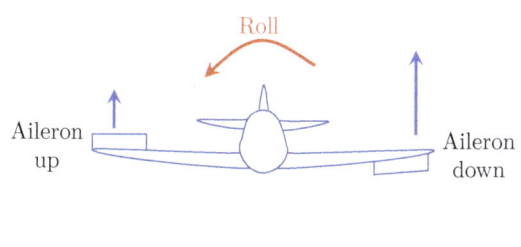

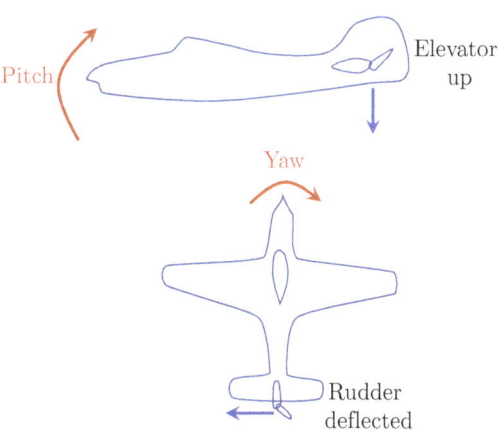

Figure 4.31: Effect of control surface deflections (Aileron, elevator and rudder) on roll, pitch and yaw.

is important to know the stability associated with these axes. An axis is an imaginary line stretching from wing tip to wing tip and rotation about this axis which causes 'nose up' or 'nose down' movement is called *pitching*. The horizontal line stretching from the extreme nose of the aircraft to the extreme tail is called the *roll* axis. A vertical axis that passes through the intersection of the other two and pointed downward is called *yaw* axis. As shown in fig. 4.31, there are control surfaces which affect the movements about the three axes. Pitching is controlled by the elevators, roll is controlled by the ailerons and yaw is controlled by the rudder.

Stability

If the forces and moments on the body caused by a disturbance tend initially to return the body toward its equilibrium position, the body is statically stable.

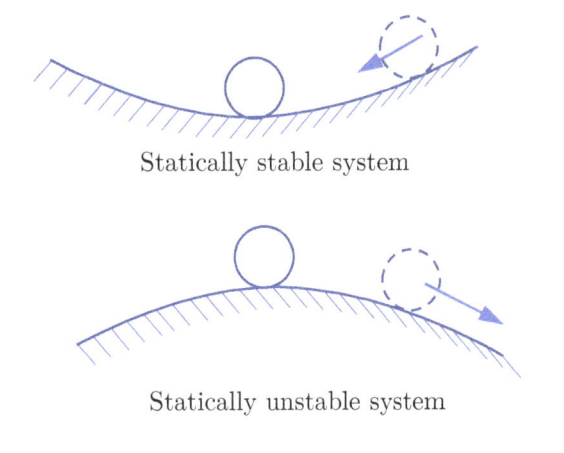

Figure 4.32: Illustration of static stability.

- Statically stable system
- Statically unstable system
- Neutrally stable system.

If the forces and moments are such that the body continues to move away from its equilibrium position after being disturbed, the body is statically unstable. Dynamic stability deals with the time history of the vehicle's motion after it initially responds to its static stability. A body is dynamically stable if, of its own accord, it eventually returns to and remains at its equilibrium position over time. It is important to observe that a dynamically stable airplane must always be statically stable. However, static stability is not sufficient to ensure dynamic stability. Nevertheless, static stability considerations is usually the first stability characteristic to be designed into an airplane. The stability of the airplane about the pitch axis is called the longitudinal stability. The stability about roll axis is called lateral stability and that about the yaw axis is called the directional stability. The longitudinal stability is determined by the characteristics of the wing, the horizontal stabilizer and their relative location with respect to the center of mass of the airplane. In a general sense, if the center of mass of the airplane is ahead of the aerodynamic center of the wing-horizontal tail combination, the airplane is considered to be longitudinally stable. For a conventional configuration airplane, the longitudinal stability is decoupled from the lateral and directional stability. The directional and lateral stability are coupled to each other. The vertical stabilizer is the primary contributor to directional stability. Providing dihedral angle and placing it in a high-wing configuration gives lateral stability to the airplane.

4.5 Flight Structures

The loads experienced by aerospace vehicles are significantly different from that experienced by other types of vehicles and the environment in which they operate can also be very harsh. Its weight has a strong bearing on the performance and economics of aerospace vehicles. So, unlike other types of vehicles, weight comes at a premium for aerospace vehicles. The design engineers spend many hours (months!) on the design and analysis so as to make vehicles as light-weight as possible. During design and analysis, it is important to estimate all the different types of loads that the vehicle will bear and then calculate the structural response to those loads. The structural analysis involves calculating the response of structures carefully to estimated/measured loadings. Due to the penalty associated with weight, the factor of safety of aerospace vehicles are much lower than other type of load bearing structures. The factor of safety is a relative measure of the load a structure can withstand to the load actually applied. In the design of aerospace vehicles, it is quite common to see the structural engineers push to the limits the load carrying capacity of structural components by the right choice of materials and ingenious design solutions. In the present section, basics of mechanics of materials is introduced in which types of loading, identifying different load carrying members, types of materials and manufacturing techniques are introduced.

In general, the mechanical structure is subjected to complex loads that produce a multi-axial deflection. Such complex loads can be approximated to several simple types of loading as shown in figure 4.33. This loading is the combined effect of complicated external forces acting on a structure. *Tensile Load* measures the pulling action (tensile force) perpendicular to the section that tends to elongate the member. *Compressive Load* measures the pushing action (compressive force) perpendicular to the section that tends to shorten the member. *Shearing Load* involves applying forces to a structure parallel to a plane which caused the material on one side of the plane to want to slide across the material on the other side of the plane. *Bending Load* involves applying a load in a manner that causes a structure to curve and results in compressing the material on one side and stretching it on the other. *Torsional Load* is the application of a force that causes twisting in a structure.

If a material is subjected to a constant force, it is called static loading. If the loading of the material is not constant but instead fluctuates, it is called dynamic or cyclic loading. The way a material is loaded greatly affects its mechanical properties and largely determines how, or if, a component will fail; and whether it will show warning signs before failure actually occurs.

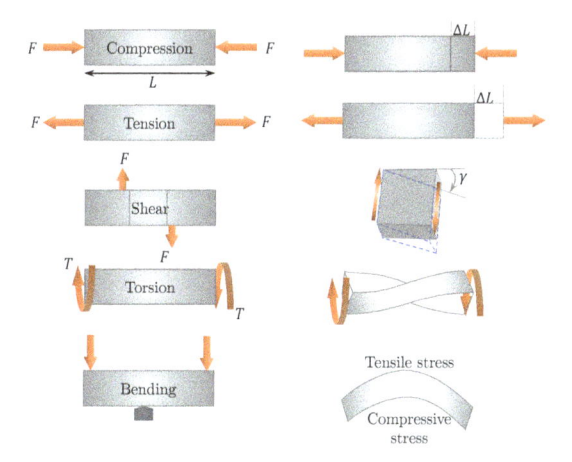

Figure 4.33: Types of structural loading: compression, tension, shear, torsion and bending.

That brings us to the very important question, what is stress and why do we need to study it? For deformable continuous bodies, the concept called *stress* is introduced to model force transmission through the body. Stress is the measure of the state of force transmission in the interior of the body. The body subjected to a complex load produces some combination of rigid motion (the distance between two particles does not change) and stretching (the distance between two particles changes). Materials resist stretching in the sense that the distance between two points in a body can change, but it takes force to do so (i.e., to stretch the molecular bonds) which gives rise to internal force. The internal resistance force per unit area is the *stress*. Depending on the direction of the force and the orientation of area, there are different components of the stress. When the force acts perpendicular to the reference surface, it is called normal stress. When the force acts parallel to the surface, it exerts a shear stress. Strain is a measure of the deformation due to the stress. Normal strain quantifies the elongation or compression of the material and is the ratio of change in length to the reference length. Shear strain measures the angular deformation of the body. The stress-strain relation for a given material is obtained by careful experiments.

Some of the terminologies related to the mechanical properties of structures are introduced. Tensile tests are used to determine properties like the modulus of elasticity, elastic limit, elongation, proportional limit, reduction in area, tensile strength, yield point and yield strength. A typical load versus elongation curve, which is then converted into a stress versus strain curve by dividing the load and elongation by constant values related to specimen geometry information, is shown in Fig. 4.37. For small loads, the stress is directly proportional to strain. In this linear region, the line obeys the relationship defined by Hooke's Law where the ratio of stress to strain is a constant, which is called the modulus of elasticity or Young's modulus (a measure of the material stiffness). Axial strain is always accompanied by lateral strains of opposite sign in the two directions mutually perpendicular to the axial strain. Poisson's ratio is defined as the negative ratio of the lateral strain to the axial strain for

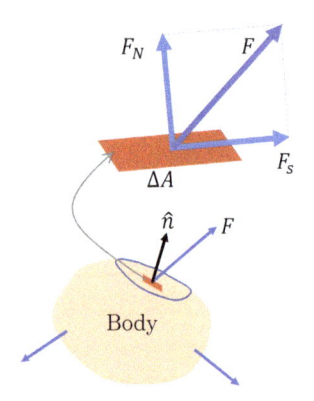

Resultant stress $\sigma = \lim\limits_{\Delta A \to 0} \dfrac{\Delta F}{\Delta A}$

Normal stress $\sigma_N = \lim\limits_{\Delta A \to 0} \dfrac{\Delta F_N}{\Delta A}$

Shear stress $\sigma_s = \lim\limits_{\Delta A \to 0} \dfrac{\Delta F_s}{\Delta A}$

Three dimensional element

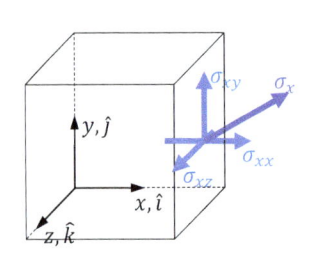

On a face whose normal is along x-axis

$$\sigma_x = \sigma_{xx}\hat{i} + \sigma_{xy}\hat{j} + \sigma_{xz}\hat{k}$$

σ_{xx} is a normal stress

σ_{xy} and σ_{xz} are shear stresses

Stress tensor at a point

$$\sigma = \begin{bmatrix} \sigma_{xx} & \sigma_{xy} & \sigma_{xz} \\ \sigma_{xy} & \sigma_{yy} & \sigma_{yz} \\ \sigma_{xz} & \sigma_{yz} & \sigma_{zz} \end{bmatrix}$$

Figure 4.34: Stress at a point

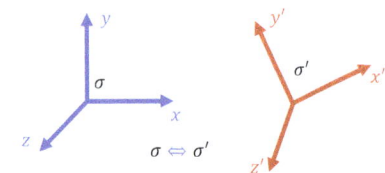

$\sigma \Leftrightarrow \sigma'$

For different coordinate system, the values will be different but still represent same stress state

$\sigma_{xy} = \dfrac{Tr}{J}$

$$\begin{bmatrix} 0 & \sigma_{xy} & 0 \\ \sigma_{xy} & 0 & 0 \\ 0 & 0 & 0 \end{bmatrix} \quad \begin{bmatrix} \sigma_{xy} & 0 & 0 \\ 0 & \sigma_{xy} & 0 \\ 0 & 0 & 0 \end{bmatrix}$$

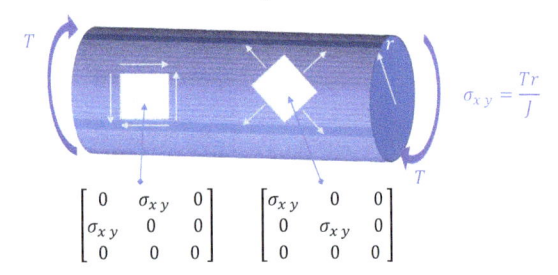

Figure 4.35: The stress state is not absolute

$$\sigma = \begin{bmatrix} \sigma_{xx} & \sigma_{xy} & \sigma_{xz} \\ \sigma_{xy} & \sigma_{yy} & \sigma_{yz} \\ \sigma_{xz} & \sigma_{yz} & \sigma_{zz} \end{bmatrix} \quad \sigma^P = \begin{bmatrix} \sigma_1 & 0 & 0 \\ 0 & \sigma_2 & 0 \\ 0 & 0 & \sigma_3 \end{bmatrix}$$

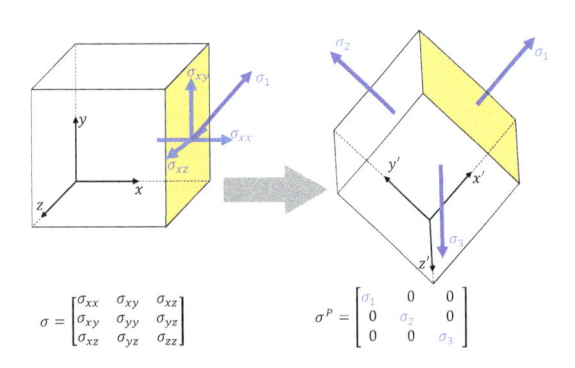

Figure 4.36: Principal stresses

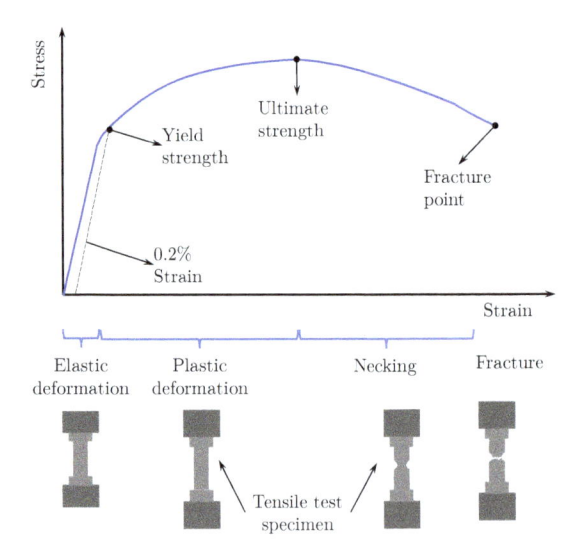

Figure 4.37: Stress-strain curve for a ductile material

a uniaxial stress state. For a perfectly isotropic elastic material, Poisson's ratio is 0.25, but for most engineering materials, the value lies in the range of 0.28 to 0.33. For ductile materials like Aluminum, when the load is increased, at some point, the stress-strain curve deviates from the straight-line relationship and Hooke's law no longer applies. Beyond that, the structure starts yielding. Further increase in the applied force results in some permanent deformation occuring in the specimen and the material is said to react plastically to any further increase in load or stress. The material will not return to its original, unstressed condition when the load is removed. In brittle materials like glass, little or no plastic deformation occurs and the material fractures near the end of the linear-elastic portion of the curve.

The *yield strength* is the stress required to produce a small, amount of plastic deformation. The yield strength is used for most engineering designs. Since the transition from elastic to plastic behavior is gradual, it is hard to determine the exact yield point. The offset yield strength is the stress corresponding to the intersection of the stress-strain curve and a line parallel to the elastic part of the curve offset by a specified strain, typically 0.2% for metals. *Proportional limit* is the maximum stress up to which stress remains directly proportional to strain; the upper end of the straight-line portion of the stress-strain or load-elongation curve. *Elastic limit* is the maximum stress to which a material may be subjected with no permanent deformation after the release of the applied load. The *ultimate tensile strength* is the maximum stress level reached in a tension test. The strength of a material is its ability to withstand external forces without breaking. In brittle materials, the ultimate tensile strength will be close to the elastic limit. In ductile materials, the ultimate tensile strength will be well outside of the elastic portion into the plastic portion of the stress-strain curve.

Strength is the ability of a material to resist the externally applied load. The stronger the materials the greater the load it can withstand before it fails. Yield strength or ultimate tensile strength is a measure of the strength of the material.

Stiffness is the ability of a material to resist deformation or deflection under stress. The higher the value of Young's modulus, the stiffer the material in tensile and compressive stress.

Elasticity is the property of a material to regain its original shape after removal of the externally applied load.

Plasticity is the ability of a material to undergo some degree of permanent deformation before it fails.

Ductility is the property of a material that enables it to be drawn into the thin wire with the application of tensile load. Ductility is measured in terms of percentage elongation and percent reduction in the area. In general, materials that possess more than 5% elongation are called as ductile materials. Examples include mild steel, copper, aluminum, nickel, zinc, tin, and lead.

Brittleness, opposite of ductility, is a property of material that enables breaking with little permanent distortion (less than 5% elongation). Glass, cast iron, brass and ceramics are considered brittle materials.

Malleability is the ability of the material which enables it to be flattened into thin sheets under applications of heavy compressive forces without cracking. A malleable material should be plastic but it is not essential to be so strong. Copper, aluminum, lead and steel are considered as highly malleable metals.

Hardness is the ability of a material to resist localized deformation. It embraces many different properties such as resistance to wear, resistance to indentation, resistance to scratches, resistance to deformation and machine mobility, etc. Diamond is the hardest known material naturally.

Toughness is the ability of a material to absorb energy and deform plastically before it fails. Ductility is a measure of how much something deforms plastically before fracture, but just because a material is ductile does not make it tough. Toughness is a good combination of strength and ductility.

Resilience is the ability of the material to absorb energy without permanent deformation.

Creep refers to material undergoing a slow and permanent deformation (time-dependent deformation) when subjected to a load that is below its yield strength for a longer period of time. It most often occurs at elevated temperatures, but for some materials creep can happen at room temperature also. Creep terminates in a rupture if steps are not taken to bring it to a halt.

Factor of Safety (FoS) is the ratio of the actual load-bearing capacity of a structure and working stress. FoS of 1 means that a structure fails exactly when it reaches the design load, and cannot support any additional load. Structures with FoS < 1 are not viable; basically, FoS of 1 is the minimum required for a structure. A FoS of 2 means that a component will fail at twice the design load. Typical FoS for aircraft components lies in the range 1.5-2.5. Due to the absence of human payload, missile and satellite launch vehicle structures have even lower values for factor of safety.

Fatigue is the failure of a material due to cyclic or repeated loading. The intensity of the applied load may be very less than the ultimate tensile stress. The fatigue life of a component can be expressed as the number of loading cycles required to initiate a fatigue crack and to propagate the crack to a critical size. Hence, fatigue failure occurs in three stages – crack initiation; slow, stable crack growth; and rapid fracture.

Stress-Strain diagram in Figure 4.38 shows the difference between stiff and flexible (high or low Young's modulus (E)) material; soft and rigid (low or high yield stress) material; weak and strong (low or high ultimate tensile stress (UTS)) material and brittle and ductile material (low or high plastic strain).

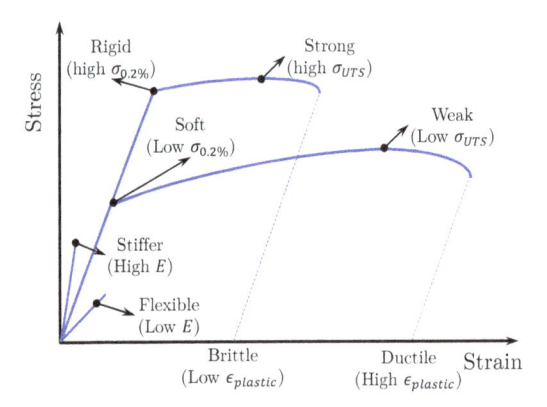

Figure 4.38: Stress-Strain diagram showing different mechanical properties of materials

4.5.1 Material Types

Aerospace structures require materials that have a high strength-to-weight ratio which means the materials have good mechanical properties but with a low density. The materials used for aerospace structures can fall into any of the following categories: *metal alloys, polymers, composites or ceramics*. The optimization of the aerospace structures is based on the material type, structural shape and manufacturing process.

When metals are alloyed with other metals, their properties are improved significantly. For example, adding a few percent of copper and magnesium to aluminum increases its yield and ultimate strengths by a factor of 4 to 6. Metal alloys have been used extensively for airframes, skins, and other stressed components. The choice of alloy is based on properties such as strength (yield and ultimate stress), ductility, ease of manufacture, resistance to corrosion and amenability to protective treatment, fatigue strength, freedom from liability to sudden cracking due to internal stresses, and resistance to fast crack propagation under load. Different types of aerospace vehicles have differing requirements. For example, a military aircraft having a relatively short life measured in hundreds of flying hours does not demand the same degree of fatigue and corrosion resistance material as a commercial airplane with a life $> 30,000$ hours.

Pure Aluminum with relatively low-strength, and extreme flexibility has little structural applications in aerospace vehicles. But, its mechanical properties are improved significantly when alloyed with other metals. Many alumimum alloys are used quite extensively for aerospce structural components. One of the most extensively used aluminum alloy in the aerospace industry is duralumin. Though steel has high tensile strengths but other properties such as ease of manufacturing or getting finished components are sacrificed. Maraging steels, alloys without Carbon, has found a wide range of applications in aerospace vehicles as some of the undesirable properties including carbon such as brittleness and distortion are improved significantly. Steel in its stainless form has found applications primarily in the construction of super- and hypersonic experimental and research aircraft, where temperature effects are considerable. Stainless steel formed the primary structural material in the *Bristol 188*, built to investigate kinetic heating effects, and also in the American rocket aircraft, the *X-15*, capable of speeds of the order of Mach number 5–6. Titanium alloys were used in the airframe and engines of *Concorde*, while the *Tornado* wing carry-through box is fabricated from a weldable medium-strength titanium alloy. Titanium alloys are also used extensively in the *F15* and *F22* American fighter aircraft and are incorporated in the tail assembly of the *Boeing 777* civil airliner. Other uses include forged components, such as flap and slat tracks and undercarriage parts. In aircraft engine design, titanium alloys are used in the early stages of a compressor, while nickel-based alloys or steels are used for the hotter later stages.

Due to low strength and stiffness, *polymers* are generally not employed as structural materials. Mechanical properties are temperature dependent, and at low temperatures they become brittle. Elastomers like polymers such as rubber which is flexible and exhibits a very large strain before failure occurs, has been used in tires and seals. Plain plastic materials which has specific gravities less than half those of the aluminum alloys are used for windows or lightly stressed parts whose dimensions are established by handling requirements rather than strength. Plastics are also particularly useful as electrical insulators and as energy-absorbing shields for delicate instrumentation and even structures where severe vibration, such as in a rocket or space shuttle launch, occurs. Other polymers such as natural fibers, synthetic fibers, and nylon are generally used to reinforce a low-strength polymer to form a composite structure.

Glass in the form of a plain or laminated plate or heat-strengthened plate is frequently employed in pressurized cabins for flight at high altitudes and also windscreens which are subjected to loads normal to their mid-planes.

Due to extreme brittleness and inability to sustain strain, *ceramics* are not used as structural materials in aerospace vehicles. However, they are used in thermal protection systems for an actual structure that would be otherwise susceptible to mechanical degradation at high temperatures.

Composites are engineering materials, in which two or more distinct and structurally complementary substances with different physical or chemical properties are combined, to produce structural or functional properties not present in any individual component. An example of a composite is the fiber-reinforced polymer composite, which consists of two distinct and complement materials, namely fibers and polymer. The function of the fibres is to reinforce the polymer providing strength and stiffness to the material and, by doing so, to carry the main portion of the load. The function of the polymer is to support the fibres and to transfer the load to and from the fibres in shear. A composite (fiberglass and aluminum) is used in the tail assembly of the *Boeing 777*, while the leading edge of the *Airbus A310–300/A320* fin assembly is of conventional reinforced glass fiber construction, reinforced at the nose to withstand bird strikes. A complete composite airframe was produced for the *Beechcraft Starship* turboprop executive aircraft, which, however, was

Table 4.6: Eamples of metal alloys used in aerospace vehicles

Metal	Alloyed with	Important property
Aluminum	4% copper, 0.5 % magnesium, 0.5 % manganese, 0.3 % silicon, and 0.2 % iron	Moderately high strength
	1–2 % of nickel and also magnesium, copper, silicon, and iron	Retention of strength at high temperatures
	2.5 % copper, 5 % zinc, 3 % magnesium, and up to 1 % nickel	High strength
Steel	*Maraging steels* (17–19 % nickel, 8–9 % cobalt, 3–3.5 % molybdenum, 0.15–0.25 titanium%)	Higher fracture toughness and notched strength, simpler heat treatment, much lower volume change and distortion during hardening, very much simpler to weld, easier to machine, and better resistance to stress corrosion and hydrogen embitterment
Titanium	*Ti6Al-4V (5)* - 6% aluminum, 4% vanadium, 0.25% (max) iron and 0.2% (max) oxygen	Good fatigue strength/tensile strength ratio with a distinct fatigue limit, and retain considerable strength at temperatures up to 400–500 °C. Good resistance to corrosion and corrosion fatigue

not a commercial success. Metal matrix composites, such as graphite–aluminum and boron–aluminum, are lightweight and retain their strength at higher temperatures than aluminum alloys but are expensive to produce. However, a fiber metal laminate is used as the upper fuselage skin of the *Airbus A380*. The use of composites in aircraft construction has increased steadily over the past two decades. For example, composites in the *Boeing 787* airliner comprise nearly 50 percent of its structural weight, while a carbon-epoxy composite features significantly in the *Boeing-McDonnell-Douglas F/A-18E* jet fighter. Usage of the composite over the year for *Airbus* passenger aircraft is shown in Fig. 4.39.

4.5.2 Aircraft Structures

The aircraft from the first decade of aviation were built with *truss* structures in which the bars, tubes and wires carried all loads and transfer the loads primarily via tension and compression. Structural rigidity is obtained by placing wires or rods diagonally at 45^0. The skin or fabric coverage did not contribute to the load-bearing function. Some of the later applications introduced tailored truss structures that has led to significant weight savings. As the metal was introduced into the aircraft structure, structural rigidity is obtained from the sheet metal. In such configurations, the sheet metal bears a shear load, whereas the wires and rods only bear tension and compression load. Further, the truss structure was replaced by the thin-walled shell structure in combination with sheet metal to increase the strength.

Fuselage structures: Fuselage is a thin-walled pressure vessel as shown in Figure 4.41 which has different elements such as fuselage skin, frames, stringers, bulkheads, splices and joints that together provide strength and rigidity. The stringers that are employed with skin panels along with the closed cylindrical shell structure provide longitudinal stiffness to the fuselage. Stringers are combined with longerons which are heavy longitudinal stiffeners that carry large loads to create structural rigidity. This concept then has only few longitudinal stiffeners that have a larger cross-section. Pressurized fuselages require the application of bulkheads at the front and rear end of the fuselage to create a pressure vessel and to maintain the forces related to the pressurization.

Wing Structures: The wings are the main part of the aircraft that is designed to generate lift. Wing design, which is a full cantilever so that no external bracing is needed, for an aircraft depends on size, weight and use of the aircraft, desired speed in flight and at landing, and desired rate of climb. The wing contains similar stiffened shell concepts similar to the fuselage. However, wing structures can be divided into wing skin panels, ribs, stringers and spars. Some wings contain straight spars running from the wing root to the wing tip that connects the wing fuselage while other wings contain kinks in the spars. Such a kink in the spar can be

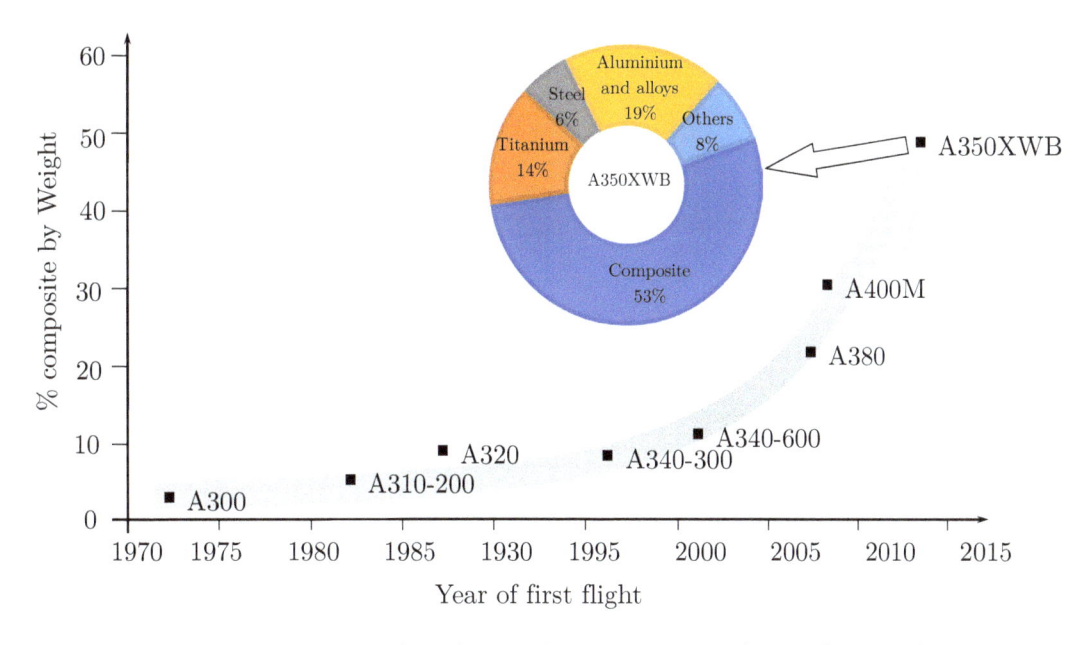

Figure 4.39: Trends in the use of composite materials in *Airbus* aircraft.

achieved by connecting two spar elements at that particular location. Where some wings contain ribs that are placed in flight direction, other wings contain ribs that are placed more or less perpendicular to the spars and wingspan direction. Despite the advantages of sectioning large structures with respect to manufacturing and assembly, a potential disadvantage may be related to the highly loaded joints as a result. The ribs in wing structures maintain the aerodynamic profile of the wing. Because the aerodynamic profile of the wing is defined in flight direction, this may be an important reason to place the ribs in flight direction. Ribs also transfer the aerodynamic and fuel loads acting on the skin to the rest of the wing structure. The aerodynamic loads are related to the under pressure above the upper skin, while the fuel loads act directly on the lower wing skins. These loads are brought into the structures via the ribs. Finally, the ribs provide stability against panel buckling.

4.5.3 Manufacturing

The metallic components in aircraft are either sheet-metal parts or machined parts which may be made from extrusion, forgings or castings. Machined parts are those which have been worked on using a machine tool such as the drilling machine or the lathe. Lathes produce parts that have circular cross sections, such as bolts, bearing bushes, pistons, etc. The milling process is used to make smooth flat faces, grooves, and shaped recesses or slots. This is done by cutting chips of metal from a surface by rotating a tool with one or more cutting teeth and driving it through the metal to be removed. The tools for working sheet metal in the aerospace industry are enlarged and power-operated versions of the hand tools used by typical sheet metal workers.

Cutting can be done either by guillotine, blanking or routing. The guillotine makes long, straight cuts, whereas blanking is a punch and die process. Any irregular shape can be cut out as a hole in the thick steel plate to make the die. Routing also produces irregular sheet-metal shapes. In this method, a special milling cutter run at very high speed cuts through a pack of five to ten sheets of material. *Forming* sheet metal is probably a major activity in aircraft manufacturing. Straight bending of flanges or single-curvature skin panels, such as those on the

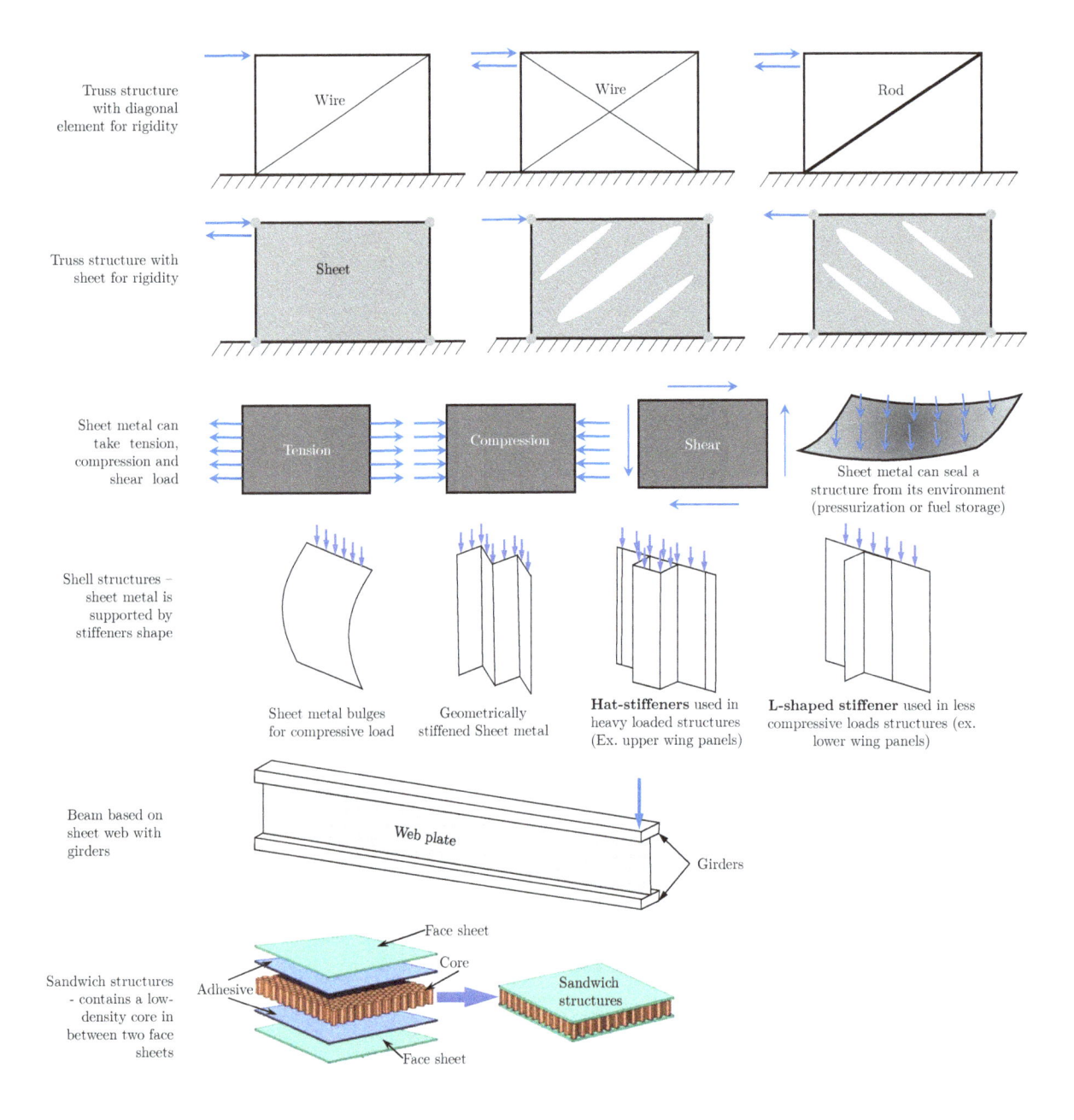

Figure 4.40: Structural concepts in aviation

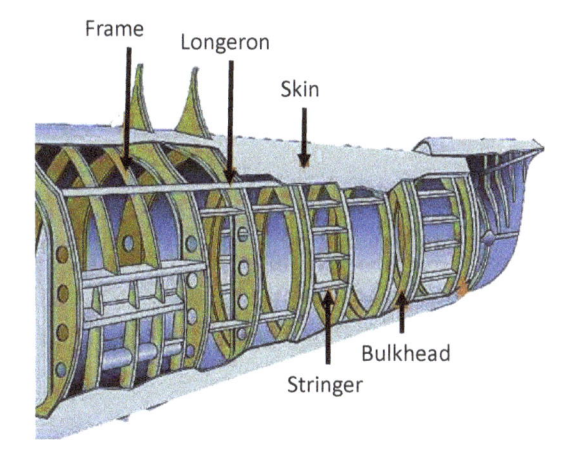

Figure 4.41: Fuselage structure.

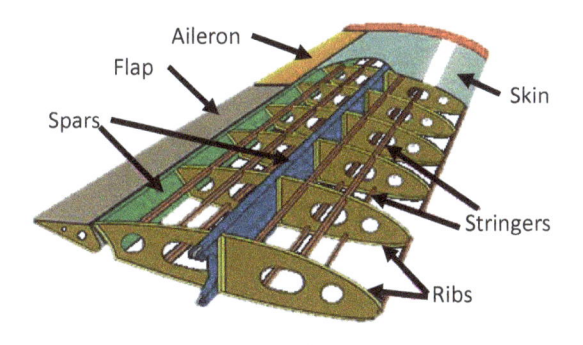

Figure 4.42: Wing structure.

parallel central part of a fuselage or the surface of a conventional wing, is relatively easy. Forming parts with double curvature, such as the skin panels at the nose of an aircraft, is considerably more difficult, but the processes are well understood in the industry. Bending or folding of straight flanges is carried out in a folding machine or a press brake. Double curvature forming may be carried out in a press with punch and die tools, but arranged with sufficient clearance so that the sheet is not cut between the two parts of the tool but is formed to the shape of the punch.

Casting is the production process in which liquid molten metal is poured into a mould with the required shape of the product. The shape of the inside cavity of the mould defines the outer shape of the product. After cooling, the part is retrieved by breaking the mould. Machining is a solid state cutting and milling processes, which means that the process is applied to materials in their solid state at room temperature. However, machining often required the use of a cooling agent, because cutting and milling create significant heating of the material due to friction.

Jointing: *Welding* is the process whereby components made of metal or some plastics are joined by being made so hot in an area local to the joint, that the material of both components melts and runs together so that when the heat is removed the material is continuous across the joint. Usually, to compensate for gaps between the joint faces, a filler of similar material is melted into the weld. There are two other similar processes, soldering and brazing, in which the components being joined are not melted but are made hot enough to fuse a softer material such as tin, which runs between them and acts like an adhesive. Advanced welding techniques using lasers or electron beams are used for some specialized work. From the structural strength viewpoint, welding is not a process to be used too freely. Welding can significantly change the material properties in the heat-affected area, not to mention the presence of residual stresses and local distortion.

These problems can be minimized by the use of special welding techniques and reheat treating and/or stress relieving after welding but all these add to the cost and complexity of the process.

Adhesive bonding of metal structures is another technique widely used by some aircraft manufacturers. All designers of airframe structures use adhesives for glass-reinforced plastic components and sandwich structures in various combinations of wood, paper, plastic and metal, but for the bonding of stringers to skins and other metal-to-metal joints in primary structure it is not so widely used.

Rivets and *bolts* and some *special fasteners* are widely used joining methods in aerospace structure. One structural disadvantage of riveted construction is that every rivet hole is potentially the start of a tear or a crack, so the apparently simple drilled hole in fact requires care in its preparation, and an additional reserve of strength in its surrounding structure, with the consequent weight disadvantage.

4.6 Orbital Mechanics

Orbital mechanics is the study of the motions of satellites and space vehicles moving under the influence of forces such as gravity, atmospheric drag, thrust, etc. The engineering applications of orbital mechanics include ascent trajectories, atmospheric entry from outer space and landing, rendezvous computations, and lunar and interplanetary trajectories.

4.6.1 Basic Laws

Kepler's[2] three *Laws of Planetary Motion*, *Newton's Law of Universal Gravitation*, and *Newton's* three *Laws of Motion* provide the building blocks upon which orbital mechanics is built. These basic laws use the concepts of mass and force extensively.

Kepler's Laws of Planetary Motion

Kepler's first law *A satellite describes an elliptical path around its center of attraction.* Every planet's orbit about the Sun is an ellipse. The Sun's center is located at one foci of the orbital ellipse. *Kepler's second law In equal time intervals, the areas swept out by the radius vector of a planet from the Sun are the same.* The imaginary line joining a planet and the Sun sweeps equal areas of space during equal time intervals as the planet orbits. The point of nearest approach of the planet to the Sun is termed perihelion and the farthest point in the planet's orbit is the aphelion. Hence, according to Kepler's second Law, a planet moves fastest when it is at perihelion and slowest at aphelion.

Kepler's third law The periods($\tau_{1,2}$) of any two satellites about the same planet are related to their semi-major axes($a_{1,2}$) as $\frac{\tau_1^2}{\tau_2^2} = \frac{a_1^3}{a_2^3}$.

Though *Kepler* hadn't known about gravitation when he came up with these three laws, they were instrumental in Isaac Newton proposing the theory of universal gravitation, which explains the unknown force behind *Kepler's* third Law. In 1686, *Isaac Newton* presented the three laws which governs the motion of bodies. *Newton's* laws along with *Kepler's* laws explained why planets move in

---

[2]*Johannes Kepler* (1571 - 1630), a German mathematician proposed laws of planetary motion. He had the opportunity to work for the famous astronomer *Tycho Brahe* (1546 - 1601) in Prague. *Tycho Brahe* was credited with the most accurate astronomical observations of his time. *Kepler* started working as an assistant in 1600 for *Brahe* to continue the work on observations and mapping of planetary motion. Later, *Kepler* proved inconsistency in *Brahe's* geocentric model according to which the Earth is the center of the solar system and the planets Mercury, Venus, Mars, Jupiter, and Saturn all orbit the Sun, which in turn orbits the Earth. *Kepler* believed in the heliocentric model which, unlike the geocentric model, correctly placed the Sun at its center. After much struggle, *Kepler* realized that the orbits of the planets are not circles but were instead the elongated or flattened circles called ellipses. *Kepler* published the three laws of planetary motion, obtained from an exhaustive examination of astronomical data from *Brahe*.

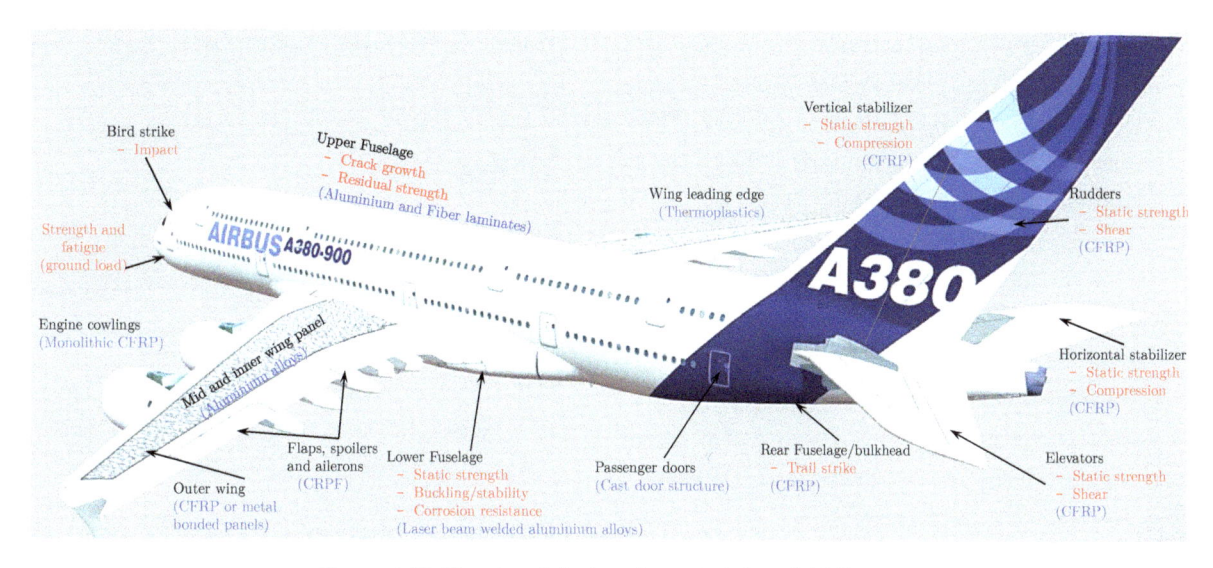

Figure 4.43: Structural design characteristics of *A380*.

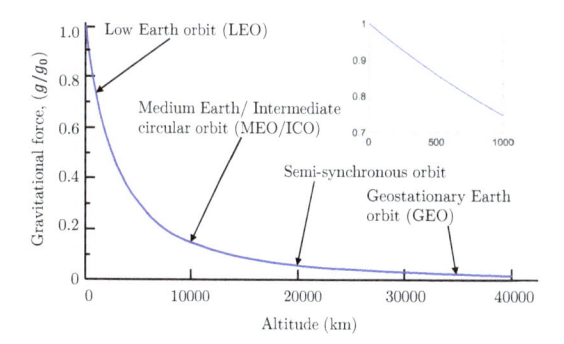

Figure 4.45: Variation of the acceleration of gravity with altitude.

elliptical orbits rather than in circles. These are the laws which govern the motion of satellites around the Earth. The gravitational force due to Earth expressed as acceleration due to gravity is plotted as a function of altitude in Fig. 4.45.

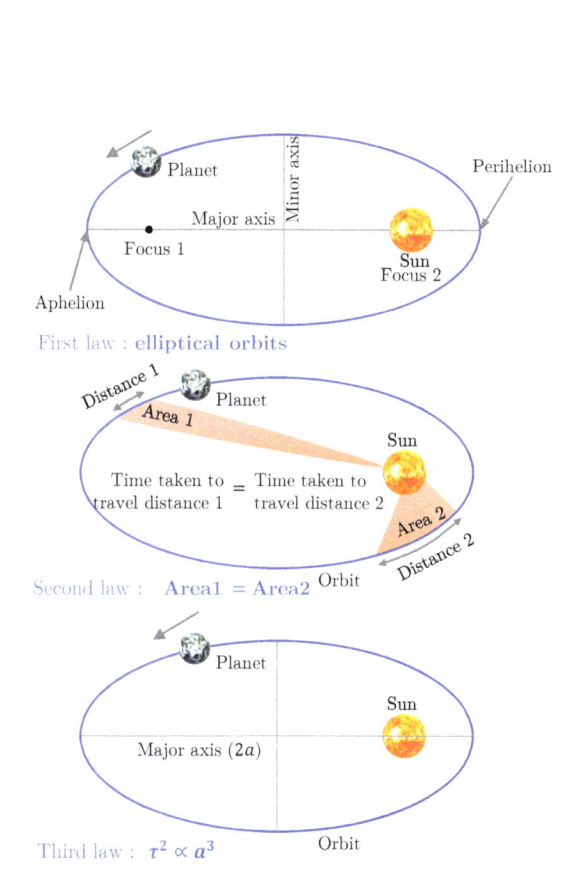

Figure 4.44: Kepler's three laws of planetary motion

4.6.2 Space Vehicle Trajectories

The equation of the path (orbit equation) of the trajectory of a space vehicle moving under the gravitational field of Earth or an artificial satellite in orbit about the earth is given by

$$r = \frac{h^2/k^2}{1 + A\left(h^2/k^2\right)\cos(\theta - C)},$$

where, h^2 is constant angular momentum per unit mass, A and C are constants to be determined from the location at a given instant in its path. These constants are fixed by conditions at the instant of burnout of the rocket booster. $k^2 = GM$ product of the universal gravitational constant and the mass of the Earth. r and θ are the geometric coordinates. θ is the angular orientation of r which is the distance between the space vehicle and the center of the Earth. The orbit equation is the standard form of a conic section in polar coordinates. The path of a body of mass m is determined by it's kinetic energy ($\frac{1}{2}mV^2$) and the gravitational field as

- if $\frac{1}{r}mV^2 = \frac{GMm}{r^2}$, the path is a circle.
- if $\frac{1}{2}mV^2 < |\frac{GMm}{r}|$, the path is an ellipse.
- if $\frac{1}{2}mV^2 = |\frac{GMm}{r}|$, the path is a parabola.
- if $\frac{1}{2}mV^2 > |\frac{GMm}{r}|$, the path is a hyperbola.

The family of curves represented by the orbit equation called *conic sections* (i.e., circle, ellipse, parabola and hyperbola) represents the only possible paths for an orbiting object in the two-body problem. The circle and ellipse are closed-loop conics, while parabola and hyperbola are open conics. The focus of the conic must be located at the center of mass of the central body. The specific mechanical energy of a satellite which is the sum of the kinetic and potential energies does not change as the satellite moves along its orbit. However, there is an exchange of energy between the two forms which means that the satellite must slow down as it gains altitude (i.e., r increases) and speeds up as r decreases so that the total energy remains constant. The orbital motion takes place in a plane fixed in the inertial space. Also, the specific angular momentum of a satellite remains constant as there is no external torque. As r and v change along the orbit, the flight path angle (θ) must change so that the specific angular momentum remains constant.

In this chapter, a brief introduction of the engineering aspects of aerospace vehicles was presented. In the next part of the book, we will look at the different types of aerospace vehicles.

Bibliography

J. D. Anderson Jr. (2005) *Introduction to Flight*, McGraw-Hill Higher Education.

M. Drela (2005) *Flight Vehicle Aerodynamics*, MIT press.

E. L. Houghton and P. W. Carpenter (2003) *Aerodynamics for engineering students*, Elsevier.

J. N. Nielsen (1960) *Missile Aerodynamics*, McGraw-Hill.

N. Cumpsty and A. Heyes (2015) *Jet Propulsion*, Cambridge University Press.

Y. A. Cengel, M. A. Boles and M. Kanoğlu (2011) *Thermodynamics: an Engineering Approach*, McGraw-Hill New York.

T. A. Ward (2010) *Aerospace Propulsion Systems*, John Wiley & Sons.

J. D. Mattingly (1996) *Elements of Gas Turbine Propulsion*, McGraw-Hill New York.

J. D. Mattingly, K. M. Boyer and H. von Ohain (2006) *Elements of Propulsion: Gas Turbines and Rockets*, AIAA.

H. I. H. Saravanamuttoo, G. F. C. Rogers, H. Cohen, P. V. Straznicky and A. C. Nix (2017) *Gas Turbine Theory*, Pearson Education London, UK.

A. F. El-Sayed (2017) *Aircraft Propulsion and Gas Turbine Engines*, CRC press.

G. P. Sutton and O. Biblarz (2001) *Rocket Propulsion Elements*, John Wiley & Sons.

S. M. Yahya (2003) *Fundamentals of Compressible Flow: SI units with Aircraft and Rocket Propulsion*, New Age International.

R. F. Stengel (2005) *Flight Dynamics*, Princeton University Press.

A. Miele (2016) *Flight Mechanics: Theory of Flight Paths*, Courier Dover Publications.

D. G. Hull (2007) *Fundamentals of Airplane flight Mechanics*, Springer.

L. V. Schmidt (1998) *Introduction to Aircraft Flight Dynamics*, AIAA.

T. R. Yechout (2003) *Introduction to Aircraft Flight Mechanics*, AIAA.

W. F. Phillips (2004) *Mechanics of Flight*, John Wiley & Sons.

D. F. Anderson and S. Eberhardt (2001) *Understanding Flight*, McGraw-Hill New York.

P. T. Kabamba and A. R. Girard (2014) *Fundamentals of Aerospace Navigation and Guidance*, Cambridge University Press.

G. M. Siouris (2004) *Missile Guidance and Control Systems*, Springer Science & Business Media.

R. M. Lloyd (1998) *Conventional Warhead Systems Physics and Engineering Design*, Progress in astronautics and aeronautics.

P. Jerome (2001) *Composite materials in the airbus A380-from history to future*, Proceedings 13th International Conference on Composite Materials (ICCM-13).

J. Cutler (2005) *Understanding Aircraft Structures*, Blackwell Publishing.

R. Alderliesten (2018) *Introduction to Aerospace Structures and Materials*, Delft University of Technology.

T. H. G. Megson (2016) *Aircraft Structures for Engineering Students*, Butterworth-Heinemann.

A. P. Boresi and R. J. Schmidt (2002) *Advanced Mechanics of Materials*, John Wiley & Sons.

L. M. Celnikier (1993) *Basics of Space Flight*, Atlantica Séguier Frontières.

H. D. Curtis (2010) *Orbital Mechanics for Engineering Students*, Elsevier Butterworth Heinemann.

Part III

Types of Aerospace Vehicles

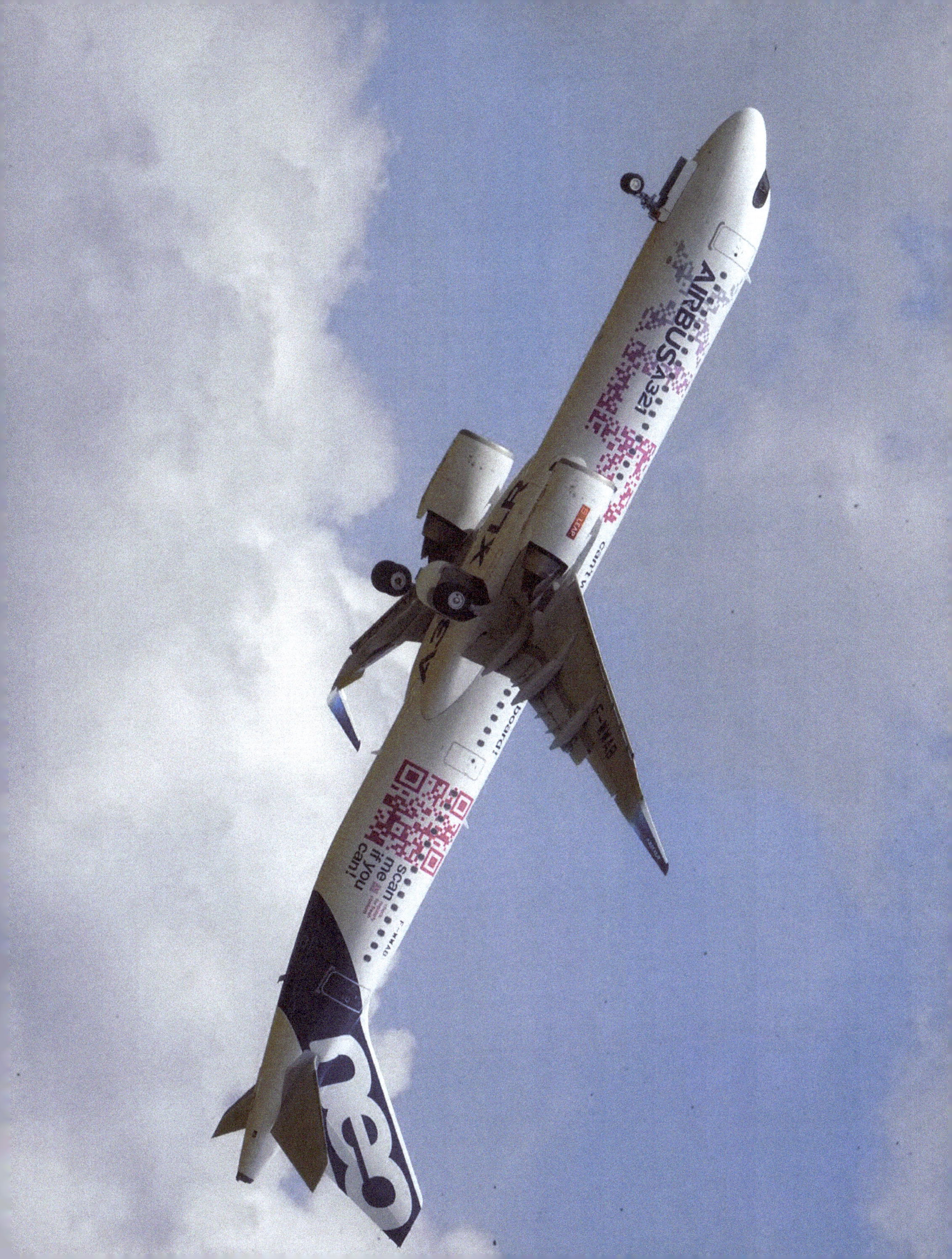
AIRBUS
A321
XLR
scan me if you can!
neo

Chapter 5

Civilian Airplanes

Contents

5.1	Introduction		**243**
5.2	**Commercial Airliners**		**247**
	5.2.1	Basic Terminologies	248
	5.2.2	Parts of Airliners	248
	5.2.3	Notable Airliners	254
	5.2.4	Identifying the Airliners	261
	5.2.5	Supply and Services of Airliners	265
	5.2.6	Airliner Operations	268
5.3	**Safety and Comfort in Commercial Aviation**		**270**
	5.3.1	Aviation Safety Management System (SMS)	271
	5.3.2	Airliner Noise	276
5.4	**Business Aviation**		**277**
	5.4.1	Business Aircraft	278
5.5	**Airports**		**282**
	5.5.1	Airport Types	283
	5.5.2	Airport Services	284
5.6	**Governing Bodies**		**287**
5.7	**Civil Aviation in India**		**287**

In the realm of human achievement and progress, few innovations have transformed our world as profoundly as civil aviation. Let's commence the chapter on civil aviation with two stories: the first one exemplifies remarkable progress due to airplanes, and the second one is about the profound and lasting impact aviation has on humanity, showcasing the significance of air travel.

Story 1: Wings of Progress
Picture this scenario in the present day in India: you decided to pursue your education in London, a city that promised a world of opportunities and knowledge. You have prepared, collecting every document and making travel plans to ensure you make the most of your time abroad. Your journey began on a bright morning, much like the one Mahatma Gandhi experienced over a century ago. However, today, the mode of transportation is strikingly different. Instead of a long sea voyage, you are taking to the skies, traveling by air. The contrast between your experience and Gandhi's is striking.

As you step onto the airplane, you can't help but imagine what Gandhi's journey must have been like–a lengthy sea voyage marked by the rhythm of the waves, the vast expanse of the ocean, and the endless horizon. Experiences like motion sickness where one stands still but still feels like moving, dizziness, nausea, increased saliva production, loss of appetite, and pale skin are additional suffering of a long sea voyage. Some individuals may also encounter headaches, fatigue, or shallow breathing after a long sea voyage. The 25 days (from 4th to 29th September 1888), it took for Mahatma Gandhi to reach Tilbury (UK) seem like an eternity compared to your mere hours in the air.

The airplane amazes you with its high speed, comfort, and convenience. At this moment, you can't help but feel a profound sense of gratitude for the incredible advancements of modern air travel. You realize that the sky is no longer a limit; it's a gateway to the world.

Story 2: Wings of Hope (Mission of the century)
The horror of the global COVID-19 pandemic cast a long shadow over the world, bringing bustling cities to a standstill, leaving streets eerily deserted, and turning masks into an enduring symbol of our shared struggle. Amid the pandemic, the world was racing against time to develop, produce, and distribute vaccines. Scientists had worked tirelessly to create a solution and finally, Margaret Keenan, a 91-year-old woman in the UK, became the first person to be vaccinated against COVID-19 on December 8, 2020.

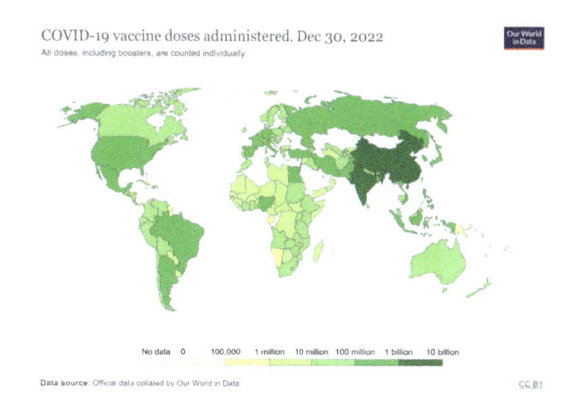

COVID-19 vaccine doses administered, till December 30, 2022 (source:*ourworldindata.org*)

The real challenge lay in ensuring that the vaccines reached every corner of the globe and that's where the marvel of air travel played a pivotal role. In less than three months from the first dose, approximately 240 million people worldwide had been vaccinated against COVID-19. By the end of 2022, roughly 13.2 billion COVID-19 vaccine doses were administered, and the logistics of the vaccine distribution was made possible by the efficiency of air transportation.

The challenge at hand was monumental: to administer a single vaccine dose to the global population of 7.8 billion requires enough vaccines to fill 8,000 Boeing 747s. However, with two-dose vaccines, the delivery of roughly 10 billion doses needed was accomplished in an astonishing 15,000 flights. The global distribution of COVID-19 vaccines was described as "the mission of the century for the global air cargo industry and largest single transport challenge ever"

5.1 Introduction

From a historical perspective spanning several decades, the air transport industry has displayed remarkable growth, consistently doubling in size approximately every fifteen years, surpassing the expansion rates of most other sectors. Since 1960, increasing demand for passenger and freight services, technological progress and associated investment have combined to multiply the output of the aviation industry by a factor of more than 30.

One of the pivotal catalysts behind this burgeoning growth in passenger traffic is the steadily increasing accessibility of air travel, largely driven by its affordability. For five decades, the actual cost of air travel has decreased by an impressive margin, with a reduction of over 70% since 1970.

Furthermore, **the aviation sector has established itself as the safest mode of long-distance transportation**, further propelling its rapid expansion. The safety of air travel has witnessed substantial improvements, exemplified by the data depicted in the Figure. In 2019, the fatal accident rate in commercial air transport declined to a mere 0.18 fatal accidents per million flights, down from 0.28 accidents per million flights in 2018. This translates to one fatal accident occurring for every 5.58 million flights in 2019. Aviation stands as the world's safest and most efficient means of long-range transportation, making it an unparalleled contributor to global connectivity and commerce.

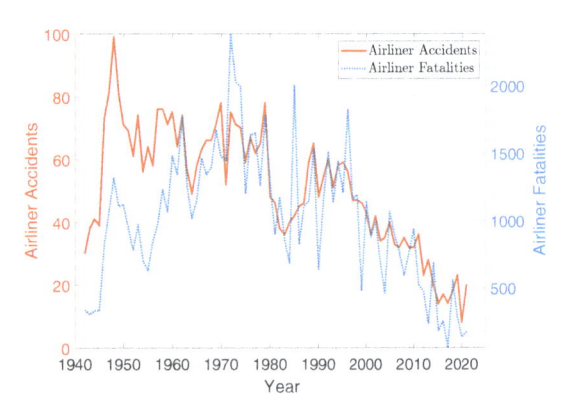

Trends in aviation accidents and fatalities over the year show that flight safety has improved dramatically, especially last two decades

As per the Air Transport Action Group (ATAG) report *Aviation: Benefits Beyond Borders 2020*, the aviation industry supported 87.7 million (8.77 crores) jobs worldwide and generated $3.5 trillion economy which is 4.1% of global GDP. The industry served 48,044 routes globally transporting 4.5 billion passengers (8.68 trillion revenue passenger kilometers) and 61 million tons of freight (254 billion cargo ton kilometers) in 46.8 million scheduled commercial flights. The aviation industry is truly a global industry connecting all parts of the world seamlessly with 1,478 commercial airlines operating 33,299 commercial aircraft in 3780 airports. 58% of international tourists travel by air. With the disruptions brought about by the pandemic over, the aviation sector is growing at a steady pace, especially in countries like India and China.

Civil aviation is one of two major categories of flying (the other being military aviation), representing all non-military and non-state aviation, both private and commercial. Civil aviation includes three major categories: (1) Commercial air transport, including scheduled and non-scheduled passenger and cargo flights (2) Aerial work, in which an aircraft is used for specialized services such as agriculture, photography, surveying, search and rescue, etc. (3) General aviation (GA), including all other civil flights, private or commercial. Some of the terminologies associated with civil aviation are mentioned below.

Commercial air transport operation: An aircraft operation involving the transport of passengers, cargo or mail for remuneration or hire. Commercial air transport includes scheduled and non-scheduled passenger and cargo flights

Scheduled commercial air transport includes a series of flights (1) it is performed by aircraft for the transport of passengers, mail or cargo for remuneration, in such a manner that each flight is open to use by members of the public (2) it is operated, so as to serve traffic between the same two or more points according to a timetable.

A non-scheduled air service is a commercial air transport service performed as other than a scheduled air service. A charter flight is a non-scheduled

5.1. INTRODUCTION

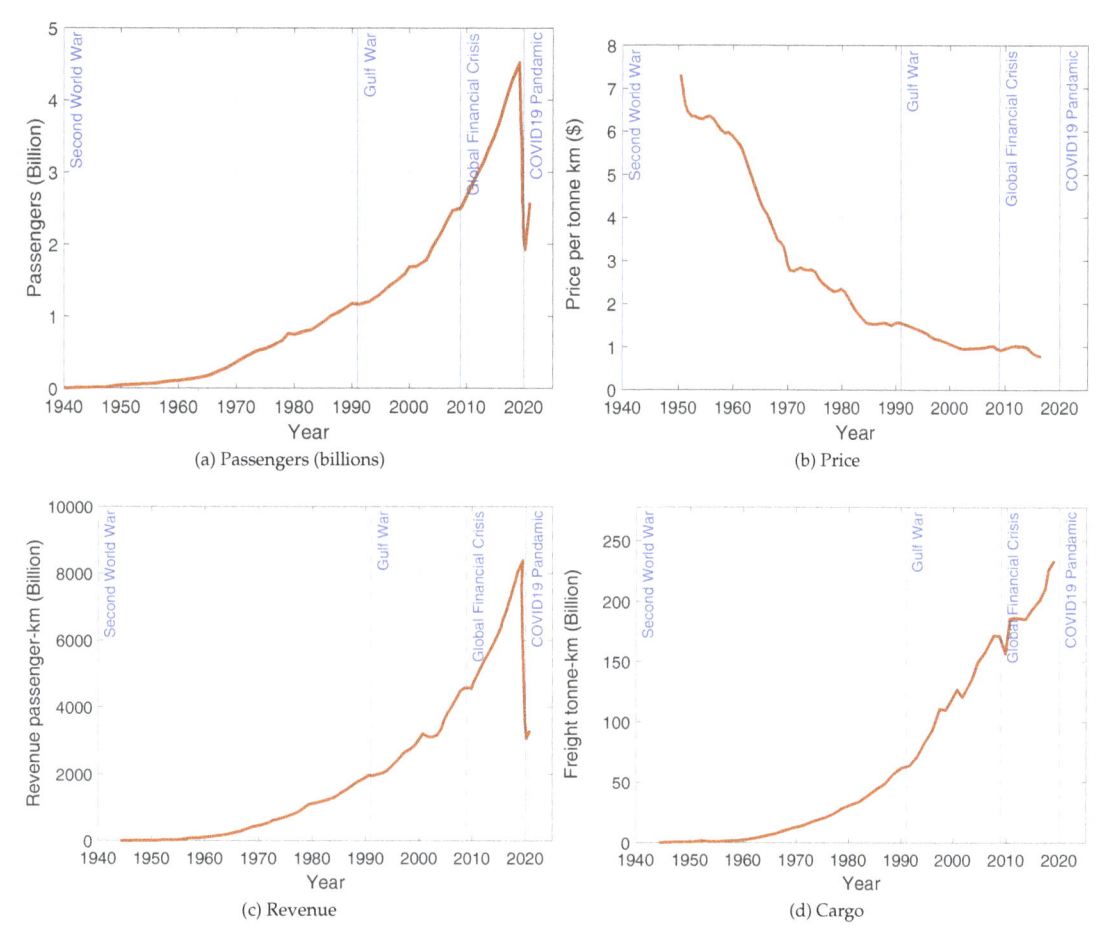

(a) Passengers (billions)

(b) Price

(c) Revenue

(d) Cargo

Evolution of commercial aviation. *Price per tonne km* is the cost of carrying a tonne of cargo over a distance of one kilometer. *Freight tonne-km* measures a metric tonne of freight carried one km. *Revenue passenger km (RPK)* is a product of the number of paying passengers by the distance traveled in km. For example, an airplane with 100 passengers that flies 250 km has generated 25,000 RPK. A large passenger airplane operated by an airline is generally referred to as an airliner.

operation for carrying passengers or cargo using a chartered aircraft (the business of renting an entire aircraft).

On-demand air taxi service is a non-scheduled non-charter flight for the carriage of individually ticketed or individually waybilled traffic. These are flights not operated according to a published schedule but sold to individual members of the public.

General aviation includes all civil aviation operations other than scheduled air services and non-scheduled air transport operations for remuneration or hire. The general aviation activities are classified into instructional flying, business flying, pleasure flying, aerial work and other flying.

Instructional flying is the use of an aircraft for purposes of formal flight instruction with an instructor. The flights may be performed by aero clubs, flying schools or commercial operators.

Pleasure flying is the use of an aircraft for personal or recreational purposes not associated with a business or profession.

Business aviation is that sector of aviation which concerns the operation or use of aircraft by companies for carrying passengers or goods as an aid to the conduct of their business, flown for purposes generally considered not for public hire and piloted by individuals having a valid commercial pilot license.

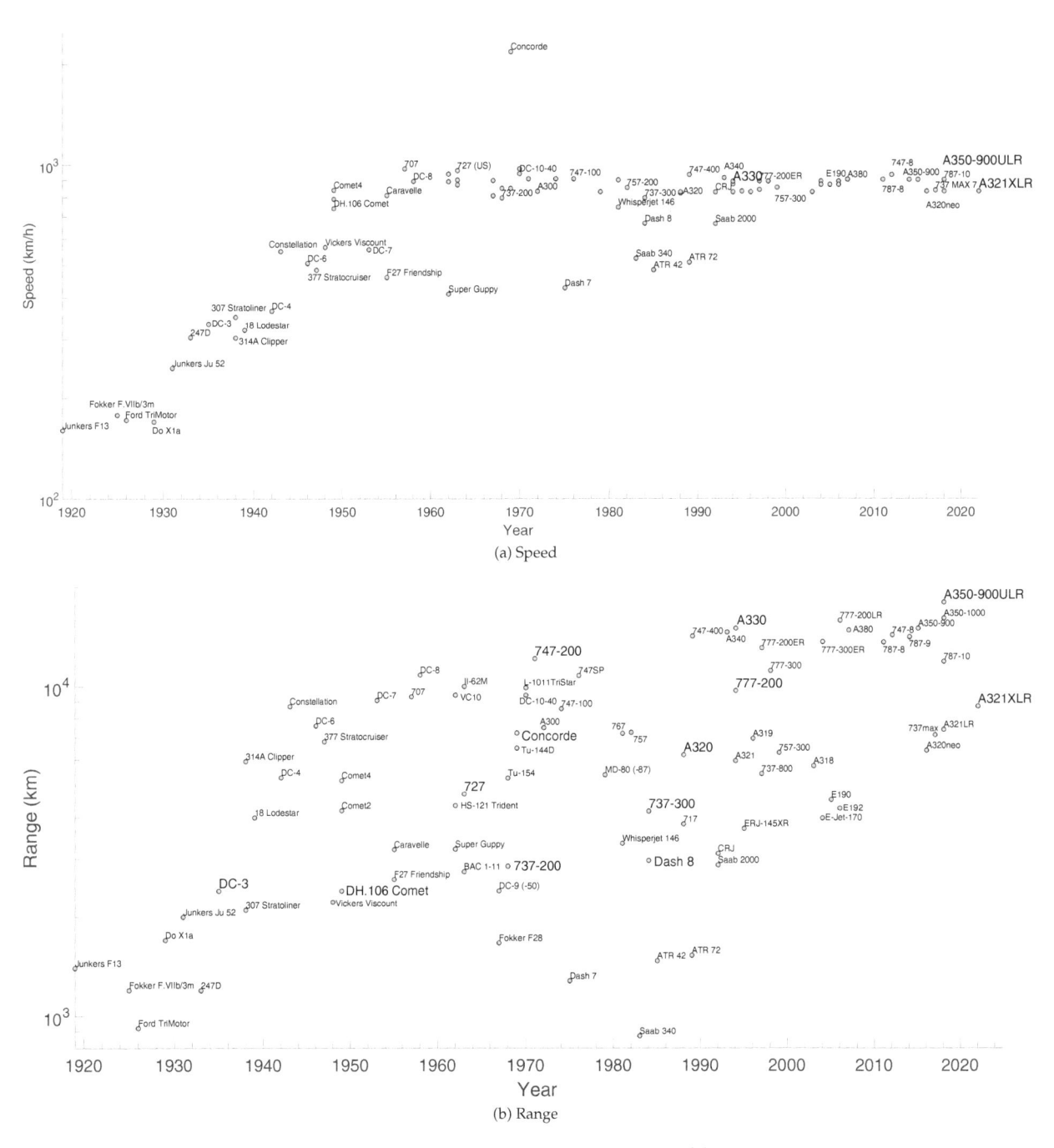

Evolution of commercial aviation in terms of Range and Speed.

Aerial work. An aircraft operation in which an aircraft is used for specialized services such as agriculture, construction, photography, surveying, observation and patrol, search and rescue, aerial advertisement, etc.

Airport or Aerodrome: A defined area on land or water intended to be used for the arrival, departure and surface movement of aircraft.

International airport: Any airport designated by the Contracting State in whose territory it is situated as an airport of entry and departure for international air traffic, where the formalities incident to customs, immigration, public health, agricultural quarantine and similar procedures are carried out

5.2. COMMERCIAL AIRLINERS

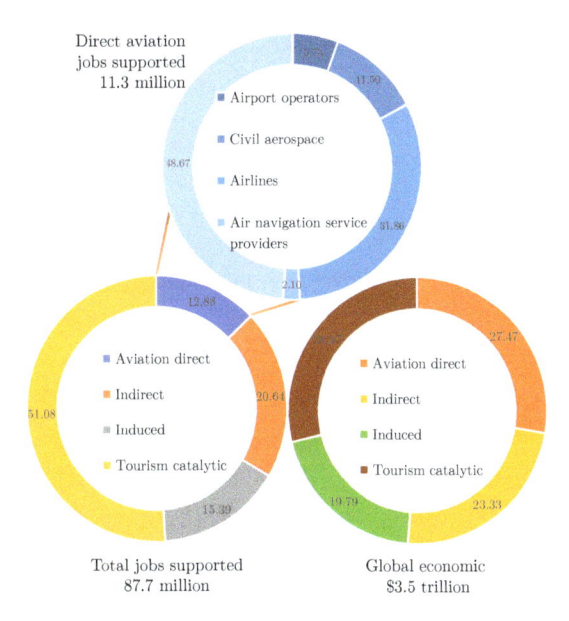

Total jobs supported and global economy generated by the aviation industry in 2019

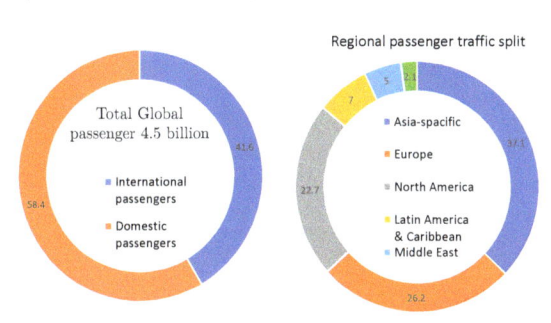

Global passengers and regional traffic split

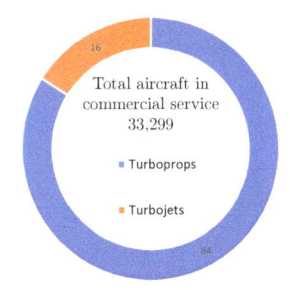

Total aircraft in commercial service in 2019

Air navigation services include air traffic management (ATM), communications, navigation and surveillance systems (CNS), meteorological services for air navigation (MET), search and rescue (SAR) and aeronautical information services (AIS). These services are provided to air traffic during all phases of operations (approach, aerodrome and en route)

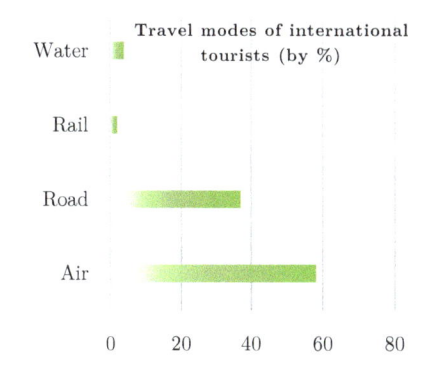

Travel modes of international tourists in 2019

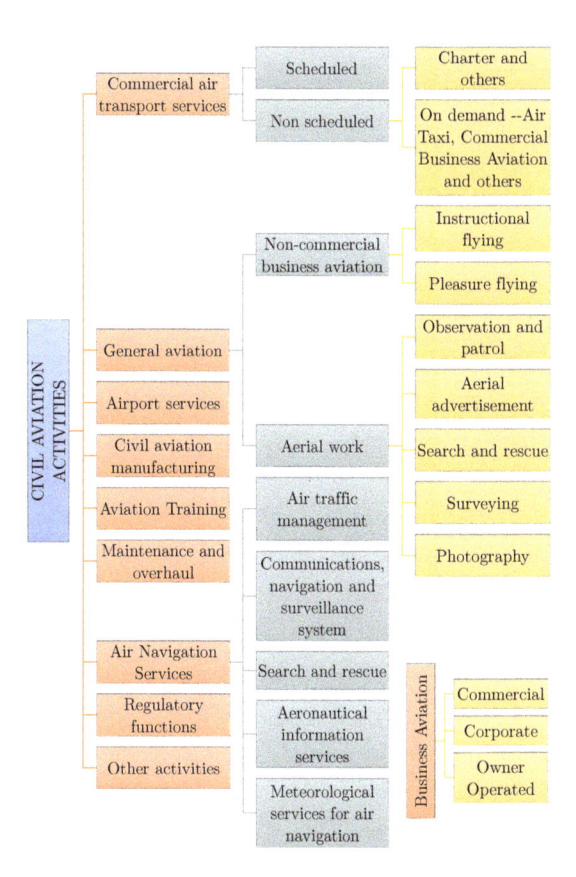

ICAO classification of civil aviation activities

5.2 Commercial Airliners

An airline's fleet is described by the total number of aircraft and the specific types of aircraft that it operates. The aircraft can have different technical and performance characteristics such as range and size.

5.2.1 Basic Terminologies

The *range* is the maximum distance an aircraft can fly with the same payload between takeoff and landing without stopping for additional fuel. Based on the range or time of flight, the flight can be a short, medium and long haul.

Short-haul flights lasting anywhere from 30 minutes to 3 hours, usually, have low volumes of around 100-200 passenger capacity. The most common short-haul aircraft are the variants of the *Boeing 737* (series – *737NG, 737 MAX*) and *Airbus A320* (series – *A319, A320* and *A321*). Some other popular short-haul aircraft are the *Embraer E-Jet* family (*E170, E175, E195* and *E190*) and the *Bombardier CRJ200, CRJ700,* and *CRJ900*. The *Boeing 717* and *ATR 72* are also short-haul aircraft, but they're much less common in the industry. *Airbus A220* has recently entered the market which caught the attention of several airlines.

Medium-haul is defined by flights lasting between 3-6 hours. One of the most common medium-haul aircraft is the *Boeing 787* series (*787-8, 787-9*, and *787-10*). However, *787-9* and *787-10* variants can be used for long-haul routes. More airplanes that are designed for medium-haul flights include the *Boeing 757* and *Boeing 767*.

Lastly, *long-haul flights* are those that extend beyond 6 hours. The most common long-haul aircraft include the *Boeing 777* variants, larger versions of the *787*, larger versions of the *767*, the *Airbus A330* variants, the *Airbus A340* variants, the *Airbus A350* variants, the *Airbus A380*, and the iconic *Boeing 747*. These are constantly modified to improve one over another and upcoming advanced long-airliners are *A330neo, A350XWB,* and *777X*.

The size can be represented by the amount of payload (passenger or cargo) that an aircraft can carry. The size of the aircraft, in turn, as the maximum takeoff and landing weights (MTOW and MLW) determine the feasibility of the airport for operating the aircraft. For higher takeoff and landing weights, the airport needs to have a longer minimum run-way. Also, type of aircraft imposes limitations on taxiways and gate space, or ground equipment at different airports can impose constraints on the airline's choice of aircraft.

Based on the width of the fuselage and seating arrangement, the airliner can be narrow-body or wide-body. *Narrow-body airliners* have one aisle (a passage between rows of seats) and up to six passenger seats across. A typical narrow-body plane has a diameter of 3-4 m accommodating between 100 and 240 passengers. A *wide-body* or *twin-aisle* aircraft has a twin-aisle and can seat up to ten passengers across. a typical wide-body plane has a diameter of 5-6 m accommodating a total capacity of 200 to 850. They tend to operate on medium, long-haul routes.

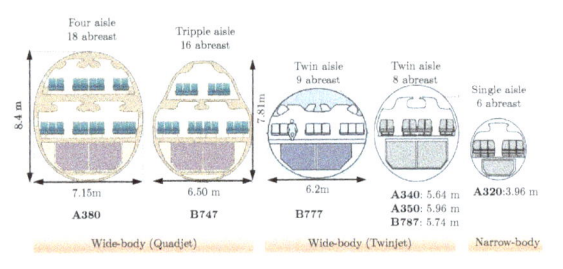

Comparison of narrow-body and wide-body airliner

5.2.2 Parts of Airliners

Airliners are complex machines composed of numerous parts and components that work in harmony to ensure the safe and efficient transportation of passengers and cargo. Here is an overview of the key parts of an airliner:

Fuselage is the main body of the aircraft, housing the cockpit, passenger cabins, cargo holds, and other essential systems. It provides structural integrity and aerodynamic shape to the aircraft. The wings and tail assembly are attached to the fuselage, with the wings typically attached to the mid-fuselage and the tail assembly at the rear. The fuselage is designed to withstand the stresses and loads experienced during flight, including turbulence, pressurization, and structural forces. The fuselage's shape is carefully designed to minimize aerodynamic drag, which affects fuel efficiency and overall flight performance.

Wings of an airliner are one of its most distinctive and critical features, playing a central role in the aircraft's ability to generate lift and maintain stable flight. The primary function of an airliner's wings is to generate lift, which is the force that counteracts the weight of the aircraft, allowing it to stay aloft. Airliner wings can have various configurations, including swept-back wings for high-speed flight and straight wings for slower aircraft. The choice of wing design is influenced by the aircraft's intended use and performance requirements.

Many modern airliners are equipped with winglets at the tips of their wings. Winglets improve aerodynamic efficiency by reducing drag and enhancing lift. They are a common feature on commercial aircraft, as they increase fuel efficiency.

Airliner wings are equipped with flaps and slats that can be extended or retracted. Flaps are deployed during takeoff and landing to increase lift and reduce landing speed. Slats are deployed at lower speeds to enhance lift and stability.

Many airliners incorporate integral fuel tanks within the wings. These tanks store the aircraft's fuel supply, which is used for propulsion during flight. The location of fuel tanks in the wings helps distribute weight and maintain balance.

Most of the airliners feature engines mounted on the wings, which is a common configuration for many modern jet aircraft. These engines are attached either below the wings (underwing) or above the wings (overwing), with specific advantages related to aerodynamics and ground clearance.

Wings are equipped with navigation lights and beacons to enhance the aircraft's visibility during night and low-visibility conditions. These lights include position lights, strobe lights, and landing lights.

Engines: Airliners typically have two or more jet engines mounted under the wings or on the tail. These engines provide thrust to propel the aircraft forward. The turbofan engine is the most common type of propulsion system used in airliners. Turbofan engines can be mounted on the aircraft's wings or fuselage, with two common configurations: underwing engines or tail-mounted engines. Many airliners have engines mounted under the wings. This location offers advantages in terms of ground clearance, maintenance access, and simplified fuel and hydraulic connections. Some aircraft, such as regional jets and smaller airliners, have engines mounted at the rear of the fuselage. Tail-mounted engines provide flexibility in aircraft design and can help improve aerodynamics.

Turbofan engines are mounted on the aircraft using sturdy structures called engine pylons. These pylons attach the engine to the wing or fuselage and provide support, stability, and the necessary connections for fuel, hydraulic systems, and electrical systems.

Many turbofan engines are equipped with thrust reversers, which redirect engine exhaust forward upon landing to assist with braking. The design and installation of thrust reversers are crucial factors in aircraft safety and performance.

Landing Gear of an airliner is a critical component responsible for supporting the aircraft during takeoff, landing, and while on the ground. It consists of a complex system of wheels, struts, and hydraulics designed to withstand the immense forces and stresses encountered during these phases of flight.

The typical landing gear assembly includes multiple components: Main Landing Gear and Nose Landing Gear. The main landing gear is the large, sturdy wheels that support the majority of the aircraft's weight. The main landing gear is typically located beneath the wings or fuselage. The nose landing gear is the wheel located under the aircraft's nose. It supports the front portion of the aircraft and aids in steering during taxiing. Each landing gear strut is equipped with shock absorbers, which absorb and dissipate the impact forces experienced during landing. They provide a cushioning effect to protect the aircraft structure and passengers from abrupt jolts. The wheels are fitted with aircraft tires, specially designed to handle the high speeds and weight of the aircraft. Braking systems are also installed to slow down and stop the aircraft during landing and taxiing on the runway.

Hydraulic systems play a crucial role in the extension and retraction of the landing gear. They are operated by the flight crew to lower the landing gear before landing and retract it after takeoff.

After takeoff, the landing gear is retracted into the aircraft's fuselage or wings to reduce drag and improve fuel efficiency. Airliners are equipped with emergency systems that allow the landing gear to be extended manually or under gravity if hydraulic systems fail.

Tail Assembly of an airliner, also known as the empennage, is a critical component located at the rear of the aircraft. It serves several key functions to ensure the aircraft's stability, control, and safety during flight. The tail assembly typically consists of the following components:

Vertical Stabilizer (Fin): The vertical stabilizer, or fin, is the upright tail component located at the back of the aircraft. It plays a vital role in maintaining directional stability, preventing yaw (side-to-side) oscillations, and helping the aircraft stay on a straight flight path.

Horizontal Stabilizer (Tailplane): Positioned horizontally, the horizontal stabilizer, or tailplane, is located in the tail assembly's upper section. It provides longitudinal stability, ensuring the aircraft maintains a steady pitch (nose-up or nose-down) attitude during flight.

Elevator: The elevator is part of the horizontal stabilizer and is attached to its trailing edge. It can move up or down to control the aircraft's pitch and, subsequently, its altitude. Elevators are crucial for controlling the aircraft's climb, descent, and level flight.

Rudder: The rudder is a movable surface attached to the vertical stabilizer. It is responsible for controlling the aircraft's yaw, allowing it to make coordinated turns. By deflecting the rudder, the pilot can adjust the aircraft's direction.

Trim Tabs: Trim tabs are small, adjustable surfaces on the control surfaces, such as the elevator and rudder. They help the pilot fine-tune the aircraft's control forces and maintain desired attitudes with minimal effort.

Antennas and Lighting: The tail assembly often houses various antennas and lighting systems, such as anti-collision lights, navigation lights, and communication equipment to ensure safe and effective aircraft operation.

Cockpit of an airliner, often referred to as the flight deck, is the nerve center of the aircraft, where the flight crew, including the captain and first officer, operate and control the aircraft. The aircraft's cockpit houses essential flight instruments displayed on the instrument panel, as well as the controls necessary for piloting the aircraft. In the case of most airliners, a secure door serves as a barrier between the cockpit and the passenger cabin. In the wake of the September 11, 2001 attacks, all major airlines reinforced their cockpit security measures to prevent unauthorized access by potential hijackers.

Cockpit of an *A380*. Most *Airbus* cockpits are glass cockpits (digital flight instrument displays, typically large LCD screens rather than the traditional style of analog dials and gauges) featuring fly-by-wire technology.

Passenger Cabins: Airliners are equipped with passenger cabins that include seats, overhead bins, lavatories, and in-flight entertainment systems.

These cabins are designed to provide comfort and safety to passengers.

Cargo Holds: Airliners have cargo holds, both in the fuselage and beneath the passenger cabins, for transporting cargo, luggage, and freight.

Avionics: Airliners are equipped with advanced avionics systems that include navigation, communication, and control systems, as well as weather radar, collision avoidance systems, and autopilots.

Fuel Systems in airliners are intricate and critical components that ensure the aircraft's engines receive a steady supply of aviation fuel, a necessity for propulsion. These systems are designed with precision, safety, and redundancy to guarantee that the aircraft can operate efficiently and securely.

Airliners are equipped with multiple fuel tanks, which are typically located within the wings and fuselage. These tanks store the aircraft's fuel supply and are designed to minimize the risk of fuel leaks or fires, even in the event of damage. Advanced fuel quantity sensors and indicators are installed in the tanks to provide real-time information about the amount of fuel on board. The flight crew relies on these indications to monitor fuel consumption and plan for refueling or alternative landing options in the case of an emergency. Fuel transfer pumps and systems are integrated into the aircraft's design to ensure that fuel is distributed evenly between the tanks. This balance is essential to maintain proper weight and balance throughout the flight.

Specialized fueling procedures are followed by ground crews to ensure that the correct type and quantity of fuel are loaded into the aircraft. Fueling is a meticulously controlled process to prevent contamination and ensure safety. Airliners are equipped with emergency fuel systems that can provide fuel to the engines even in the event of certain system failures. These systems offer redundancy and safety measures to cope with unexpected situations.

In rare situations where an aircraft must return to the airport shortly after takeoff due to an issue, it may need to reduce its weight quickly. Some airliners are equipped with a fuel jettison system that can rapidly release excess fuel to reach a safe landing weight.

Aviation fuel, often referred to as jet fuel, is a specialized type of fuel formulated for use in aircraft, ranging from small general aviation planes to large commercial airliners. Jet fuel is a colorless, combustible, straight-run petroleum distillate liquid.

The most commonly used aviation fuels are Jet A and Jet A-1 (unleaded kerosene), adhering to international quality standards. Jet B (naphtha-kerosene blend), with its improved cold-weather attributes, is the sole alternative jet fuel typically employed in civilian turbine-powered aviation. Avgas (aviation gasoline) finds its application in small aircraft, light helicopters, and vintage piston-engine planes. Its composition sets it apart from the typical gasoline used in land vehicles. Despite the availability of various grades, Avgas boasts a higher octane rating than the standard motor gasoline.

Aviation fuels are designed to provide a high energy content to power aircraft engines efficiently. Jet fuels are formulated to remain liquid at extremely low temperatures encountered at high altitudes. Aviation fuels have low vapor pressure to minimize the risk of fuel vapor igniting during storage, handling, and flight. This property enhances safety. Aviation fuels are typically color-coded for easy identification. *Jet A* and *Jet A-1* are straw-colored, while *Avgas* is often dyed blue or green.

Emergency Equipment on civil airliners is a critical component of ensuring passenger safety in the event of unexpected situations. Airliners are equipped with an array of emergency tools and systems designed to address a variety of scenarios, from rapid evacuations to in-flight medical emergencies.

Escape Slides: One of the most visible pieces of emergency equipment is the emergency exit, which includes doors and window exits that can be quickly opened in case of an evacuation. These exits are equipped with inflatable slides or ramps that allow passengers to quickly exit the aircraft during

Airframe
Specially formed damage resistance aluminium alloy panel is riveted over the airframe made of longerons (stretched from nose to tail) and vast network of stringers, intercostals and subframes.

Wind shield and cabin windows made of strong glass panels. An additional acrylic panel in passenger side has a small hole and air gap for pressure equalization.

winglet

Static discharger on winglet of the wingtip are flexible metal rods that discharge built up static electricity

Engine Pylon Strong and flexible connection used to attach engine to the wing

Fuel stores in center box and internal wing area

Vertical stabilizer

Horizontal stabilizer

Wing

Fuselage

Nose landing gears operate rack-pinion steering system

Engine

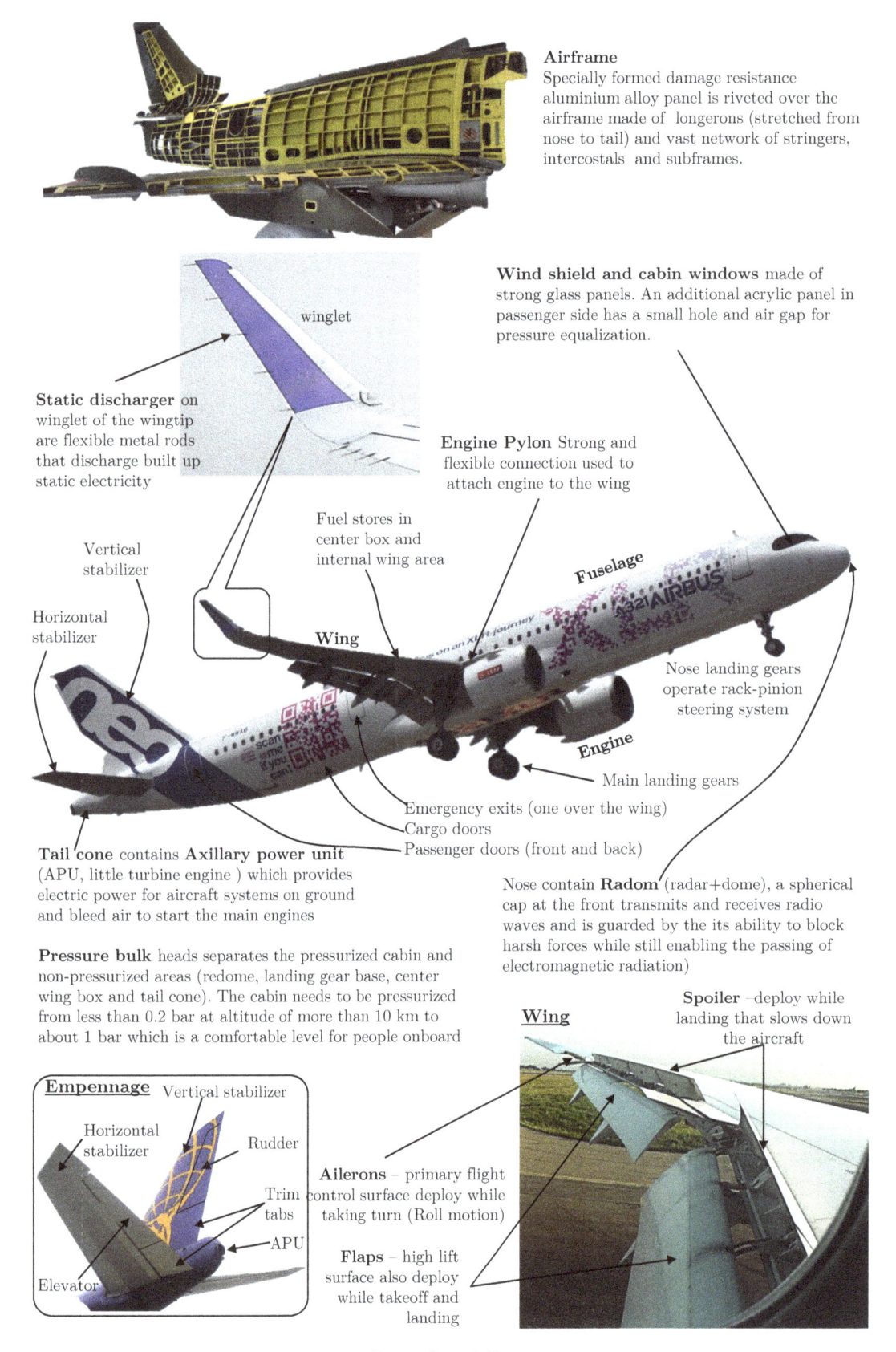

Main landing gears

Emergency exits (one over the wing)
Cargo doors
Passenger doors (front and back)

Tail cone contains **Axillary power unit** (APU, little turbine engine) which provides electric power for aircraft systems on ground and bleed air to start the main engines

Nose contain **Radom** (radar+dome), a spherical cap at the front transmits and receives radio waves and is guarded by the its ability to block harsh forces while still enabling the passing of electromagnetic radiation)

Pressure bulk heads separates the pressurized cabin and non-pressurized areas (redome, landing gear base, center wing box and tail cone). The cabin needs to be pressurized from less than 0.2 bar at altitude of more than 10 km to about 1 bar which is a comfortable level for people onboard

Wing

Spoiler –deploy while landing that slows down the aircraft

Empennage Vertical stabilizer
Horizontal stabilizer
Rudder
Trim tabs
APU
Elevator

Ailerons – primary flight control surface deploy while taking turn (Roll motion)

Flaps – high lift surface also deploy while takeoff and landing

Parts of an airliner

an emergency landing. In addition to exits, airliners also carry emergency oxygen masks that deploy if there is a sudden loss of cabin pressure, ensuring passengers and crew can breathe normally at high altitudes.

Another crucial component of emergency equipment is fire suppression systems. Aircraft are equipped with fire extinguishers and fire-resistant materials in the cabin and cargo holds to contain and control fires.

Life Vest, also known as life jackets, are essential safety devices found on all commercial aircraft. These vests are designed to ensure the safety and survival of passengers and crew members in the event of an emergency water landing. Life vests are strategically located under passenger seats, cabin attendant seats, and behind every cockpit seat. A limited quantity of non-functional life jackets is also present on board to assist flight attendants in providing clear, visible instructions to passengers.

First Aid Equipment: First-aid kits and defibrillators are on board to address medical emergencies during flights. Flight attendants are trained to use this equipment, ensuring they can respond effectively to passenger needs. Overall, the presence of emergency equipment on civil airliners underscores the industry's commitment to passenger safety and preparedness for unforeseen events.

Escape Rope for Cockpit: An escape rope for the cockpit is a critical safety feature in certain aircraft, such as the *Airbus A320*. Within the *Airbus A320* aircraft, there is a rope stored in compartments above the sliding windows on both sides of the overhead panel. When the cabin is depressurized, the sliding windows can be manually opened. This rope serves as an emergency means of exit for the cockpit crew in such situations, providing them with an escape option.

Universal Precaution Kit (UPK): Every aircraft is equipped with a universal first-aid kit that includes tools for the removal of foreign objects from the cabin. This practice reduces the risk of potential contamination for both the crew and passengers.

The APU exhaust in the tail cone of an *Airbus A380*.

Emergency Locator Transmitter (ELT) and Underwater Locator Beacon (ULB) An ELT and ULB are a vital piece of equipment installed in aircraft to enhance safety and facilitate search and rescue operations in the event of an emergency. ELTs play a crucial role in situations where an aircraft may be difficult to locate after an accident or crash. ELTs are designed to transmit distress signals on designated emergency frequencies. ULBs are designed to aid in locating and recovering aircraft that have ditched or crashed into bodies of water. This transmission continues for a period exceeding 90 days.

Auxiliary Power Unit (APU) is a crucial component found in most commercial aircraft and is typically located at the rear of the aircraft. Placing the APU in the aircraft's tail, away from the main engines, enhances safety during ground operations and maximizes storage space for cargo and fuel. The high-pressure air it supplies to the engines remains efficient over this distance. The APU is a compact turbine engine that operates similarly to the aircraft's main engines but on a smaller scale. Unlike the main engines, the APU doesn't produce thrust for propulsion. Instead, its primary functions are to generate electrical power and provide compressed air for various essential systems, both when the aircraft is on the ground and during flight preparations. The other main function of the APU is to start the main engines. The APU plays a vital role in aircraft operations. While on the ground, it allows the aircraft to have its electrical systems powered without relying on ground-based equipment. This is particularly important for activities such as pre-flight checks, passenger boarding, and air conditioning while the aircraft's main engines are not running. In the air, the APU can serve as a backup power source in case of an engine failure and can provide air pressure for functions like environmental control systems.

Environmental Control Systems (ECS) in civil airliners are essential components responsible for maintaining a comfortable and safe cabin environment for passengers and crew throughout a flight. ECS is responsible for supplying conditioned air, regulating temperature, and maintaining cabin pressure. It also serves additional roles, such as cooling avionics, detecting smoke, and suppressing fires.

One primary function of the ECS is *temperature control*. It regulates the cabin temperature to provide a comfortable and consistent environment. Air from the engines is often used to heat or cool the air within the cabin through a special heat exchanger. Additionally, cabin temperature is controlled using thermostats and sensors that monitor and adjust the temperature to maintain a set range.

Another critical aspect of ECS is *cabin pressurization*. Airliners typically cruise at altitudes where the external air pressure is significantly lower than at sea level. ECS systems maintain a controlled cabin pressure to ensure passengers and crew receive enough oxygen to breathe comfortably. This is achieved by compressing bleed air from engines and regulating the mix of fresh and recirculated air within the cabin. The ECS is responsible for controlling how much-compressed air enters the cabin. Although the majority of airplanes utilize engine-derived air for their cabins, this is not a standard practice. For instance, the *Boeing 747* employs a different approach, drawing in external air through inlets situated on its wings. Subsequently, this externally sourced air is employed to pressurize the cabin.

Moreover, ECS systems manage *cabin air quality*. This includes filtering out contaminants, ensuring a supply of clean, fresh air, and regulating humidity levels to prevent the air from becoming too dry, which can cause discomfort to passengers. Filtering systems remove airborne particles, and in some cases, even potential pathogens. These systems help maintain a clean and healthy cabin environment for the duration of a flight, enhancing the overall travel experience for passengers and crew alike.

ECS also integrates *smoke detection* and *fire suppression systems* to enhance passenger safety. These systems can detect the presence of smoke or fire in the cabin and take appropriate actions to control or extinguish the fire and direct passengers to safety.

5.2.3 Notable Airliners

Throughout the history of civil aviation, several iconic airliners have left a lasting mark on the industry and the way people travel. One such legendary aircraft is the *Boeing 747*, often referred to as the "Queen of the Skies." Introduced in 1970, the *Boeing 747* revolutionized long-haul air travel with its unprecedented capacity, enabling airlines to carry more passengers efficiently. It played a pivotal role in making air travel more accessible to a broader demographic, marking a significant milestone in aviation history.

Another renowned airliner is the *Concorde*, a supersonic passenger jet introduced in the late 1960s. The *Concorde* represented the epitome of speed and luxury in air travel, flying at twice the speed of sound and reducing transatlantic flight times significantly. While the *Concorde*'s operational history was relatively short, it remains an enduring symbol of innovation and prestige in the aviation world. These famous airliners, along with others like the *Douglas DC-3*, *Airbus A380*, and *Lockheed Constellation*, have each contributed to the advancement and evolution of civil aviation, leaving an indelible legacy in the skies and in the hearts of travelers worldwide.

5.2. COMMERCIAL AIRLINERS

Notable airliners in the history of civil aviation. EIS: Entry into service

Company Airliner	Airlines	EIS	Range (km)	Max. Seats	Speed (km/h)	Number built	Remarks
Junkers F13		1919	1,400	4-6	160	345	The first airliner made of metal with enclosed cabin
Fokker F.VIIb/3m	Dutch airline KLM	1925	1,200	8-10	178		Also known as Fokker Trimotor as it is driven by three engines
Ford TriMotor-4ATE	Military transport (US)	1926	920	11-14	172	199	Henry Ford believed mix of Fokker's three-engine and Junker's all-metal construction was the future for aircraft construction
Dornier Do X1a		1929	1700	100	170	3	Driven by 12 Siemens engines, known for an epic transatlantic tour and largest aircraft then
Junkers Ju 52/3m g3e	Swissair and Deutsche Luft Hansa	1931	1998	17-19	246	4845	Following the rise of Nazi Germany, thousands of Ju 52s were were used in military transport
Boeing 247D	Boeing Air Transport	1933	1199	10-13	304	75	All-metal semi-monocoque construction, an aerodynamically efficient airframe, retractable landing gear and autopilot and de-icing boots for the wings and tailplane
Douglas DC-3	American Airlines, United Airlines, Trans World Airlines, Eastern Air Lines	1935	2400	21-32	333	607	The most successful twin-engine airliner before WWII and the foundation of the world's resurgent airliner after the war. It was the first to fly profitably without government subsidy. A larger, improved 14-bed sleeper version of the Douglas DC-2.
Lockheed Model 14 Super 18 Lodestar	Northwest Airlines	1937	1370	10	400	354	Too expensive to operate and failed to even dent the Douglas dominance despite the aircraft's excellent performance
		1939	400	18	400	625	
Boeing 314A Clipper	Pan American World Airways	1938	5930	68	303	12	One of the largest aircraft of its time, it had the range to cross the Atlantic and Pacific oceans
Boeing 307 Stratoliner	Pan American Airways and TWA	1938	2100	33-60	350	10	The first pressurized airliner to fly at a higher altitude of 6100 m compared to DC-3 at 2400 m for smoother and more comfortable services. Driven by four supercharged engines.
DC-4	United Air Lines	1942	5300	44	365	80	
DC-6		1946	7600	89	507	704	
DC-7		1953	9069	105	557	338	Arguably the ultimate expression of four-piston-engined airliner just before the jet age.
Lockheed Constellation (Connie)	Trans World Airlines	1943	8700	95	550	856	The most elegant airliners ever built featuring the distinctive triple-tail and dolphin-shaped fuselage. first widespread airliner with pressurization in its cabin flying at a record altitude of 3,810 m. Various models such as Super Constellation, Warning Star, and Starliner were developed.
Boeing 377 Stratocruiser	Pan American World Airways	1947	6800	115	484	56	Advanced and complex with a five-man flight crew, twin decks, turbo-compound engines, cabin pressurization and heated, pressurized cargo hold. Developed from the C-97 Stratofreighter military transport that itself is a derivative of the B-29 Superfortress
Vickers-Viscount	Capital Airlines, Trans-Canada Air Lines, Air Canada	1948	2220	75	566	445	Known for its smooth ride and large cabin windows and remains the UK's most widely produced airliner type. Later, Vickers Vanguard was developed in 1959 for 139 passengers flying at 679 km/h in a range of 2950 km
De Havilland Comet	BOAC, British European Airways, Dan-Air	1949	2400	44	740	114	The world's first commercial jet airliner offering a relatively quiet, comfortable passenger cabin. Aerodynamically clean design with four turbojet engines buried in the wing roots,
Comet 2			4200	44	790		A pressurized cabin, and large windows are other features. The revolutionary Comet transformed the airliner performance by flying above 13000 m.
Comet 3			4300	76	840		But a series of accidents tarnished its reputation.

Company Airliner	Airlines	EIS	Range (km)	Max. Seats	Speed (km/h)	Number built	Remarks
Comet 4			5190	81	840		The constant stress of pressurization and depressurization at high altitudes weaken an area of the fuselage around the Comet's rectangular windows and the skin of the aircraft had been made as thin as possible to save weight
Fokker F27 Friendship	New Zealand National Airways Corporation; Trans Australia Airline, East-West Airlines; and Turkish Airlines	1955	2600	56	460	586	Most successful European turboprop airliner of its era
Boeing 707	Pan Am, Trans World Airlines, American Airlines, Air France	1957	9300	141	972	865	Narrow-body jetliners are credited with beginning the Jet Age. A swept-wing, quadjet with podded engines. Its larger fuselage cross-section allowed six-abreast economy seating, retained in the later 720, 727, 737, and 757 models. It established Boeing as a dominant airliner. The variant 707-120B (1961) can accommodate 174 in one class with a 6700 km range and the 707-320B (1962) can fly 9300 km with 141 passengers
McDonnell Douglas DC-8	United Airlines, UPS Airlines, Delta Air Lines, Trans Air Cargo Service	1958	10843	189	892	556	A low-wing, quad jet airliner, the narrow-body airliner was designed as a competitor to Boeing's 707. DC-8-50 was the most produced variant (built 142) and could fly non-stop from the U.S. West Coast to London
Hawker Siddeley Trident	British European Airways, British Airways, CAAC Airlines, Cyprus Airways	1962	4350	115	937	117	The jetliner is powered by three rear-mounted low-bypass turbofans, it has a low swept wing and a T-tail. Advanced avionics allowed it to be the first airliner to make a blind landing
Vickers-Armstrongs, Vickers VC10	BOAC, East African Airways, Ghana Airways, Royal Air Force	1962	9410	151	890	54	Designed to operate on long-distance routes from the shorter runways of the era and commanded excellent hot and high performance for operations from African airports. The fastest crossing of the Atlantic by a subsonic jet airliner until February 2020 when the Boeing 747 broke the record. The VC10 is often compared to the larger Soviet Ilyushin Il-62, the two types being the only airliners to use a rear-engined quad layout
Aero Spacelines Guppy and Super Guppy	Aero Spacelines, NASA, Airbus, Aeromaritime	1962	3211		410	1	a large, wide-bodied cargo aircraft. The design inspired later designs, such as the jet-powered Airbus Beluga and Boeing Dreamlifter. Derived from Boeing 377 Stratocruiser
Boeing 727	American Airlines, Lineas Aéreas Suramericanas, Kalitta Charters, Total Linhas Aereas	1963	4720	189	960	1832	Six-abreast, narrow-body airliner designed for shorter flight lengths from smaller airports, the only trijet aircraft to be produced by Boeing. one Low-bypass turbofans engine below a T-tail fed through an S-duct, one on each side of the rear fuselage.
BAC One-Eleven (BAC 1-11)	British Airways; American Airlines; Braniff Airways; British United Airways	1963	2744	119	871	244	The short-haul, narrowbody aircraft was powered by aft-mounted low-bypass turbofans, a configuration similar to the earlier Sud Aviation Caravelle and later Douglas DC-9 and also competed with early Boeing 737 variants
Ilyushin Il-62M Classic	Air Koryo, Aeroflot, Russian VIP transport, LOT Polish Airlines	1963	10000	200	900	292	The world's largest in its time and the USSR's first long-haul jet airliner, successor to the popular turboprop Il-18, proved capable and reliable in service but never matched the efficiency of Western equivalents

5.2. COMMERCIAL AIRLINERS

Company Airliner	Airlines	EIS	Range (km)	Max. Seats	Speed (km/h)	Number built	Remarks
McDonnell Douglas DC-9 (-50)	USA Jet Airlines, Aeronaves TSM, Everts Air Cargo, Northwest Airlines	1967	2400	139	897	976	Five-abreast single-aisle airliner powered by two rear-mounted low-bypass turbofan engines under a T-tail for a cleaner wing aerodynamic with built-in airstairs to better suit smaller airports. Smaller variants of DC-9 competed with the BAC One-Eleven, Fokker F28, and Sud Aviation Caravelle, and larger ones with the Boeing 737.
Fokker F28-4000 Fellowship (Netherlands)	Garuda Indonesia, AirQuarius Aviation. Linjeflyg, Biman Bangladesh Airlines	1967	1668	85	808	241	A twin-engined, short-range jet airliner.
Tupolev Tu-154	Air Koryo	1968	5,280	180	850	1026	A three-engined, medium-range, narrow-body airliner
Tupolev Tu-144D Charger	Aeroflot	1969	6500	150	2200	16	The world's first commercial supersonic transport aircraft offered scintillating performance, but its service career was cut short after two fatal accidents. Only 13 aircraft flew and services were canceled on 1 June 1978
McDonnell Douglas MD-80 (-87)	Aeronaves TSM, World Atlantic Airlines, LASER Airlines, European Air Charter,	1979	5400	155	830	1,191	The second generation, the MD-80 series, a lengthened DC-9-50 with a larger wing and a higher MTOW was introduced that competed Boeing 737 Classic and Airbus A320ceo family. This was further developed into the third generation, the MD-90, as the body was stretched again, fitted with high-bypass turbofans, and an updated flight deck.
Boeing 717	Delta Air Lines, Hawaiian Airlines, QantasLink, AirTran Airways	1988	3820	134	822	156	The shorter and final version, the MD-95, was renamed as the Boeing 717 after McDonnell Douglas's merger with Boeing in 1997
Boeing 737 family							
Original Generation 737-200	Southwest Airlines, Ryanair, United Airlines, American Airlines	1968	2850	130	796	1988	Powered by low-bypass engines and offered seating for 85 to 130passengers.
Classic Generation 737-300	UTair, Jet2.com, Southwest Airlines, US Airways	1984	4176	149	796	1988	Was re-engined with a high-bypass turbofan, for better fuel economy and had upgraded avionics. It competed with the McDonnell Douglas MD-80 series, and then with the Airbus A320 family
Next Generation 737-800	Southwest Airlines Delta Air Lines United Airlines American Airlines	1997	5436	189	842	7102	Redesigned wing with a larger area, a wider wingspan, greater fuel capacity, and higher MTOW and longer range with a glass cockpit. The 737NG's primary competition is the Airbus A320 family. As of April 2023, a total of 7,124 737NG aircraft had been ordered
MAX Generation 737 MAX 7	Southwest Airlines, Ryanair, United Airlines, flydubai	2017	7130	153	839	1161	Driven by more efficient LEAP-1B (high bypass) engines with efficient aerodynamic including distinctive split-tip (Scimitar) winglets. As of April 2023, the 737 MAX has 4,174 unfilled orders. The 737 MAX suffered a recurring failure in the Maneuvering Characteristics Augmentation System (MCAS), causing two fatal crashes. the company estimated a loss of $18.4 billion for 2019, and it reported 183 canceled MAX orders for the year
Aerospatiale/BAe Concorde	Air France and British Airways	1969	7,222	128	2,180	20	The supersonic airliner had a delta wing in a canard configuration with a droop nose for landing visibility and a narrow fuselage design permitting a 4-abreast seating. It had four turbojets with variable engine intake ramps and a reheat option. The first airliner to have analogue fly-by-wire flight controls
McDonnell Douglas DC-10-40		1970	9400	270	940	386	McDonnell Douglas entered the widebody market with the DC-10, a trijet wide-body aircraft. Later MD-11 was derived from DC-10 with updated turbofan, wider wing with winglets, 15% increase in MTOW and stretched by 11% to accommodate 298 passengers

Company Airliner	Airlines	EIS	Range (km)	Max. Seats	Speed (km/h)	Number built	Remarks
Lockheed L-1011-500 TriStar	British Airways, Delta Air Lines, Eastern Air Lines	1970	9899	330	972	250	Widebody trijet airliner. The aircraft has an autoland capability, an automated descent control system, and available lower deck galley and lounge facilities.
Airbus A300	UPS Airlines, European Air Transport Leipzig, Mahan	1972	7,500	247	833	561	The first twin-engine widebody
Boeing 747-100	Pan Am, United Airlines, British Airways	1974	8,560	366	907	205	Boeing 747-100 with six upper deck windows was the first variant. 747SR (seating 498-550) was built and entered service in 1973 as a short-range version of the 747-100 with lower fuel capacity and greater payload capability. Further, it was stretched as 747-100B in 1979 to accommodate 452 passengers in a range of 9300 km
747-200	Japan Airlines, Lufthansa, Air France	1971	12,150	366	907	393	Improved version of 747-100 featuring more powerful engines, increased MTOW, and greater range. The 747-200 was produced in passenger (-200B), freighter (-200F), convertible (-200C), and combi (-200M) versions
747-300	Singapore Airlines, Kalitta Air, Cathay Pacific	1971	11,720	400	907	81	Features a longer upper deck and increase in Mach number to 0.85 with a minor change in aerodynamics compared to Mach number 0.84 of 747-200
747-400	British Airways, United Airlines, Rossiya Airlines, Atlas Air	1989	14,205	416	933	694	An improved model with increased range and fuel efficiency with wingtip extensions and winglets. A new glass cockpit was designed for a flight crew of two instead of three. The -400 was offered in passenger (-400), freighter (-400F), combi (-400M), domestic (-400D), extended-range passenger (-400ER), and extended-range freighter (-400ERF) versions.
747-8	Lufthansa, Korean Air, China Airlines	2012	14,320	467	933	155	Advanced propulsion and cockpit technology, quieter, more economical, and more environmentally friendly. The passenger version was named 747-8 Intercontinental or the cargo version was 747-8I and the 747-8 Freighter, or 747-8F. The 747-8 has surpassed the range of Airbus A340-600, a record it would hold until the 777X broke it in 2020
747SP	Pan Am	1976	10,800	276	907	45	Boeing needed a smaller and long-range aircraft to compete with the McDonnell Douglas DC-10 and Lockheed L-1011 TriStar tri-jet wide-bodies
De Havilland Canada Dash 7		1975	1280	50	428	113	A turboprop-powered regional airliner with short take-off and landing (STOL) performance.
Boeing 767	Delta Air Lines, UPS Airlines, United Airlines	1981	7200	214	900	1,273	A seven-abreast cross-section aircraft has a conventional tail and a supercritical wing for reduced aerodynamic drag.
BAe 146-300 / Avro RJ Whisperjet	Mahan Air, National Jet Express, Pionair Australia	1981	3340	112	747	394	Regional and short-haul airliner was a high cantilever wing, T-tail airliner with four jet engines. BAe 146, a low-maintenance, low-operating cost, feeder airliner, is arguably Britain's most successful jet program. An improved version known as the Avro RJ began operation in 1992. Operates very quietly hence the name Whisperjet.
Boeing 757-200	American Airlines, Delta Air Lines, Northwest Airlines	1982	7,250	239		913	Narrow-body airliner and a twinjet successor for the 727 (a trijet)
Boeing 757-300	Continental Airlines, United Airlines	1999	6,295	243	854	55	The stretched version of 757-200 and the longest (54.5 m) single-aisle twinjet ever built
Saab 340	Regional Express Airlines, Silver Airways, Loganair	1983	870	33-34	524	459	A Swedish twin-engine turboprop aircraft,

5.2. COMMERCIAL AIRLINERS

Company Airliner	Airlines	EIS	Range (km)	Max. Seats	Speed (km/h)	Number built	Remarks
Saab 2000	Eastern Airways, Loganair	1992	2,870	50	665	63	A Swedish twin-engined high-speed turboprop airliner derived from Saab 340
De Havilland Canada Dash 8	Porter Airlines, Air Canada Express, Horizon Air	1984	2,963	39-90	667	1,258	Dash 8 is a series of turboprop-powered regional airliners introduced by de Havilland Canada (DHC) in 1984 which was later bought by Boeing in 1988, then by Bombardier in 1992; then by Longview Aviation Capital in 2019.
ATR 42-400	Air Tahiti, Air Corsica, Air New Zealand	1985	1470	42	484	497	A high-wing, turboprop regional airliner by Franco-Italian manufacturer ATR. The number 42 in its name is derived from the aircraft's original standard seating capacity of 42 passengers. Later variants are upgraded with new avionics, a glass cockpit, and newer engine versions.
ATR 72	Azul Linhas Aéreas, FedEx, Alliance Airlines	1989	1,528	68-78	510	1,000	The stretched version for ATR 42
Airbus A318	Avianca Brasil, Mexicana de Aviación	2003	5741	136	829	80	Airbus A320 family is a series of narrow-body airliners that pioneered the use of digital fly-by-wire and side-stick flight controls in airliners. The Airbus A318 is the smallest member of the Airbus A320 family
Airbus A319	US Airways, Northwest Airlines, America West Airlines	1996	6950	156	829	1486	In October 2019, the A320 family surpassed the Boeing 737 to become the highest-selling airliner (16,874 ordered and 10,891 delivered as of May 2023). The direct Boeing competitor is the 737-700
Airbus A320	Air Berlin, Virgin America	1988	6200	180	829	4763	The A320 series has two variants, the A320-100 and A320-200 with different wingtip fences and increased fuel capacity. The closest competitor is the 737-800
Airbus A321	Swissair, Thomas Cook	1994	5950	220	829	1791	The heavier and longer range compared to A320. A321's closest competitors are the 737-900/900ER
Airbus A319 neo	American Airlines, United Airlines	2018	6950	160	833	92	Airbus A320neo (new engine option) family was improved version with re-engined and sharklets as standard. By 2019, the A320neo had a 60% market share against the competing Boeing 737 MAX. As of May 2023, a total of 8,754 A320neo family aircraft had been ordered by more than 130 customers.
Airbus A320 neo	IndiGo, Frontier Airlines, China Eastern Airlines	2016	6400	194	833	3995	20% efficiency gain per passenger compared to its successor
Airbus A321 neo	IndiGo, American Airlines	2017	6482	244	833	4667	Lengthened fuselage variant has structural strengthening in the landing gear and wing, increased wing loading, and other minor modifications due to higher MTOW. By July 2022, the A321neo represented over 53% of all A320neo family orders
Airbus A321LR	TAP Air Portugal, Air Transat, Aer Lingus	2018	7400	169			Long Range
A321XLR		2022	8700	220			eXtra Long Range
Airbus A340	Lufthansa, Air France, Virgin Atlantic	1993	12,400-14,600	239-440	913	377	A long-range, wide-body, eight-abreast, quadjet airliner with a similar airframe to the A330. A340 is the first derivative of A300
Airbus A330	AirAsia X, Delta Air Lines, Qantas	1994	7,200-15,000	246-440	871-913	1,567	A wide-body aircraft. In July 2014, Airbus announced the re-engined A330neo (new engine option) comprising A330-800/900, which entered market in 2018
Bombardier CRJ	American Eagle, Delta Connection, United Express	1992	2,420-3,110	50-100	828	1,858	The most successful family of regional jets powered by turbofan engines
Boeing 777-200	United Airlines, Japan Airlines, Air China	1994	9700	313	892	88	*The world's largest twinjet*, commonly referred to as the *Triple Seven*, is a long-range wide-body airliner. 777-200 made its maiden flight in 1994,
777-200ER (Extended Range)	United, American, Airlines, British Airways	1997	13,080	313			422	Has additional fuel capacity compared to 777-200 and an increased MTOW enabling transoceanic routes. It competed with the A340-300 and was then replaced by 787-10

Company Airliner	Airlines	EIS	Range (km)	Max. Seats	Speed (km/h)	Number built	Remarks
777-300	Cathay Pacific, Singapore Airlines, Emirates	1998	11,165	550		60	Launched at the Paris Air Show in 1995, with stretched version (about 20% compared to earlier model)
777-300ER	Emirates, Qatar Airways, Air France, Cathay Pacific	2004	13,649	550		832	Higher MTOW and increased fuel capacity compared to 777-300. The 777-300ER features raked and extended wingtips, a strengthened fuselage and wings (an aspect ratio of 9.0), and a modified main landing gear.
777-200LR (Long Range)	Emirates, Delta Air Lines, Qatar Airways	2006	15,843	440		61	A long range commercial airliner and nicknamed *Worldliner* as it can connect almost any two airports. It holds the record for the longest nonstop flight by a commercial airliner.
Embraer ERJ(-145XR)	American Eagle, SkyWest Airlines, Air Canada Express	1995	3700	50	833	1231	The twinjet with a swept wing is powered by two rear-fuselage-mounted turbofans. The family includes the ERJ135 (37 passengers), ERJ140 (44 passengers), and ERJ145 (50 passengers), as well as the Legacy 600 business jet and the R-99 family of military aircraft
Embraer E-Jet-170	Republic Airways, S7 Airlines, Air France Hop	2004	3,982	78	871	191	Embraer E-Jet family: four-abreast, narrow-body, short- to medium-range, twin-engined jet airliners. The E170 is the smallest aircraft in the E-Jet family and was the first to enter service in 2004.
E175	SkyWest Airlines, Republic Airways, Envoy Air, Mesa Airlines	2005	4,074	88		730	The E175 is a slightly stretched version of the E170. Initial winglets were changed to wider, angled as part of an efficiency improvement package in 2014.
E190	JetBlue, Aeroméxico Connect, KLM Cityhopper, Tianjin Airlines	2005	4,537	114		568	Larger stretches of the E170/175 models fitted with a new, larger wing, a larger horizontal stabilizer, two emergency overwing exits, and a new engine.
Airbus A380-800 (Super Jumbo)	Emirates, Singapore Airlines, Qantas	2007	14,800	555	903	254	The world's largest passenger airliner and only full-length double-deck jet airliner developed to challenge the dominance of the 747 in the long-haul market. It has 40% more usable space than the 747.
Boeing 787-8	American Airlines, ANA, Qatar Airways, Air India	2011	13621	220	903	386	Wide-body jet airliners developed aiming 20% less fuel burn than aircraft like the 767. It is recognizable by its four-window cockpit, raked wingtips, and chevrons on its engine nacelles. The 787-8 is the base model of the 787 family and was the first to enter service in 2011 to replace the 767-200ER and 330-200.
Boeing 787-9	ANA, Riyadh Air, United Airlines, Air Canada	2014	14140	259		580	787-9 is a lengthened and strengthened variant with a longer fuselage and higher MTOW. It features active boundary-layer control on the tail surfaces, reducing drag. It competes with the A330-900.
Boeing 787-10	United Airlines, Korean Air, Singapore Airlines, British Airways	2018	11,910	310		79	Stretching the 787-9 further with similar capacity as that of Airbus A350-900, and the Boeing 777-200ER. It is targeted to replace the Boeing 767-400ER and Airbus A330-300. It competes with the A330-900
Airbus A350	Singapore Airlines, Qatar Airways,	2015		530			Airbus A350 family is a long-range, wide-body, nine-abreast economy cross-section, twin-engine jet airliner
A350-900	Cathay Pacific, Delta Air Lines	2015	315	15,372		294	An eXtra Wide Body design developed in response to the Boeing 787 Dreamliner. The first Airbus aircraft largely made of CFRP
A350-900ULR	Singapore Airlines	2018	315	18,000			The longest range of any airliner in service as of 2022. With a MTOW of 280 tonnes, the A350-900ULR can fly more than 20 hours non-stop, combining the highest levels of passenger and crew comfort with unbeatable economics for such distances. As of 2022, the A350-900ULR is used on the longest flight in the world, Singapore Airlines Flights 23 and 24 from Singapore to New York JFK.
A350-1000		2018	369	16,100		104	Designed to replace the A340-600 and compete with the Boeing 777-300ER and 777-9.

5.2.4 Identifying the Airliners

If you are an aviation enthusiast and an avid plane spotter, there's nothing quite like the thrill of spotting aircraft at an airport and successfully identifying them. In the following section, we provide you with certain guidance for identifying commercial airliners. Note that this guide is specific to commercial jet airliners and excludes piston or turboprop aircraft and business jets. More specifically, we are focusing on *Boeing* and *Airbus* passenger airplanes.

There are several things to look for when differentiating aircraft – including the number of engines, the landing gear, the tail fins, nose shape, cockpit windows, and fuselage layout/exit doors.

Boeing vs. Airbus

The airliners from *Airbus* are in A3XX format (for example *A320, A350*, etc.). However, *Boeing*s don't start with a B – they're just numbers in a 7X7 format (examples are *737* and *747*) The cockpit windows and nose shape are easy ways to distinguish whether an aircraft is a *Boeing* or an *Airbus*. The classic *Boeing* airliner has a pointed nose with a V-shaped windshield. However, the typical narrow-body *Airbus* airliner features a rounded nose with a windshield straight across the bottom, and the rear with a windshield window notched.

Wide body, four-engine aircraft (Quadjets)

The *Airbus A380* is the largest widebody passenger aircraft, and in a three-class configuration, it holds around 540 passengers. The number of engines is probably the easiest start to identify the airliner. Many of the first initially built jet airliners had four engines, among which stands the *De Havilland Comet*, the world's first commercial jetliner. Currently, there are only a few aircraft with four engines – the *Airbus A380* and *A340* and the *Boeing 747*. The only full two-deck aircraft flying today, the *A380* can be easily identified.

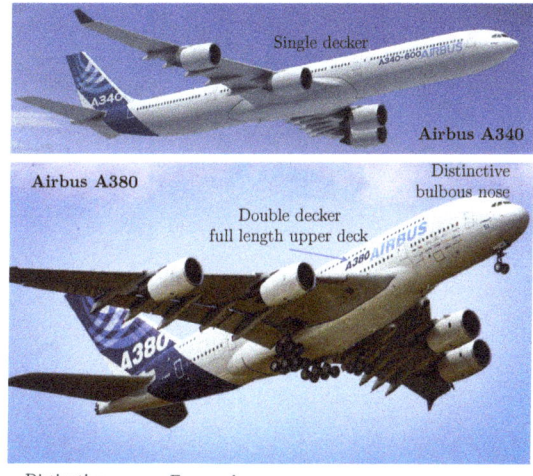

The larger 747-8 can be identified by its longer upper deck, with more windows.

Wide-body quad jet airliners.

Wide-body twin-jets

With improvements in safety and the permitted range of two-engine flights, widebody-twinjets replaced four-engine aircraft operating long-haul over-ocean flights. Further, Airbus has introduced *A321XLR*, the long-range narrow-body airliner, that may change the direction of aviation. Commercial wide-body twin-jets today include the *Boeing 767, 777*, and *787*. And from *Airbus*, the *A300, A330*, and *A350* are in this category.

Narrow-body airliners

The largest narrow-body airliner is the *Boeing 757-300* which can seat up to 295 passengers. These are designed to serve short to medium-range routes and offer greater fuel efficiency and lower operating costs compared to their wide-body counterparts. Prominent examples include the *Boeing 737, Airbus A320*, and *Embraer E-Jet* families. Other examples include the Airbus *A220 series, Boeing 757*, and *Bombardier CRJ* series. The two best-selling commercial jetliners in history are the *Boeing 737* and *Airbus*

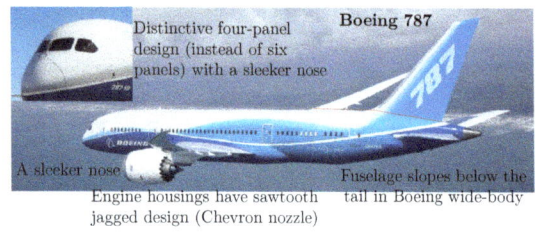

Latest wide-body twinjet airliners

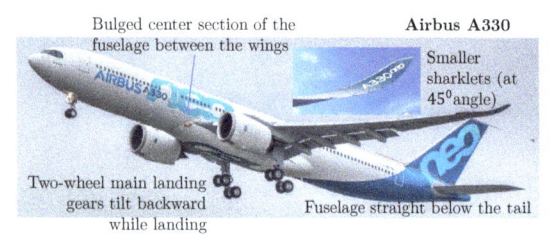

Wide-body twinjet airliners (exception is that *Boeing 757* is a narrow-body airliner)

A320 families, both narrow-body aircraft. The new generation of narrow-body aircraft, such as the *Airbus A321LR* and *A321XLR*, is transforming the industry by covering longer distances, even transatlantic routes traditionally dominated by wide-body aircraft.

Although no longer in production, the *757*, with its impressive range of capabilities, still operates on many routes worldwide. The Brazilian manufacturer *Embraer*'s *E-Jet* series includes the *E170*, *E175*, *E190*, and *E195*. These are smaller narrow-body aircraft often used for regional operations. Although *Bombardier* has exited the commercial aviation business, their *CRJ* (Canadair Regional Jet) series aircraft continue to operate extensively for regional air travel.

Popular narrow-body jetliners: *Boeing 737* and *Airbus A320*.

Boeing and Airbus Family of Airliners

Boeing 737 family are one of the most recognizable narrow-body, six-abreast seating airliners. As of April 2023, 15,591 *737*s have been ordered and 11,395 delivered. On average, there are 1,250 *737*s airborne at any given time with 2 landing or departing every 5 seconds. The best-selling jet airliner of all time and its latest variants offer unthinkable performance compared to the original. It is a low-cost solution derived from the *Boeing 727* (trijet) and *Boeing 707* (quad-jet) with two under-wing turbofans. The twelve variants of the *Boeing 737* are split into four generations of aircraft as follows:

5.2. COMMERCIAL AIRLINERS

- Original Generation (1968) - *Boeing 737-100* and *737-200*, powered by low-bypass engines and offered seating for 85 to 130 passengers. They had very small but long pod engines nestled under each wing. Wings are much shorter than any other *737* with no wingtip structures. They are not in operation now

- Classic Generation (1984) - *Boeing 737-300*, *737-400*, and *737-500* powered by high-bypass engines and offered seating for 110 to 168 passengers.

- Next Generation (1997) - *Boeing 737-600*, *737-700*, *737-800*, and *737 900ER* powered by high-bypass engines with a larger wing and an upgraded glass cockpit offering seating for 110 to 215 passengers.

- MAX Generation (2017) - *737 MAX 7/8/9/10* offering seating for 138 to 204 passengers.

Figure 5.1: Boeing's popular narrow-body jetliners: *Boeing 737 Next generation (NG)* and *Boeing 737 MAX*.

Winglet design of Boeing's two popular narrow-body jetliners: *737 Max* and *737-Next generation*.

The only narrow-body aircraft being produced by *Boeing*, the *737* has replaced the *Boeing 707, 727,* and *757* as well as the *DC9, MD80* and *MD90*. The *737 MAX*, designed to compete with the *A320neo*, was grounded worldwide between March 2019 and November 2020 following two fatal crashes.

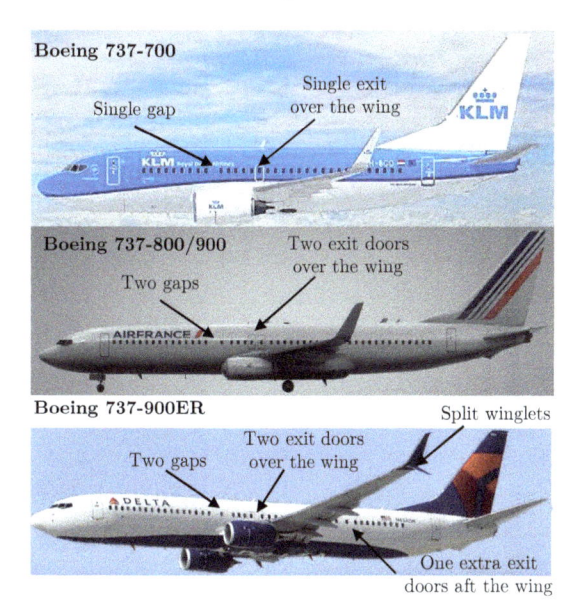

Variants of *Boeing*'s popular narrow-body jetliner, *737-Next generation*. Note that the older *Boeing 737-900ER* variant had blended winglets. The figure shows a newer variant which has split scimitar winglets.

Boeing 777 family referred as *Triple Seven*, a wide-body airliner with a 9/10 abreast seating, is the world's largest twinjet designed to replace older *DC-10*s and *L-1011* trijets and compete with *Airbus A340* and the *McDonnell Douglas MD-11*. *777* family is recognizable for its large-diameter turbofan engines, six wheels on each main landing gear, fully circular fuselage cross-section, and a blade-shaped tail cone and have extended raked wingtips. It is the first *Boeing* aircraft with fly-by-wire controls. *Boeing* announced the *777X* development with the -8 and -9 variants, both featuring composite wings with folding wingtips and *General Electric GE9X* engines.

Airbus A320 Family are the world's most popular narrow-body, twinjet airliners with six-abreast seating. As of April 2023, a total of 10,840 A320 family aircraft had been delivered and total orders of

16,874 were placed from 300+ customers. A forerunner in adopting fly-by-wire technology, it has four different variants: *A318* (2003), *A319* (1996), *A320* (1988), and the *A321* (1994). *A318* was the smallest (136 seats, 5741 km range) and *A321* was the largest (236 seats, 5926 km range) of the family. *A320* was Airbus' first narrow-body offering in the single-aisle, short to medium-range market. *A320* was set to compete with *Boeing*'s *717*, *757*, and *737* as well as *McDonnell Douglas*' *MD80* and *MD90*. It was the first narrow-body airliner to use an appreciable amount of composite materials in its construction. The *A320neo* or *A320E* (A320 Enhanced) was introduced in 2006 featuring 14% lower cash operating costs per seat, 20% fuel-saving with advanced *LEAP-X* engine, better passenger comfort with 50% noise footprint reduction and enhanced aerodynamic performance with wingtip devices known as *Sharklets*. *A321neo* (6482 km rage) improved long range capability in *A321LR*(7400 km) which was then further improved in the extra long range *A321XLR* (8700 km).

Boeing 757 family is a narrow-body airliner, designed for short and medium-length routes to replace aging trijet *727s* and challenge the *A321* which offered operator exceptionally efficient service and demonstrated long range. It has three variants: *757-200/200F* and *300* of which, *757-200* accounted for 913 of the type's 1,049 commercial deliveries. **Boeing 787 Dreamliner** is a wide-body, twinjet airliner, powered by high-bypass turbofans, that burns 20% less fuel than replaced aircraft like the *Boeing 767*. The first airliner with an airframe primarily made of composite materials and made extensive use of electrical systems. It is recognizable by its four-window cockpit, raked wingtips, and noise-reducing chevrons on its engine nacelles. Its variants seat 242 to 335 passengers in typical three-class seating configurations. It is *Boeing*'s most fuel-efficient airliner and is a pioneering airliner with the use of composite materials as the primary material in the construction of its airframe. The shortest *Dreamliner* variant, the *787-8* (248 passengers, 13530 km range) was the first variant to fly in 2009, then the longer *787-9* (296 passengers, 14010 km range) began operation in 2013, followed by the longest variant, the *787-10* (336 passengers, 11730 km range), in 2017. *787-9* which competes with the Airbus *A330-900*, features active boundary-layer control on the tail surfaces, reducing drag. On March 16, 2020, an *Air Tahiti Nui 787-9* achieved the longest commercial flight of 15,715 km. The *Dreamliner*'s distinguishing features include swept wings with no winglets, two 4-wheel main landing gear, a sleek pointed nose, and noise-reducing chevrons on its engine nacelles.

Boeing 787 Dreamliner.

Boeing 747 (Queen of the Skies or Jumbo Jet): As the first twin-aisle, ten-abreast economy seating, long-range wide-body airliner, *Boeing 747* revolutionized air travel and enabled more people to fly farther, faster, and more affordable. It held the size record for more than 35 years until it was surpassed by the *Airbus A380*. However, the *An-225* cargo transport remains the world's largest aircraft in service, while the *Hughes H-4 Hercules* has an even larger wingspan. More than 100 customers have purchased 1,574 Jumbo Jet aircraft. The hump created by the upper deck has made the *747* a highly recognizable icon of air travel. In 1989, a *Qantas 747-400* flew non-stop from London to Sydney, 18001 km in 20 hours and 9 minutes with no passengers or freight aboard.

Boeing 747-400 vs. 747-8: Identifying the final two jumbo jet series

Airbus A380-800 (Super Jumbo jet): This quad-jet airliner is the only full-length double-decker airliner and can carry up to 853 passengers on transoceanic flights. It challenged the dominance of the *Boeing 747* in the wide body, quadjet segment. With its high aspect ratio wings, the *A380* improves fuel efficiency and lift performance. To decrease drag and increase fuel efficiency, the wing incorporates advanced wingtip devices called winglets.

A summary of the airliners with the longest range capabilities is given in Table 5.2. A summary of the most produced civilian airplanes is mentioned in Table 5.3.

5.2.5 Supply and Services of Airliners

The aviation industry is a complex ecosystem with several key players contributing to the successful operation of airliners. Supply (airframe, engine and components) and services(Maintenance, Repair and Overhaul) play a crucial role in keeping these aircraft in the air.

Boeing and *Airbus* stand as the globe's leading suppliers of commercial aircraft, engaged in designing, manufacturing, selling, and maintaining aircraft. Joining them are *Embraer* and *Sukhoi*, although the aviation industry predominantly depends on a select few airframe suppliers. A pivotal role for these airframe suppliers is integrating components from various sources into their self-designed aircraft, hence often earning them the title of integrator due to their role in merging engines and components crafted by other manufacturers into their proprietary airframes. An aircraft's engine serves as its core, shaping the aircraft's operational capabilities and playing a crucial role in its commercial

Table 5.2: The longest-range passenger aircraft. The original long-haul aircraft was the Boeing 747SP, with a range of 12,320 km. This was later beaten by the Boeing 777-200LR at 15,843 km (became the longest-range two-engined aircraft until the A350-900ULR broke the record) and the Airbus A340-500 at 17,000 km. Today the A350-900ULR is the longest-range aircraft in the world, with the latest competition from Boeing (the 777-8) still some way behind. With no extra long-haul aircraft in development currently, the A350 could hold this record for a few more years.

Airliner	Range (km)
Airbus A350-900ULR	18,000
Airbus A340-500	16,670
Boeing 777-8	16,170
Airbus A350-1000	16,100
Boeing 777-200LR	15,843
Airbus A350-900	15,372
Airbus A330-800 (Neo)	15,094
Airbus A380-800	14,800
Boeing 747-8i	14,320
Boeing 787-9	14,140

Table 5.3: The most produced passenger aircraft. Total delivered and orders are as per May 2023 data. Douglas DC-3 (16,079 between 1936-1950) was originally intended as an airliner but after just 607 airliner variants were built, it was re-tasked to serve as a military transport. *Cessna 172* take the crown being the single most produced aircraft of all time. More than 44,000 172s have been produced since production first began in 1956

Aircraft	Production	Delivered	Order
Boeing 737	1967–	11,431	15,646
Airbus A320	1987–	10,891	16,874
Bombardier CRJ	1991-2020	1945	–
Boeing 727	1962-1984	1,832	-
Boeing 777	1993–	1,709	2,135
Embraer E-Jet	2001–	1,661	1,746
Boeing 747	1968–2023	1,574	–
Airbus A330	1992–	1,571	1,775
Antonov An-26	1969–1986	1,403	-
Antonov An-24	1959–1979	1,367	-

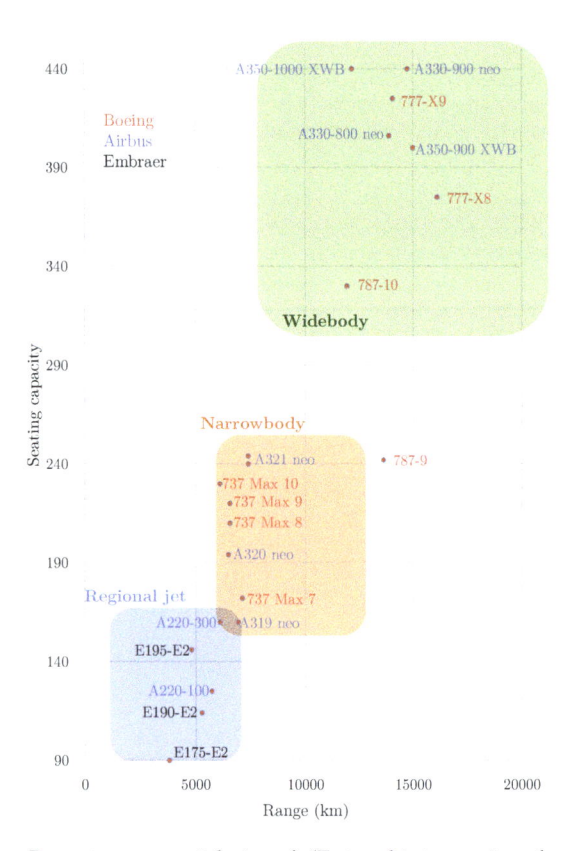

Recent commercial aircraft (Entered into service after 2015) with passenger capacity and the range (Source: *IATA's Aircraft Technology Roadmap to 2050*)

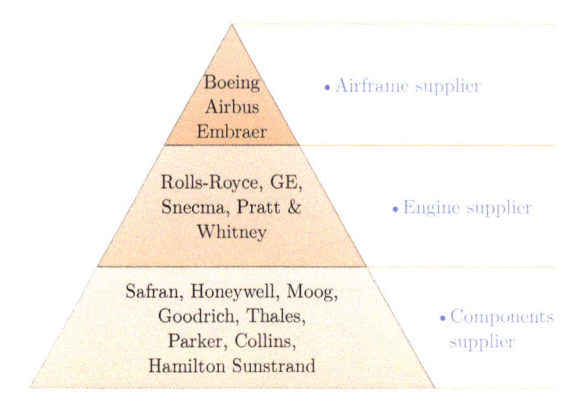

Commercial aviation supply hierarchy

Both airframe and engine suppliers integrate an extensive array of components, encompassing landing gear, avionics, auxiliary power units, hydraulic and pneumatic systems, interiors, and in-flight entertainment, into their final products. This intricate process contributes to the seamless assembly of aircraft. Some components are exclusively tailored for a particular aircraft type. The landing gear is a high-cost, technologically advanced component exclusively designed for a specific aircraft type. This critical part is primarily provided by two dominant suppliers: *BF Goodrich* in the US and *Messier-Dowty*, a French subsidiary of the component conglomerate *Safran*. Avionics include a comprehensive array of computers, antennas, and sensors crucial for communication, navigation, and surveillance (CNS) during aircraft operation. Key industry leaders in this domain include *Rockwell Collins* in the US, a division of *UTC*, *Honeywell*, and the French company *Thales*, known for their advanced electronic systems integrated into aircraft worldwide.

Aircraft require hydraulic power for functions such as landing gear retraction, wheel brakes, and flight controls, as well as electrical power to operate various electrically driven systems. Specific suppliers cater to each of these requirements. Furthermore, an aircraft's pneumatics play a crucial role in pressurizing the cabin and regulating oxygen levels. The environmental control system (ECS) of an aircraft comprises numerous components, often tailored to specific aircraft types and sourced from a

success. Aircraft design is centered around the engine, placing engine suppliers in a pivotal position in aircraft development. In the aviation industry, just four engine suppliers hold significant influence: *General Electric (GE)* and *Pratt & Whitney* in the US, *Rolls-Royce* in the UK, and *Snecma* in France. Engine suppliers often collaborate, leading to the joint development and production of various engine models, with a 50/50 partnership in some cases. For instance, the highly successful *CFM-56*— engine, used in the *Boeing 737* and *Airbus A320*, is the result of a joint venture between *GE* and the *Snecma*, with production facilities on both sides of the Atlantic. Similarly, the alternative engine for the *A320*, the *V2500*, is a product of a joint venture between *Pratt & Whitney* and *Rolls-Royce*. In a notable example, *Pratt & Whitney* and *GE* jointly developed the engine *Alliance 3200* for the *A380* as an alternative to the *Rolls Royce Trent 900*.

variety of suppliers. On the ground, these systems are powered by the aircraft's auxiliary power unit (APU). Aircraft interiors encompass a unique blend of components tailored to specific aircraft types, such as wall panels, interior lighting, and overhead bins, along with non-type specific elements like aircraft seats, inflight entertainment (IFE) systems, and internet connectivity.

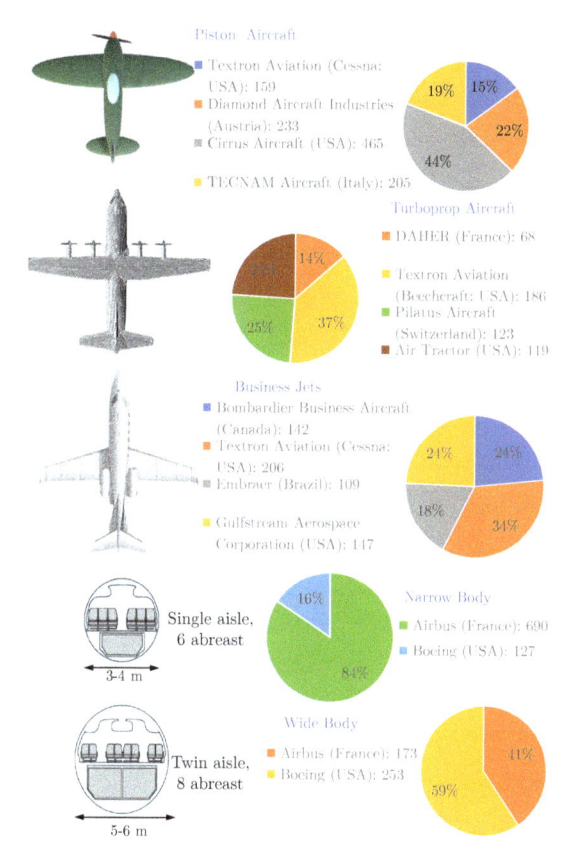

General aviation aircraft supplier by type (data is for 2019, source: GAMA's General Aviation Aircraft Shipment Report 2019).

Maintenance, Repair and Overhaul (MRO)

Aircraft Maintenance, Repair, and Overhaul (MRO) is a crucial aspect of the aviation industry that encompasses a wide range of services aimed at ensuring the safety, airworthiness, and longevity of aircraft.

What is MRO? MRO involves routine inspections, repairs, and maintenance tasks that cover various aircraft components, including engines, avionics, airframes, landing gear, and interiors. MRO comprises three distinct yet interconnected elements: maintenance, repair, and overhaul.

Maintenance is done to maintain the aircraft's technical integrity and is carried out regularly, making it predictable and planned. Manufacturers define the necessary maintenance tasks based on time-related inspection intervals.

Repair: The components need to be repaired when they are failed. Such failures are inherently unpredictable and, as a result, cannot be planned. Repairing a malfunctioning component of an aircraft is rarely performed while the aircraft is in operation but it is not ruled out. A component that can be replaced on an aircraft while in operation is called a Line Replaceable Unit (LRU).

Therefore, not every failed part requires immediate replacement. The aircraft manufacturer provides a document known as the Minimum Equipment List (MEL). The MEL outlines which components may be unserviceable under what conditions to allow the aircraft to operate. If a component is not listed in the MEL and is unserviceable, it results in an Aircraft On Ground (AOG) situation.

With advanced sensors and the rapid progress in digital data management, today, there's a significant shift toward predictive maintenance. This approach utilizes real-time performance data from a unit to anticipate its impending failure, enabling a cost-effective replacement just before the failure occurs.

Overhaul involves the complete disassembly of a unit, followed by the replacement of any parts that are worn or damaged. After an overhaul, the unit is restored to a "zero-hour" condition, akin to a brand-new state.

Why is MRO important? Aircraft MRO are essential for ensuring the aircraft's continued airworthiness for several key reasons: (1) Legal obligation: To maintain airworthiness, aircraft must undergo MRO procedures as specified in the manufacturer's manuals. Compliance with these standards is legally required and crucial for airworthiness approval. (2) Operational integrity: Proper MRO is vital for ensuring an airline's punctuality and safety, reducing the risk of in-flight technical issues. (3) Market value: Aircraft that receive thorough maintenance retain higher market values compared to those with less maintenance.

What is the frequency of MRO? MRO intervals are typically based on flight hours, which account for the total time an aircraft spends in the air. However, components exposed to high temperatures, such as engines, or those subject to repeated stress, like the wing and fuselage structures, require consideration of flight cycles which involves starting up a cold engine to its operating temperature and back. A cycle also includes the landing based on the number in which the landing gear and wheel brakes MRO is scheduled (approximately 1,500 landings).

Airframe MRO involves inspections based on the number of flight hours. As aircraft age, especially during their mid-life phase, the airframe must undergo routine checks to identify material fatigue. For engines, performance is continually monitored, focusing on thrust and temperature. An engine is only removed from the aircraft for a comprehensive overhaul in an engine shop when its performance falls below a specified minimum threshold. Following this overhaul, the engine's performance is restored to like-new condition. Consequently, engines require minimal in-service maintenance. All components within the aircraft are categorized as either *hard-time items* (crucial for flight safety and need to be replaced after a certain usage in terms of flight hours or cycles) or *on-condition items* (secondary components and remains in the aircraft until it fails). Not all failed components can be repaired. Those that fail or need replacement without repair are termed *consumables*, while repairable components are referred to as *rotables*.

How is MRO performed? Aircraft MRO is divided between *Line maintenance* and *Base maintenance*. Line maintenance is performed at the ramp during the ground turnaround process or at night when the aircraft is in service that involves light checks and servicing like oil replenishment. Line maintenance also includes repairing an AOG (Aircraft On Ground) with a technical failure in operation. If the failed component is a Line Replaceable Unit (LRU), line maintenance can perform on-site repairs at the ramp.

Base maintenance encompasses all procedures conducted within a hangar when the aircraft is not in service. When the replacement component is not an LRU, the repair necessitates an indoor hangar environment, often accompanied by post-repair testing.

Who performs MRO? Before the 1980s, airlines used to handle their aircraft MRO internally, but they later started outsourcing some or all of their MRO needs. At the same time, some large airlines transformed their in-house MRO departments into separate profit centers, offering MRO services to airlines that outsourced them. Major MRO providers like *LHT (Lufthansa Technik)*, *AFI (Air France Industries/KLM E&M)*, or *Delta Ops* are among the industry leaders who have origin in an airline or are still part of an airline company.

MRO is a substantial business in the aviation sector. For Original Equipment Manufacturers (OEMs), the main profit source lies in parts sales. Some MRO organizations maintain close relationships with OEMs, collaborating on the exclusive use of OEM parts and repairs. In such cases, these MRO organizations can become key partners in the OEM's global support network, conducting services on behalf of the OEM.

5.2.6 Airliner Operations

Certificate of Airworthiness

Before any airliner can commence operations, it must obtain a Certificate of Airworthiness (CoA) from the Civil Aviation Authority (CAA) of the state or nation in which the aircraft is registered, granting it permission to fly in strict compliance with ICAO's guidelines. The Federal Aviation Administration (FAA) is responsible for certifying aircraft manufactured in the United States while its EU counterpart is the European Union Aviation Safety Agency (EASA).

The CoA signifies an aircraft's adherence to specific type design standards, validated through requisite maintenance and inspections. It reflects the rigorous assessments conducted to ensure safe operations for passengers and those on the ground. In addition to the airworthiness certificate, each aircraft must hold an Airworthiness Review Certificate (ARC), mandating annual renewal for flight eligibility. For a new aircraft type, its engines, and all installed components, an Original Type Certificate is mandatory for flight authorization. In cases of subsequent aircraft type modifications, such as engine changes, an Amended Type Certificate is required. When changes pertain only to specific systems without altering the aircraft type, a Supplemental Type Certificate is needed.

Most modern aircraft models are equipped with two engines, ensuring the capability to continue flight in the event of an in-flight engine failure. These aircraft are indeed designed to operate effectively on a single engine. The permissible duration for single-engine operation is determined by its Extended-range Twin Operational Performance Standards (ETOPS) certification. For instance, ETOPS-90 means that the aircraft, its engines, and the operator are collectively certified to fly with one engine for a maximum of 90 minutes. Present-day twin-engined aircraft often possess ETOPS180 or even higher certifications. However, regardless of the specific ETOPS standard an aircraft holds, operators must adhere to a comprehensive set of regulations and guidelines to utilize this capability.

Weight of the Airplane

Aircraft productivity relies on its capacity to transport a given payload over a specified distance. The aircraft's performance and productivity are inherently tied to its weight.

The first weight is the *Manufacturer's Empty Weight (MEW)* which is the weight of the base aircraft in certified flying condition without interior, livery (insignia comprising color, graphic, and typographical identifiers) and eventual optional equipment that the operator has specified on the aircraft.

The *Operating Empty Weight (OEW)* or *Dry Operating Weight (DOW)* of an aircraft is weight without payload and fuel that forms the basis for calculating any further weight in aircraft operation. When delivered, a manufacturer will provide its OEM to the operator. The OEW includes the MEW as well as the aircraft's interior, livery, optional equipment, safety equipment, catering inserts and the entire flight crew with their luggage. Interestingly, the OEW of two of the same aircraft can differ significantly for large airliners as the different operator has different requirements in terms of luxurious interiors apart from minimum comfort amenities.

MTOW (ZFW + Fuel)	The weight when aircraft is loaded with payload and fuel. The absolute maximum weight at which the aircraft can take-off and climb out safely.
ZFW (OEW + payload)	A certified weight includes Operating Empty Weight (OEW) and loaded payload such as passenger, luggage or cargo
OEW (MEW + safety and luxury equipment)	Flying aircraft with interior, livery, optional equipment, safety equipment, catering inserts and the entire flight crew with their luggage.
MEW	Weight of the base aircraft in certified flying condition

Weight of the aircraft

Zero Fuel Weight (ZFW) is the weight of the plane with all passengers and cargo loaded. This means ZFW is the sum of DOW and the payload. This weight is defined by the structural integrity and maximum payload capability of the aircraft

The **Fuel Weight** for a trip has components such as contingency fuel (to account for additional en-route fuel consumption caused by wind, routing changes, usually 5% of the trip fuel), alternate fuel (flying to an alternate airport in emergency), reserve fuel (holding at the destination airport approximately to hold for 45 minutes) and taxiing fuel (prior to takeoff, normally include pre-start APU consumption, engine start and taxi fuel).

Cargo loading and passenger boarding into Airbus 380 *Super Jumbo Jet* at Hamad International Airport, Doha, Qatar.

While ZFW provides the aerodynamic insight to determine the amount of lift that the wings must generate to support the fuselage, TOW is used to calculate the thrust setting and runway distance requirements for takeoff. TOW is limited by the structural capacity of the airplane. Hence the structural takeoff weight limit is the legally certified maximum takeoff weight (MTOW) for a particular aircraft. In addition to the aircraft structural limit, the maximum take-off weight also restricts the runway length, the climb rate, and the landing limit.

Each aircraft is assigned a *Maximum Landing Weight* (MLW, well below the MTOW) by the manufacturers, any weight above MLW during landing may cause severe structural damage to the aircraft and an overweight landing leads to structural integrity will be impaired. MTOW limits landing because if any extra fuel or cargo is added to the aircraft that is above its MLW when it arrives at the destination, the aircraft needs to dump the extra fuel to land. In some emergency landing situations, the pilot needs to dump the extra fuel into the air to bring the total weight of the aircraft well within the limit of MLW before attempting to land.

Fuel Jettison is the procedure to dump the fuel from the aircraft in certain emergencies before a return to the airport shortly after takeoff, or before landing short of the intended destination (emergency landing) to reduce the aircraft's weight. Large airliners have a maximum landing weight that is lower than the maximum take-off weight. For instance, the *Boeing 777-300* has a maximum takeoff weight of 330 tonnes and a maximum landing weight of 262 tonnes and a difference of 68 tonnes. A *Delta Air Line*'s *Boeing 777-200* flying from

Fuel Jettison

Atlanta (USA) to Tokyo (Japan) dumped around 55 tonnes of fuel on October 13, 2017, as it was advised to divert to Minneapolis/Saint Paul (USA) because of a medical emergency.

Aircraft life cycle: The technical lifespan of an aircraft frame is determined by 60,000 cycles, with each cycle representing a single flight regardless of its duration. The cumulative impact of taking off, ascending to high altitudes in low pressure conditions, descending, and landing on the higher pressure surface induces metal fatigue in the airframe. This fatigue of the metal serves as the critical factor in determining an aircraft's technical longevity. The typical commercial airplane lifespan is approximately 25 years. Nevertheless, specific aircraft models can deviate significantly from this average.

5.3 Safety and Comfort in Commercial Aviation

In the early days of aviation, numerous significant accidents underscored the inherent risks of this mode of transportation. As an example, during the early 1920s, the U.S. Army began tracking aviation accidents and documented around 467 accidents per 100,000 hours of flight in 1921. To provide a modern context for this exceptionally high accident rate, it would be equivalent to the daily loss of several thousand airliners worldwide in today's operations.

During the early years of commercial aviation, pilots and the occasional passenger used to fly in open cockpits, exposing themselves to the harsh and uncomfortable elements. The air travel across Europe and USA remained rugged, noisy, and uncomfortable. During this period, the primary source of income for most airlines came from carrying mail for the government.

A century of dedicated effort has transformed commercial aviation into an exceptionally safe and comfortable mode of transportation. The journey began early 20th century but fast forward over a century, and the global commercial aviation fleet of around 27,400 aircraft carried around 32.4 million passengers safely to their destinations in 2023. The detailed design and testing procedures practiced by the aircraft and engine manufacturers, the uncompromising certification procedures followed by the governing bodies, the rigorous and extensive training for the pilots and staff in the industry, and the use of sophisticated technologies have all contributed in making commercial aviation the safest mode of travel.

The safety and comfort of air travel saw remarkable advancements with the introduction of airliners operating at higher altitudes. Prior to that, airliners lacked pressurization, necessitating low-altitude flights that frequently subjected passengers to the whims of wind and weather, resulting in frequent turbulence and air sickness. While airlines made efforts to provide various comforts to alleviate passenger discomfort, the journey continued to be a demanding adventure, persisting well into the 1940s. However, despite the high costs and discomforts of piston engines and unpressurized cabins, commercial aviation managed to entice thousands of new passengers annually who were eager to experience the advantages and adventure of flight. The American airline industry experienced remarkable growth, with passenger numbers skyrocketing from a mere 6,000 in 1929 to over 450,000 by 1934 and further to 1.2 million by 1938.

After World War II, passenger air travel experienced an unprecedented surge. As wartime travel restrictions were lifted, airlines found themselves inundated with passengers. Emerging carriers entered the scene, and innovative technology commenced a transformative era in civil aviation. In 1955, for the first time, more people in the United States traveled by air than by train. As air travel gained popularity and became more commonplace, the nature of the flying experience started to evolve. By the end of the 1950s, *American Airlines* was introducing a higher degree of speed, comfort, and efficiency to the traveling public. Yet, as passenger numbers continued to rise steadily, the level of personalized service diminished. The anxieties associated with air travel started to supplant the excitement. Flying ceased to be a novelty or an adventure; it was evolving into a necessity.

The introduction of jet engines marked a pivotal moment in the evolution of air travel. These powerful and durable engines empowered aircraft manufacturers to build bigger, faster and safer aircraft. As planes gained greater speed and passenger volumes soared, travelers began grappling with physiological challenges caused by traversing multiple time zones in a matter of hours. The altered duration of days and nights disrupted natural circadian rhythms, making sleep a challenging endeavor. Although later coined as "jet lag", this phenomenon was initially observed following extended journeys on fast piston-engine and turboprop airliners.

In-flight entertainment (IFE) including personal screens with a wide range of entertainment options keeps passengers entertained during the flight.

5.3.1 Aviation Safety Management System (SMS)

Aircraft Safety Management Systems (SMS) represent a structured and proactive approach to enhancing aviation safety. SMS is a comprehensive and systematic framework implemented by aviation organizations, including airlines, maintenance providers, and airports, to manage and continually improve safety performance. It is a critical part of the broader concept of safety culture within the aviation industry, emphasizing a commitment to identifying and mitigating risks to ensure the safety of passengers, crew, and the public.

One of the primary functions of an SMS is to assess and analyze potential safety risks within an aviation organization. This involves identifying hazards, evaluating the likelihood and severity of these hazards leading to accidents, and prioritizing them. The assessment often includes inputs from employees, safety reporting systems, incident investigations, and data analysis. By understanding these risks, aviation organizations can make informed decisions and allocate resources to mitigate them effectively.

SMS establishes clear safety policies, standards, and procedures that are communicated to all employees within the organization. These policies outline the commitment to safety and the roles and responsibilities of various personnel in maintaining and promoting safety. By providing a framework for safety, SMS ensures that everyone involved in aviation operations understands and adheres to the safety practices and guidelines in place.

Safety assurance within SMS involves monitoring and measuring the effectiveness of the safety management system. This includes regular safety audits, inspections, and performance assessments to ensure that safety objectives are being met. If deviations or issues are identified, corrective actions are taken to address them. Safety assurance activities are designed to ensure the ongoing effectiveness of safety measures and to identify areas for improvement.

SMS is a crucial part of aviation safety. By systematically identifying and managing safety risks, setting clear policies and procedures, and continuously monitoring safety performance, SMS contributes to a proactive safety culture within the aviation industry, reducing the likelihood of accidents and incidents and enhancing overall safety for passengers, crews, and everyone involved in aviation operations.

Instrument Landing System (ILS) is a standard ICAO precision landing aid introduced in 1949. It is a vital component of modern aviation, playing a pivotal role in enhancing aviation safety by providing accurate guidance to pilots when visibility is compromised due to adverse weather conditions or low-visibility situations. ILS is a ground-based radio navigation system comprising of two primary components: the localizer and the glide slope. The localizer provides lateral guidance, ensuring that the aircraft lines in the runway, while the glide slope provides vertical guidance, facilitating a precise descent path. ILS functions through a network of transmitters situated at the end of the runway. The localizer transmits a specific radio frequency signal that helps the pilot align the aircraft with the runway's centerline. Simultaneously, the glide slope transmitter sends another signal, indicating the correct descent path. Aircraft equipped with ILS receivers interpret these signals, guiding the pilots through their instruments, such as the Horizontal Situation Indicator (HSI) and Vertical Speed Indicator (VSI). Pilots follow these indications to maintain the correct approach angle and alignment with the runway.

Precision Approach Path Indicator (PAPI) is a critical visual aid used at airports to assist pilots during the landing phase of a flight. It is a lighting system consisting of a series of lights arranged in a line on the side of the runway to assist pilots during their approach to landing through a visual glide path. When an aircraft is on the correct glide path for landing, the pilot observes two white lights and two red lights, signifying that they are at the ideal altitude. If the pilot perceives more red lights than white, it indicates that the aircraft is too low, while more white lights than red imply that the aircraft is too high. This visual feedback is an indispensable tool for maintaining the precise approach slope, ensuring a safe and accurate landing. PAPI's simplicity, effectiveness, and universal application make it an essential component in the toolbox of aids that support safe and efficient air travel.

Table 5.4: Aviation Safety Milestones

Year	Aviation Safety Milestones
1944	the International Civil Aviation Organization (ICAO) was established to create regulations for aviation safety, security, efficiency and regularity and environmental protection.
1945/9	Standards and Recommended Practices (SARPs) to support aviation safety were adopted
1949	The Instrument Landing System (ILS) was included for precision landing aid under normal or adverse weather conditions.
1951	Aircraft Accident Inquiry and Aerodromes were adopted
1955	Voice transmission, air traffic control centers and the use of more advanced navigation aids began
1964	Fully automatic landing using ILS occurred
1969	The provisions on bird strike hazard reduction were introduced
1971	SARPs for Aircraft noise were first adopted
1974	The precision approach path indicator (PAPI) system which provides guidance information by visual aid to help a pilot acquire and maintain the appropriate approach to a runway
1976	Introduction of the runway end safety area (RESA) intended to reduce the risk of damage to an airplane undershooting or overrunning the runway
1978	Ground proximity warning systems (GPWS) was fitted to certain airplanes as a major mitigation for controlled flight into terrain (CFIT)(alert pilots for an immediate danger or an obstacle).
1981	SARPs for the safe transport of dangerous goods by air were adopted
1982	The introduction of the glass cockpit, combined with electronic cockpit displays and improved navigation systems as well as the introduction of terrain awareness and warning systems (TAWS), significantly reduced the rate of CFIT accidents
1986	Introduction of provisions for helicopters to be equipped with flight recorders
1987	Established the International Airways Volcano Watch (IAVW) commission that defines international protocols for the monitoring and provision of warnings to aircraft in the presence of volcanic ash in the atmosphere. Also, aircraft using Fly-by-Wire (FBW) technology with Flight Envelope Protection functions were introduced. This helps to protect against Loss of Control In-Flight (LOC-I) accidents
1993/4	The Global Positioning System (GPS) and Global Navigation Satellite System (GNSS) became operational leading to a number of safety and efficiency-related enhancements to air navigation.
1994	An Emergency Locator Transmitter (ELT) was introduced. ELT is an emergency beacon used in aircraft to alert rescue authorities and to indicate the location and the identity of an aircraft in distress.
1995	First operational use of controller-pilot data link communications (CPDLC) automatic dependent surveillance
1996	Pressure-altitude reporting transponders and the carriage of airborne collision avoidance systems (ACAS) were introduced
1997	Global Aviation Safety Plan (GASP) was introduced to continually reduce fatalities, and the risk of fatalities, by guiding the development of a harmonized aviation safety strategy.
2001	SARPs on the certification of aerodromes were introduced. Also, SARPs supporting Global Navigation Satellite System (GNSS) operations based on augmenting core satellite constellation signals was adapted to meet safety and reliability requirements.
2006	Safety Management Manual published including guidance on Safety Management Systems (SMS) and specific reference to safety culture.
2014	Use of digital formats for volcanic ash and tropical cyclone advisories
2014	Develop a comprehensive flight tracking system, Global Aeronautical Distress and Safety System (GADSS)
2016	Global Reporting Format (GRF) a harmonized methodology for assessing and reporting runway surface conditions to help mitigate the risk of runway excursion. Establishment of an accident investigation authority (AIA). Introduced the autonomous runway incursion warning system (ARIWS) that provides autonomous detection of a potential incursion or of the occupancy of an active runway and a direct warning to a flight crew or a vehicle operator

Runway End Safety Area (RESA) is a vital component of airport infrastructure designed to enhance the safety of aircraft operations during takeoff and landing. It is essentially an extended area at the end of a runway, engineered to provide a buffer zone that can be used in the event of an aircraft overshooting the runway during landing or a rejected takeoff. RESA is a critical safety feature, as it helps mitigate the potential consequences of runway excursions, which can lead to accidents or incidents involving aircraft overrunning the pavement. In cases where it's not possible to establish a full-length RESA, airports may implement alternative solutions like an Engineered Material Arresting System (EMAS), which uses specialized materials to decelerate and safely stop an aircraft that overshoots the runway. These safety measures are particularly crucial at airports where challenging conditions, such as short runways or proximity to bodies of water or urban areas, increase the risk of runway excursions.

The Ground Proximity Warning System (GPWS) is a crucial safety tool in aviation designed to prevent accidents caused by an aircraft's proximity to the ground. Sophisticated avionics systems, GPWS continuously monitor an aircraft's altitude and position in relation to the terrain during flight. The primary purpose of GPWS is to provide advance warnings to flight crews when the aircraft's altitude, rate of descent, or approach profile poses a potential risk of collision with the ground, obstacles, or other terrain features. If GPWS detects a deviation from the safe flight profile or potential danger, it issues audible and visual warnings to the flight crew. One of the most significant contributions of GPWS to aviation safety is its ability to prevent Controlled Flight Into Terrain (CFIT) accidents. CFIT accidents occur when an aircraft, under the pilot's control, unintentionally collides with terrain, often due to factors like poor visibility, pilot error, or improper navigation. GPWS has significantly reduced the occurrence of such accidents by providing timely and effective warnings, allowing pilots to take evasive action and avoid potential collisions with the ground or obstacles.

Terrain Awareness and Warning Systems (TAWS) is another advanced avionics system that continuously monitors an aircraft's position, altitude, and approach profile, providing real-time awareness of the surrounding terrain and obstacles. The primary objective of TAWS is to alert flight crews to potential dangers, allowing them to take corrective action and avoid CFIT accidents, where an aircraft, under the pilot's control, inadvertently collides with terrain or obstacles. The implementation of TAWS has been a significant milestone in aviation safety, substantially reducing the occurrence of CFIT accidents.

GPWS is a more basic system primarily concerned with ground proximity and descent-related warnings. It uses simpler sensors and offers fewer alerts. TAWS, on the other hand, is a more advanced and comprehensive system that provides a broader range of warnings related to terrain and obstacles. TAWS has largely supplanted GPWS due to its more advanced capabilities and its effectiveness in preventing CFIT accidents.

Emergency Locator Transmitter (ELT) is a critical equipment in aviation safety. It serves as a potentially life-saving device designed to transmit distress signals in the event of an emergency, such as an aircraft crash. ELTs are typically mounted onboard aircraft and are activated automatically upon impact or can be manually activated by the crew or passengers. When triggered, an ELT emits a radio signal on a designated frequency that can be detected by search and rescue authorities, significantly enhancing their ability to locate and assist distressed or missing aircraft.

ELTs play a crucial role in saving lives by expediting the response of search and rescue teams. These devices are equipped with built-in GPS technology, allowing for more accurate pinpointing of the aircraft's location, which is particularly essential in remote or challenging environments. Over the years, advancements in ELT technology have led to increased reliability and reduced false alarms, further bolstering their role in ensuring the safety and security of those engaged in air travel.

Controller-Pilot Data Link Communications (CPDLC) is a significant advancement in aviation safety that enhances the efficiency and reliability of communication between air traffic controllers and pilots. Unlike traditional voice communication, CPDLC relies on digital data transmission, allowing for more precise and streamlined exchanges of critical information, especially during long-haul or oceanic flights.

One of the primary safety benefits of CPDLC is the reduction of potential miscommunications and misunderstandings between pilots and air traffic controllers. By using standardized, pre-formatted messages, CPDLC minimizes the risk of misinterpreted instructions or misunderstandings due to language barriers or radio interference. This, in turn, helps prevent incidents that can occur from miscommunications, such as runway incursions or navigational errors.

CPDLC also supports the sharing of important information related to weather updates, route changes, or emergency situations in a concise and efficient manner. This real-time data exchange enhances situational awareness for both pilots and controllers, enabling them to make well-informed decisions promptly. Moreover, CPDLC is especially valuable in remote or oceanic airspace, where traditional voice communication can be unreliable due to distance or radio congestion. The ability to maintain a reliable data link with ground control significantly contributes to the safety and efficiency of long-haul flights.

Airborne Collision Avoidance Systems (ACAS) is designed to prevent mid-air collisions between aircraft. These systems are a testament to the aviation industry's commitment to enhancing safety and mitigating the risk of accidents in congested airspace. ACAS operates on the principles of active surveillance and cooperative communication between aircraft to provide timely collision avoidance guidance.

When an aircraft equipped with TCAS detects another aircraft within a certain proximity, the system issues advisories and resolution advisories to both flight crews. These advisories can include audible alerts and visual displays to notify pilots of the impending collision risk and recommend specific avoidance maneuvers.

The effectiveness of ACAS in preventing mid-air collisions has been significant. By providing real-time collision avoidance guidance, ACAS has played a pivotal role in reducing the risk of accidents in congested airspace, where aircraft frequently operate in close proximity.

Global Navigation Satellite System (GNSS) GNSS is a constellation of satellites that provides precise global positioning and navigation services to users worldwide. GNSS is a broad term that encompasses various satellite navigation systems, with the Global Positioning System (GPS) being the most well-known. Other major GNSS systems include the Russian GLONASS, the European Galileo, the Indian IRNSS and the Chinese BeiDou. These systems work collectively or independently to provide highly accurate positioning and timing information for a wide range of applications. GNSS satellites orbit the Earth, continuously transmitting signals that are received and processed by GNSS receivers on the ground, in the air, or at sea. These signals contain data that allows the receiver to calculate its exact position, velocity, and time. The accuracy of GNSS is remarkably high, with modern systems capable of providing position information down to a few centimeters.

The GNSS plays a pivotal role in enhancing aviation safety by providing accurate and reliable positioning, navigation, and timing information to aircraft and air traffic management systems. GNSS enables aircraft to determine their precise position, altitude, and velocity, allowing for more accurate navigation. This precision is particularly beneficial during all phases of flight, including takeoff, en-route, and landing. It minimizes the risk of navigational errors and helps pilots maintain safe and efficient flight paths. In low-visibility conditions or during challenging instrument approaches, GNSS aids pilots in executing precision approaches and landings, such as the Instrument Landing System (ILS) or RNAV (Area Navigation) procedures. These approaches improve safety by guiding aircraft along precise paths to the runway, reducing the risk of accidents caused by poor weather or human error. GNSS, when integrated with terrain databases, provides terrain awareness and warning systems (TAWS) with accurate position information. Air traffic control (ATC) systems also rely on GNSS to track aircraft positions accurately, allowing for better traffic management and conflict resolution.

Global Aeronautical Distress and Safety System (GADSS) is a comprehensive framework in aviation safety designed to improve the tracking and response to aircraft distress situations. In response to high-profile aviation incidents, such as the disappearance of *Malaysia Airlines Flight MH370* in 2014, the aviation industry recognized the need for enhanced measures to prevent and manage distress situations more effectively. It mandates the use of certain technologies, communication protocols, and standards that enable real-time aircraft tracking, especially in cases where aircraft deviate from their intended flight path or encounter unexpected events.

GADSS introduces a two-fold approach to aviation safety. First, it enhances the tracking and situational awareness of aircraft by requiring real-time position reporting. This involves automatic position reporting at least once every 15 minutes, and more frequent reporting in distress situations. Second, GADSS mandates the development and implementation of Autonomous Distress Tracking (ADT) systems that autonomously transmit critical aircraft information, such as location and status, in distress scenarios. These measures, collectively, aim to minimize response times in emergencies, facilitate search and rescue operations, and ultimately improve aviation safety by reducing the likelihood of extended periods during which aircraft are unaccounted for, which can be critical in incident resolution and saving lives.

Autonomous Runway Incursion Warning System (ARIWS) is a critical component of aviation safety designed to prevent runway incursions at airports. A runway incursion occurs when any aircraft, vehicle, or person improperly enters a runway that is actively being used for takeoffs or landings, which can lead to dangerous situations and accidents. The ARIWS is a technological system that provides real-time alerts and warnings to help prevent these incursions, enhancing safety on the airfield.

ARIWS relies on a combination of sensors and surveillance technologies to continuously monitor the movements and positions of aircraft, vehicles, and personnel on and around the airport's runways and taxiways. This information is then processed and analyzed to identify potential incursion scenarios. If the system detects an incursion risk, it automatically issues warnings to relevant parties, such as air traffic controllers, pilots, or ground vehicle operators. These warnings may include audible alerts, visual notifications, and advisories through radio communication, aiming to alert those involved to take immediate corrective action.

The ARIWS serves as a critical layer of safety at airports, helping to prevent runway incursions, which can lead to dangerous conflicts and potential accidents. By providing timely alerts and improving situational awareness, ARIWS contributes to the overall safety of aviation operations and reduces the risk of runway-related incidents, enhancing the safety of both passengers and personnel on the ground.

5.3.2 Airliner Noise

Civil aviation has made significant advancements in passenger comfort features to enhance the overall flying experience. These innovations aim to provide travelers with a more pleasant and stress-free journey. Improved sound insulation and quieter engines reduce cabin noise, creating a more peaceful environment for passengers. Over time, advancements in both engine technology and aircraft aerodynamics have led to a significant reduction in noise levels. High-bypass turbofan engines, in particular, are known for their reduced noise levels.

Airframe noise is a significant noise source for modern large aircraft. The noise primarily stemming from landing gear stands out as the predominant source. The possibility of reducing airframe noise depends significantly on the aircraft category, design, and operational factors, and the realization of such reductions will be influenced by a multitude of constraints.

The aviation industry is committed to mitigating aircraft-related noise. Aircraft noise has witnessed a remarkable reduction equivalent to 30 decibels, over the past six to seven decades. Replacing an *A320* with a new, quieter *A320neo*, for example, halves the noise at takeoff, exemplifying the industry's commitment to noise reduction. The graph

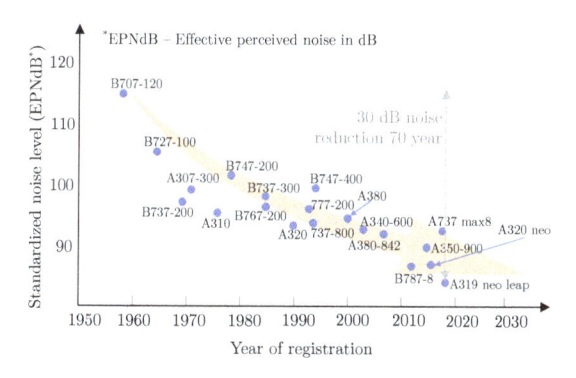

Noise reduction achieved in the past. The calculation of EPNdB is given in ICAO's Environmental Technical Manual (2004)

illustrates noise levels measured in Effective Perceived Noise in decibels (EPNdB). Among older aircraft like the *Boeing 707* and *Boeing 747-200*, which are less common these days, noise levels exceed 100 decibels. In the medium-sized aircraft category, such as the *Airbus A320* and *Boeing 737-800*, which are prevalent worldwide, noise levels typically range around 93-94 decibels. More modern passenger aircraft like the *Airbus A350*, *Boeing 787*, *Airbus A320neo*, and *Boeing 737 MAX* have reduced noise levels to approximately 90 decibels. The *Airbus A320neo*, in particular, has achieved noise levels below 90 decibels. The newer *A319neo* with *LEAP* engines have achieved noise much less than 90 decibels.

5.4 Business Aviation

Business aviation is the use of any general aviation aircraft (all flights not operated by the military or scheduled airlines) for a business purpose. That means that business aviation is a part of general aviation that focuses on the use of aircraft (mainly small, including helicopters, propeller-driven airplanes, and jets) for business transportation. Most business aircraft typically accommodate 6-20 individuals and operate flights that cover distances that can be covered in a few hours.

Business aviation serves as a vital resource for companies and organizations of various types and scales, encompassing universities, nonprofits, and hospitals. Additionally, firefighters, law enforcement personnel, and government agencies rely on these aircraft daily. Companies of all sizes utilize business aircraft.

Business aviation offers significant advantages to its users, such as saving time for employees, increasing employee productivity, improving safety and security, efficiently reaching multiple destinations, and accessing communities with limited or no airline service. Nearly half of business aviation missions are destined for locations with limited or no scheduled airline service. In addition, more than a third of these missions go to destinations that have never been served by commercial air travel.

As an industry, business aviation creates employment opportunities for more than a million people, including aircraft manufacturing jobs and airport-related jobs. In 2018, general aviation contributed $128.3 billion to the U.S. GDP. In 2021, the worldwide market for business jets reached a value of $ 29 billion, with a projected compound annual growth rate (CAGR) of 4.17% anticipated throughout the forecast period.

Some interesting facts about business jets include:

- Business jets can operate at higher altitudes and speeds compared to commercial jets, allowing faster travel times and more direct routes. For example, the *Bombardier Global 8000* boasts a maximum speed of Mach 0.94 (999 km/h) and a ceiling height of 15.5 km, offering a range of 14,800 km. The *Cessna Citation X+* can reach a maximum speed of Mach 0.935 and a ceiling height of 15.5 km. The *Gulfstream G500* is capable of flying at Mach 0.925, covering a range of 9,816 km at a ceiling height of 15.545 km.

- Many business jets are equipped with luxurious amenities such as plush seating, private bedroom, fully stocked kitchen on board, dedicated workspace, personalized service, luxurious bathrooms, and entertainment systems.

- Business jets can access smaller airports that are not served by commercial airlines, making it easier to reach remote or less accessible locations. 80% of business airplane flights are directed toward airports located in small towns and communities where scheduled airline service is not available.

- Some business jets are equipped with advanced technology, such as satellite communication systems and Wi-Fi onboard, allowing passengers to stay connected during their flight.

- Business jets are often used by high-level executives, celebrities, and government officials for private and secure travel.

5.4.1 Business Aircraft

According to forecasts, the size of the global business jet market is expected to increase from $43.97 billion in 2023 to $62.66 billion by 2030, with a compound annual growth rate (CAGR) of 5.19%.

Fleet: As of May 2023, the global fleet of business jets comprised 23,369 aircraft, with the top 20 countries (the largest being the United States with 64%, making it 14,999 jets) markets collectively representing 89% of this total fleet.

Business: In 2022, a total of 712 business jets were delivered (488 in the USA and 80 in Europe). *Gulfstream* led the market with $ 6.60 billion in revenue from 120 aircraft, followed by *Bombardier* with $ 6.04 billion from 123 jets. *Textron Aviation* secured $ 3.62 billion, *Dassault Aviation* earned $ 1.76 billion from 239 aircraft, *Embraer* generated $ 1.36 billion from 102 jets, and *Pilatus* garnered $ 900 million from 123 aircraft.

Table 5.5: Biggest business jet companies as per 2022 data

Company	Delivered business jets(2022)	(B$)	Popular models
Gulfstream Aerospace (USA)	120	6.6	G280 G500, G600, G650, and G650ER models
Bombardier (Canada)	123	6.04	Challenger 350/650, Global Express, Global 7500 Learjet 70 / 75 / 75 Liberty
Textron Aviation (USA)	568	3.62	Cessna Citation family, Skyhawk, Skylane
Dassault Aviation (French)	239	1.76	Falcon 6X, Falcon 900LX, Falcon 2000LXS, Falcon 8X, and Falcon 10X
Embraer (Brazil)	102	1.36	Phenom 100EV, Phenom 300E, Praetor 500, and Praetor 600

A variety of business aircraft ranging from propeller-driven planes to jets to helicopters are available. The fleet encompasses everything from piston aircraft, only slightly larger than a car and capable of covering a few hundred km before refueling, to jets accommodating over a dozen passengers and capable of conducting non-stop international flights. However, the majority of business aircraft typically seat six passengers in a cabin approximately the size of a large SUV and operate with an average stage length of less than 1,600 km. Depending on their capabilities, these aircraft may fly at altitudes below commercial airlines (below 6.1 km) or above them (above 12.2 km).

Business jets can be categorized according to their size as (1) very light jets (2) light jets (3) mid-size jets (4) super-mid-size jets (5) large jets (6) long-range jets (7) VIP airliners.

Very Light Jets (VLJs) are the smallest category of private jets available in the market, typically accommodating between 4 and 6 passengers. They offer cost-effective operation and maintenance compared to larger jets. Their compact size enables access to small airfields with short runways and narrow taxiways, particularly in remote areas. VLJs are ideal for short-haul flights, typically lasting up to three hours maximum. Most commonly used for weekend getaways, these jets can cover distances between destinations ranging from 1,500 to 2,000 kilometers without interruption. Despite their limited size, they feature a small luggage compartment of less than 70 cubic feet. Many VLJs are certified for single pilot operations and have a maximum take-off weight of less than 4,500 kilograms.

Light jets offer enhanced passenger capacity, comfortably accommodating up to eight individuals, which is particularly favored by business travelers. With an average flight distance of up to 2000 km, these jets are well suited for two to three hour flights, including intercontinental routes. Similar to their smaller counterparts, light jets boast the capability to access small airports and runways, providing business travelers with increased flexibility to utilize less congested airports rather than busy commercial ones. Although most light jets may not feature cabin attendants, they can be equipped with lavatories, distinguishing them from most VLJs

Midsize jets are the ideal choice for travelers

Cessna Citation Mustang, most-sold very light jet. The jet, with MTOW of 3,921 kg, is equipped with two *Pratt & Whitney Canada PW615F* turbofans, each producing 6.5 kN of thrust. It can achieve a speed of 630 km/h and has a range of 2,161 km.

The *Cessna CitationJet* (photo: Cessna CitationJet/M2) is a family of light business jets produced by *Cessna Aircraft Company*. It was first introduced in 1993 and has since become one of the most popular models in the *Citation* series. The *CitationJet* features a sleek design, efficient engines, and advanced avionics systems, making it a popular choice for corporate and private jet owners. It offers a range of up to 2800 km and can accommodate up to 7 passengers. More than 2,000 (as of June 8, 2017) *Cessna CitationJets* have been delivered.

who need longer flight capacity. With range up to 4000km, these jets can handle both short-haul and long-haul flights, making them perfect for transcontinental travel.

These jets have a larger cabin, providing more headroom, standing capacity, and luggage space. They also offer stylish interiors and can comfortably accommodate up to 12 passengers.

Midsize jets typically have room for two pilots, a

Table 5.6: Business jets category according to their size

Category	Seat	Range (km)	MTOW (ton)	Popular models
Very Light Jets (VLJ)	4-6	up to 2000	4.5-7	Cessna Citation Mustang, Eclipse 500, Embraer Phenom 100, HondaJet HA-420, Eclipse 400
Light Jets	6-8	up to 2500	7-10	Hawker 400 XP, Dassault Falcon 10, Cessna Citation CJ2, Cessna Citation CJ3, Embraer Phenom 300, Learjet 45, Cessna Citation CJ4
Mid-Size Jets	9-12	up to 4000	10-14	Gulfstream G150, Cessna Citation Latitude, Hawker 900XP, Bombardier challenger 300, Learjet 60, Challenger jets
Super Mid-Size Jets	10-19	5500	14-18	Bombardier Challenger 350, Cessna Citation Sovereign, Gulfstream G280, Gulfstream G200, Dassault Falcon 2000, Cessna Citation X
Heavy/Long-Range Jets	13-19	up to 6000	18-40	Bombardier Challenger 605, Cessna Citation Sovereign, Gulfstream G350, Dassault Falcon 900
Ultra-Long-Range Heavy Jets	13-19	10000	40-50	Gulfstream V, Dassault Falcon 7x, Bombardier Global 6000, Bombardier Global 7500, Embraer Lineage 1000
Executive/VIP Airliners	20-50	13000	50-100	Boeing Business Jet BBJ, Airbus Corporate Jet ACJ 380, Airbus Corporate Jet ACJ 319, Lineage 1000E

flight attendant, a service galley, and an on-board lavatory. Some may even have an enclosed shower and fold-out divans. Despite their larger size, mid-size private jets can still use smaller airports and are more cost-efficient to operate compared to heavy jets.

The *British Aerospace 125*, originally known as the *DH.125 Jet Dragon*, is a popular twinjet mid-size business jet that was developed by *de Havilland*. The aircraft had more than 60% of its total sales to North American customers.

Super-mid-size jets offer increased cabin space and higher passenger capacity compared to standard mid-size private jets. These jets can fly up to seven hours, covering an average distance of 5500 km.

They provide ample standing and walking room, along with space for a private lavatory and service galley. Equipped with advanced avionics for quieter operation, super mid-size jets ensure a more comfortable travel experience with improved speed and range for passengers. They seamlessly blend transatlantic capabilities with the speed and luxury associated with wide-body high-altitude aircraft.

Ultra-long-range heavy jets are considered more luxurious compared to regular heavy jets due to their extended range capabilities, allowing for approximately 12 hours of continuous flight at cruising speeds. These jets feature spacious cabins that can accommodate up to 19 passengers.

Designed for trans-oceanic flights between continents, ultra-long-range heavy jets cater to frequent long-distance travelers seeking comfort and convenience. These jets are equipped with luxurious amenities such as lie-flat beds, retractable tabletops, refrigeration units, and plush seating, appealing to those with expensive tastes.

However, the opulence and comfort offered by

Bombardier Challenger 300, the most widespread super mid-size jet. With a range of 5,700 km, manufactured by the Canadian aircraft company *Bombardier Aerospace*, the *Challenger 300* entered service in 2004 and over 500 aircraft have been delivered worldwide.

The *Bombardier Challenger 600* is a popular family of large business jets developed by the Canadian manufacturer *Bombardier Aerospace*. The Challenger 600 series is known for its spacious cabin, long range capabilities, and high level of comfort. It has been a popular choice among business executives and charter companies for private and corporate travel.

A *Bombardier Global Express*, a large cabin, long-range business jet produced by *Bombardier Aviation*.

these aircraft come at a premium cost, making them more expensive to charter, rent, or purchase compared to smaller jet categories. For many travelers, opting for an ultra-long-range heavy jet signifies a desire for extravagance and luxury during air travel.

The *Dassault Falcon 6X*, a business jet with a large cabin and long-range capabilities, designed by *Dassault Aviation* in France. The *Falcon 6X* features turbofan engine with the most expansive purpose-built cabin in the business jet category. This configuration enables an impressive range of 10,200 km and a top speed of Mach 0.90.

The *Embraer Legacy 600*, a business jet derivative of the *Embraer ERJ* family.

Executive airliners, also known as *bizliners*, are specifically built for business travel which are converted commercial airline planes. They are the largest and most expensive private jets to charter or operate. They have a broader cabin that can accommodate up to 50 passengers in one flight.

Executive jets are equipped with special business amenities such as conference rooms, lavish sleeping

The *Bombardier Global Express (Global 6000)* is a large cabin, long-range business jet designed and manufactured by *Bombardier Aviation*.

The *Bombardier Global 5000* is a long-range business jet produced by Bombardier Aerospace.

BBJ 737-9, a variant of the *Boeing 737 MAX 9* and is the largest member of the *BBJ MAX* family, offering a cabin floor area of 104.1 m^2. *Boeing Business Jets (BBJ)* are customized versions of Boeing's commercial jetliners, designed to cater to the private, governmental, and corporate jet market. These jets can fly over 11000 km without stopping and provide more cabin space compared to conventional long-range business jets. The inaugural *BBJ*, developed from the 737-700 model, was unveiled in 1998 and completed its maiden flight on September 4, 1998.

quarters, and luxurious bathrooms with hot showers to ensure that passengers are refreshed during the flight. These aircraft are often used by heads of states, with *Air Force One* being a prime example. *Air Force One*, a *Boeing 747* operated for US presidents, not only features a luxurious private jet interior, but also boasts advanced military systems designed to safeguard the president from potential threats.

5.5 Airports

An airport is any area of land or water used for buildings or facilities required for the landing or takeoff of aircraft. Airports also include special types of facilities such as seaplane bases, heliports, and facilities to accommodate tilt-rotor aircraft. Across the globe, approximately 40,000 airfields have been designated as civil aerodromes, each identified by a specific ICAO aerodrome code. Among these airfields, around 4,000 function as airports equipped to accommodate commercial aircraft, serving passengers and/or cargo transport needs. Interestingly, a subset of roughly 100 airports manages a significant traffic volume of over 15 million passengers annually.

An airport serves as the connecting point between aviation and ground transportation. This is the juncture where the world of aeronautics (airside), governed by its stringent rules and regulations, operational procedures, and protocols, converges with the non-aeronautical domain (land side).

The *air side* of the airport encompasses areas where aircraft taxi, take off, land, and park, all of which fall under the jurisdiction of air law. Conversely, the *land side* of the airport comprises spaces where passengers navigate, congregate, or await their journeys, and where national or local authorities hold administrative control. In a precise sense, one can assert that in a modern airport equipped with jet bridges connecting the terminal to the air-

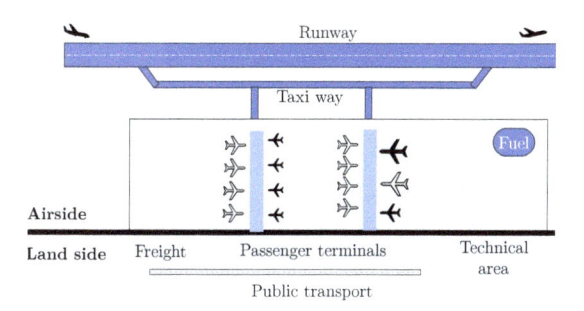

Layout of an airport

craft, the boundary between the air side and land side for passengers is demarcated by the threshold of the aircraft's entrance door. Once passengers cross this threshold and board the aircraft, the air side prevails, with the pilot in command or an appointed representative from the airline assuming authority over the aircraft's operations.

5.5.1 Airport Types

The primary distinguishing feature of airports lies in their functionality such as cargo airport, refueling airport, and hub airport (used by home airlines to connect their flights, for example, Atlanta Hartsfield, USA). Airports such as *Liege* in Belgium and *Memphis* in Tennessee serve as classic illustrations of airports whose primary significance lies in cargo operations. DXB (Dubai) in the Middle East is a rational choice as a refueling airport. It's crucial to recognize that many airports exhibit a blend of functions.

Another pivotal contrast among airports is whether they are categorized as *international* or *domestic*. An international airport boasts customs and immigration facilities, as well as the capability to conduct security checks in adherence to the regulations of the arrival country. Typically, international airports tend to be larger than their domestic counterparts, with many major airports naturally falling into the international category. Furthermore, there is a growing trend of regional airports evolving into international ones.

The airport can be *towered* (has an operating control tower and pilots are required to maintain two-way radio communication with the tower, to acknowledge and comply with their instructions) and *nontowered*. The airport without an operating control tower communicates based on CTAF (Common Traffic Advisory Frequency), a frequency designated to carry out airport advisory practices while operating to or from an airport without an operating control tower. The CTAF may be a Universal Integrated Community (UNICOM, a nongovernment air/ground radio communication station), MULTICOM, Flight Service Station (FSS), or tower frequency and is identified in appropriate aeronautical publications. On pilot request, UNICOM stations may provide pilots with weather information, wind direction, the recommended runway, or other necessary information.

Airport Surface Detection Equipment, a form of ground surveillance radar that, in conjunction with other data, allows airport ground controllers to see traffic on the ground, even during poor weather or when visibility is limited. It is useful for stopping runway incursions by vehicles or aircraft.

An aircraft must have two-way radio communication capability when operating in or out of towered airports and in many parts of the airspace system. Therefore, a pilot should be familiar with radio station license requirements and radio communications equipment. Pilots need to know which direction the wind is blowing. In a towered airport, an operating control tower provides this information. At airports that can receive and broadcast on this frequency, information may also be provided by personnel either located at that airport or remotely through a remote communication outlet. In the absence of any of these services, visual wind indicators such as wind cones, wind socks, tetrahedrons, or wind tee may be used to determine wind direction and runway usage. The pilot needs to check these wind indicators even when information is provided.

Further, airports can be classified as *Civil Airports* (open to the general public), *Military airports* (operated by the military or other agencies of the Government) and *Private Airports* (designated for private or restricted use only, not open to the general public).

Table 5.7: Busiest airport for international flights in the world during June 2023. However, Atlanta Hartsfield–Jackson International Airport is the world's busiest airport (domestic and international) in June 2023 with 5.3 million seats. Tokyo International Airport (Haneda Airport), Dallas Dallas/Fort Worth International Airport and Istanbul Airport come next with 4.3, 4,2 and 4 million seats, respectively. The rankings are based on scheduled airline capacity during the month.

2023	Airport Name	Seats (Million)
1	Dubai International	4.64
2	London Heathrow	3.96
3	Amsterdam	3.25
4	Paris Charles de Gaulle	3.24
5	Frankfurt International	3.06
6	Singapore Changi	3.03
7	Istanbul	3.01
8	Seoul Incheon International	2.79
9	Hamad International, Doha	2.36
10	London Gatwick	2.15

5.5.2 Airport Services

Airports are required to adhere to a comprehensive set of rules and regulations for their operational integrity, and they must also deliver a range of services to their users (airlines). Typical services are Air Traffic Control (ATC), fueling services, flight information services, rescue and firefighting services, customs and immigration, and security at the airport. These services are typically furnished by Aeronautical Navigational Service Providers (ANSPs), an organization and legitimate holder of responsibility for managing the aircraft in flight or in the maneuvering area in an airport. The extent of services required at an airport is defined by the airport categorization system by ICAO.

Air Traffic Control (ATC)

ATC, the ground-based air traffic controller, is responsible for providing the safe, orderly, and expeditious flow of air traffic in airspace around an airport and designated to traffic to and from that airport. The type of operations and/or volume of traffic in non-controlled airspace requires various advisory services from ATC. The primary purpose of ATC worldwide is to prevent collisions, organize and expedite the flow of air traffic, and provide information and other support for pilots. ATC monitors the location of aircraft in their assigned airspace by radar and communicates with the pilots by radio.

Air traffic control (ATC) tower at RGIA Hyderabad, India.

The ATC may have a Primary Radar, Radar Beacon System, Transponder, Automatic Dependent Surveillance–Broadcast (ADS-B) and Radar Traffic Advisories. *Radar* is a device that provides information on the range, azimuth and elevation of objects in the path of the transmitted radio waves. The ATC *Radar Beacon System* (ATCRBS) or *secondary surveil-*

lance radar helps in reducing some of the limitations associated with primary radar such as degradation of radar over distance and the inability to penetrate through solid objects. The advantages of ATCRBS include the reinforcement of radar targets, rapid target identification, and a unique display of selected codes.

The second most crucial ground service at an airport is the capability to fuel an aircraft. Even when an aircraft doesn't need to land for passenger or cargo operations, it might still require a stop at an airport for refueling. It's evident that an airport's role in aviation is fundamentally contingent on its ability to supply fuel. Therefore, ensuring the availability of fuel services at an airport is an essential aspect, albeit one that can be quite challenging to establish as aircraft consume large amounts of fuel.

Each National Aviation Authority (NAA) is mandated to furnish up-to-date aeronautical information and data regarding their airspace and aerodromes as per the mandate by the ICAO regulations. This data pertains to the utilization of the country's airspace and the operational status of navigational aids and runways. Flight Information Services (FIS) encompasses two critical components vital for airport operations: Flight Planning services and meteorological services. A comprehensive flight plan must include essential aircraft data including the fuel required, the aircraft's weight, passenger count, and cargo description, and the entire route, commencing from the departure airport and concluding at the destination airport. The flight plan should also specify the anticipated overflight times for each of these locations based on the aircraft's speed. Fuel requirements include the fuel needed for the journey to the destination along with the reserve fuel needed for holding, overshooting the arrival airport and flying to an alternate airport. The alternate airport is specified in the flight plan, as well as alternate airports en route.

Ground Operations

The majority of airports have a variety and type of lighting systems for night operations. Airport lighting is standardized so that all the airports use the same light colors for runways and taxiways. *Airport beacons* (an omnidirectional capacitor-discharge device, or it may rotate at a constant speed, that produces the visual effect of flashes at regular intervals) help a pilot to identify an airport at night. Some of the most common beacon types are:

- Flashing white and green for civilian land airports,
- Flashing white and yellow for a water airport,
- Flashing white, yellow, and green for a heliport,
- Two quick white flashes alternating with a green flash for a military airport

There are various types of signs that one may observe in airports: *Mandatory instruction signs* (red background with white inscription signs denoting an entrance to a runway, critical area, or prohibited area), *Location signs* (black with yellow inscription and a yellow border, no arrows signs used to identify a taxiway or runway location, to identify the boundary of the runway, or identify an instrument landing system (ILS) critical area), *Direction signs* (yellow background with black inscription), *Destination signs* (yellow background with black inscription and arrows), *Information signs* (yellow background with black inscription) and *Runway distance remaining signs* (black background with white numbers sign to indicate the distance of the remaining runway in thousands of feet).

The safe flight begins with a careful visual inspection of the airplane to determine that the airplane is legally airworthy (airworthiness certificate, Registration certificate, Federal Communications Commission(FCC) radio station licensee or Airplane operation limitations) and is in a condition for safe flight (pre-flight inspection including cockpit, outer-wing surfaces and tail section, fuel and oil, landing gear, tyre and brakes or engine and propeller). The pilot needs to be familiar with hand signals to operate the airplane safely on the ground.

The pilot needs to know the specific procedure for engine starting including hand propping in case the airplane is not equipped with electric starters. The pilot must understand and be proficient in taxiing which is the controlled movement of the airplane under its power while on the ground. The position of the control surface during taxiing, especially under downwind (wind in the direction of airplane motion) and upwind (wind in the direction opposite to airplane motion), is very crucial information for the pilot.

A pushback tug with a towbar on the apron.

The pushback tractor or tug is pushing back an airport at RGIA Hyderabad, India. *Pushback* is the procedure for the rearward moving of an aircraft from a parking position to a taxi position by use of specialized ground support equipment.

A basic runway may only have centerline markings and runway numbers.

There are *markings and signs* used at airports that provide directions and assist pilots in airport operations. All airport markings are painted on the surface, whereas some signs are vertical and some are painted on the surface. Since aircraft are affected by the wind during takeoffs and landings, runways are laid out according to the local prevailing winds.

Certain procedures are followed by pilots for collision avoidance under various circumstances.

Before takeoff: Before taxiing onto a runway or landing area in preparation for takeoff, pilots should scan the approach area for possible landing traffic, executing appropriate maneuvers to provide a clear view of the approach areas.

Climbs and descents: During climbs and descents in flight conditions that permit visual detection of other traffic, pilots should execute gentle banks left and right at a frequency that permits continuous visual scanning of the airspace.

Straight and level: During sustained periods of straight-and-level flight, a pilot should execute appropriate clearing procedures at periodic intervals.

Traffic patterns: Entries into traffic patterns while descending should be avoided.

Traffic at very high-frequency omnidirectional range (VOR) sites: Due to converging traffic, sustained vigilance should be maintained in the vicinity of

VORs and intersections.

Training operations: Vigilance should be maintained and clearing turns should be made prior to a practice maneuver. During instruction, the pilot should be asked to verbalize the clearing procedures (call out "clear left, right, above, and below"). High-wing and low-wing aircraft have their respective blind spots. The pilot of a high-wing aircraft should momentarily raise the wing in the direction of the intended turn and look for traffic prior to commencing the turn. The pilot of a low-wing aircraft should momentarily lower the wing and look for traffic prior to commencing the turn.

Heads-up, eyes outside. The pilot may request progressive taxi instructions, including step-by-step taxi routing instructions at an unfamiliar airport. The pilot should have a current airport diagram, remain *heads-up with eyes outside*, and devote full attention to surface navigation as per ATC clearance.

5.6 Governing Bodies

India's Directorate General of Civil Aviation (DGCA), the Federal Aviation Administration (FAA) of the United States, Transport Canada's Civil Aviation Directorate (TCCA), the United Kingdom's Civil Aviation Authority (CAA), the Civil Aviation Administration of China (CAAC), and the Europea Aviation Safety Administration (EASA) are some of the civil aviation authorities in the respective countries. Most of these civil aviation authorities are responsible for setting, monitoring, and enforcing safety standards across the local aviation environment and the standards apply not just to airlines based in that country but also to airlines who fly into (and often over) the country.

In 1944, representatives from 54 nations gathered at the International Civil Aviation Conference in Chicago, at the invitation of the United States Government. By the conference's conclusion on December 7, 1944, a total of 52 of these nations had affixed their signatures to the Convention on International Civil Aviation, more commonly recognized today as the "Chicago Convention". This historic accord served as the cornerstone for the creation of the *International Civil Aviation Organization (ICAO)* and the formulation of Standards and Recommended Practices (SARPs) and protocols for worldwide air navigation, all aimed at fostering the safe and organized advancement of aviation. The organization sets standards and regulations necessary for aviation safety, security, efficiency, and regularity, as well as for aviation environmental protection. ICAO defines the protocols for air accident investigation that are followed by transport safety authorities in countries signatory to the Convention on International Civil Aviation.

The International Air Transport Association (IATA) is a trade association of the world's airlines founded in 1945. Consisting of 300 airlines in 2023, the IATA's member airlines account for carrying approximately 83% of total available seat miles air traffic. IATA supports airline activity and helps formulate industry policy and standards. It is headquartered in Montreal, Canada with executive offices in Geneva, Switzerland.

5.7 Civil Aviation in India

The Indian aviation market is experiencing remarkable growth, ranking as the world's third-largest market after USA and China and fastest-growing market. Over the past nine years (2014 - 2023), domestic air travel in India has jumped by an impressive 130%, showcasing a remarkable surge in people taking to the skies for their travels. India has a fleet of over 700 airliners, showcasing a remarkable increase of more than 100% over the past nine years. From April 2023 to January 2024, domestic passenger traffic reached 254.44 million, showing a 15.3% rise, while international passenger traffic hit 57.57 million, marking a 23.5% increase compared to the previous year's same period. This significant increase highlights the growing popularity and accessibility of air travel in India as a preferred mode of transportation for many individuals. Domestic flights in India account for 69% of total airline traffic in South Asia.

By 2042, *Airbus* predicts that 685 million people will fly in India, a significant increase from the 254 million passengers in 2024. Additionally, a report from *Barclays* revealed that Indian airlines hold almost 7% of the total industry orders backlog, making it the second largest in the world after the United States.

As of November 2023, India has around ten scheduled airlines providing passenger services, in addition to other cargo carriers. In the Paris Airshow 2023, *Airbus* received an order for 500 *A320neo* aircraft from *IndiGo Airlines*, making it the largest deal ever made by a single carrier in civil aviation history. In February 2023, the newly privatized *Air India* placed an order for 470 narrow and widebody jets from both *Boeing* and *Airbus*. When the largest two India airliners, *IndiGo* and *Air India* placed huge orders, they collectively hold a dominant 81.3% share of the domestic market, with *IndiGo* transporting 56.2% of all passengers in the country and *Air India* accounting for 25.1% of the market.

The Ministry of Civil Aviation (MoCA) of the Government of India is in charge of creating national policies and programs for the development and regulation of civilian aviation. It is also responsible for designing and executing plans to ensure the organized growth and expansion of civilian air transport. The ministry oversees aviation-related autonomous organizations such as the *Airports Authority of India* (AAI) and the *Bureau of Civil Aviation Security* (BCAS).

The Directorate General of Civil Aviation (DGCA) is the regulatory body in India that oversees civil aviation activities. It is responsible for regulating and supervising air transport services to ensure that safety and security standards are met. The DGCA also formulates and enforces rules and regulations related to civil aviation, issues Private Pilot Licenses (PPLs), Student Pilot Licenses (SPLs) Commercial Pilot Licenses (CPLs), licenses for aircraft maintenance engineers, and air traffic controllers, and conducts inspections and audits to ensure compliance with aviation regulations. DGCA conducts investigations of aviation accidents and incidents.

The Airports Authority of India (AAI) is a government agency responsible for creating, upgrading, maintaining, and managing the civil aviation infrastructure in India. AAI oversees a network of 140+ airports, comprising 34 international airports, 10 Customs Airports, 81 domestic airports, and 23 Civil enclaves at Defence airfields. AAI's operational presence extends to all airports and 25 additional sites to ensure aircraft safety. Across India, AAI manages key air routes with 29 Radar installations at 11 locations and 700 VOR (Very High Frequency Omnidirectional Range Station, a short-range radio navigation system for aircraft) installations co-located with Distance Measuring Equipment (DME). Additionally, AAI has implemented an Automatic Message Switching System at 15 Airports and installed Instrument Landing System (ILS) facilitating night landings at most of these airports.

Bibliography

M. Gert (2020) *Fundamentals of Aviation Operations*, Routledge.

S. K. Kearns (2018) *Fundamentals of International Aviation*, Routledge.

H. A. Kinnison and Others (2013) *Aviation Maintenance Management*, McGraw-Hill Education.

U. Michel (2014) *Correlation of aircraft certification noise levels EPNL with controlling physical parameters*, 19th AIAA/CEAS Aeroacoustics Conference.

C. C. Rodrigues and Others (2012) *Commercial Aviation Safety*, McGraw-Hill Education.

A. J. Stolzer and R. L. Sumwalt and J. J. Goglia (2023) *Safety Management Systems in aviation*, CRC Press.

T. Filburn (2019) *Commercial Aviation In The Jet Era And The Systems That Make It Possible*, Springer.

S. Eriksson and H. Steenhuis (2015) *The Global Commercial Aviation Industry*, Routledge.

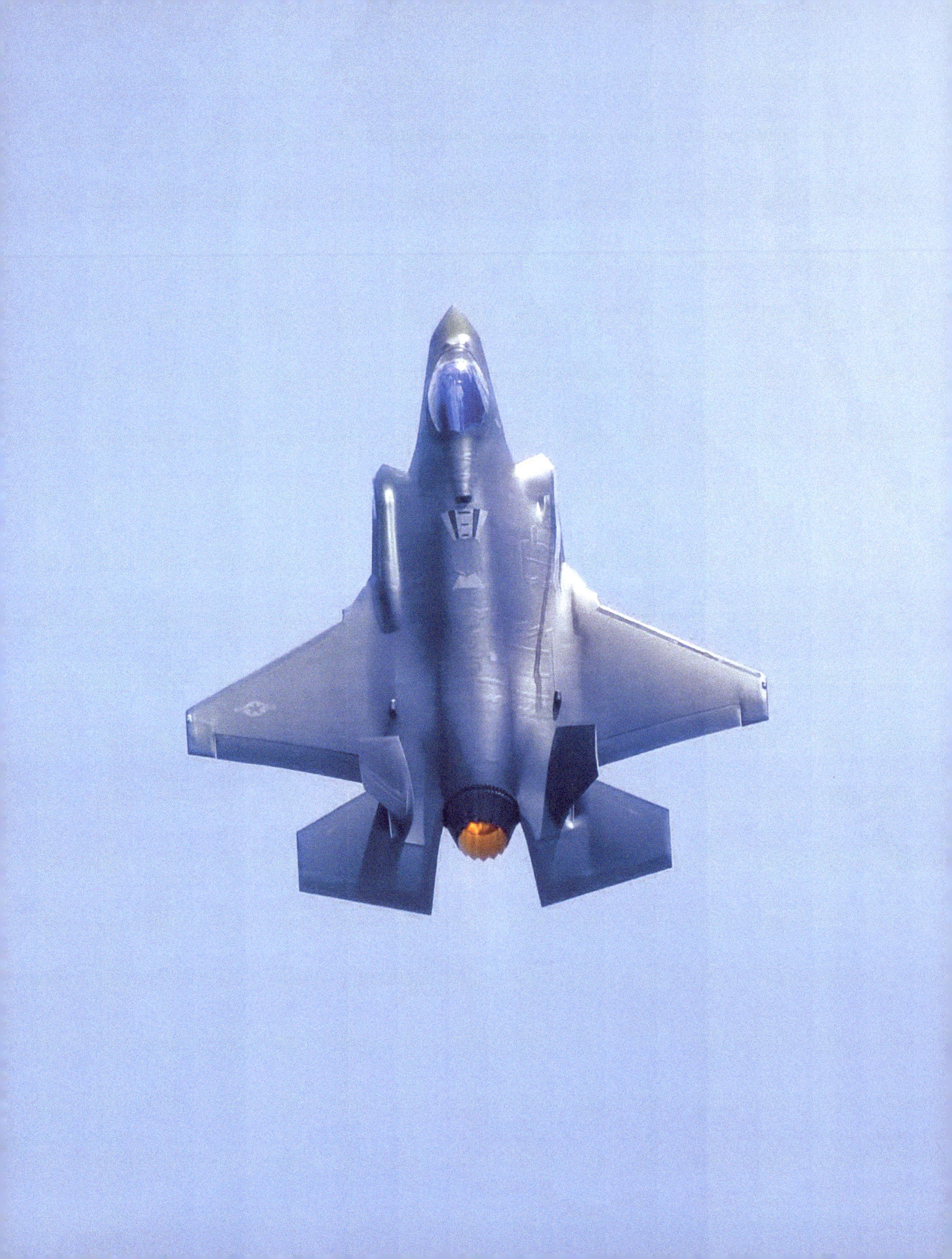

Chapter 6

Military Airplanes

Contents

6.1	**Introduction**		**292**
	6.1.1	Military Airplane Generations	292
	6.1.2	Next Generation Military Airplanes	296
6.2	**Capabilities and Performance Parameters**		**298**
	6.2.1	Maximum and Minimum Speeds	298
	6.2.2	Range and Endurance	299
	6.2.3	Climb Performance	299
	6.2.4	Turning Performance	300
	6.2.5	Takeoff and Landing	300
	6.2.6	Stealth Technology	301
	6.2.7	Airborne Armaments	301
6.3	**Roles of Military Airplanes**		**302**
	6.3.1	Aerobatics	302
	6.3.2	Attack	306
	6.3.3	Bomber	309
	6.3.4	Fighter and Interceptor	312
	6.3.5	Reconnaissance	312
	6.3.6	Tanker	314
	6.3.7	Trainer	316
	6.3.8	Transport	317
	6.3.9	Multirole Airplanes	321
	6.3.10	Airborne Early Warning & Control (AEW&C) System	326
6.4	**Military Aviation in India**		**328**

6.1 Introduction

Ever since the *World War II*, air forces have played a significant role in most of the military conflicts around the world. During these conflicts, military airplanes are used for aerial bombardment, aerial combat, reconnaissance, transportation of military infrastructure, interception of enemy airplane and providing aerial shield against enemy attack. Even in peaceful times, these airplanes and the associated infrastructure and personnel are kept ready for some potential conflicts that can happen in the future. The strength of the Air Force forms a critical component in the military strength of any country or coalition. The outcome of many military conflicts is determined by the air power deployed. For instance, during the *Gulf War* of 1991 fought after the invasion of Kuwait by Iraq, the coalition forces used aerial bombardment to such devastating effect in the *Operation Desert Storm* that the liberation of Kuwait was achieved in a matter of few weeks with significantly lower losses on the coalition side.

Military airplanes are specially designed and utilized for various military purposes, including combat, reconnaissance, training, transportation, and logistical support. These airplanes play a crucial role in modern warfare, providing nations with strategic and tactical advantages in both offensive and defensive operations. The key difference between military airplanes and civilian airliners is in their design and operation. Military combat airplanes are often built with a focus on performance, agility, and survivability in combat situations. They incorporate advanced aerodynamic designs, higher maneuverability, and specialized features like stealth technology, armored structures, and weapon systems. While civilian aircraft prioritize efficiency, comfort, and safety for passengers. They have more spacious cabins, larger windows, and amenities such as lavatories and galley areas.

Military airplanes are equipped with advanced avionics, communication systems, and radar technologies tailored for military operations. They also feature specialized equipment such as weapon pylons, targeting systems, electronic warfare capabilities, and self-defense mechanisms. However, civilian airliners have navigation and communication systems optimized for commercial air traffic control, passenger entertainment systems, and safety features like emergency exits and oxygen masks. Military combat airplanes are designed for high-speed flight, quick acceleration, and rapid climb rates to carry out military missions effectively. They often have powerful engines, superior maneuverability, and the ability to sustain high G-forces during aerial combat. However, civilian airliners prioritize fuel efficiency, long-range capability, and passenger comfort. They are designed for cruising at relatively lower speeds and altitudes.

Military airplanes operate in diverse and often hostile environments, including combat zones, remote areas, and unprepared airfields. They are designed to handle rough landing conditions, short takeoff and landing distances, and even operate from aircraft carriers. Civilian airliners operate in controlled environments such as airports with long runways and dedicated facilities for passenger services. It is important to note that some airplanes can be used in both military and civilian roles, such as cargo transport or reconnaissance. In such cases, their configurations and equipment can be modified to suit the specific operational requirements.

6.1.1 Military Airplane Generations

Military airplane generations are a way to classify and describe the evolution and advancements in military aviation technology over time. Although there are different interpretations and variations of these generations and there is no universally agreed-upon definition for each generation, the following overview provides a general understanding.

6.1. INTRODUCTION

First Generation

The first-generation military airplanes emerged during and after World War I and were primarily biplanes made of wood and fabric with open cockpits. They were armed with machine guns and mainly used for reconnaissance and limited bombing roles. Notable examples include the *Sopwith Camel (United Kingdom), Curtiss Jenny (United States), Fokker D.VII (Germany), SPAD S.XIII (France)*.

Some even consider that the first generation of fighters comprised those that emerged during the Jet Age, commencing towards the end of World War II and extending into the Korean War. The *Messerschmitt Me-262*, a German aircraft, was the first operational jet fighter. These fighters saw further advances in aerodynamics, structure, and power and were the first to be powered by turbojet engines. The design was aimed at the air superiority interceptor role. They were monoplanes and had advanced engine technology. They introduced features such as enclosed cockpits, retractable landing gear, and metal construction. These airplanes were employed in *World War II*, serving in both air-to-air combat and ground-attack roles. These first-generation fighters were not equipped with radar systems and relied primarily on older technologies such as conventional guns, dumb bombs, and rockets, since guided missiles were primarily in the experimental stage at that time.

Iconic examples include the *Messerschmitt Bf 109 (Germany), Supermarine Spitfire (United Kingdom), Yakovlev Yak-3 (Soviet Union) and North American P-51 Mustang (United States), Mikoyan-Gurevich MiG-15, the North American F-86 Sabre and Hawker Hunter*.

North American P-51D Mustang, among the best and most well-known fighters used by the U.S. Army Air Forces during World War II. Possessing excellent range (2,660 km) and maneuverability, it operated primarily as a long-range escort fighter and also as a ground attack fighter-bomber

Supermarine Spitfire, a highly maneuverable aircraft that could out-turn and out-climb many opponents. It has an innovative design including a distinctive elliptical wing shape, retractable landing gear and a powerful Rolls-Royce Merlin engine.

Mikoyan-Gurevich *MiG-15*, the most representative fighters of first generation. *MiG-15*, a high-altitude (Service ceiling: 15.500 km, Mach 0.87 at sea level) day interceptor first flew in late 1947, featured the first production swept wing, pressurized cockpit, and ejection seat on a Soviet aircraft.

Second Generation

Second-generation airplanes emerged after the *Korean War* in the 1950s and saw significant technological advances. Post-war interceptor aircraft adopted afterburning engines to achieve supersonic speed, while advances in radar and infrared homing missiles significantly improved their accuracy and firepower. These aircraft were equipped with radar

Best representing Second Generation fighters is the American *Century series*. clockwise from the bottom: *F-104 Starfighter, F-100 Super Sabre, F-102 Delta Dagger, F-101 Voodoo,* and *F-105 Thunderchief.* The series, the first successful supersonic aircraft design in the United States, was a mix of fighter-bombers (F-100, F-101A, F-105) and pure interceptors (F-101B, F-102, F-104, F-106). F-102 was the first aircraft in the world to utilize area rule in its design. F-104 was the first combat aircraft capable of Mach 2 flight, and holds the world speed and altitude records

and used the first guided air-to-air missiles. These jets had swept-wing designs and more advanced avionics systems of their time. These airplanes became the primary fighters of the Cold War era, capable of supersonic speeds and engaging in air-to-air combat. Examples include *Century Series* (popular name for a group of US fighter aircraft, F-100 - F-106), *Mikoyan-Gurevich MiG-21 (Soviet Union), Dassault Mirage III (France),* and *English Electric Lightning (United Kingdom).*

Third Generation

Third-generation fighters started appearing in the Vietnam War and represented a significant leap in technology with capabilities of performing both air defense and ground attack missions. Third-generation fighter's design was focused on multi-role capabilities. During this era, aircraft were tasked with carrying an extensive array of weaponry and ordnance, including air-to-ground missiles and laser-guided bombs, while also possessing the capability for beyond-visual-range air-to-air interception. This period saw significant advancements in supporting avionics, including pulse-doppler radar, off-sight targeting, and terrain-warning systems

McDonnell Douglas F-4 Phantom II, the best representative of third-generation fighters. One of the premier fighter jets of the USA, the F-4 was a tandem two-seat, twin-engine, all-weather, long-range supersonic interceptor and fighter bomber that saw combat in Vietnam, Israel, Iran, and Turkey.

Other examples of third-generation fighters are *McDonnell Douglas F-4 Phantom, Mikoyan-Gurevich MiG-23, Sukhoi Su-17, Shenyang J-8, and Hawker Siddeley Harrier, British Aerospace Harrier, Dassault Mirage F.1, Dassault Super Etendard, Shenyang J-8II*

Fourth Generation

Fourth-generation aircraft were advanced multi-role fighters equipped with progressively advanced avionics and weaponry. These fighters shifted focus towards prioritizing maneuverability over speed to excel in air-to-air combat.

They featured advanced avionics, fly-by-wire, composite materials, thrust-to-weight ratios greater than unity (enabling the plane to climb vertically),

F-16 Fighting Falcon, representative of fourth-generation aircraft. F-16 is a compact, multi-role fighter aircraft that can reach a maximum speed of over Mach 2. Innovations include a frameless bubble canopy for better visibility, a side-mounted control stick, and a reclined seat to reduce g-force effects on the pilot under high 9-g maneuvers.

hyper maneuverability, advanced digital avionics and sensors such as synthetic radar and infrared search-and-track, and stealth.

Prominent examples include the *General Dynamics F-16*, *Saab 37 Viggen*, *MiG-29 Fulcrum*, Anglo-American *Harrier II*, *McDonnell Douglas F-15 Eagle* (United States), *Sukhoi Su-27* (Soviet Union/Russia) or *Saab JAS 39 Gripen* (Sweden), *Lockheed F-117*.

4.5 Generation

These are more recent fourth-generation fighters. These fighters typically have the same fundamental traits as fourth-generation planes but incorporate augmented capabilities through advanced technologies reminiscent of fifth-generation fighters such as Supercruise ability (sustained supersonic flight of a supersonic aircraft without using afterburner), high-capacity digital network communications(link 16, a military tactical data link network), *An electro-optical targeting system (EOTS)* employed to track and locate targets in aerial warfare, *low-probability-of-intercept radar (LPIR)* (Employed to avoid detection by passive radar detection equipment while it is searching for a target or engaged in target tracking). They make use of advanced avionics to improve mission capability and limited

A member of 4.5 generation fighter, *Eurofighter Typhoon*. This is a European multinational twin-engine, supersonic, canard delta wing, multirole fighter. Typhoon can simply switch from air-to-air into the air-to-ground role and back within the same mission.

F-18 Super Hornet, a good example of 4.5 generation fighter. It is a supersonic, twin-engine, carrier-capable (can operate from aircraft carriers), multi-role fighter

stealth characteristics to reduce visibility.

Notable examples include *F-18 Super Hornet*, *Eurofighter Typhoon, HAL Tejas MK 1A, Sukhoi Su-30/33/35* and *Dassault Rafale*.

Fifth Generation

The fifth-generation airplanes began development in the 1990s and are currently in service. They incorporate advanced stealth technology, advanced sensor fusion, advanced avionics, and highly networked systems. These aircraft are designed to operate in highly contested environments, provide superior situational awareness, and possess enhanced capabilities for air-to-air and air-to-ground mis-

The Fifth Generation, *Lockheed Martin F-22 Raptor*. The unique combination of stealth, speed, agility, and situational awareness, combined with lethal long-range air-to-air and air-to-ground weaponry, makes the F-22 the best air dominance fighter in the world.

sions. They have advanced radar systems, high-performance engines, and significant network-centric warfare capabilities.

Notable fifth-generation airplanes include the *Lockheed Martin F-22 Raptor (United States), Lockheed Martin F-35 Lightning II (United States), Chengdu J-20 (China), Sukhoi Su-57 (Russia)*. India has been working on its own fifth-generation aircraft program called the *Advanced Medium Combat Aircraft (AMCA)*.

An *F-16 Fighting Falcon* (leading), *P-51D Mustang* (bottom), *F-86 Sabre* (top), and *F-22 Raptor* (trailing) fly in a formation representing four generations of American combat aircraft.

Sixth Generation

Furthermore, discussions around *sixth-generation airplanes* are already underway, with expectations of incorporating technologies like directed energy weapons, hypersonic speed, dual-mode engines, adaptive shapes and improved autonomy. Several countries and aerospace companies are actively researching and working on concepts and technologies that could potentially be incorporated into future sixth-generation airplanes replacing the pilot in an autonomous or semi-autonomous command aircraft. A few examples of initiatives and concepts being explored for sixth-generation airplanes are given below.

United States (*Next Generation Air Dominance - NGAD*) Program: The United States Air Force is actively pursuing the Next Generation Air Dominance program, which aims to develop a future air superiority platform. While specific details are classified, it is expected to incorporate advanced capabilities such as improved stealth, enhanced situational awareness, directed energy weapons, and increased autonomous capabilities.

United Kingdom (Tempest): The *Tempest* program is a joint effort by the United Kingdom and industry partners to develop next-generation combat airplanes. It aims to integrate advanced technologies like artificial intelligence, unmanned systems, and networked capabilities. The Tempest is envisioned as a highly adaptable, stealthy, and versatile platform.

The classification of generations helps provide a historical context and a general understanding of the technological advancements and capabilities of military airplanes throughout history. The more general term used for fifth and lately developed airplanes is next-generation airplanes.

6.1.2 Next Generation Military Airplanes

Next-generation military airplanes refer to the latest advancements in aircraft technology and capabilities that are being developed to replace or augment

6.1. INTRODUCTION

Table 6.1: Five generations of fighter aircraft

Type	Features	Examples
First generation (1940s–1950s)	High subsonic/Transonic, swept wing, conventional armaments such as machine guns or cannons, unguided bombs and rockets. Mostly interceptors	F-86 Sabre and MiG-15, Me 262, DH Vampire, P-80
Second Generation (1950s-1960s)	Early supersonic, advances afterburner engine design, air-to-air radar-guided missiles, interceptors (e.g. MiG-21F, SU-9, F-106) and fighter-bombers (e.g. F-105, SU-7), air-to-air combat within visual range	F-100, MiG-19, F-104, MiG-21, Mirage III
Third Generation (1960s - 1970)	Multirole fighters capable of performing air defense and ground attack missions. air-to-air interception beyond visual range, turbofan engines with thrust vectoring, advanced avionics	F-4 Phantom II, F-5, MiG-23/25, Sukhoi Su-15/17/20/22, Tupolev Tu-28P, Dassault Mirage F.1, Dassault Super Etendard, Shenyang J-8II
Fourth Generation (1970 - 1990)	Supersonic aerodynamics, especially area ruling; radar for search and fire control, Fly-by-wire control systems, multirole aircraft with the ability to switch and swing roles between air-to-air and air-to-ground, Use of composite materials	F-14, F-15 Eagle, F-16 Falcon, F-18 Hornet, MiG-29/31, Su-27, Yak-38, Mirage 2000, Saab Viggen, J-10, HAL LCA
4.5 Generation (1990 - 2000)	Enhanced capabilities, advanced digital avionics based on microchip technology and highly integrated systems, advanced avionics and super maneuverability, stealth, radar absorbent materials, thrust vector controlled engines, greater weapons carriage capacity and extended range and endurance.	F-18 Super Hornet, Sukhoi Su-30/33/35, Typhoon, Saab Gripen, Rafale
Fifth Generation (2000 - present)	Stealth, low-probability-of-intercept radar (LPIR), agile airframes with supercruise performance, advanced avionics features, and highly integrated computer systems capable of networking with other elements within the battle space for situation awareness and C3 (command, control and communications) capabilities.	Lockheed Martin F-22 Raptor, Lockheed Martin F-35, Sukhoi Su-57, J- 20

existing platforms. They incorporate cutting-edge technologies, enhanced performance characteristics, and improved mission capabilities. *Next Generation Military Airplanes* feature advanced stealth technology, integrate advanced avionics and sensor systems, possess enhanced performance characteristics, incorporating autonomous capabilities, are designed to be part of network-centric warfare systems, have the capability to conduct long-range strikes with precision-guided munitions, enabling them to engage targets from extended distances accurately. However, the specific features and capabilities of next-generation military airplanes will depend on the requirements and objectives of the countries and organizations developing them, as well as advancements in technology and evolving threats.

F-35 Lightning II (Joint Strike Fighter) is a fifth-generation multi-role fighter airplane developed by *Lockheed Martin Corporation*. It features advanced stealth capabilities, advanced avionics, sensor fusion, and network-centric warfare capabilities. The *F-35* is designed to replace various aging fighter airplanes in the inventories of the United States and its allies.

Su-57, Prospective Airborne Complex of Frontline Aviation(PAK FA) is a fifth-generation fighter jet developed by Russia's *Sukhoi Company*. It incorporates stealth technology, advanced avionics, supercruise capability, and advanced sensor systems. The *Su-57* is designed to provide air superiority and perform a range of missions in complex environments.

Chengdu J-20 is a fifth-generation stealth fighter airplane developed by China's *Chengdu Aerospace Corporation*. It features advanced stealth characteristics, integrated avionics, and a long-range air-to-air and air-to-surface strike capability. The *J-20* is designed to enhance China's air superiority and strike capabilities.

Northrop Grumman B-21 Raider is an American next-generation strategic bomber under development for the United States Air Force (USAF) by *Northrop Grumman Corporation*. It aims to provide long-range strike capabilities with advanced stealth characteristics, improved range, and payload capacity. The *B-21* is intended to replace the aging *B-1B* Lancer and *B-2 Spirit* bombers.

Shenyang FC-31 Gyrfalcon, developed by China's

Shenyang Aircraft Corporation, is a fifth-generation stealthy fighter airplane. It features advanced avionics, reduced radar cross-section, and improved stealth characteristics. The *FC-31* is designed for both air-to-air and air-to-surface missions and is intended to complement China's existing *J-20* stealth fighter.

FCAS (Future Combat Air System) is a collaborative effort between France, Germany, and Spain to develop a next-generation combat airplane system development by *Dassault Aviation*, *Airbus* and *Indra Sistemas*. It aims to deliver a combination of manned and unmanned aircraft, as well as advanced sensors and weapons, to achieve air superiority and conduct a wide range of missions in future conflicts. The *FCAS* will consist of a Next-Generation Weapon System (NGWS) as well as other air assets in the future operational battlespace.

FCAS (Future Combat Air System) at Paris Air Show 2023

HAL Advanced Medium Combat Aircraft (AMCA) is an indigenous program by India's Hindustan Aeronautics Limited (HAL) to develop a fifth-generation stealth multirole fighter airplane. It is expected to incorporate advanced stealth features, supercruise capability, advanced avionics, and advanced weapons systems.

TAI TF-X is a next-generation indigenous fighter jet program led by Turkey's Turkish Aerospace Industries (TAI). The TF-X aims to develop a twin-engine, fifth-generation fighter airplane with advanced stealth capabilities, advanced avionics, and high maneuverability. It is intended to replace Turkey's aging fleet of F-16s.

Mitsubishi X-2 Shinshin, also known as the *ATD-X*, is a technology demonstrator developed by Japan's Mitsubishi Heavy Industries. It serves as a testbed for Japan's future indigenous stealth fighter program. The X-2 incorporates advanced stealth technology.

6.2 Capabilities and Performance Parameters

The performance of an airplane refers to its ability to accomplish certain objectives during its operation, making it useful for certain tasks. Aircraft that can land and take off in a very short distance, for instance, are important to pilots who operate on short, unimproved airfields. For transport airplanes to perform well, they must be able to carry heavy loads, fly at high altitudes at high speeds, and/or travel long distances. At equilibrium condition (or trimmed condition) the forces and the moments acting on the airplane are balanced and the airplane is not accelerating. During steady, level flight, the lift produced by the airplane balances its weight and the thrust produced by the engine(s) balances the drag. Therefore, the aerodynamic drag defines the thrust required to maintain steady, level flight. More details of aerodynamic forces are covered in Chapter 4. Many important performance parameters like maximum speed, minimum speed, range, endurance, absolute ceiling, service ceiling and climb characteristics can be estimated for a given airplane and engine configuration by assuming the airplane is flying in equilibrium condition.

6.2.1 Maximum and Minimum Speeds

The *maximum speed* of an airplane at trimmed condition is obtained when the power/thrust required to fly at the given speed at the given altitude equals the maximum power/thrust available from the powerplant. Many military airplanes use afterburners for a short duration to enhance their maximum speed. Some airplanes are capable of supercruise which is the ability to cruise at supersonic speeds without the need for use of afterburner. For a given airplane, its maximum speed is determined by both the aerodynamic and engine characteristics. For a given engine setting and altitude, there is a lower limit on the speed with which the airplane can fly in a steady, level flight. The *minimum speed* of an airplane is determined by this or the stall velocity, whichever is higher.

6.2.2 Range and Endurance

The range and the endurance are two other important performance parameters. Range deals with flying distance and endurance involves consideration of flying time. The *specific endurance* is the ratio of the flight hours to the kg of fuel consumed. The *specific range* is the ratio of the distance traveled to the kg of fuel consumed. For a maximum specific range, the flight condition must provide a maximum of speed per fuel flow. The total range is dependent on both the fuel available and the specific range. When range and economy of operation are the principal goals, the pilot must ensure that the airplane is operated at the recommended long-range cruise condition. There are certain terms associated with the range in military applications. Thus, the total range is the maximum distance an airplane can fly between takeoff and landing, as limited by fuel capacity in a powered airplane. *Ferry range* is the maximum distance an airplane can fly with maximum fuel and without any payload. *Combat range* is the range calculated for the airplane with ordinance loaded. *Combat radius* is a related measure based on the maximum distance a warplane can travel from its base of operations, accomplish some objective, and return to its original airfield with minimal reserves. The limitations imposed by range and endurance can be augmented by mid-air refueling.

6.2.3 Climb Performance

Climbing is when an airplane gains potential energy by increasing its altitude which is required for (1) the airplane to avoid hitting obstacles by flying over them and (2) flying in better weather that enhances the fuel economy. This can be achieved either using excess power above that required to maintain level flight or converting kinetic energy (airspeed) to altitude. In the first scene, the airplane climbs while maintaining its speed. However, if there is no excess power available in the powerplant, the altitude is gained by decreasing the speed. *Maximum Angle of Climb (AOC)* provides climb performance to ensure an airplane clears obstacles, useful in a short runway or a short airfield surrounded by high obstacles, such as trees or power lines. AOC is the inclination (angle) of the flight path to the local horizontal. Maximum AOC occurs at the airspeed and angle of attack (AOA) combination which allows the maximum excess thrust. *Maximum Rate of Climb (ROC)* provides climb performance to achieve the greatest altitude gain over time. Maximum ROC occurs at an airspeed and angle of attack combination that produces the maximum excess power.

The ROC and AOC measure the altitude gained in relation to the time and distance traveled to reach that altitude, respectively. Both maximum ROC and AOC profiles use the airplane's maximum throttle setting but the difference lies in the velocity and angle of attack combination the airplane manual specifies. Climb performance is directly dependent upon the ability of the engine to produce either excess thrust or excess power. Altitude is one of the important parameters that influence the Climb Performance. An increase in altitude decreases the power available from the engine(s). Hence, the climb performance of an airplane diminishes with altitude. At a certain altitude called *the absolute ceiling*, there is no excess of power and only one speed allows steady, level flight and the airplane has zero ROC. There is a safe altitude called the *service ceiling* above which the maximum ROC drops below a specific value.

During takeoff, landing and turning, the velocity of the airplane is continuously changing and hence these performance characteristics are analyzed assuming the airplane is accelerating. These performance parameters also form part of the characteristics that are evaluated to decide the capabilities of an airplane.

6.2.4 Turning Performance

Maximum rate turns occurs when the airplane is changing direction at the highest possible rate, i.e., maximum degrees turned through in minimum time. To turn at maximum rate we need maximum centripetal force and that is contributed by a component of lift in a conventional configuration airplane. In some aircraft, vectoring the thrust is also used for making sharper turns. Hence, for maximum turn rate, the airplane flies at $C_{L,max}$. The increased angle of attack means increased drag, so full power is used. As the rate of turn is proportional to velocity, the limiting factor in a maximum rate turn is power. *A minimum radius* turn achieves a change of direction using less space and is usually done at a lower speed. The turning performance of an airplane is limited by the maximum load factor. The load factor is the ratio of lift produced by the airplane to its weight. The maximum load factor is determined based on the structural strength of the airplane to withstand the loads. Turning characteristics are important performance parameters for military combat airplanes as it is directly related to the maneuverability of the airplane.

6.2.5 Takeoff and Landing

The takeoff and landing performance of an airplane, a condition of accelerated motion, is an important aspect as the majority of pilot-caused accidents occur during the take-off and landing phases. To improve flight safety, the pilot must be familiar with all the variables that influence the takeoff and landing performance. The takeoff or landing speed is generally a function of the stall speed or minimum flying speed. Runway conditions affect takeoff and landing performance. The condition of the surface affects the braking ability of the airplane. The amount of power that is applied to the brakes without skidding the tires is referred to as braking effectiveness. Water on the runways reduces the friction between the tires and the ground and can reduce braking effectiveness. The minimum takeoff distance is of primary interest in the operation of any airplane because it defines the runway requirements. The minimum takeoff distance is obtained by taking off at some minimum safe speed that allows sufficient margin above stall and provides satisfactory control and initial ROC. To obtain minimum takeoff distance at the specific lift-off speed, the forces that act on the airplane must provide the maximum acceleration during the takeoff roll. Vertical takeoff and landing (VTOL) refers to taking off and landing vertically thereby avoiding the requirement of a long runway. This is achieved by orienting the direction of thrust vertically downwards and thereby making VTOL possible. Military airplanes like *F-35* and *Harrier* are capable of VTOL.

Carrier Takeoff and Landing: One of the most interesting features of some military jets is their ability to takeoff and land on an extremely short runway aboard an aircraft carrier. The landing/takeoff distance available on aircraft carriers is only a few hundred meters compared to the kilometers available at conventional airstrips. Takeoff from an aircraft carrier is typically achieved by the use of a catapult mechanism. While landing on an aircraft carrier, after the clearance to land from Air Traffic Control is granted, the pilot begins the final approach ensuring that the angle of descent is correct and the jet is lined up with the runway. The pilot gets aid from Landing Signal Officers (LSOs) and the onboard Fresnel Lens Optical System. LSOs use a combination of radio communication and lights on the deck to either confirm the preconditions for a landing or tell the pilot to abort and try again. Approaching landing, the pilot lowers the tailhook that engages the wires on deck made by weaving together high-tensile steel. Multiple rows of wires are placed in order to increase the size of the touchdown area. During the touchdown, the pilot is expected to give full power to the engines so that they can take off again if the tailhook fails to catch a wire.

6.2.6 Stealth Technology

Stealth refers to the capability to be less observable to the detection methods used by the enemy. With the missiles used in the air defense systems capable of flying faster and higher and being more maneuverable than military combat airplanes, designing for stealth may be the viable option for the survival of a military airplane. Hence, one of the most desirable features of a modern warplane is *Stealth*. Since radars are the most common way of tracking airplanes, having a reduced *Radar Cross-Section (RCS)* is one of the design objectives of advanced military airplanes. RCS or radar signature is a measure of how detectable an object is by radar. A larger RCS indicates that an object is more easily detected. Therefore, a stealth aircraft, which is designed to have low detectability, will have design features that give it a low RCS such as absorbent paint, flat surfaces and surfaces specifically angled to reflect the signal somewhere other than toward the source.

Since the 1960s, investigation on integrated air defense systems started where ground- and air-based radars were tied into command-and-control systems, which in turn could give orders to surface-to-air missile batteries and air bases with fighter jets ready to take off. As a result, offensive air power was forced to innovate new tactics–such as airborne command-and-control, electronic warfare, air-defense suppression, and more–all to ensure that relatively few airplanes were able to pierce the defenses and reach their targets. The foundation of the air defense system was *Radar* which can detect airplanes from hundreds of km away by sending out streams of radio waves, and intercepting them when they return and gives information such as relative size, speed, altitude, and heading. *Radio waves* that strike objects in their path will be reflected back, giving defenders a warning that intruders are en route. The implications of understanding how to build radar-evading planes were the crux of stealth technology.

Stealth is a key aspect of modern airplanes involving careful design to minimize the effectiveness of enemy radar, hence reducing one's detectability by radar. The *F-117A Nighthawk* and the *B-2 Spirit* stealth bomber are a few examples. The *F-22 Raptor* was the first stealthy fighter jet capable of air-to-air combat. More recent advanced fifth generation airplane with stealth technology are *F-35 Lightning II*, *J-20*, *Su-57* and *B-21 Raider*.

6.2.7 Airborne Armaments

Military airplanes carry weapons for offensive attacks or for defense against enemy aircraft and ground targets. Bombs, missiles, torpedoes, mines, and other stores are suspended internally or externally from the airplane by bomb racks, which carry, arm and release stores. Airborne Armament Equipment is made up of items that support the expending or release of ordnance from airplanes such as bomb racks and launchers. Bomb racks are aircraft armament equipment items that provide for the suspension, carriage, and release of ordnance items from the aircraft. Ejector racks serve the same purpose as bomb racks but differ in that they use electrically fired impulse cartridges to eject the weapon/stores free of the bomb rack. Guided missile launchers provide for the carriage and release of guided missiles from an airplane. They provide the mechanical and electrical interface between the airplane and the air-launched missile. Early warplanes useed to carry either a small-caliber weapon capable of a high number of strikes per second or a larger, slower-firing weapon capable of high penetration and destruction. As technology advanced, airplanes often began to be outfitted with both.

Alpha Jet, light attack and advanced trainer airplane manufactured by *Dassault Aviation* and *Dornier Flugzeugwerke*, taking off at Paris Air Show 2023. The military cargo *Airbus A400M Atlas* is in the background.

6.3 Roles of Military Airplanes

Military airplanes are used for a variety of purposes and have a lot of designs based on their utility. Here, we look at some of the significant roles of military airplanes.

6.3.1 Aerobatics

Aerobatic planes are specifically and carefully engineered for complicated aerial maneuvers in aerobatic competitions as an entertaining event in airshows worldwide. These airplanes are designed to have exceptional maneuverability, control response, and structural strength to withstand the stresses of high-G maneuvers and dynamic flight routines. To be specific, aerobatic planes are often equipped with systems of devices such as *inverted fuel and oil systems* (fuel injection instead of a carburetor), symmetrical wings (helps in inverted flight and stunt planes), aerobatic propellers (either a fixed-pitch or constant-speed propeller in which the blade pitch is adjusted by a governor) and aileron spades. *Aileron Spades* are shovel-shaped surfaces, rigidly mounted on arms forward of the ailerons, providing aerodynamic balance, and reducing the effort needed to roll the plane. Aerobatic planes are used in aerobatic competitions, flight demonstrations in airshows, pilot training, experimental and aircraft de-

French Air and Space Force Aerobatics Display Team flying *Alpha Jets* at Paris Air Show 2023.

velopment and promotion of aviation. Some of the famous aerobatic planes and the teams are given below.

Patrouille de France of the *Armée de l'Air*(French Air and Space Force), the oldest and considered one of the best precision aerobatics demonstration units in the world, has its historical beginnings in 1931 and officially commissioned in 1953. Currently, it is comprised of 9 pilots and 35 mechanics and 10 *Alpha Jets*. The *Alpha Jet* airplanes are painted in blue, white and red, the national colors of France, and carry *Armee de l'air* inscriptions under the wing. The team uses eight airplanes that all fly at the same

The Blue Angels F/A-18 Hornets fly in a tight diamond formation.

Two *Thunderbirds* perform a calypso pass. (image:https://dod.defense.gov)

The RAF Red Arrows.

time. A ninth plane is held in reserve, but ready to take off if needed.

U.S. Navy Blue Angels is a flight demonstration squadron of the United States Navy. This is the second oldest formal aerobatic team in the world. The Navy Flight Exhibition Team performed its first flight demonstration on June 15, 1946, at their home base, Naval Air Station in Florida, flying the *Grumman F6F-5 Hellcat*. The team was introduced as the *Blue Angels* at a show in Nebraska, in July of 1946. Today, the team, composed of six Navy and one Marine Corps demonstration pilot, flies Boeing *F/A-18 Super Hornets*.

Thunderbirds is the air demonstration squadron of the United States Air Force (USAF), created in 1953 as the third-oldest formal flying aerobatic team in the world. The team consists of 12 officers, 132 enlisted support personnel and 3 civilian support personnel. It has flown a variety of jets during its history, including the *F-84G Thunderjet, F-84F Thunderstreak, F-100 Super Sabre*, and the *F-4E Phantom II*. Today, the *Thunderbirds* fly the *F-16 Fighting Falcon*.

Royal Air Force Red Arrows, the RAF aerobatic team, is one of the world's premiere demonstration units. The team was formed in late 1964 as an all-RAF team, replacing a number of unofficial teams that had been sponsored by RAF commands. The *Red Arrows* have flown almost 5,000 presentations in 57 countries. *Red Arrows* consists of 11 pilots 100 engineering & support staff. Initially, they were equipped with seven *Folland Gnat trainers* inherited from the RAF Yellowjacks display team. In late 1979, they switched to the *BAE Hawk trainer*.

Surya Kiran is the Indian Air Force aerobatics display squad founded in 1996 and is a member of the IAF's 52nd Squadron. The crew of 14 Pilots with 20 airplanes has conducted several demonstrations, generally with nine planes. Until 2011, the squadron was made up of *HAL HJT-16 Kiran Mk.2* military training airplane. In February 2011, the team was suspended but was re-established in 2017 using *Hawk Mk-132 airplane*.

Frecce Tricolori aerobatic demonstration team of the Italian Air Force, created in 1961 as a permanent group for the training of Air Force pilots in air acrobatics, consists of 13 Italian-made *Aermacchi MB-339-A/PAN* two-seat fighter-trainer jet, only 10 of which take part in airshows. With ten air-

A *BAE Hawk Mk.132* of *Surya Kiran* flying in Aeroindia 2023.

planes, nine in close formation and a soloist, they are the world's largest acrobatics patrol, and their flight schedule, comprising about twenty acrobatics and about half an hour, made them the most famous in the world.

Patrouille Tranchant is a French civilian aerobatic team, based in Rennes. It was created in 2006 by two aviation enthusiasts, *Benjamin Tranchant* and *Hugues Duval*, and is the only patrol flying the *Fouga CM-170 Magister*, a 1950s French two-seat jet trainer airplane, developed and manufactured by French aircraft manufacturer *Fouga*. Demonstrations are typically performed using four airplanes.

Other famous aerobatic team includes *Patrouille Suisse* Swiss Demonstration Team, *Snowbirds* of the Royal Canadian Air Force, *Patrulla Águila* of the Spanish Air Force and *46 Aviation*. The *Breitling Jet Team* of France flies seven *Aero L-39 Albatros* jets and is the largest civilian aerobatic team in Europe.

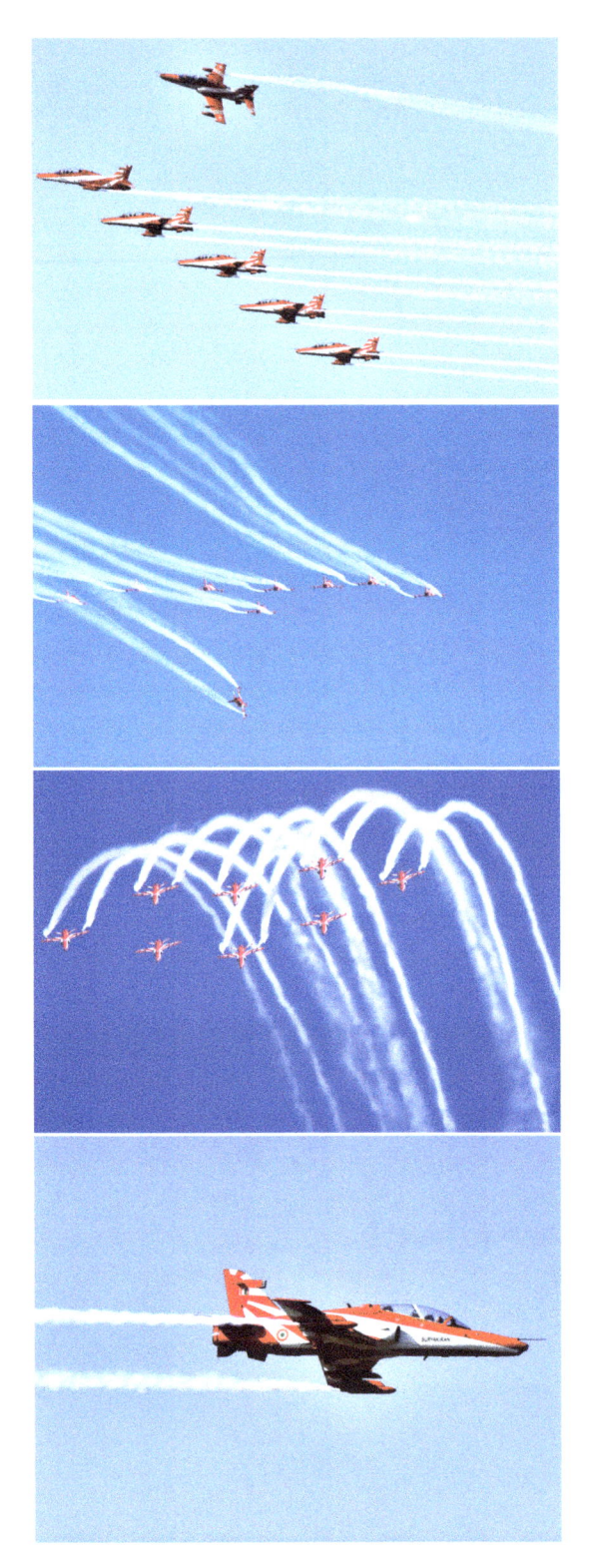

The *Surya Kiran Aerobatic Team* flying *BAE Hawk Mk.132* at *Aeroindia 2023*.

6.3. ROLES OF MILITARY AIRPLANES

Close formation of *Frecce Tricolori* at the Royal International Air Tattoo held in Fairford in 2005.

Flight demonstration La patrouille Tranchant at Paris Air show 2023

Patrouille Tranchant, a French civilian aerobatic team flying the *Fouga Magister*, a 1950s French two-seat jet trainer airplane.

The *46 Aviation*, Aerobatics Display Team perform *wing walking* in its the 1940's *Boeing Stearman*, at Paris Air Show 2023.

6.3.2 Attack

An attack aircraft or an attacker is designed with specific features and capabilities to effectively engage ground targets and provide close air support. Some key features of attack airplanes are

Robust Airframe: Attack airplanes are built with rugged airframes capable of withstanding the stresses of low-level flying and potential enemy fire. They often incorporate features such as reinforced structures, durable landing gear, and additional armor to protect critical components and the pilot.

Weaponry: Attack airplanes are equipped with a wide range of weapons to engage ground targets effectively. These include bombs, missiles, rockets, and autocannons. The airplane's design allows for the carriage of various ordnance configurations, enabling the pilot to select the appropriate weapons for specific mission requirements.

Precision Targeting Systems: Attack airplanes are equipped with advanced avionics and targeting systems to accurately locate and engage ground targets. This includes radar systems, electro-optical sensors, and laser targeting pods that provide situational awareness, target acquisition, and the ability to deliver precision-guided munitions.

Close Air Support Capabilities: Attack airplanes excel at providing close air support to friendly ground forces. They have the ability to operate at low altitudes and slow speeds, allowing pilots to visually identify targets and engage them with high precision. Attack airplane pilots work closely with ground troops to ensure effective coordination and minimize the risk of friendly fire.

Survivability Features: Attack airplanes are designed with survivability in mind. They often feature redundant systems, advanced electronic warfare suites, and defensive countermeasures to protect against threats such as anti-aircraft missiles and enemy fighters. Some attack airplanes also have additional armor plating to protect critical components and the pilot.

Long Range and Endurance: Many attack airplanes have extended range and endurance capabilities, allowing them to operate over vast distances and remain on station for extended periods. This ensures they can provide sustained support to ground forces and engage targets effectively over large areas.

Low-Speed Maneuverability: Attack airplanes are designed to operate at low speeds, enabling them to fly at the same pace as friendly ground forces. This facilitates accurate target identification and engagement, as well as close coordination with ground units.

Fairchild Republic A-10 Thunderbolt II	
Specifications	**A-10C**
Maximum speed	706 km/h,
Ferry/Combat range	4,150/463 km
Wing loading	482 kg/m^2
Thrust/weight	0.36
Rate of climb	30 m/s
Service ceiling	13,700 m
Maximum payload	7.2 tons (weapons)
MTOW/Fuel capacity	20.8/5 tons
Wingspan	17.53 m
Engines	2 × General Electric TF34-GE-100A turbofans, 40.32 kN thrust each
Armament (Guns)	1 × 30 mm GAU-8/A Avenger rotary cannon
Rockets	4 × LAU-61/LAU-68, 6× LAU-131 rocket pods
Missiles	2 × AIM-9 Sidewinder air-to-air missiles, 6 × AGM-65 Maverick air-to-surface missiles
Bombs	Mk 77 incendiary bombs or Mark 80 series of unguided iron bombs, Paveway series of Laser-guided bombs
Over 716 A-10s were built during 1972 – 1984	

These airplanes are optimized for close air support (CAS), interdiction, and precision strikes against enemy forces and infrastructure. Attack airplanes are typically characterized by their robust airframe construction, powerful engines, and heavy weapon-carrying capabilities. They are designed to operate at low altitudes, often flying close to the battlefield, and are equipped with advanced avionics and weapons systems to accurately engage ground targets.

Some examples that represent a range of attack military airplanes designed for engaging ground targets, providing close air support, and conducting precision strikes are *Sukhoi Su-25, Chengdu J-10, Sukhoi Su-34, Embraer EMB 314 Super Tucano, Ilyushin Il-2 Shturmovik, Douglas AC-47 Spooky* or *Lockheed AC-130*.

The *Fairchild-Republic A-10 Thunderbolt II*, a single-seat, twin-turbofan, straight-wing, subsonic attack airplane, is arguably one of the most famous and iconic attack airplanes in the world. It has gained widespread recognition and popularity due to its unique design, exceptional close air support capabilities, and its role in various conflicts. Its distinctive features, such as the prominent GAU-8 Avenger 30mm Gatling gun mounted in the nose, heavily armored cockpit, and high-survivability design, have made it a favorite among military enthusiasts and the public alike. The *A-10* has earned a reputation for its remarkable ability to loiter over the battlefield, deliver precise strikes against armored vehicles, and provide close air support to ground troops. Its durability, maneuverability, and dedicated ground attack capabilities have made it an indispensable asset for the United States Air Force. Additionally, the *A-10*'s participation in various conflicts, including the Gulf War and the Balkan conflicts has further cemented its fame and reputation as an exceptional attack airplane. Its effectiveness and reliability in these operations have solidified its status as one of the most renowned attack airplanes in the world.

The *Sukhoi Su-34* is a Russian twin-engine, twin-seat, all-weather supersonic fighter-bomber designed for tactical bombing and strike missions.

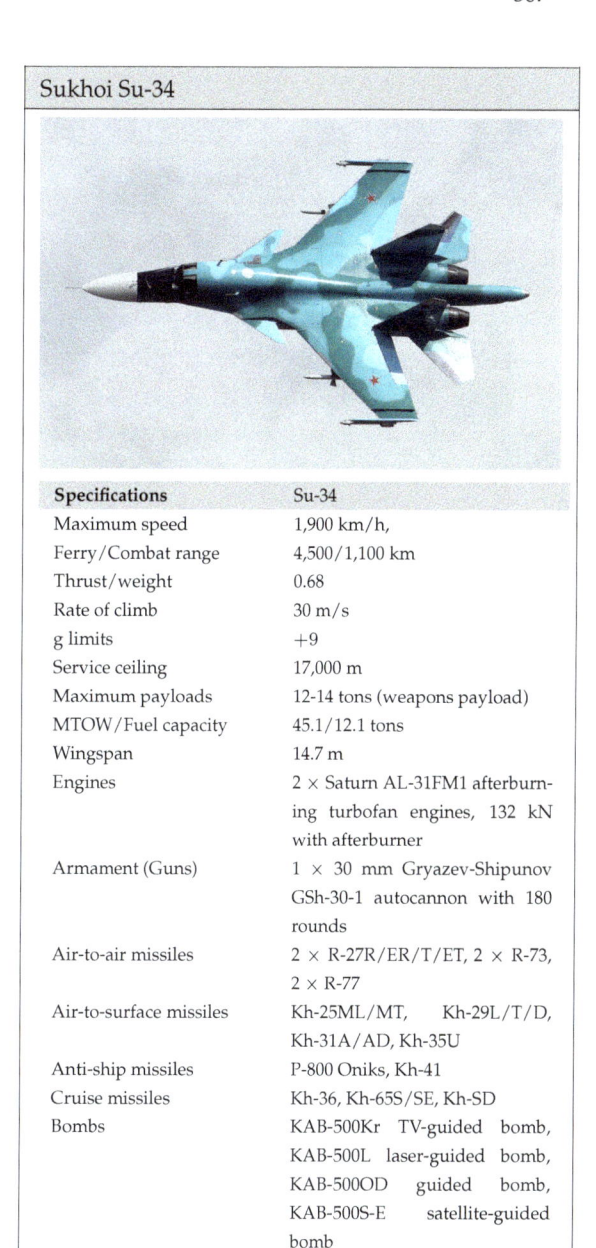

Sukhoi Su-34

Specifications	Su-34
Maximum speed	1,900 km/h,
Ferry/Combat range	4,500/1,100 km
Thrust/weight	0.68
Rate of climb	30 m/s
g limits	+9
Service ceiling	17,000 m
Maximum payloads	12-14 tons (weapons payload)
MTOW/Fuel capacity	45.1/12.1 tons
Wingspan	14.7 m
Engines	2 × Saturn AL-31FM1 afterburning turbofan engines, 132 kN with afterburner
Armament (Guns)	1 × 30 mm Gryazev-Shipunov GSh-30-1 autocannon with 180 rounds
Air-to-air missiles	2 × R-27R/ER/T/ET, 2 × R-73, 2 × R-77
Air-to-surface missiles	Kh-25ML/MT, Kh-29L/T/D, Kh-31A/AD, Kh-35U
Anti-ship missiles	P-800 Oniks, Kh-41
Cruise missiles	Kh-36, Kh-65S/SE, Kh-SD
Bombs	KAB-500Kr TV-guided bomb, KAB-500L laser-guided bomb, KAB-500OD guided bomb, KAB-500S-E satellite-guided bomb

Over 155 Su-34s have been built since their entry into service in 2014

One of the distinctive characteristics of the *Su-34* is its side-by-side seating arrangement for the pilot and co-pilot/navigator. This configuration provides excellent communication and coordination between the crew members during missions. The cockpit is equipped with advanced avionics, including modern displays, navigation systems, and weapon control systems, ensuring superior situational awareness and operational effectiveness. The *Su-34* has a sleek aerodynamic design, with swept wings and canards located in front of the cockpit. These canards enhance the airplane's maneuverability and stability, especially at high angles of attack. It is powered by two afterburning turbofan engines, which enable it to achieve supersonic speeds and operate at high altitudes. One of the key features of the *Su-34* is its extensive weapon-carrying capacity. It has a large internal weapons bay capable of housing various air-to-surface missiles, bombs, and guided munitions. Additionally, it has multiple hardpoints on its wings and fuselage, allowing for the attachment of additional ordnance, including air-to-air missiles for self-defense. The *Su-34* is equipped with an advanced radar system and electronic warfare suite, enabling it to detect and engage targets in complex environments. It also has a robust defensive countermeasures system to protect against enemy threats, including radar-guided missiles and anti-aircraft artillery.

The *Embraer EMB 314 Super Tucano* is a highly versatile and widely acclaimed light attack and advanced training aircraft manufactured by the Brazilian aerospace company *Embraer*. It is specifically designed for counterinsurgency, close air support, and aerial reconnaissance missions. The *Super Tucano* has gained a reputation as a reliable and cost-effective platform for various military forces around the world. The *Super Tucano* features a rugged and aerodynamically efficient design, with a tandem-seat cockpit. The airplane's advanced avionics and systems enable it to operate effectively in both day and night missions, as well as in adverse weather conditions. It is equipped with various weapon stations, allowing for the integration of a wide range of armaments. It can carry air-to-air missiles, precision-guided munitions, unguided rockets, and machine guns, providing it with substantial firepower for combat missions. The airplane's accurate weapons delivery capabilities make it highly effective in engaging ground targets with great precision.

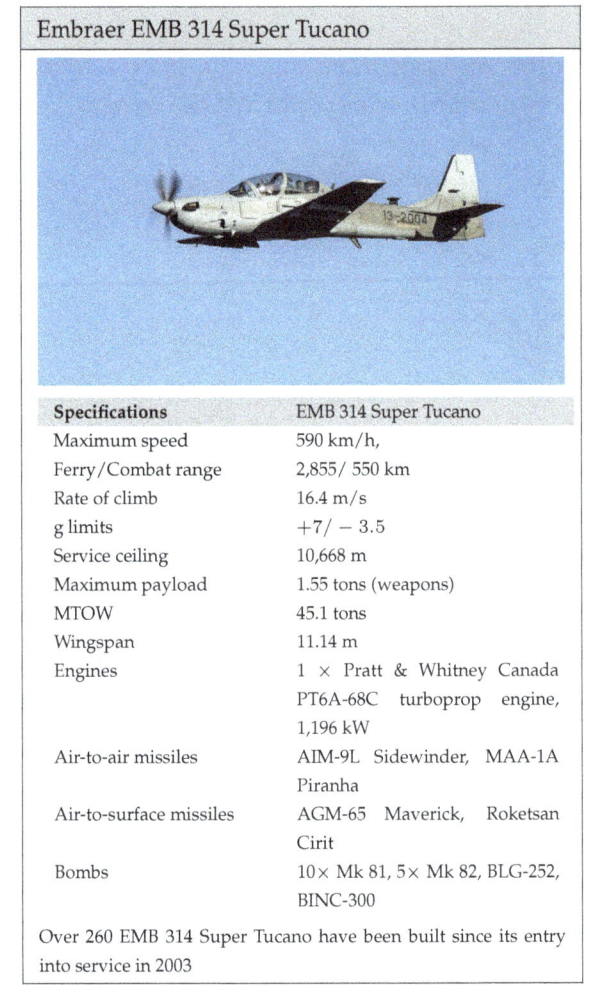

Embraer EMB 314 Super Tucano

Specifications	EMB 314 Super Tucano
Maximum speed	590 km/h,
Ferry/Combat range	2,855/ 550 km
Rate of climb	16.4 m/s
g limits	+7/ − 3.5
Service ceiling	10,668 m
Maximum payload	1.55 tons (weapons)
MTOW	45.1 tons
Wingspan	11.14 m
Engines	1 × Pratt & Whitney Canada PT6A-68C turboprop engine, 1,196 kW
Air-to-air missiles	AIM-9L Sidewinder, MAA-1A Piranha
Air-to-surface missiles	AGM-65 Maverick, Roketsan Cirit
Bombs	10× Mk 81, 5× Mk 82, BLG-252, BINC-300

Over 260 EMB 314 Super Tucano have been built since its entry into service in 2003

The *Super Tucano*'s versatility extends beyond combat operations. It can perform a variety of roles, including border patrol, intelligence gathering, and aerial reconnaissance. The airplane is equipped with advanced sensors, including an electro-optical/infrared (EO/IR) sensor and a laser designator, enabling it to conduct surveillance missions and provide real-time situational awareness.

The *Super Tucano* has achieved considerable success in the international market. It has been selected by various air forces worldwide, including those of Brazil, Colombia, Afghanistan, and the United States. The airplane's reliability, cost-effectiveness, and performance in challenging environments have made it a popular choice for countries seeking a capable light attack and training airplane.

The *Sukhoi Su-25* is a subsonic, single-seat, twin-engine, highly capable and heavily armored close air support (CAS) airplane developed by the Soviet Union and manufactured by *Sukhoi Design Bureau*. It is specifically designed to provide effective air support to ground forces in combat operations. The *Su-25* has a robust and durable airframe, built to withstand the rigors of combat and low-altitude operations. Its design includes a straight-wing configuration, twin engines, and a single-seat cockpit. The airplane's structure is reinforced with armor protection to enhance its survivability against enemy ground fire. It carries a wide array of weapons, including bombs, rockets, missiles, and cannons, allowing it to engage a variety of ground targets such as armored vehicles, bunkers, and enemy positions. The airplane's armament can be customized to suit specific mission requirements. The *Su-25* is renowned for its substantial armor protection, which enables it to operate in high-threat environments. The cockpit is heavily armored, offering the pilot increased survivability against ground fire. Additionally, vital airplane systems and critical areas are reinforced to withstand damage from enemy anti-aircraft weapons.

Avionics and Navigation Systems: The *Su-25* is equipped with advanced avionics and navigation systems to enhance situational awareness and mission effectiveness. It features a modern cockpit with multifunction displays, providing the pilot with critical flight information, sensor data, and weapon system control. The airplane's navigation systems enable it to operate accurately in all weather conditions, day or night.

Operational Flexibility: The *Su-25* has exceptional low-altitude maneuverability, allowing it to operate effectively in complex and challenging environments, including mountainous terrain. It can take off and land on unprepared runways, which enhances its operational flexibility and enables it to be deployed to remote areas with limited infrastructure.

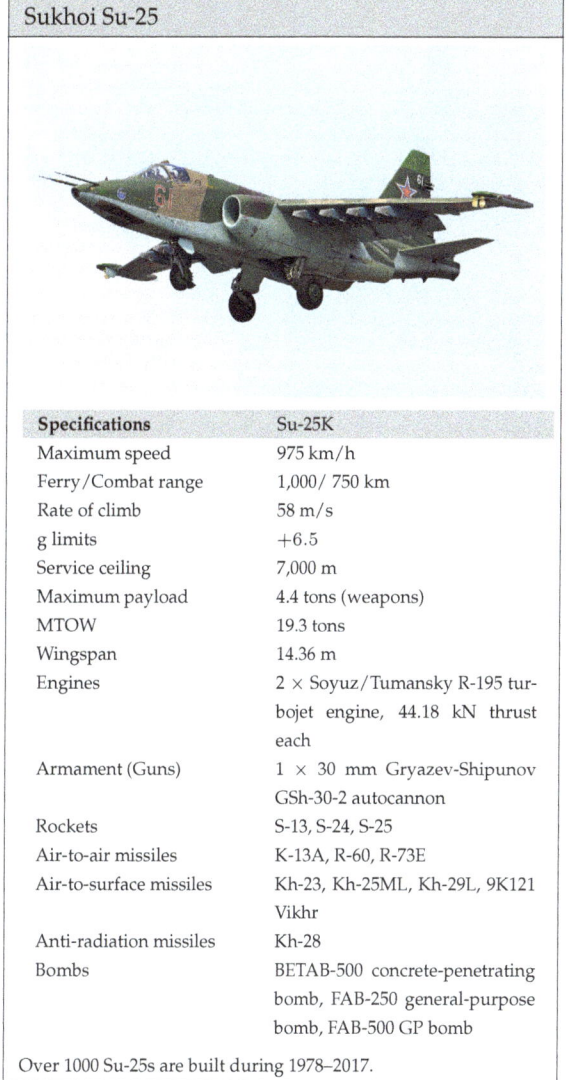

Sukhoi Su-25

Specifications	Su-25K
Maximum speed	975 km/h
Ferry/Combat range	1,000/ 750 km
Rate of climb	58 m/s
g limits	+6.5
Service ceiling	7,000 m
Maximum payload	4.4 tons (weapons)
MTOW	19.3 tons
Wingspan	14.36 m
Engines	2 × Soyuz/Tumansky R-195 turbojet engine, 44.18 kN thrust each
Armament (Guns)	1 × 30 mm Gryazev-Shipunov GSh-30-2 autocannon
Rockets	S-13, S-24, S-25
Air-to-air missiles	K-13A, R-60, R-73E
Air-to-surface missiles	Kh-23, Kh-25ML, Kh-29L, 9K121 Vikhr
Anti-radiation missiles	Kh-28
Bombs	BETAB-500 concrete-penetrating bomb, FAB-250 general-purpose bomb, FAB-500 GP bomb

Over 1000 Su-25s are built during 1978–2017.

International Deployment: The *Su-25* has been widely exported to various countries worldwide. It has been utilized in numerous conflicts and combat operations, including the Soviet-Afghan War, the Balkan conflicts, and the war in Syria. Its combat-proven performance and reliability have made it a trusted asset for air forces in need of a dedicated close air support platform.

6.3.3 Bomber

Bombers are a specific class of military airplanes primarily designed to carry and deliver large quantities of ordnance, including cruise missiles, bombs, and even torpedoes and other munitions to targets on the ground. They play a crucial role in strategic

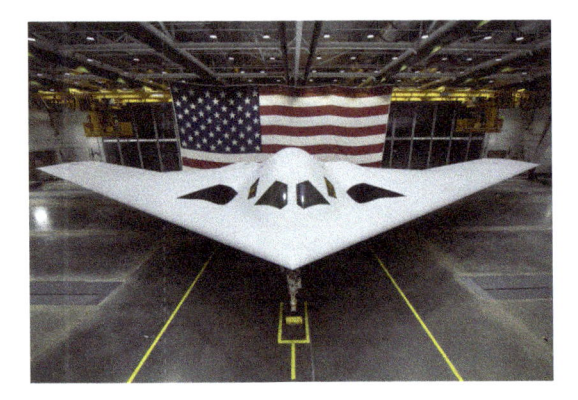

The *Northrop Grumman B-21 Raider* is an American strategic bomber.

The *Boeing B-52 Stratofortress*, an American long-range, subsonic, jet-powered strategic bomber

The *B-2 Spirit*, an American heavy strategic stealth bomber.

and tactical bombing missions, and their design and capabilities are optimized for long-range, high payload, and the ability to penetrate enemy defenses. Military bombers, mostly used for ground attacks, are neither fast nor agile and, therefore, they are unable to face the enemy head-on. Based on the type and size of the payload they carry, the military bombers can be light, medium, and heavy bombers, torpedo bombers, and dive bombers. Bombers are designed for long-range missions, enabling them to operate at extended distances from their bases. They have larger fuel capacities and fuel-efficient engines to achieve extended flight durations. Bombers often feature sleek, streamlined shapes, swept-back wings, and advanced aerodynamic features to minimize drag and maximize fuel efficiency. Some military bombers even have stealth capabilities, for instance, the *Northrop Grumman B-2 Spirit*. This makes them difficult to be detected by enemy radar. Other examples of bombers are the *B-52 Stratofortress* and the World War II *B-17 Flying Fortress*. Bombers may be equipped with electronic warfare systems, such as jammers or decoy dispensers, to disrupt or deceive enemy radar and missile systems. Many modern bombers have in-flight refueling capabilities, allowing them to receive fuel from tanker airplanes, thereby extending their range and mission endurance. Some bombers, such as the *Tupolev Tu-22M*, are equipped with advanced surveillance and reconnaissance systems,

enabling them to gather intelligence and provide situational awareness.

The United States *B-21 Raider* is a next-generation long-range strategic bomber developed by *Northrop Grumman Corporation* for the United States Air Force. It is intended to replace the aging *B-1* and *B-2* bombers. The *B-21 Raider* is expected to incorporate advanced stealth capabilities, long-range strike capabilities, and advanced avionics. *Boeing B-52 Stratofortress:* A long-range strategic bomber with a large payload capacity, known for its long service history and versatility. It has been in service since the 1950s and remains operational even today.

Northrop Grumman B-2 Spirit: A stealth bomber designed for penetrating deep into enemy territory undetected. Its advanced stealth technology and large payload capacity make it a formidable strategic bomber.

6.3. ROLES OF MILITARY AIRPLANES

The *Tupolev Tu-160*, a supersonic, variable-sweep wing heavy strategic bomber.

Rockwell B-1 Lancer, a supersonic variable-sweep wing, heavy bomber.

Tupolev Tu-160: A supersonic strategic bomber capable of carrying a variety of weapons, including cruise missiles. It is one of the largest and heaviest bombers ever built.

Rockwell B-1 Lancer: A long-range, multi-mission bomber featuring variable-sweep wings. It has the capability to operate at high speeds and low altitudes, making it effective for both strategic and tactical missions.

During *World War II*, several famous bombers played significant roles in the conflict. These bombers, among others, played critical roles in strategic bombing campaigns, contributed to the

Boeing B-17 Flying Fortress, a four-engined heavy bomber.

The *Consolidated B-24 Liberator*, an American heavy bomber.

North American B-25 Mitchell, an American bomber.

overall air superiority of their respective nations, and significantly impacted the course of *World War II*. Some notable examples are

Boeing B-17 Flying Fortress: The *B-17* was a four-engine heavy bomber used primarily by the United States Army Air Forces (USAAF). It gained a reputation for its resilience and ability to sustain heavy damage and still return to base. It was used extensively in daylight precision bombing raids over Europe.

Consolidated B-24 Liberator was another four-engine heavy bomber, employed by both the USAAF and the Royal Air Force (RAF). It had a longer range and higher top speed than the *B-17*, making it suitable for long-range bombing missions in the European and Pacific theaters.

North American B-25 Mitchell was a medium

The *Junkers Ju 88*, a German World War II *Luftwaffe* twin-engined multirole bomber.

Advanced fighter jets (from right-left– two *F-18 Hornet*s, two *F-35 Lightning* and one *F-16 Fighting Falcon*, on static display at Aeroindia 2023.

bomber widely used by the USAAF, primarily in the Pacific theater. It gained prominence for its role in the Doolittle Raid, a retaliatory bombing mission on Japan following the attack on Pearl Harbor.

Heinkel He 111 and *Junkers Ju 88* were medium bombers employed by the German *Luftwaffe*. *Heinkel He 111* played a significant role in the early stages of the war, particularly during the Blitz on London and the Battle of Britain. *Junkers Ju 88* served in various roles, including bombing, reconnaissance, and as a night fighter. It was known for its speed and maneuverability.

6.3.4 Fighter and Interceptor

We have seen attack airplanes in the previous section, the fighter airplanes can have some overlap in capabilities of attack airplanes. But, there are key differences between the two types. Fighter airplanes, air superiority fighters and interceptors, are primarily designed for air-to-air combat engagements and gaining control of the airspace. Fighter airplanes excel in speed, agility, and maneuverability to engage and defeat enemy airplanes. They are built for dogfighting, with the ability to perform tight turns, high-G maneuvers, and rapid acceleration.

Fighter airplanes are equipped with a variety of air-to-air missiles, guns, and advanced radar and avionics systems. They prioritize air-to-air combat capabilities, with superior radar systems for detecting and tracking enemy airplanes.

Fighter airplanes are typically smaller and lighter than attack airplanes, emphasizing speed and maneuverability over payload capacity. They carry a limited amount of ordnance, often focused on air-to-air missiles and a small number of air-to-ground munitions.

Differences between Attack and Fighter airplanes:

Primary Mission: Fighter airplanes are primarily designed for air-to-air combat and achieving air superiority. Attack airplanes are primarily focused on engaging ground targets and providing direct support to ground forces.

Payload and Armament: Fighter airplanes have a smaller payload capacity and are typically optimized for air-to-air missiles and air-to-air combat. Attack airplanes have a larger payload capacity and carry a wider range of ground attack weapons.

Design Priorities: Fighter airplanes prioritize speed, maneuverability, and advanced avionics for air-to-air engagements. Attack airplanes prioritize durability, protection, and versatility for ground attack missions. Some of the well-known military fighter planes include the *F-15 Eagle, Su-27* and the *Lockheed Martin F-22 Raptor*.

6.3.5 Reconnaissance

Reconnaissance airplanes are specialized airplanes designed for intelligence gathering and surveillance missions. They play a vital role in military operations, providing valuable information about en-

Sukhoi Su-27SKM fighter airplane.

The *Lockheed U-2 (Dragon Lady)*, an American single-engine, high-altitude reconnaissance airplane operated during the Cold War.

emy positions, movements, and terrain. Reconnaissance airplanes are equipped with advanced sensors, cameras, and communication systems to collect and relay real-time data to military commanders. Let's explore the key features and capabilities of reconnaissance airplanes.

Reconnaissance airplanes are equipped with a range of sensor systems to gather intelligence. These may include electro-optical/infrared (EO/IR) cameras, synthetic aperture radar (SAR), signals intelligence (SIGINT) equipment, and other specialized sensors. These sensors enable the airplanes to capture high-resolution imagery, detect enemy radar signals, and monitor communication networks.

Reconnaissance airplanes are designed for extended endurance to maximize their time on station. They typically have fuel-efficient engines and large fuel capacities, allowing them to remain airborne for prolonged periods and cover vast areas during surveillance missions.

Reconnaissance airplanes can operate at various altitudes depending on the mission requirements. They may fly at high altitudes to avoid detection or fly at lower altitudes for detailed surveillance. They are often equipped with systems that enable them to fly at low speeds, maintaining stability for accurate imaging and data collection.

Reconnaissance airplanes have advanced communication systems to transmit collected data in real time to command centers. This enables military commanders to receive immediate intelligence updates and make informed decisions based on the gathered information.

Advanced reconnaissance airplanes may incorporate stealth technology to minimize their radar cross-section and reduce the chances of detection. Stealth features enable the airplanes to operate covertly, gathering intelligence without alerting the enemy to their presence.

Lockheed U-2 was a high-altitude reconnaissance airplane known for its long-duration flights and ability to gather intelligence deep into enemy territory. It provided day and night, high altitude up to 21,300 meters, and all-weather intelligence gathering. The *U-2* has been used by the United States for several decades, including during the Cold War and recent conflicts. It remains in service today and is regarded as one of the most iconic and enduring spy planes in history.

Uninhabited Aerial Vehicles (UAVs)

In recent years, UAVs have become increasingly prevalent in the field of reconnaissance. These remotely piloted or autonomous airplanes offer advantages such as reduced risk to human personnel, extended endurance, and the ability to operate in environments hostile to manned airplanes.

General Atomics MQ-9 Reaper is a modern UAV utilized for reconnaissance and surveillance missions. The *MQ-9 Reaper* has long endurance, the ca-

General Atomics MQ-9 Reaper, an unmanned aerial vehicle capable of remotely controlled or autonomous flight operations.

A *Beechcraft RC-12N*, an airborne signals intelligence collection platform.

An *RQ-4 Global Hawk* UAV.

Flying boom refuelling – *F-15C Eagles* is refueled by a *KC-135R Stratotanker* during joint training.

pability to carry a range of sensors, and the ability to operate in both contested and permissive environments.

Northrop Grumman RQ-4 Global Hawk is another high-altitude, long-endurance UAV designed for surveillance and reconnaissance. The *Global Hawk* can fly at high altitudes for extended periods, collecting vast amounts of imagery and data.

Beechcraft RC-12 Guardrail is a reconnaissance variant of the *Beechcraft King Air* series airplane, employed by the United States Army for signals intelligence (SIGINT) missions. The *RC-12 Guardrail* is equipped with specialized equipment to intercept and analyze enemy communication signals. These examples demonstrate the diversity and importance of reconnaissance airplanes in gathering critical intelligence for military operations. Their advanced sensor systems, endurance, and ability to operate in various environments make them in-

valuable assets in modern warfare. There is a category of UAVs that will take up combat roles and are called Uninhabited Combat Aerial Vehicles (UCAVs). Future warfare will see the increased role of UCAVs with performance characteristics even better than their inhabited counterparts.

6.3.6 Tanker

Tanker airplanes (aerial refueling airplanes) are specialized airplanes designed to provide in-flight refueling capabilities to other aircraft. They play a critical role in extending the operational range, endurance, and flexibility of fighters, bombers, and other military airplanes so that they remain airborne for longer periods, reach distant destinations, and execute missions that would otherwise be limited by fuel constraints.

Tanker airplanes are equipped with specialized systems that allow them to transfer fuel to receiver

Using a probe-and-drogue, a Royal Air Force *C-130K* crew practice refuelling from an *RAF VC10* tanker.

KC-135 Stratotanker refuels an *F-16 Fighting Falcon*.

An IAF *Ilyushin Il-78MKI* provides mid-air refueling to two *Mirage 2000*.

airplanes while in flight. The most common method is the use of a *refueling boom*, which extends from the rear of the tanker airplanes and connects to a receptacle on the receiver airplane. Other methods include the use of *drogue systems*, which deploy a flexible hose and drogue basket for probe-equipped receiver airplanes to connect with.

Tanker airplanes have significantly larger fuel capacities than regular combat airplanes. This enables them to carry and transfer substantial amounts of fuel to the receiver airplane, thereby extending their range and mission endurance.

Boeing KC-135 Stratotanker is a long-serving aerial refueling airplane operated by the United States Air Force (USAF) and several other air forces. The *KC-135* is based on the *Boeing 707* airliner and has been a crucial asset for US military aerial refueling since the 1950s.

Airbus A330 MRTT is a multi-role tanker-transport airplane developed by *Airbus* based on the *A330* airliner. It is capable of performing both aerial refueling and airlift missions. It has been adopted by several air forces around the world.

Boeing KC-46 Pegasus is the latest tanker airplane in the United States Air Force inventory, based on the *Boeing 767* airliner. The *KC-46* provides advanced aerial refueling capabilities along with transport and cargo capabilities.

Ilyushin Il-78 is a Russian/Soviet tanker airplane based on the *Il-76* transport airplane. The *Il-78* serves as an aerial refueling platform for the Russian Air Force and has also been exported to other nations.

Beechcraft T-6 Texan II, a single-engine turboprop trainer airplane.

The Pilatus PC-9, a single-engine, low-wing tandem-seat turboprop training airplane.

6.3.7 Trainer

Trainer airplanes, also known as training airplanes or primary trainers, are specialized airplanes designed to facilitate pilot training and skill development for aspiring aviators. These airplanes serve a crucial role in the initial stages of pilot training by providing a safe and controlled platform for students to learn fundamental flying skills, including takeoff, landing, maneuvering, and emergency procedures. Let's explore the key features and examples of trainer airplanes.

Dual-Control Configuration: Trainer airplanes are typically designed with a dual-control setup, allowing an instructor pilot to have full control of the airplane from the second seat. This configuration enables the instructor to monitor and intervene when necessary, ensuring safe and effective training.

Stability and Forgiving Flight Characteristics: Trainer airplanes are designed to be stable and

The *Grob G 120*, a two-seat training and aerobatic low-wing trainer airplane with a carbon composite airframe, built by *Grob airplane* manufacturer.

forgiving, making them more manageable for student pilots. They often have features such as swept wings, tricycle landing gear, and low stall speeds to enhance stability and ease of handling.

Instrumentation and Avionics: Trainer airplanes have basic instrumentation and avionics systems that mimic those found in more advanced airplanes. This allows students to become familiar with instruments, controls, and procedures they will encounter in their future careers as pilots.

Relatively Low Performance: Trainer airplanes generally have modest performance characteristics, emphasizing safety and ease of operation. They typically have lower maximum speeds, simpler engine systems, and reduced complexity compared to high-performance combat airplanes.

Some examples of trainer airplanes are:

Beechcraft T-6 Texan II is a turboprop-powered trainer airplane used by the United States Air Force and several other air forces around the world. The *T-6 Texan II* is a versatile and modern training platform known for its reliability and performance.

Pilatus PC-9 is a Swiss-built turboprop trainer airplane used by numerous air forces globally. The *PC-9* is recognized for its robust construction, versatility, and excellent handling characteristics.

Grob G 120 is a modern, single-engine turboprop trainer airplane used by several air forces worldwide. The *G 120* features a glass cockpit and advanced avionics, providing a realistic training envi-

The *Diamond DA40 Diamond Star*, an Austrian four-seat, single-engine, light trainer airplane.

ronment.

Diamond DA40 is an Austrian four-seat, single-engine, light airplane constructed from composite materials. A piston-powered trainer airplane known for its fuel efficiency and advanced composite construction, the *DA40* is used for flight training as well as civilian recreational flying.

6.3.8 Transport

Military transport planes are used mostly to transport other aircraft, equipment, and personnel from one location to another. The cargo is either loaded onto pallets first to make it easier to load and unload or dropped from the planes on parachutes, which means the planes don't even have to land. Aerial tankers are transport planes that refuel other planes while in the air. Fleets of transport airplanes are an important aspect of the air-combat during the war. While the combat aircraft fleet faces the most daunting challenges from the enemy troop, the transport airplanes support the transportation of personnel and materials. Furthermore, the fleet of military transport airplanes is also employed primarily for peacetime tasks related to logistic support missions both at the national and international levels.

Airbus A400M Atlas

Specifications	
Cruise speed	781 km/h
Maximum speed	M 0.72
Range	8900 km
Wing loading	637 kg/m^2
Cruising altitude	11,300 m
Cargo volume	340 m^3
Maximum payload	37 tons
MTOW and MLW	141 and 123 tons
Wingspan	42 m
Engines	4 × Europrop TP400-D6 turboprop, 8,200 kW each

There are a total 178 orders from ten different countries, 118 are delivered and operational as of June 2023.

Over the years, the development of military transport airplanes by the leading aerospace and defence manufacturers around the world has focused on the following factors:

- On the larger size to increase payload capacity, capable of transporting oversized and heavy cargo such as battle tanks or large-size vehicles

- Enhanced operating range

- Advanced design for strategic or tactical roles or both.

Military airplanes are also designed to withstand

more extreme conditions and are built or modified to fly at higher speeds and altitudes than civilian airplanes. One of the key differences between commercial and military airplanes is the equipment onboard. Military airplanes often have specialized avionic equipment for navigation and communication, including encrypted radios, advanced radars, and in-flight refueling capabilities. Engines can also differ between versions as military missions prioritize response time, whereas commercial airliners focus on reducing fuel consumption and noise.

C-27J Spartan Next Generation

Specifications	
Cruise speed	583 km/h
Range	5,852 km
Wing loading	637 kg/m^2
Service ceiling	9,144 m
Maximum payload	11.3 tons
MTOW and MLW	32.5 and 30.5 tons
Wingspan	28.7 m
Engines	2 × Rolls-Royce AE2100-D2A turboprop, 3,458 kW each
Rate of climb	13 m/s

The **Airbus A400M ATLAS**, the proven and certified European four-engine turboprop military transport airplane designed by *Airbus Military*, combines the capability to carry strategic loads (the transport of strategic assets like outsized and heavy vehicles or equipment) with the ability to deliver even into tactical locations (transport and delivery of personnel and goods directly into theaters of operation) with small and unprepared airstrips and can act as a frontline-tanker. In addition to its transport capabilities, the *A400M* can perform aerial refueling and medical evacuation when fitted with appropriate equipment.

C-5M Super Galaxy

Unloading one of two Chinook helicopters from a C-5M Super Galaxy

Specifications	C-5M Super Galaxy
Cruise speed	830 km/h
Maximum Range	13,000 km
Service ceiling	12,000 m
Wing area	580 m^2
Maximum payload	130 tons
MTOW	381 tons
Wingspan	67.89 m
Engines	4 × General Electric F138-100 turbofan engines, 230 kN thrust each
Rate of climb	11 m/s
Thrust/weight	0.26

Crew: 7 typical (Aircraft Commander, First Pilot, 2 Flight Engineers, 3 Loadmasters). A total of 131 C-5 aircraft have been produced since it entered into service in 1970.

C-27J Spartan, developed and manufactured by *Leonardo's Aircraft Division* of Italy, is a high-wing, multi-mission airlifter military airplane powered by twin turboprop engines from Roll-Royce. The *C-27J* can be configured to carry out tactical transport missions including transportation of troops and cargo, drop paratroopers, undertake medical evacuation as well as provide humanitarian assistance and disaster relief, fire fighting, intelligence surveillance and reconnaissance missions, maritime patrol and tactical support missions.

KC-10 Extender

KC-10 Extender refuels F-16 Fighting Falcon

Specifications	KC-10A Extender
Maximum speed	866 km/h
Range	18,500 km
Service ceiling	13,000 m
Wing area	367.7 m^2
Maximum payload	77 tons
MTOW	267 tons
Wingspan	50.4 m
Engines	3 × General Electric F103 (GE CF6-50C2) turbofan engines, 234 kN thrust each
Rate of climb	34.9 m/s

Crew: 4 (Aircraft Commander, copilot, flight engineer, and boom operator). A total of 60 KC-10 Extender airplanes have been produced since it entered into service in 1981.

The Lockheed Martin made **C-5M Super Galaxy** is a heavy intercontinental-range strategic military transport airplane used by the US Air Force. The *C-5M* can transport a typical load of six Mine-Resistant Ambush Protected vehicles or five helicopters. The *C-5M Super Galaxy* is an upgraded version with new engines and modernized avionics designed to extend its service life to 2040 and beyond. *C-5M* allows for quick loading/unloading from the front and rear simultaneously.

The McDonnell Douglas **KC-10 Extender** is an American tanker and cargo airplane in service with the US Air Force and the Royal Netherlands Air Force with primary roles of aerial refueling and transport. *KC-10 Extender* is developed from a *DC-10* trijet airliner.

The **Antonov An-124 Ruslan** is a large, high wing, regular tail, mid-set, strategic airlift, four-engined airplane designed by the *Antonov design bureau* in the USSR. The *An-124* is the world's second heaviest gross weight production cargo airplane and heaviest operating cargo airplane, behind the destroyed one-off *Antonov An-225 Mriya* and the *Boeing 747-8*. The *An-124* remains the largest military transport airplane in service. The airplane features a pressurized cargo compartment that is 20% larger than that of *C-5 Galaxy*. The front and rear cargo doors of the airplane ensure rapid loading/off-loading of heavy cargo with ease.

Antonov AN-124 Ruslan

Specifications	
Cruise speed	800-850 km/h
Maximum range	15700 km
Wing loading	640.1 kg/m^2
Service ceiling	12,000 m
Maximum payload	150 tons
MTOW and MLW	402 and 330 tons
Wingspan	73.3 m
Engines	4 × Progress D-18T high-bypass turbofan engines, 229 kN each
Thrust/weight	0.23

Crew: Eight (pilot, copilot, navigator, chief flight engineer, electrical flight engineer, radio operator, two loadmasters). A total of 55 airplanes were produced in the period 1982 to 2004.

The **C-390 Millennium** is the new generation military multi-mission airplane with improved mobility, high productivity and operation flexibility at low operational costs on a single and unique modern platform.

Embraer C-390 Millennium

Specifications	C-390 Millennium
Cruise speed	870 km/h
Range	6,130 km
Service ceiling	11,000 m
Maximum payload	26 tons
MTOW and Fuel capacity	87 and 23 tons
Wingspan	35.05 m
Engines	2 × IAE V2500-E5 turbofan, 139.4 kN thrust each

Crew: Three (two pilots, one loadmaster). A total of 9 airplanes have been produced since the entry into service in 2019.

Boeing C-17 Globemaster III

Specifications (C-17A)	
Cruise speed	830 km/h
Range	11,540 km
Wing loading	730 kg/m^2
Service ceiling	14,000 m
Maximum payload	77.5 tons
MTOW and MLW	265.3 and 330 tons
Wingspan	51.7 m
Engines	4 × Pratt & Whitney F117-PW-100 turbofan engines, 179.9 kN thrust each
Thrust/weight	0.277

Crew: 3 (2 pilots, 1 loadmaster). A total of 279 C-17s were built in the period 1991 to 2015.

The *C-390* is capable of transporting and launching cargo and troops and performing a wide array of missions including medical evacuation, search and rescue, humanitarian search and rescue, aerial refueling, aerial firefighting and humanitarian assistance.

The **Boeing C-17 Globemaster III** is a large military tactical and strategic airlift airplane designed and developed by *McDonnell Douglas*. *C-17* performs other roles such as transporting troops and cargo throughout the world, medical evacuation and airdrop duties. A unique feature of this airplane is that its engine reverse thrust is operable in flight as well.

Boeing KC-46 Pegasus

Specifications	
Cruise speed	851 km/h
Range	11,830 km
Service ceiling	12,200 m
Maximum payload	29 tons
MTOW and MLW	415 and 310 tons
Wingspan	47.5 m
Engines	2 × Pratt & Whitney PW4062 turbofan, 289 kN thrust each

Crew: 3 (2 pilots, 1 boom operator). A total of 72 KC-46 are built since its inception to service in 2019.

The **Boeing KC-46 Pegasus** (winner of *KC-X* tanker competition by USAF to replace older *Boeing KC-135 Stratotankers*) is a multi-mission aerial refueling and strategic military transport airplane developed by *Boeing* from its *767* jet airliner, a widebody, low wing platform powered by two turbofan engines. With greater refueling, cargo and aeromedical evacuation capabilities compared to the *KC-135*, the *KC-46A* will provide aerial refueling support to the United States Air Force, Navy and Marine Corps.

Ilyushin Il-76

Specifications	(Il-76TD)
Maximum speed	900 km/h
Range	9,300 km
Service ceiling	13,000 m
Maximum payload	48 tons
MTOW	190 tons
Wingspan	50.5 m
Engines	4 × Soloviev D-30KP turbofans, 117.7 kN thrust each
Thrust/weight	0.252

Crew: 5. A total of 960+ Il-76s are built since its entry into service in 1974.

Lockheed Martin C-130J Super Hercules

Specifications (C-130J)	
Cruise speed	644 km/h
Range	3,300 km
Service ceiling	8,500 m
Maximum payloads	19 tons
MTOW	70 tons
Wingspan	40.41 m
Engines	4 × Rolls-Royce AE 2100D3 turboprop engines, 3,458 kW each

Crew: 3 (two pilots and one loadmaster are minimum crew). A total of 500 C-17s were built in the period 1996 to 2023.

The **Ilyushin Il-76** is a multi-purpose, fixed-wing, four-engine turbofan strategic airlifter designed (as competitor to American *Lockheed C-141A Starlifter*) by the Soviet Union's *Ilyushin* design bureau as a commercial freighter in 1967, to replace the *Antonov An-12*.

The **Lockheed Martin C-130J Super Hercules** is a four-engine turboprop military transport airplane. The *C-130J* is a comprehensive update of the Lockheed *C-130 Hercules*, with new engines, flight deck, and other systems. The *C-130J* airlifter is a choice for 22 nations and 26 operators around the world.

6.3.9 Multirole Airplanes

Multirole fighter or *combat airplane*, is an airplane designed to perform multiple missions and roles effectively. Unlike specialized airplane that are tailored for specific tasks, such as air superiority or ground attack, a multirole airplane is designed to have a wide range of capabilities, allowing it to adapt to various operational requirements.

The concept of a multirole airplane emerged as a response to the need for increased flexibility and efficiency in military operations. By incorporating features and systems that can handle different missions, a multirole airplane offers the advantage of versatility and cost-effectiveness. It reduces the need for multiple specialized airplanes, streamlines logistics, and maximizes the utilization of resources.

The primary advantage of a multirole airplane is its ability to engage both air and ground targets. It typically possesses air-to-air combat capabilities, allowing it to engage enemy aircraft, enforce air superiority, and provide aerial defense. At the same time, it can also carry and deliver a variety of air-to-ground weapons, such as bombs, missiles, and rockets, enabling it to conduct precision strikes against ground targets.

In addition to air-to-air and air-to-ground capabilities, multirole airplanes often incorporate other mission capabilities. These may include electronic warfare capabilities to disrupt enemy communications and radar systems, reconnaissance capabilities to gather intelligence, and even aerial refueling capabilities to extend operational range and endurance.

Multirole airplanes typically have advanced avionics and sensor suites that enhance situational awareness and enable effective target acquisition and tracking. They often feature advanced radar systems, electronic warfare systems, integrated communications systems, and modern cockpit displays, allowing pilots to manage and process large amounts of information quickly and efficiently.

Examples of renowned multirole military airplanes include *Eurofighter Typhoon*, the *F-16 Fighting Falcon*, the HAL (India) *Tejas*, the Sukhoi (Russia) *Su-57*, *Su-34*, *Su-35*, *Su-30*, the Mikoyan's (Russia) *MiG-29*, *MiG-35*, the Dassault *Rafale*, and the *F/A-18 Hornet/Super Hornet*. These airplanes have proven their effectiveness in a variety of missions, demonstrating the advantages of having a single platform capable of performing multiple roles.

Swing-Role

Some airplanes like the *Saab JAS 39 Gripen* have the capability to **swing-role** which means they have the capability to quickly switch between different roles or missions during the course of a single mission or sortie. The concept of swing-role airplanes was developed to enhance operational flexibility and efficiency. It allows an airplane to perform different tasks within a single mission, reducing the need for multiple specialized airplanes and maximizing the use of available assets. Swing-role airplanes are often categorized as a subset of multirole airplanes but with the added capability of rapidly transitioning between roles.

While both multirole and swing-role airplanes can perform multiple missions, there is a subtle distinction between the two concepts. Multirole airplanes are designed to perform a wide range of missions but typically require a ground crew to reconfigure the airplanes and load different weapon systems before transitioning between roles. In contrast, swing-role airplanes are designed with built-in systems and features that facilitate quick role changes during flight, often with minimal physical modifications.

Dassault Rafale

Specifications	
Maximum speed	1,912 km/h
Ferry/Combat range	3,700/ 1,850 km
Wing loading	328 kg/m^2
Thrust/weight	0.988
Rate of climb	304.8 m/s
g limits	$+9/-3.6$
Service ceiling	15,835 m
MTOW/Fuel capacity	24.5/4.7 tons
Wingspan	10.90 m
Engines	2 × Snecma M88-4e turbofans, 50.04 kN thrust each dry, 75 kN with afterburner
Armament (Guns)	1 × 30 mm (1.2 in) GIAT 30/M791 autocannon with 125 rounds
Armament (Missiles)	ASMP-A (nuclear missile), MBDA Apache, MBDA Storm Shadow/SCALP-EG, AASM-Hammer, GBU-12 Paveway II, AS-30L, Mark 82 (Surface-to-air), MBDA MICA, MBDA Meteor Magic II (Air-to-air).

Total of 239 are delivered and operational as of 2021 since its entry into service in 2001.

Examples of swing-role airplanes include the *Panavia Tornado*, which was specifically designed with the swing-role concept in mind. The *Tornado* is capable of performing air defense, ground attack, and reconnaissance missions within a single sortie. The airplane features modular weapon systems and interchangeable avionics, allowing for quick reconfiguration between roles. The *SEPECAT Jaguar* is a swing-role airplane developed by the United Kingdom and France. It was primarily designed as a ground attack airplane but also possesses limited

IAF Tejas landing during *Aero India 2023* with airbrakes and parachute open.

air-to-air capabilities. The *Jaguar* was optimized for low-level, close air support missions, including deep interdiction, reconnaissance, and nuclear strike roles.

The *Dassault Rafale* is a versatile multi-role fighter airplane manufactured by the French company *Dassault Aviation*. It is one of the most advanced and capable fighter jets in the world, known for its exceptional performance, cutting-edge technology, and wide range of combat capabilities. The *Rafale* is a twin-engine, delta-winged airplane with a canard configuration, giving it excellent maneuverability and high agility. Its airframe is constructed using composite materials, which makes it lighter and more durable. The airplane features advanced avionics, an integrated electronic warfare system, a state-of-the-art glass cockpit, and a fly-by-wire flight control system, providing the pilot with superior situational awareness and control.

The development of the *Rafale* began in the late 1970s as a joint project between France, Germany, Spain, and the United Kingdom. It is designed to perform a wide range of missions, including air superiority, ground attack, reconnaissance, and nuclear deterrence. The airplane can carry an extensive payload of air-to-air and air-to-ground weapons, including missiles, bombs, guided munitions, and anti-ship missiles. It can also conduct in-flight refueling, extending its range and endurance.

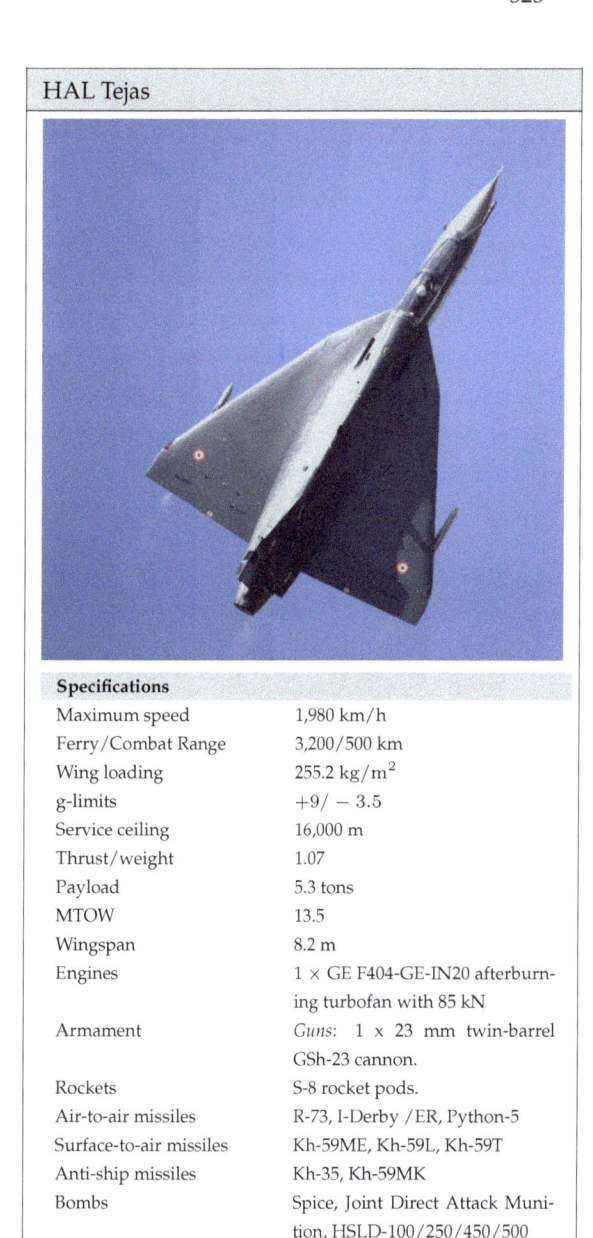

HAL Tejas

Specifications

Maximum speed	1,980 km/h
Ferry/Combat Range	3,200/500 km
Wing loading	255.2 kg/m^2
g-limits	+9/ − 3.5
Service ceiling	16,000 m
Thrust/weight	1.07
Payload	5.3 tons
MTOW	13.5
Wingspan	8.2 m
Engines	1 × GE F404-GE-IN20 afterburning turbofan with 85 kN
Armament	*Guns*: 1 x 23 mm twin-barrel GSh-23 cannon.
Rockets	S-8 rocket pods.
Air-to-air missiles	R-73, I-Derby /ER, Python-5
Surface-to-air missiles	Kh-59ME, Kh-59L, Kh-59T
Anti-ship missiles	Kh-35, Kh-59MK
Bombs	Spice, Joint Direct Attack Munition, HSLD-100/250/450/500

There are a total of 123 orders from the Indian Air Force and a total of 40 are delivered and operational as of October 2021.

Tejas, the Light Combat Aircraft (LCA), is a multi-role supersonic fighter airplane developed by the *Aeronautical Development Agency (ADA)* in collaboration with *Hindustan Aeronautics Limited (HAL)* in India. *Tejas* is a single-engine, delta-winged airplane with a tailless compound delta platform, giving it a distinctive appearance. It is one of the most significant indigenous aviation projects in India's aerospace history. The development of *Tejas* began in the 1980s intending to replace the aging fleet of Indian Air Force (IAF) airplanes. The primary objective was to create a lightweight, agile, and technologically advanced fighter airplane that could meet the requirements of the IAF and operate in various combat roles. The airplane features an advanced fly-by-wire flight control system, advanced avionics, and a modern glass cockpit, providing the pilot with enhanced situational awareness. It is equipped with a wide range of weapons systems, including missiles, bombs, rockets, and guns. The airplane can carry a diverse payload and has the capability to integrate various weapon systems.

The *F-16 Fighting Falcon*, often referred to as the *Viper*, is a highly versatile and widely used multirole fighter airplane developed by *General Dynamics* (now *Lockheed Martin*) in the United States. It has become one of the most successful and iconic fighter jets in the world, known for its agility, performance, and adaptability. The development of the *F-16* started in the 1970s, intending to create a lightweight, cost-effective, and high-performance fighter airplane that could excel in both air-to-air and air-to-ground missions. The airplane was designed to be highly maneuverable, capable of flying at high speeds, and easy to maintain. The *F-16* features a compact and streamlined design with a delta wing and a single-engine configuration. This design, combined with its high thrust-to-weight ratio, gives the airplane exceptional agility and maneuverability. The *F-16* is known for its ability to perform tight turns, high-G maneuvers, and quick acceleration, making it a formidable opponent in aerial combat. The avionics and systems of the *F-16* have evolved over the years to incorporate the latest technological advancements. The airplane is equipped with a sophisticated radar system, advanced electronic warfare systems, a Heads-Up Display (HUD), and a modern glass cockpit with multifunction displays. These features enhance the pilot's situational awareness, provide precise targeting information, and streamline the management of various systems and sensors. The armament capabilities of the *F-16* are extensive and diverse. It can carry a wide range of air-to-air and air-to-ground weapons, including guided missiles, bombs, rockets, and a 20mm *M61 Vulcan Gatling* gun.

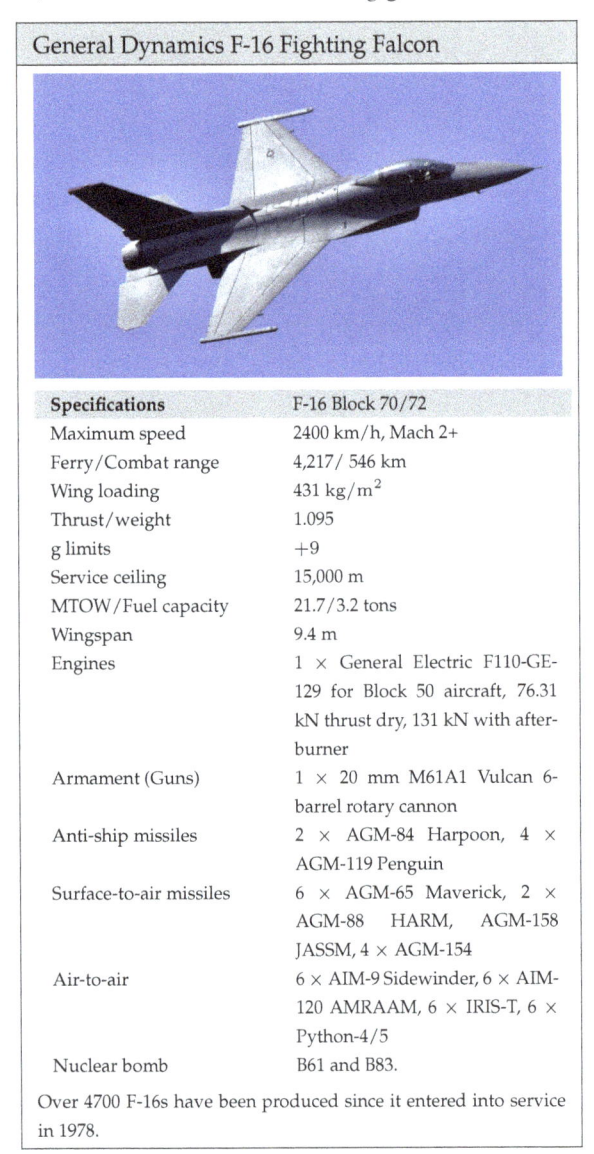

F-16 Fighting Falcon take off at Paris Air Show 2023

General Dynamics F-16 Fighting Falcon

Specifications	F-16 Block 70/72
Maximum speed	2400 km/h, Mach 2+
Ferry/Combat range	4,217/ 546 km
Wing loading	431 kg/m^2
Thrust/weight	1.095
g limits	+9
Service ceiling	15,000 m
MTOW/Fuel capacity	21.7/3.2 tons
Wingspan	9.4 m
Engines	1 × General Electric F110-GE-129 for Block 50 aircraft, 76.31 kN thrust dry, 131 kN with afterburner
Armament (Guns)	1 × 20 mm M61A1 Vulcan 6-barrel rotary cannon
Anti-ship missiles	2 × AGM-84 Harpoon, 4 × AGM-119 Penguin
Surface-to-air missiles	6 × AGM-65 Maverick, 2 × AGM-88 HARM, AGM-158 JASSM, 4 × AGM-154
Air-to-air	6 × AIM-9 Sidewinder, 6 × AIM-120 AMRAAM, 6 × IRIS-T, 6 × Python-4/5
Nuclear bomb	B61 and B83.

Over 4700 F-16s have been produced since it entered into service in 1978.

Eurofighter Typhoon

Specifications	Eurofighter Typhoon
Maximum speed	2,125 km/h, Mach 2.0
Ferry/Combat range	3,790/1,389 km
Wing loading	312 kg/m^2
Thrust/weight	1.15
Rate of climb	315 m/s
g limits	+9/ − 3
Service ceiling	19,812 m
MTOW/Fuel capacity	23.5/5 tons
Wingspan	10.95 m
Engines	2 × Eurojet EJ200 afterburning turbofan engines, 60 kN thrust each dry, 90 kN with afterburner
Armament (Guns)	1 × 27 mm Mauser BK-27 revolver cannon with 150 rounds
Anti-ship missiles	Marte ER
Surface-to-air missiles	Storm Shadow, Brimstone, AGM-88 HARM, AGM-65 Maverick, Taurus KEPD 350
Air-to-air missiles	AIM-120 AMRAAM, MBDA Meteor, IRIS-T, AIM-132 ASRAAM, AIM-9 Sidewinder
Nuclear bomb	Paveway II/III/Enhanced Paveway series of laser-guided bombs (LGBs), 500-lb Paveway IV, Small Diameter Bomb (planned for P2E), Joint Direct Attack Munition (JDAM), work started in 2018, HOPE/HOSBO, in the future, Spice 250

As of November 2022, over 581 Typhoons have been produced since it entered into service in 2003.

The *F-16* can engage multiple targets simultaneously and possesses a robust sensor suite that allows for effective target acquisition and tracking. About 4700 F-16s have been produced over the past 49 years and 3,000+ of them are still in service with 25 nations around the world. The most recent version is called the *Block 70/72* and includes a new cockpit, computers, and an advanced AESA radar based on the one installed in the *F-35*.

The *Eurofighter Typhoon* is a high-performance, multirole, twin-engine, canard delta wing fighter airplane designed and developed through a collaborative effort between several European countries (the UK, Germany, Italy and Spain). It is regarded as one of the most advanced and capable combat airplanes in the world. The development of the *Eurofighter Typhoon* began in the 1980s with the aim of creating a next-generation fighter airplane to replace the aging fleet of airplanes in the respective countries' air forces. The project aimed to produce a versatile, agile, and highly maneuverable airplane capable of excelling in air-to-air combat, air superiority, and ground-attack missions. The *Typhoon* features a delta wing design with canards, giving it excellent maneuverability and agility. Its aerodynamic design allows it to fly at supersonic speeds without using afterburners, providing it with increased fuel efficiency and extended range. In terms of its avionics and systems, the *Eurofighter Typhoon* is equipped with state-of-the-art technology. It incorporates a sophisticated radar system, called the European Common Radar System (ECRS), which provides superior situational awareness, detection capabilities, and tracking of multiple targets. The airplane also features advanced electronic warfare systems, communication systems, and a Helmet-Mounted Display (HMD) that enhances the pilot's ability to track and engage targets. The *Typhoon*'s armament capabilities are diverse and extensive. It can carry a wide range of air-to-air and air-to-surface missiles, such as the *Meteor* beyond-visual-range air-to-air missile and the *Storm Shadow* and *Brimstone* precision-guided air-to-surface missiles. In addition to the partner nations, the *Typhoon* is operated by Austria, Kuwait, Oman, Qatar and Saudi Arabia.

Lockheed Martin F-35 Lightning II

Specifications	F-35A
Maximum speed	2000 km/h, Mach 1.6
Ferry/Combat range	2,800/1,239 km
Wing loading	526 kg/m^2
Thrust/weight	0.87
g limits	+9
Service ceiling	15,000 m
MTOW/Fuel capacity	29/8.2 tons
Wingspan	11 m
Engines	1 × Pratt & Whitney F135-PW-100 afterburning turbofan, 125 kN thrust dry, 191 kN with afterburner
Armament (Guns)	1 × 25 mm GAU-22/A 4-barrel rotary cannon, 180 rounds
Hardpoints	4 × internal stations, 6 × external stations on wings
Anti-ship missiles	AGM-158C LRASM, Joint Strike Missile
Surface-to-air missiles	AGM-88G AARGM-ER, AGM-158 JASSM, AGM-179 JAGM
Air-to-air missiles	AIM-9X Sidewinder, AIM-120 AMRAAM, AIM-132 ASRAAM
Nuclear bomb	B61 mod 12 nuclear bomb, Joint Direct Attack Munition, Paveway, Precision-guided glide bomb

Over 945+ (as of July 2023) F-35s have been built since it entered into service in 2015.

The Lockheed Martin *F-35 Lightning II* is a fifth-generation, single-seat, multirole fighter airplane that represents a significant leap forward in advanced fighter technology. It is designed to meet the operational requirements of the United States and its allies, offering unparalleled capabilities for air superiority, strike missions, and advanced reconnaissance.

The *F-35 Lightning II* program was initiated in the late 1990s as a joint effort between the United States, the United Kingdom, and several other partner nations. The aim was to develop a family of airplanes that shared common components and systems while meeting the unique needs of each participating country. The result was three primary variants: the *F-35A* for conventional takeoff and landing, the *F-35B* with short takeoff and vertical landing capability, and the *F-35C* optimized for carrier operations. The *F-35 Lightning II* features advanced stealth technology, giving it a low radar signature and enabling it to operate undetected in hostile environments. The airplane's aerodynamic design, combined with internal weapon storage, helps minimize its radar cross-section, enhancing its survivability and lethality.

One of the key strengths of the *F-35 Lightning II* is its advanced sensor suite and integrated avionics. The airplane is equipped with the Active Electronically Scanned Array (AESA) radar, which provides superior situational awareness, long-range detection, and precision targeting capabilities. Additionally, it incorporates advanced electro-optical targeting systems, an integrated communications suite, and a helmet-mounted display that allows pilots to see through the airplane.

The *F-35 Lightning II*'s armament capabilities are highly diverse and flexible. It can carry a wide range of air-to-air and air-to-ground ordnance, including missiles, guided bombs, and internally mounted guns. The airplane's internal weapon bays help maintain its stealth characteristics while providing ample storage for various munitions.

6.3.10 Airborne Early Warning & Control (AEW&C) System

An AEW&C system is an airborne radar system designed to detect aircraft, missiles, and other incoming projectiles at long ranges and perform command and control of the battle space by directing airplanes and fighter strikes. AEW&C units are also used to carry out Battle/Post Strike Damage Assessment, surveillance over air or ground targets, frequently perform battle management command and control and even perform functions similar to an air traffic controller. When used at altitude, the radar on the airplane allows the operators to detect and track targets and distinguish between friendly and hostile airplanes much farther away than a similar ground-based radar AEW&C systems are typically network-centric and use secure two-way data links augmented with satellite communication to receive surface inputs and transmit to receivers located on the ground. The system integrates and synthesizes the input information derived from onboard/surface sensors, as well as commands and control inputs before transmitting it to receivers. Some of the military planes used for this work include the *Lockheed EC-121 Warning Star*, the *Fairey Gannet AEW.3*, and the *Northrup Grumman E-2 Hawkeye*. They can be used for both offensive and defensive operations. For example, they can direct fighters to their target locations and counterattack enemy forces both on the ground and in the air. They are also used by both the Air Force and the Navy. Many countries have developed their own AEW&C systems. The American *Boeing E-3 Sentry* and *Northrop Grumman E-2 Hawkeye* are the most common systems worldwide. Other AEW&C systems are Russia's *Beriev A-50*, *KJ-2000*, India's *Netra*, Swedish *Erieye* and *GlobalEye*.

Boeing E-3 Sentry

Boeing E-3 Sentry is an American AEW&C airplane derived from the *Boeing 707* airliner. The *E-3* has a distinctive rotating radar dome (rotodome) above the fuselage.

Northrop Grumman E-2 Hawkeye

The *Northrop Grumman E-2 Hawkeye* is an American all-weather, carrier-capable tactical AEW&C twin-turboprop airplane, first entered service in 1965.

NATO has a fleet of *Boeing E-3A* Airborne Warning & Control System (AWACS) airplane, a modified *Boeing 707* airliner, equipped with distinctive long-range radar domes mounted on the fuselage and sensors capable of detecting air and surface contacts over large distances and provides air surveillance, command and control, battle space management and communications.

It carries out a wide range of peacetime missions and the full spectrum of wartime missions. Under normal circumstances, the airplane operates for about eight and a half hours (capable of flying longer operations as it has air-to-air refueling capability) at around nine kilometer altitude and covers a surveillance area of more than three hundred thousand square kilometers.

Embraer EMB145I

DRDO, India's (AEW&CS) based on the first fully modified *EMB-145i* airplane.

Boeing E-7A Wedgetail

The *Boeing E-7A Wedgetail* AEW&C system based on the *Boeing 737 Next Generation* design.

Embraer R-99

The *Embraer R-99* is the Brazilian AEW&C based on the *ERJ 145* civil regional jet.

6.4 Military Aviation in India

One of the most remarkable achievements of the early twentieth century was the conquest of the skies by man. The first flight in India took place in

Karun Krishna Majumdar was the first Indian officer to be awarded the Distinguished *Flying Cross*. He was the first Indian to reach the rank of *wing commander*. Majumdar was among twelve of the world's most outstanding airmen featured by Life magazine in May 1944.

Allahabad in 1910, and later the first air mail in the world was launched in 1911. Subsequently, the first military flying school was set up in Sitapur (UP) in 1913. The Indian Air Force (IAF) was officially formed as the Royal Indian Air Force (RIAF) on October 8, 1932 after the recommendation by the *Skeen Committee* headed by Lieutenant General Sir *Andrew Skeen*. The RIAF was a colonial auxiliary force of the British Indian Empire. The world experienced the second most deadly war in human history, World War II. During World War II, the IAF grew from a single squadron to a nine-squadron force, gaining valuable operational experience by flying alongside the RAF in the Burma campaign and other theaters. The Indian Air Force played a crucial role in World War II in stopping the Japanese army in Burma by executing air strikes on Japanese military bases in Arakan and northern Thailand. IAF's *Karun Krishna Majumdar*, the first Indian to reach the rank of wing commander, was recognized as one of the best pilots among the Allied Air Forces. The role of IAF in various operations, including the Burma campaign where it distinguished itself and contributed to the Allied victory is commendable. The same period marked the birth of Independent India and the establishment of the Indian Air Force. In January 1950, upon India becoming a Republic, the prefix 'Royal' was removed from the *Royal Indian Air Force*. At that time, the Indian Air Force (IAF) comprised six fighter squadrons equipped with *Spitfires*, *de Havilland Vampire*, and *BAE Systems Tempest*, stationed at Kanpur, Poona (Pune), Ambala, and Palam. Additionally, there was one *B-24 bomber* squadron, one *C-47 Dakota* Transport squadron, one *AOP (Air Observation Post)* flight, a Communication squadron based at Palam, and a burgeoning training organization. Transitioning from its former supplementary role assisting the RAF prior to India's independence, the IAF now assumed full responsibility for the air defense of the country. The IAF's inherited a small force of six fighter squadrons and one transport squadron was inadequate to meet the nation's defense needs. In the 1950s and 1960s, the IAF underwent a massive expansion and modernization program, inducting a wide range of modern aircraft like Vampires, *Toofanis*, *Mysteres*, *Canberras*, *Hunters*, and *MiG-21s*. The combat strength of the IAF grew from 15 squadrons in the 1950s to around 45 squadrons by the late 1960s, equipped with increasingly capable multi-role fighters. The IAF has played a vital role in various wars and major operations since its inception. Some of the key operations include: *Kashmir War (1947-48)*: The IAF conducted extensive air operations in support of the Indian Army and provided air maintenance to the Poonch garrison. The IAF's airlift of troops to Srinagar on October 27, 1947 was a critical and historic operation that saved the state of Jammu and Kashmir.

India-China War (1962): The IAF's transport and helicopter operations were crucial in supporting the Indian Army, despite the IAF's limited capabilities at the time. Helicopters proved their worth in the inhospitable terrain and bad weather conditions.

India-Pakistan War (1965): The IAF's fighter aircraft, particularly the *Gnat* and *Hunter*, performed exceptionally well against the superior Pakistani aircraft. The IAF's ground support operations in the Chhamb sector were strategically significant in halting the Pakistani armored thrust.

India-Pakistan War (1971): The IAF achieved air dominance in the Eastern sector within the first two days, enabling it to focus on other air operations. The IAF's air superiority, interdiction of enemy targets, and support to the Army and Navy were crucial factors in India's decisive victory.

Kargil War (1999): The IAF faced the challenge of operating at high altitudes and addressing targets close to the Line of Control. The IAF employed innovative techniques and utilized available resources to effectively support the Indian Army's operations.

Over the years, the IAF has continuously modernized its fleet, acquiring a wide range of aircraft, including fighters, transport planes, and helicopters, from both domestic and international sources. The IAF kept pace with technological advancements in the global aviation industry, inducting more capable fighters like *Jaguars*, *MiG-23/27*, and later the *Mirage 2000* and *MiG-29*. It also acquired force multipliers like air-to-air refuelers, AWACS, and upgraded its air defense network with modern radars and surface-to-air missile systems. The induction of the *C-17 Globemaster*, *C-130J*, and *Mi-17 helicopters* significantly improved the IAF's strategic airlift and combat support capabilities.

India has also made significant advancements in developing indigenous military aviation capabilities, with programs such as *Light Combat Aircraft (LCA)* and *Advanced Medium Combat Aircraft (AMCA)*. In general, military aviation in India continues to evolve and strengthen, ensuring that the country is well prepared to face any challenges that may arise. India has collaborated with foreign partners for technology transfer and joint development projects. Today, the IAF is a modern, technologically advanced, and highly capable air force that plays a crucial role in India's national defense and security. It has evolved from a colonial auxiliary force to an independent, self-reliant, and formidable air power, capable of meeting the challenges of the 21st century.

Bibliography

P. Darman (2004) *Twenty-first Century War Planes & Helicopters*, Grange.

J. D. Murphy, and M. A. McNiece (2008) *Military Aircraft, 1919-1945: An Illustrated History of Their Impact*. Bloomsbury Publishing USA.

R. Jackson (2009) *The encyclopedia of military aircraft*. Parragon Publishing India.

J. D. Murphy (2005) *Military aircraft, origins to 1918: an illustrated history of their impact*. Bloomsbury Publishing USA.

Chapter 7

Missiles

Contents

7.1	Introduction	332
7.2	Brief History of Missile Technology	332
7.3	Basic Concepts	333
7.4	Missile Subsystems	338
	7.4.1 Airframe	338
	7.4.2 Propulsion System	341
	7.4.3 Navigation, Guidance and Control	343
	7.4.4 Warheads	346
	7.4.5 Missile Launchers	347
7.5	Types of Missiles	347
	7.5.1 Based on Launch Mode	347
	7.5.2 Based on Target	347
	7.5.3 Multiple Independently Targetable Reentry Vehicles (MIRVs)	350
	7.5.4 Missile Defense System (MDS)	350
7.6	Some Examples	353
7.7	Major Missile Manufacturers	356
7.8	Treaties & Agreements	357
7.9	Missile Programs in India	357

7.1 Introduction

The ingenuity of using firepower in military conflicts has shown remarkable advancement since the early days when bamboo tubes filled with gunpowder were used to propel projectiles at the enemy. During *World War I*, bombs were dropped from airplanes into enemy territory. These bombers risked being hit by enemy fire and were also limited by their weight-carrying capacity. Also, it was difficult to hit a target accurately. *World War II* saw significant aerial bombing compared to *World War I*, mainly because the airplanes' capabilities drastically improved compared to that of *World War I*. The war saw extensive use of *V-2* missiles by Germany causing significant damage to the Allied forces despite the limitations in its range and ability to hit the target accurately.

The advent and rapid development of jet aircraft following *World War II* changed the scene of air-to-air combat. The high speed and maneuverability of the jet powered aircraft marked the end of the dogfight and the necessity to engage with targets beyond visual ranges arose. The way to address this problem was by using air-to-air or surface-to-air missiles. Following World War II, many countries including USA and USSR had access to the new weapon developed in Germany which is the high-altitude rockets. Improving the guidance and accuracy of these missile systems became one of the most important research and development problems in the military context.

Since *World War II*, many countries have embarked on developing missiles with varying levels of capabilities meant for different types of targets. Missiles have been developed that can take nuclear warheads to reach any point on Earth. Some missiles can be launched from a plethora of platforms (shoulder-launched, aircraft-launched, ship-launched, submarine-launched, etc.) to hit a wide variety of targets (aircraft, missiles, ships, armored vehicles, satellites, radar systems, bunkers, runways, etc.). Some missiles can hit a target hundreds of kilometers away with a Circular Error Probability (a measure of the accuracy of hitting the target) of less than a few meters. Some missiles can track and neutralize targets that fly at hypersonic speeds. A few countries have even demonstrated the capability to hit satellites moving at speeds as fast as a few kilometers per second.

Recent military conflicts attest to the importance of missiles and rockets in winning wars and their role as a deterrent against the enemy. The Israel-Palestine conflict, which started in October 2023, saw the *Hamas* firing thousands of rockets into Israel and many of them were neutralized by the *Iron Dome* missile defense system of Israel. Countries will be pursuing technologically challenging missile development programs quite earnestly to bolster their place as a military power. Many other countries will import many of these missile systems to address their military requirements. An advantage of missile systems, compared to combat aircraft, is that since they do not have a human payload, the design can be pushed to the limits to get the best performance. The versatility with which the warheads can be configured and the advantage provided by the stand-off distance make missiles a very significant part in the arsenal of any military.

7.2 Brief History of Missile Technology

The use of missiles has become an indispensable need of modern warfare. Though the radical evolution of technology and intense use of such missiles have been observed during the past few decades, their transformation from more rudimentary weapons has taken place over many centuries. The idea of using arrows and spears has been around since prehistoric times. The earliest recorded use of rockets in warfare was in 1232 by the Chinese to repel the Mongol besiegers of the city of Pein-King. Further, rockets were used during the 14th century in the war between Venice and Genoa. Later during the 15th century, the design of rockets started in Italy. In the 18th century, Indian soldiers of various kingdoms were reported to be using incendiary rockets weighing 3 to 6 kg against the British and French forces. Colonel *William Congreve* with the British army was impressed by these Indian rockets and began to experiment with them in the 19th century. During the *World War I*, rockets were found to be useful against small but strategic targets. The French successfully air launched them against enemy balloons and German *Zeppelins* during the war. Furthermore, attempts were made to design radio-controlled rockets in the USA and Europe during this time. After World War I, the Treaty of Versailles was signed at the Paris Peace Conference (1919) imposing several restrictive conditions on German research and development activities relating to military weapons. However, the treaty did not mention an explicit restriction on rocket research. The Germans took advantage and started a top secret research establishment at *Kummersdorf* in 1932 and *Peenemunde* in 1933 under the leadership of *Werner von Braun* to carry out research on rockets without attracting much attention. This intensive research activity, prior to the *World War II*, led to the development of the famous vengeance missiles *V-1* and *V-2*. The Russian army was poised to occupy *Peenemunde* when Germany and its allies were loosing the *World War II*. But, *von Braun* and his team of scientists escaped from there and were later captured by the Americans and flew the whole team of scientists back to the USA where they continued their experiments with rocketry. The German scientists were involved in developing surface-to-air missile, (e.g., *Hermes*, range 250 km, warhead mass of 450 kg), many surface-to-surface missiles, e.g., *Corporal*, (range 130 km, length 15 m, launch mass of 5.5 tons), *Sergeant* (range 40-140 km, length 11 m, launch mass of 4.5 tons), and *Redstone* (range 400 km, length 22 m, launch mass of 30 tons). *Convair MX-774* was one of the first guided missiles to be built and tested in the US which became the stepping stone for the development of *Atlas*, the first ICBM to be developed by the US.

On the other side of the Atlantic Ocean, the Russians too used the basic *V-2* design to build *M-101*, the forerunner of many huge Russian ICBMs. From 1950 to 1970 several missile programs were launched in the USA. Some of the missiles developed by various countries are given in the following table.

7.3 Basic Concepts

What is a Missile?

Before understanding a missile, let us understand simpler weapon systems like bombs and rockets. Once dropped, a bomb is completely governed by the laws of ballistics, meaning the only forces that act upon it after its release are the force of gravity and the aerodynamic force. A bomb with a propulsion system is called a rocket which is more accurate, faster and has got better range. You must have seen the jet engine under the wing of the airplane, which sucks the air, add energy to it by burning fuel and then push the gas out of its back at high speed through the nozzle and the gas pushes the airplane forward. Unlike a jet engine, a rocket does not need ambient air for its combustion. Instead, it carries both the fuel and oxidizer and generates the thrust by expelling the combustion products at high velocities through a nozzle. Therefore, a rocket engine can work in space also where there is no air. Rocket propulsion is the most common form of propulsion used in missiles. A rocket that is capable of maneuvering toward the target can be called a missile. A generic definition of a missile can be as follows. *An aerospace vehicle with a propulsion system, typically rocket propulsion, carrying a warhead as payload launched to destroy or cause damage to the target and designed to move towards the target is a missile.* The incorporation of energy sources in the missile to provide the required force (propulsion) for it's movement, intelligence to go in the correct direction by effective maneuvering (guidance and control), the means to damage the target (warhead) and the structure which houses all these components (airframe) are the main subsystems of a missile.

Evolution of missile technology

Arrows and spears → **Flaming arrows** → **Bamboo tubes filled with explosive powder** → **First solid-propellant rockets**

- Arrows and spears: Thrown from a distance for hunting or warfare since pre-historic times
- Flaming arrows: Greeks used against the Arabs in the war of Constantinople in the
- Bamboo tubes filled with explosive powder: The Chinese used to propel the tubes to some distance in the
- First solid-propellant rockets: Used in the war between Venice and Genoa in 14th century

Rockets as unguided missiles → **Early investigations** → **Multi-stage rocket** → **Swept-back fin arrangement**

- Rockets as unguided missiles:
  - Rockets have been used by the Chinese employed them as unguided missiles to repel the Mongol besiegers of the city of Pein-King
  - The high-powered firearm was utilized in the southern provinces to thwart the Japanese marauders
- Early investigations:
  - Albert Magnus (Germany) and Roger Bacon (England) (13th century) and Leonardo da Vinci and Giovanni da Fontana (Italy), (15th century) had begun investigating the design of rockets.
- Multi-stage rocket:
  - Conrad Haas (German, 16th century) sketched a multi-stage rocket in which the first stage burns itself out
- Swept-back fin arrangement:
  - Haas also proposed a swept-back fin arrangement to improve the stability of the rocket

Incendiary rockets → **Congreve rockets - Willim Congreve, a British army Colonel (19th century)** → **Spinning rockets - William Hale, an Englishman** → **Spinning rockets - Wilhelm Unge, a Swedish engineer (19th century)**

- Incendiary rockets:
  - The armies of various kingdoms (18th century) in India used in the battle against the British and French forces
  - Weighing about 3 to 6 kgs and could be fired either in a ballistic path or in a horizontal path close to the ground
  - Simple design, could be easily transported, easily operated, and were readily producible
- Congreve rockets:
  - Stabilizes Incendiary rockets using the guide stick and increase weight to about 150 kgs, and designed them to carry a warhead of about 25 kgs at the nose-end.
  - Used extensively against the French in the Napoleonic wars in the beginning of the 19th century and War of 1812 and First Anglo-Burmese War in 1824-26
  - In 1807 the city of Copenhagen was destroyed by about 25,000 to 40,000 such rockets fired during a three-day siege
- Spinning rockets - William Hale:
  - Stability is achieved by spinning the rocket (weighing 27kg). This was done by directing the propellant exhaust through some slanted exit holes at the rear of the rocket
  - Used by United States Army in the Mexican–American War (1846-48)
- Spinning rockets - Wilhelm Unge:
  - Worked to improve the range and accuracy of Hale rockets by using improved propellant fuels. The research was financed by Alfred Nobel, the inventor of dynamite
  - His research concluded that spinning stabilizes a rocket in its flight, but in his design the spin was imparted to the rockets by spinning the launcher tube itself.
  - In 1908 the German company Krupp bought all of Unge's patents

Evolution of missile technology (1/2).

7.3. BASIC CONCEPTS

Prof. A.M. Low (1888-1956, London) (Radio-controlled rockets)
- Designed a small radio-controlled monoplane called the Flying Target
- Designed radio-controlled rockets

Charles Kettering (USA, 1918, aerial torpedo)
- Designed an unmanned airplane intended to serve as a missile with range of 100 km and speed of about 150 kmph.
- The flight direction was controlled by preset vacuum relays and after a certain time, its engine was automatically shut off and the wings were released so it fell along with its payload onto the target in a ballistic trajectory

Werner von Braun (Secret research establishment at Kummersdorf (1932))
- Treaty of Varsailles(1919) imposed restrictive conditions on German research and development activities relating to military weapons
- Rocket research was not explicitly mentioned in the treaty; hence, Germany started a top research establishment lead by Braun along with Hermann Oberth, Johannes Winkler, and Rudolf Nebel to develop rockets

Secret research establishment at Peenemunde headed by Werner von Braun
- The Peenemunde laboratories were equipped with a few wind tunnels and other elaborate test facilities which was a the first of its kind and a systematic and concerted research effort devoted to the development of a flight vehicle.
- This intensive research activity, prior to the World War II, led to the development of the famous vengeance missiles V-1 and V-2
- First instance a man-made object penetrating the atmosphere and enter space

German scientists in US after WWII
- The first US missile programme Hermes was conducted by von Braun association with the General Electric Company, it served as the base design for many ballistic missiles developed later
- The German scientists were involved in developing many surface-to-surface missiles, e.g., Corporal (Range 130kms, length 15 m, launch weight 5500kgs), Sergeant (Range 40-140) kms, length 11 m, launch weight 4500kgs), and Redstone (Range 400 kms, length 22 m, launch weight 30,000kgs)

Missile programs in US
- One of the first guided missiles to be built and tested in the US was the Convair MX-774 which was modelled on the German V-2.
- Testing data of Convair was used in the development of Atlas, the first ICBM to be built in the west

Missile programs in Russia
- They used the basic V-2 design to build a missile called M-101 (forerunner of many huge Russian ICBMs).
- The design of the German V-1 missiles were also exploited by both the Soviet Union and the US to develop cruise missiles - the JB (jet-bomb) series of cruise missiles were developed by the US, while the J-1 J-2, and J-3 cruise missiles were developed by the Russians.

Missile programs during 1950-1970
- Terrier/Tartarship based SAMs, Talos air defense SAM, Triton SSM, and the Typhon ship-based area defense SAMs), Sparrow (AAM), the Nike family of air defense SAMs, Hercules (SAM), Falcon (AAM), Phoenix (long range AAM), Maverick (ASM), Sidewinder (AAM), Chaparral (SAM), Northrop SM-62 Snark (cruise missile), SM64 Navaho; cruise missile with nuclear warhead), Pershing(SSM), Hawk(SAM), and the Patriot (SAM)

Evolution of missile technology (2/2).

Current missile programs around the world.

SRBM - Short-range ballistic missile; MRBM- Medium-range ballistic missile; IRBM-Intermediate-range ballistic missile; ICBM - Intercontinental ballistic missile; LACM -Land Attack Cruise Missiles; ASCM-Anti-ship cruise missile; ALCM - Air-launched cruise missile; LACM -Land Attack Cruise Missiles; GLCM-Ground Launched Cruise Missile; HGV-Hypersonic glide vehicle; ALCM-Air-launched cruise missile; SLBM -Submarine-launched ballistic missile; ASM -Air-to-surface missile; ATGM-Anti-tank guided missile

Missile Name	Country/Type	Range (km)
India		
Agni-I	SRBM	700 - 1,200
Agni-II	MRBM	2,000 - 3,500
Agni-III	IRBM	3,000 - 5,000
Agni-IV	IRBM	3,500 - 4,000
Agni-V	ICBM	5,000 - 6000
BrahMos	Cruise	300 - 500
Dhanush	SRBM	250 - 400
Nirbhay	Cruise	800 - 1,000
Prahaar	SRBM	150
Prithvi-II	SRBM	350
Sagarika/ Shaurya	SLBM	700 / 3,500
Israel		
Delilah	LACM	250 - 300
Gabriel	ASCM	35 - 400
Harpoon	ASCM	90 - 240
Jericho 2	MRBM	1,500 - 3,500
Jericho 3	IRBM	4,800 - 6,500
LORA	SRBM	280
Popeye	ALCM	75 - 100
Russia		
Kalibr (SS-N-30A)	LACM	1,500 - 2,500
3M-54 Kalibr	ASCM	220 - 300
Iskander	SRBM	500
9M729(SSC-8)	GLCM	2,500
Avangard	HGV	6,000+
Kh-101 / Kh-102	ALCM	2,500 - 2,800
Kh-47M2 Kinzhal	ALBM	1,500 - 2,000
Kh-55 (AS-15 Kent)	ALCM	2,500
OTR-21 (SS-21)	SRBM	70 - 120
P-800	ASCM	300
R-29 Vysota	SLBM	6,500
Shtil (SS-N-23)	SLBM	11,000
R-36 (SS-18 Satan)	ICBM	10-16,000
Granat (SS-N-21)	Cruise	2,400-3,000
RS-24 Yars (SS-27)	ICBM	10,500
RS-26 Rubezh	ICBM	2,000-5,800
RS-28 Sarmat	ICBM	10,000+
Bulava (SS-N-32)	SLBM	8,300
Topol (SS-25 Sickle)	ICBM	11,000+
Topol-M (SS-27)	ICBM	11,000
UR-100 (SS-19)	ICBM	10,000
France		
ASMP	LACM	80 - 300
APACHE AP	ALCM	140 - 400
Black Shaheen	ALCM	290
Exocet	ASCM	40 - 180
M51	SLBM	8,000
Storm Shadow	ALCM	250-400
SCALP Naval	SLCM	1000-1400
United Kingdom		
Brimstone	ASM	8 - 60
Harpoon	ASCM	90 - 240
PGM-500	ASM	15 - 50
Storm Shadow	ALCM	250-400
SPEAR 3	ALCM	120-140
Trident D5	SLBM	12,000
United States of America		
ALCM	ALCM	950 - 2,500
FGM-148 Javelin	ATGM	2.5-4.5
Harpoon	ASCM	90 - 240
Hellfire	ASM	7 - 11
JASSM / JASSM ER	ALCM	370 - 1,000
ATACMS	SRBM	165 - 300
Minuteman III	ICBM	13,000
Tomahawk	Cruise	1,250-2,500
Trident D5	SLBM	12,000

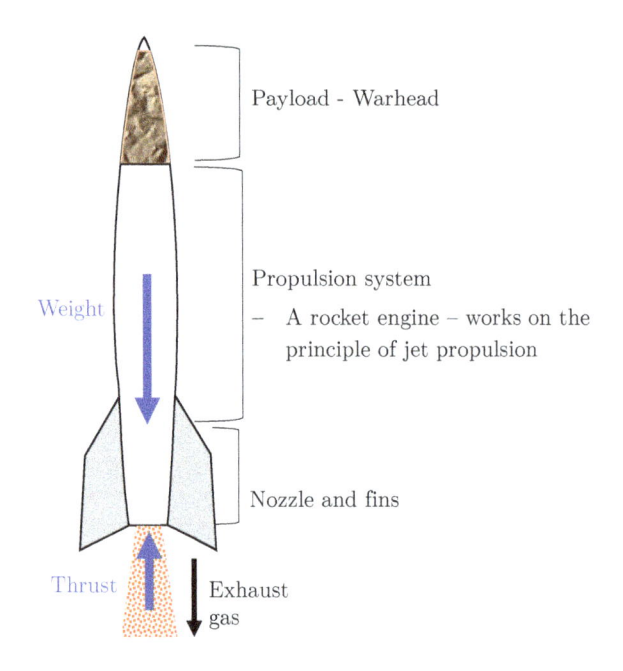

Payload - Warhead

Weight

Propulsion system
— A rocket engine – works on the principle of jet propulsion

Nozzle and fins

Thrust   Exhaust gas

Strategic vs. Tactical Missiles

Based on the distance over which missiles are used, the context in which they are employed and the warheads they can carry, the missile can be referred to as "Strategic" or "Tactical". The terms *strategic* and *tactical* are used to differentiate between the intended purpose and use of different types of missiles. Strategic missiles are designed to achieve broad, long-term military and political objectives, such as deterring or responding to large-scale threats. Tactical missiles are designed for more immediate, localized military objectives on the battlefield.

Strategic missiles have long-range, typically use nuclear warheads with yields that vary from about 100 kilotons to over a megaton of TNT which is used against bigger targets like cities. Strategic missiles are used to achieve long-term goals, the kind of goals that will make a strategic difference in the outcome of a war. By the very nature of their function, they are large in size, have long ranges, and have vast destructive capabilities. Strategic missiles are mainly offensive weapons. Long-range guided missiles need large quantities of fuel and complex guidance and control systems. These missiles are usually stored in specially designed areas. Later-generation ballistic missiles are designed for underground storage and to be launched from these silos as retaliatory measures in the event of an attack from an unfriendly nation.

Tactical missiles are short-range missiles, typically using conventional warheads with lower yields (measured in kilotons, 1-50) which are used on the battlefield. Tactical missiles are used for achieving some short-term missions. By the very nature of their function, they are small in size, have short ranges and have limited destructive power. Tactical missiles are mainly defensive weapons with limited capability for the offense.

Exceptions include short-range tactical carrying nuclear weapons and long-range missiles with conventional warheads. For example, Global Prompt Strike (PGS), the USA's precision-guided conventional weapon, is capable of airstrike anywhere in the world within one hour, similar to strategic nuclear missiles. On the other hand, some tactical nuclear weapons are capable of causing widespread destruction as 15 kilotons atomic bomb dropped by the US on Hiroshima.

Guided vs Unguided Missiles

Missiles can be divided into two categories based on the guidance system: unguided missiles and guided missiles.

Unguided missiles follow the natural laws of motion under gravity and aerodynamic forces to establish a ballistic trajectory. An unguided missile is usually called a rocket and is normally not a threat to airborne aircraft. *Guided missiles* carry within themselves the means for controlling their flight path. Guided missiles are homing missiles that include propulsion system, warhead, guidance system and sensors such as radar and lasers. Movable aerodynamic control surfaces are moved based on commands from the guidance system in order to direct the missile in flight. An alternative way of controlling the missile is by vectoring the thrust produced by the propulsion system.

Cruise vs Ballistic Missiles

Cruise missile is a dispensable, pilot-less, self-guided (requires continuous guidance), continuously powered, and air-breathing vehicle supported by aerodynamic surfaces and designed to deliver warhead to the target.

Ballistic missile is powered and hence usually guided only during the initial part of its flight, after which it follows a free-fall trajectory governed only by the gravitational and aerodynamic forces. Ballistic missiles typically belong to the strategic missile class and have a ballistic trajectory over most of their flight paths. Once the fuel is used up, the missile's direction is not able to be altered except in the case where atmospheric entry guidance and control are employed. Eventually, gravity and aerodynamic forces guide the missile - and its payload, which might be an explosive, a chemical or biological weapon, or a nuclear device - down towards its target. Ballistic missiles are not typically used against moving and maneuverable targets. A bal-

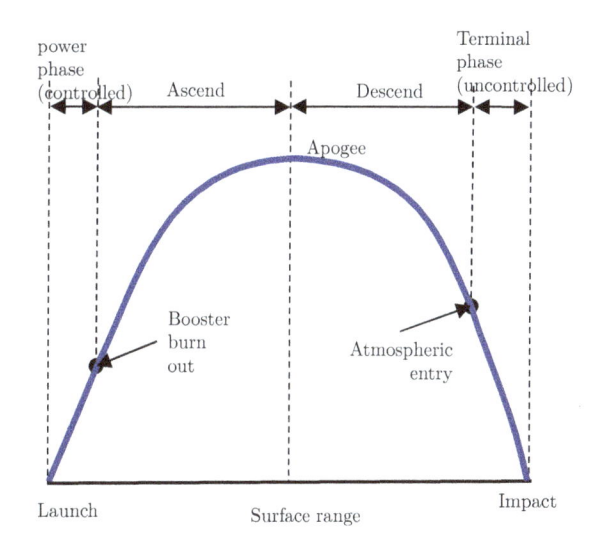

Ballistic missile trajectory

listic missile remains guided during the powered portion of the flight. Further, it follows a ballistic trajectory (free-falling body) after the thrust is terminated. Therefore, typical ballistic missile do not use aerodynamic surfaces to produce lift. The trajectory of the ballistic missile can be divided into three segments: (1) the powered portion of the flight which lasts from launch till thrust cutoff or burnout (2) the free-flight portion, which constitutes most of the trajectory and is typically outside the Earth's atmosphere where the only significant force acting is the gravitational pull of the Earth, and (3) the terminal phase, which begins at some point where the aerodynamic drag becomes a significant force in determining the missile's trajectory and lasts until the impact on the target. During the powered flight, inertial guidance is used for a ballistic trajectory in order to establish proper velocity for a hit by free fall. The ballistic missiles are roll stabilized and have no coupling between the longitudinal and the lateral motions. Due to the relatively high speeds, ballistic missiles are least likely to be intercepted. Rocket propulsion is employed in ballistic missiles such as Intercontinental ballistic missiles (ICBMs) to accelerate the missile to a position of high altitude and speed. This places it on a trajectory that meets certain guidance specifications in order to carry a warhead, or another payload, to a pre-selected target. An operational ballistic missile may be flying at speeds as high as 24000 km/hr during its atmospheric entry.

Range of Missiles

Missiles can be short-range (up to 500 km), medium-range (up to 1000 km), intermediate-range (up to 5000 km) and intercontinental ballistic missiles (above 5000 km). Russia's longest-range operational missile is the inter-continental ballistic missile (ICBM) *R-36*, which can hit targets up to 16,000 kilometers away.

7.4 Missile Subsystems

The main components of missiles are the airframe, propulsion system, guidance and control system, and warhead.

7.4.1 Airframe

The airframe is the structure that houses all the missile subsystems. By the very nature of the missile and its functionality, the construction of the missile is different from the aircraft. Missiles are expendable and uninhabited. Hence, the factor of safety used in the structural design of missiles is significantly smaller than that used in airplanes. Also, missiles are designed to have increased ranges of speed, altitude, and maneuvering accelerations, which have brought with them new aerodynamic problems. For instance, the higher allowable altitudes and maneuvering accelerations permit operation in the nonlinear range of high angles of attack. A missile may be ground-launched or air-launched and in consequence, can undergo large longitudinal accelerations, can utilize very high wing loading, and can dispense with the landing gear. Since missiles are pilot-less, sometimes they can be permitted to roll and thereby introduce new dynamic stability phenomena. Many missiles tend to be slender and utilize more than the usual two-wing panels. These trends have brought about the importance of slender-body theory and cruciform aerodynamics for missiles.

Airframe Structures and Materials

The structural integrity of the missile is crucial not only for its operation in space but also for its safe deployment and journey into space. The main structural components of the missile are primarily designed to withstand the various loads experienced during launch. The forces generated by thrust at the lower end of the vehicle during launch need to be transferred through the structure. To facilitate the transfer of loads throughout the vehicle's structure, two distinct categories of spacecraft structures can be identified: strutted structures and central cylindrical shell structures. *Strutted structures* utilize struts, which are elements designed to primarily bear compressive loads along their longitudinal axes. Strutted structures often resemble truss configurations and rely on interconnected struts to distribute loads efficiently. In contrast, *central cylindrical shell structures* feature a central cylindrical component that serves as the primary load-bearing member which is typically designed to withstand the loads imposed by thrust during launch. Other subsystems of the missile are either directly attached to this central cylinder or connected through a combination of struts, platforms, and shear webs.

Materials utilized in missile must meet specific design and quality criteria, including high strength-to-weight ratio to make lightweight structures, excellent corrosion resistance to ensure proper performance during storage, high fracture toughness, good fabricability, defect detectability, and reliability. Metallic materials such as magnesium, aluminum, titanium alloys, and maraging steel are widely utilized in conjunction with nonmetallic materials like polymers and carbon-carbon composites. *Magnesium alloys* are employed in outer shells, wings, and control surfaces because of their reduced density, leading to lighter weight, and their effective damping capabilities. *Aluminum alloys* are widely utilized in both airframe structures and propulsion systems, taking the form of sheets, ring-rolled forgings, extrusions, forgings, and castings. *Titanium alloys* are chosen for applications like high-pressure air bottles and sustainer casings because they offer exceptional strength while weighing only half as much as steel. High strength *steels* finds application in rocket motor casings and combustion chambers. The steel is preferred for its remarkable combination of high strength and fracture toughness, as well as its minimal distortion resulting from the absence of water quenching after heat treatment. Soft *magnetic alloys*, controlled expansion alloys, hard tungsten alloy spheres, nickel-based superalloys, beryllium copper, and oxygen-free high-conductivity copper are commonly employed in the manufacturing of gyroscope, accelerometer, and other control and guidance system components. Polymeric materials find diverse applications in missiles such as in propulsion systems as high-energy binders for composite propellants, as lining materials in combustion chambers with thermal stability and toughness, as thermally stable ablative polymers for protection against erosion, as adhesives for metal and carbon-FRP bonding and elastomeric sealant.

Missile Configurations

Based on the source of lift and location of the control surfaces, different basic missile airframe configurations are possible.

Cruciform missiles
(control surfaces and/or lifting surfaces are located at 90^0 from each other)

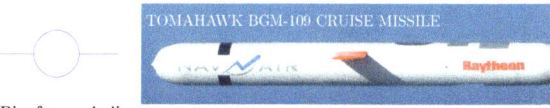

Planform missiles
(control surfaces and/or lifting surfaces are located at 180^0 from each other)

Control Mechanisms

The flight control system is responsible for stabilizing the missile, controlling the missile in its flight and ensuring that the missile airframe responds effectively to guidance commands. The control system in a missile is necessary to maintain the stability of the missile and implement the guidance commands to maneuver the missile in flight. Missiles can be further classified according to the control mechanisms as follows.

Aerodynamic control consists of external structures called control surfaces located on the missile body. These surfaces are responsible for generating deflection forces which are required to maneuver the missile in pitch, roll and yaw motion. Based on the longitudinal location of control surfaces on the body, missiles can be further classified into three types i.e., wing control (located at the center portion), tail control (at the rear end), and canard control (located at the front end). Examples: *Akash*

Various airframe configurations

Examples	Lift generating surfaces	Control surface	Comments
Sidewinder AIM-9L/9M, Chaparral MIM-72, GBU-15, Stinger FIM-92, Redeye MIM-43 and RAM	Body Tail	Canards	Deflection of canards produces the desired control force to cause a change in the angle of attack of the missile, which in turn produces the required lift on the body and tail of the missile. The servos used are small due to the large distance between the canards and the moment center.
Sparrow AIM-7F or 7M Sea Sparrow RIM-7M	Body, Tail and Wings	Wings	Deflection of the wings produces a force which moves the missile laterally and causes a change in its angle of attack which in turn causes a lift to be generated by the body, tail, and the wings. This airframe has a reasonably fast response. A large control force is required hence larger servos need to be placed near close to wings to cause the required lateral movement of the body as the wings are placed close to the moment center.
Phoenix AIM-54A AMRAAM AIM-120A Harpoon AGM/RGM Standard SM-2 Maverick AGM-65	Body Wings	Tail	Deflection of tails generates the force which causes the missile to rotate and change its angle of attack. Servos used to deflect the control surfaces are small, and the response of this type of airframe is quite slow
Patriot MIM-104	Body	Tail	Servos used to deflect the control surfaces are small, and the response of this type of airframe is quite slow

Airframe	Features
Cruciform	Four fixed wings and four movable control surfaces placed at 90° to each other. This configuration permits lateral maneuvering in any direction without first rolling. Sets of controls at right angles permit the missile to turn immediately in any plane without the necessity of its banking
Planform	A bank-to-turn missile, like an airplane, banks into the turn to bring the normal acceleration vector as close to the vertical plane of symmetry as possible

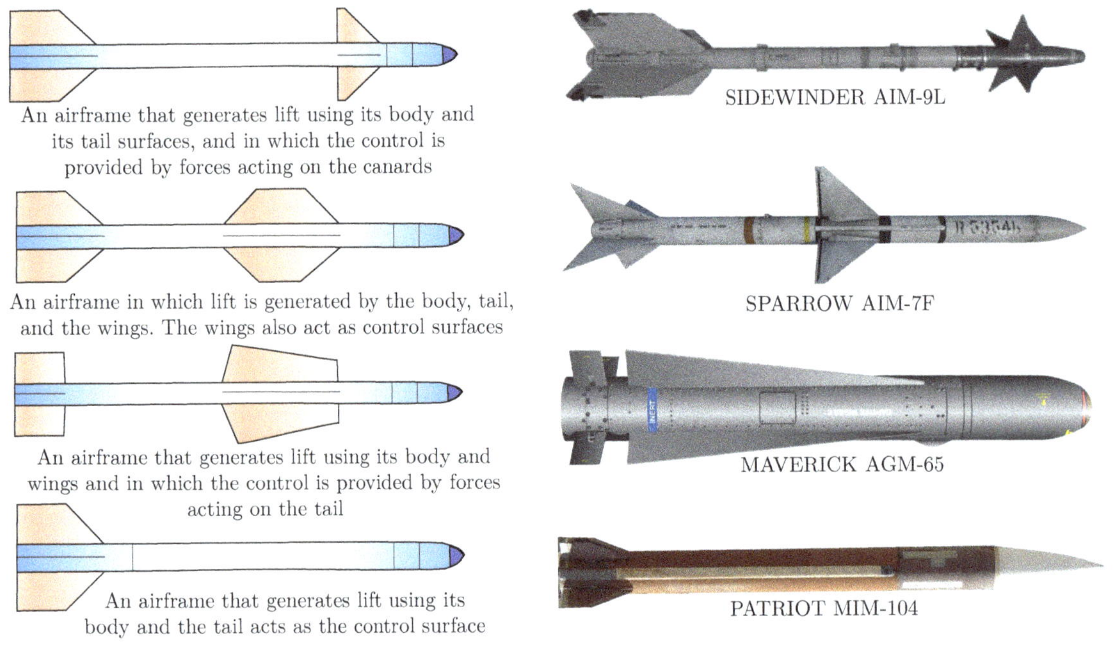

An airframe that generates lift using its body and its tail surfaces, and in which the control is provided by forces acting on the canards

SIDEWINDER AIM-9L

An airframe in which lift is generated by the body, tail, and the wings. The wings also act as control surfaces

SPARROW AIM-7F

An airframe that generates lift using its body and wings and in which the control is provided by forces acting on the tail

MAVERICK AGM-65

An airframe that generates lift using its body and the tail acts as the control surface

PATRIOT MIM-104

Figure 7.1: Airframe configurations (Drawn not to scale)

is wing-controlled SAM, and *Sparrow* missile uses wing control with tail fins for stability and aerodynamics. *Nag* is a tail-controlled ATGM, *Phoenix (Falcon)* missile uses tail control, via fins with fixed wings. *Python* is a canard-controlled air-to-air missile and a typical *Sidewinder* uses fixed tail wings, with movable nose fins (canards).

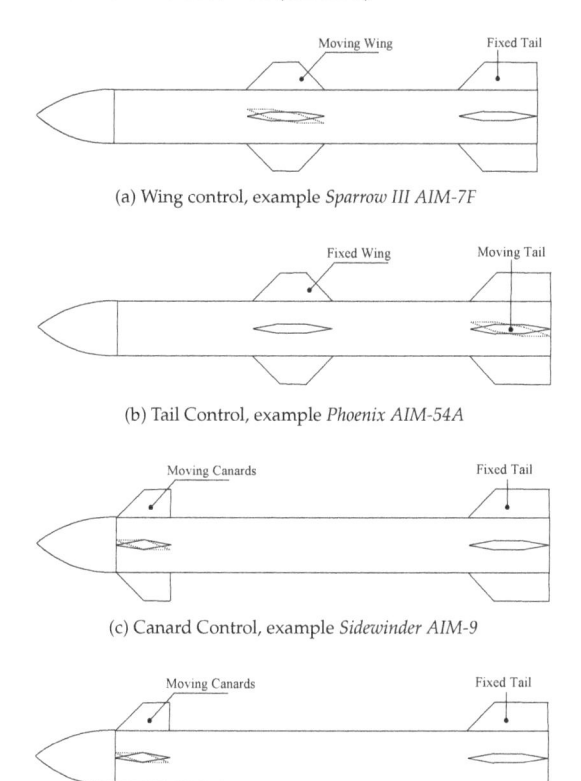

(a) Wing control, example *Sparrow III AIM-7F*

(b) Tail Control, example *Phoenix AIM-54A*

(c) Canard Control, example *Sidewinder AIM-9*

(d) TVC, example *Trident C4* gimballed nozzle

Thrust Vector Control (TVC) generates control forces similar to aerodynamic control but using deflection of the thrust (exhaust jet) axis. Deflection of the thrust causes a moment to be generated about the center of gravity since forces act at the rear end of the missile. The vehicles can be controlled in various ways, including by deflecting the main engine thrust vector through gimballing the thrust nozzle or using movable vanes in the exhaust. This system is useful for controlling the missile at very low velocity (launch phase of a huge ballistic missile) or high altitudes (atmospheric density is very low) where sufficient aerodynamic forces cannot be generated. Examples: *Trident C4* uses gimballed nozzle TVC, *Javelin* is a jet-vane controlled ATGM. Many modern missiles, like air-to-air missiles, use deflector vanes in their rocket exhaust. This helps them make sharp turns to the left or right. These missiles use thrust-vectoring augment canards to control their pitch and yaw and tail ailerons to roll control. Thrust vector control is especially useful in short-range air-to-air missiles and vertically launched intercontinental ballistic missiles (ICBMs), as well as submarine-launched missiles like the Trident, where early boost course corrections are necessary. Another method is Secondary Injection Thrust Vector Control (SITVC), where fluid injected into the thrust nozzle deflects the rocket thrust gases to steer the vehicle. This method offers fast response times compared to mechanical systems.

Reaction Control System uses multiple small rocket motors mounted on the missile body apart from the main rocket motor. These micro thrusters are set on and off as per requirements of missile position correction. RCS is mostly used for missile systems that operate at high service ceilings or exo-atmosphere. Examples: *THAAD* Anti-ballistic air defense system.

7.4.2 Propulsion System

The propulsion system is the main thrust provider for the missile. It provides forward acceleration to the missile body and helps it to reach the target. The propulsion system of the missile provides the required initial thrust to the missile to enable it to fly with sufficient velocity during the subsequent engagement period with the target. Sometimes there are two phases in the propulsion: *Boost* and *Sustain*. During boost, the propulsion system provides a high level of missile acceleration over a relatively short period (1-15 seconds). The purpose of sustained propulsion is to maintain the missile at the desired velocity for most of the remaining missile flight.

A typical propulsion system contains these important components, a pressure chamber, propellant, an igniter and a nozzle for thrust generation.

The combustion of chemicals takes place inside the combustion chamber at high temperature and pressure, and the resulting hot gases are passed through the nozzle to obtain the necessary thrust to propel the missile. The propulsion system is initiated using the igniter, which is triggered using an electrical pulse.

Solid Propellant Rocket

The rocket motor consists of propellant in the solid phase (also known as 'grain'), encased inside the combustion chamber. The solid propellant grain has both fuel and oxidizer mixed in proportion and then it is casted/extruded according to the desired dimensions of the combustion chamber. The solid propellant is placed inside a casing that is designed to withstand high pressures. The solid propellant is the most widely used propellant, with applications ranging from small tactical missiles (which contain a few kilograms) to ballistic missiles containing tonnes of solid propellant. Examples: *Nag* Anti-Tank Guided Missile. *Agni-V* Inter-Continental Ballistic Missile. The solid propellant rocket motor has a long storage life, typically more than a decade. Its simple design and ease of use make it the most suited candidate for missile applications.

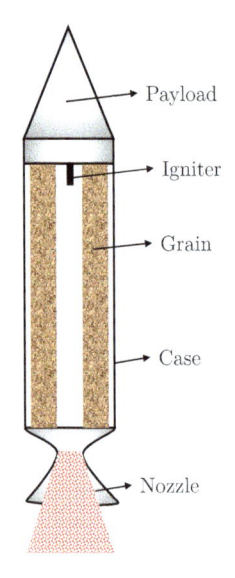

Solid rocket motor.

Liquid Propellant Rocket

The liquid propulsion system consists of liquid fuel and liquid oxidizer stored in different tanks inside the missile. Two pumps are used to pump liquid fuel and oxidizer into the combustion chamber and ignite it to generate high-temperature and pressure gases which later expand through the nozzle. The pumps are driven using a separate turbine-driven system. Instead of a pump, a pressure fed system can also be used. If the propellants are stored at cryogenic temperatures, which are liquid oxygen existing below $-183°C$ and liquid hydrogen below $-253°C$, then the liquid propulsion system is known as a cryogenic engine.

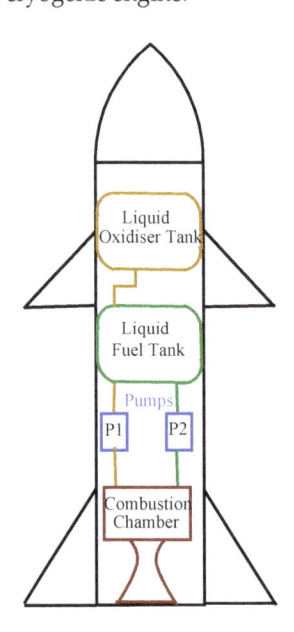

Liquid propellant rocket engine

Turbojet Engine

The missiles that typically fly at constant speeds (cruise mode) incorporate turbojet engines similar to commercial aircraft. These engines fall in air breathing class of propulsion systems. The oxidizer in solid or liquid propulsion in rockets contributes up to 75% of the total weight of the propulsion system. Hence, the missile system utilizing air-breathing engines is light in weight and thus covers a longer distance. The limitation of using a turbojet engine is additional moving parts, which is attributed to the higher cost of the missile (usually single-use systems). Examples: *Nirbhay* subsonic cruise missile and *Tomahawk* subsonic cruise missile.

Ramjet Engine

Ramjets are air-breathing engines that do not contain any rotary parts, such as turbojet engines. Ramjet's engine utilizes high-speed incoming air to increase the combustion chamber pressure, where the fuel is injected, and thrust is obtained by passing hot combustion gases through the nozzle. In a ramjet engine, the combustion process occurs at subsonic speeds within the combustor. For a vehicle traveling at supersonic speeds, the air entering the engine must be decelerated to subsonic velocities through the use of shock waves generated in the inlet. However, as the vehicle's speed increases much beyond Mach number 5, the performance losses caused by these shock waves become so significant that the engine can no longer produce net thrust. This performance limitation poses a challenge for the development of ramjet-powered vehicles intended to operate at hypersonic speeds. Another limitation of the ramjet engine is that it cannot be used at very low or zero air velocities. Therefore, they are assisted by a solid rocket booster to achieve the initial ramming speed and generate initial thrust. Examples: *Bloodhound* Surface-to-Air Missile. *BrahMos* supersonic cruise missile.

Scramjet Engine

In the 1960s, an improved ramjet was proposed, where the combustion in the burner would occur supersonically. This new design, known as the supersonic combustion ramjet or scramjet, aimed to minimize the losses associated with slowing the flow, allowing the engine to produce net thrust for a hypersonic vehicle. Unlike a rocket, which must carry all its oxygen, the scramjet uses external air for combustion, making it a more efficient propulsion system for flight within the atmosphere. Scramjets are particularly well-suited for hypersonic flight within the atmosphere.

in 2021, U.S. Defense Advanced Research Projects Agency (DARPA) developed and had a successful flight of the Hypersonic Air-breathing Weapon Concept (HAWC), a scramjet-powered hypersonic air-launched cruise missile. Unlike traditional explosive-based weapons, the HAWC is designed as a kinetic energy weapon, utilizing its immense speed and impact force to deliver a devastating blow without the need for an explosive warhead.

7.4.3 Navigation, Guidance and Control

The problem of guiding the missile towards the target poses considerable challenge to the missile guidance system. Further, the coupling of the automatic guidance system and the airframe introduces problems in stability and control. The guidance subsystem acts like the sensory organs and the brain of the missile. It acquires the position of the target relative to the missile and subsequently tracks the target. During the tracking, it decouples the seeker motion from the missile body motion and disturbances, thus improving the stability of the seeker system. With the signals from the target collected by the seeker system and using appropriate algorithms, the guidance system generates relevant commands to guide the missile for an intercept of the target.

The guidance system generates guidance commands which attempt to change the missile's flight direction. Generally, this decision has to be taken at very short intervals of time (of the order of milliseconds) during the flight of the missile. A missile guidance system has the relevant sensors that measures the position of the guided missile with respect to its target and changes the missile flight path in accordance with a guidance law. Usually, the missile guidance system includes sensing, computing, and control components. *A guidance law* is an algorithm that determines the required commanded missile acceleration needed for an intercept of the target. Some of the terminologies related to the missile guidance system are as follows:

Flight path planning refers to determining a nominal trajectory and associated control inputs for a given flight vehicle to accomplish specified objectives under specified constraints.

Navigation is the determination of a strategy for estimating the position of a vehicle along the trajectory, given the outputs from specified sensors.

Guidance is the determination of a strategy for following the nominal path in the presence of off-nominal conditions, wind disturbances, and navigational uncertainties.

Control is the determination of a strategy for maintaining the angular orientation of the vehicle during the flight that is consistent with the guidance strategy, and other constraints.

Typically, missile guidance is different during the three phases of its trajectory, viz., boost, mid-course and terminal. The initial boost phase starts from the time the missile leaves the launcher till the booster burns all of its fuel. Depending on the objectives, the missile may or may not be actively guided during this phase. The mid-course phase is usually the longest phase in its trajectory. During this phase, depending on system requirements, guidance may or may not be required to bring the missile onto the desired situation from which the terminal guidance can successfully take over. The terminal phase is the last and most critical phase of guidance and must have high accuracy and a fast reaction to ensure an intercept with the target. In this phase, the guidance seeker is locked onto the target, allowing the missile to be guided all the way to the target. Significant amount of work has been done to develop accurate equipment for use in this guidance phase.

The missile guidance system can be homing or non-homing or direct (external) guidance. In a non-homing guidance system, the missile and the target are continuously tracked from one or more vantage points, and the necessary path for the missile to intercept the target is computed and relayed to the missile by some means such as radio. A homing missile has a seeker, which sees the target and gives the necessary directions to the missile to intercept the target. The homing missile can be subdivided into classes having active, semiactive, and passive guidance systems.

Non-Homing Guidance

Inertial guidance: The sensors in this system are accelerometers that measure translational acceleration. Then numerical integrations are performed using electronic circuits to calculate the velocity and distance traveled. The accelerometer thus makes it possible to keep track of the distance traveled from the launching pad and simultaneously the distance from the target.

Homing or Seeker Guidance

The missile pursues the target making use of a target seeker and an onboard computer. Homing guidance is used to describe the guidance process that can determine the position of the target with respect to the missile and can develop the necessary commands to guide itself to the target. More specifically, a homing system involves selecting, identifying, and following a target using some distinguishing characteristic of the target. Such identifying characteristics can be radiated heat or reflections of radar waves from the target.

Homing is advantageous in tactical missiles where requirements such as "fire-and-forget" require sensing of target to be done from the missile itself. Homing guidance can be active, semi-active or passive based on the location and methods of using the components of homing guidance systems. In the active class, the missile illuminates the target and receives the reflected signals. In the semiactive class, the missile receives reflected signals from a target illuminated by means external to the missile. The passive type of guidance system depends on a receiver in the missile sensitive to the radiation of the target itself. An example of an active homing missile system is the European *Meteor* active radar-guided AAM. An example of semiactive missile guidance is found in the supersonic *Sparrow III* missile. The *Sidewinder* is an example of a passive infrared-homing guided missile.

Direct (External) Guidance

This type of guidance system relies on missile guidance commands calculated at the ground launching site and transmitted to the missile.

Beam Rider Guidance

In this approach, the target is tracked using an electromagnetic beam, which may be transmitted from a ground radar. In order to follow the beam, the missile's onboard guidance equipment includes a rearward-facing antenna, which senses the target-tracking beam. By utilizing the properties of the beam, control signals that are a function of the position of the missile with respect to the center of the target-tracking beam are computed and sent to the control surfaces. These signals produce control surface deflections which keep the missile close to the center of the target-tracking beam.

Command Guidance

Here, the guidance commands come from sources external to the missile. In this approach, a tracking system outside the missile is used to track both the missile and the target. Hence, a seeker is not required in a missile using command guidance. The target and missile positions are fed to a computer and using the information, the computer determines the flight path the missile should take that will result in a successful intercept of the target. The *Nike* family uses this type of guidance. Also, the *Patriot MIM-104* air-defense missile uses a modified version of command guidance, in which only one radar is needed. A disadvantage of command guidance is that the external energy source must illuminate the target. Hence, the target may get alerted of the radar's operation and may resort to evasive actions.

Command to Line of Sight (CLOS)

Another type of guidance is when the missile is always commanded to move along the line of sight (LOS) between the tracking unit and the target. Here, the missile is controlled to stay as close as possible on the LOS from the tracking unit to the target. In CLOS guidance, an up-link is used to transmit guidance commands from the ground based controller to the missile. This type of guidance is used mostly in short-range air defense and antitank systems.

Autopilot system

The autopilot is a component of the main guidance system that operates as a closed-loop system, ensuring the missile achieves the desired accelerations and remains stable. This control system specifically includes a roll autopilot.

The role of the autopilot is to stabilize and steer the missile by adjusting the fin positions, which result in the rotation and translation of the missile body. The main goal of designing an autopilot system for a missile is to ensure that the missile remains stable and performs effectively, while also being able to withstand a wide range of flight conditions that it may encounter during its entire flight path. This includes ensuring that the autopilot can handle changes in speed, altitude, and other variables to keep the missile on its intended course. Additionally, the autopilot must be robust enough to adapt to unexpected situations and maintain control of the missile throughout its flight. During the target acquisition phase, its primary function shifts from aligning the seeker gimbal angles (if applicable) to meeting acceleration requirements. The fin servos react to the autopilot's instructions, and the specific fin positions are determined by the balance of servo torque and aerodynamic hinge moment. These adjustments in fin position help to influence the behavior of the airframe according to its dynamic model.

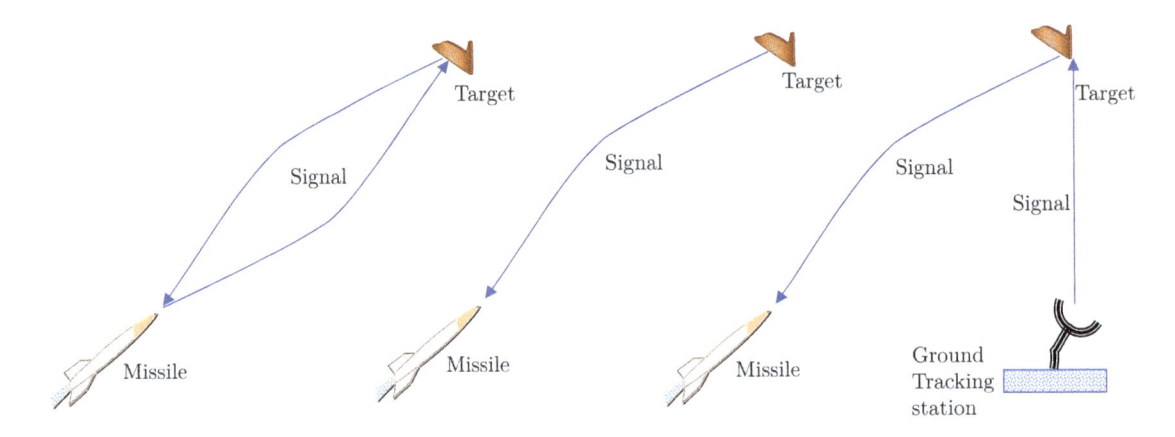

Different categories of homing guidance.

7.4.4 Warheads

The role of the missile is to deliver the warhead to the target, which in turn generates the mechanism to inflict damage to the threat target. The prediction of the amount of damage to a target is expressed as the *Probability of Damage*, which is a statistical measure of the likelihood that the target will somehow be incapacitated and possibly destroyed. The probability of damage depends on the Circular Error Probable (CEP), the vulnerability of the target, and the type of warhead. CEP is the proximity of a warhead to the target, which measures the accuracy of the delivery vehicle. More specifically, CEP is defined as the radius of a circle about the aim point inside of which there is a 50% chance that the weapon will impact and/or detonate. Target vulnerability and warhead type go hand in hand because different weapons will affect each target differently.

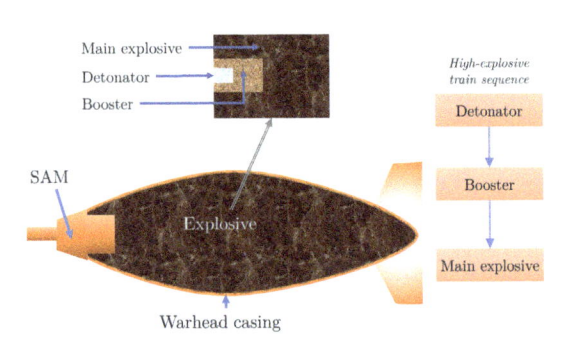

Parts of a conventional warhead.

Commonly used composites explosives which are mixtures of compounds

Composite explosives	Composition
AMATOL	80/20 Ammonium nitrate/TNT
ANFO	94/6 Ammonium nitrate/Diesel oil
COMP A-3	91/9 RDX/WAX
COMP B-3	64/36 RDX/TNT
COMP C-4	91/5.3/2.1/1.6 RDX/Di(2-ethyhexyl) sebacate/Polyisobutylene/Motor Oil
DYNAMITE	75/15/10 RDX/TNT/Plasticizers
Octol	HMX/TNT

Common explosives and their uses

Detonators or Primers	Booster	Main Explosive
Mercury fulminate, Lead azide, Lead styphnate, Tetracene, Diazodinitrophenol (DDNP)	Tetrytol, PETN, Tetryl, TNT	RDX, Comp-A, B, C, Cyclotol, HBX-1, 3, H-6, MINOL 2 Ammonium picrate

Warheads can be broadly classified as conventional or non-conventional types. In the non-conventional types, the warheads can be nuclear warheads or biologically hazardous materials. In a nuclear warhead, the energy for destruction comes from nuclear fission or fusion reactions. The 1972 Biological Weapons Convention prohibits the development of biological weapons. As of 2021, 183 countries have become party to the treaty.

The conventional warhead has three functional

parts the Safety and Arming Mechanism (SAM) or Fuze, the explosive fill and the warhead casing. The SAM contains the equipment for detecting the proximity to the target and initiating the detonation sequence. It also contains one or more safety mechanisms that prevent the inadvertent detonation of the main charge. SAM also provides the start of the high explosive train, which consists of the detonator, which is a small amount of primary high explosive, and possibly the booster charge.

Warhead of *Hunter* loitering munition.

The conventional high-explosive warheads can be further classified into directed energy warhead or the chemical energy warhead. In the directed energy warhead type, the explosive energy converges a metal liner into a high speed jet. Depending on the shape of the liner, there are shaped charge (conical liner), hemicharge (hemispherical liner) and explosively formed penetrator (curved liner). The speeds, the shape of the jet formed and its stability are distinctly different for these three types of directed energy warheads. In a shape charge warhead the speed achieved by the jet is of the order of 7-10 km/s and is used against armored vehicles. A chemical energy warhead on detonation creates both blast and fragmentation effects. Blast warheads uses explosive air shocks resulting in over-pressure to cause damage to the target. Fragmentation warheads uses explosive energy to accelerate metal fragments to high velocities to cause the damage.

7.4.5 Missile Launchers

All the missiles need certain infrastructure to launch them at specific targets. The ballistic missiles are launched from underground silos or submarines or mobile vehicle based launchers. Missile silos are underground vertical cylindrical container for storing and launching intercontinental ballistic missiles. They typically have the missile stored some distance below the ground surface, protected by a large blast door on top. Some of the relatively small missiles are launched from a launch-cum-container tube resting on a human shoulder. The launchers can be very demanding piece of engineering effort with precision in aiming the launcher at a particular target and very high rates of turning in elevation and azimuth in case of anti-aircraft missiles.

7.5 Types of Missiles

7.5.1 Based on Launch Mode

Missiles can be classified as Surface-to-Surface Missiles (SSM), Surface-to-Air Missiles (SAM), Air-to-Surface Missiles (ASM) and Air-to-Air Missiles (AAM) based on the position of the launch point and the position of the target. The details of these types of missiles are summarised in table 7.1 and Fig. 7.2.

7.5.2 Based on Target

On the basis of target, missiles can be called anti-tank, anti-personnel, anti-aircraft/helicopter, anti-ship/anti-submarine, anti-satellite or anti-missile. Examples of such missiles are given below.

Anti-tank/anti-armour: *TOW (BGM-71)* (Mach number 0.8–0.9, 3.75 km range, wire-guided optical semiautomatic CLOS and automatic IR track-

Table 7.1: Guided missiles classification based on position of the launch point and position of the target

Missile type	Definition	Target	Design requirements	Example
Surface-to-Surface Missiles (SSM)	Launched from a point on the surface of the Earth to destroy a target on the surface of the earth. *Offensive missiles*	Large and stationary target	– Size – Modularity – Range	– CSS-3 (China, ICBM, 7000 km), – R36M2 Voivode (USSR, ICBM, 12000 km), – Minuteman, (USA, ICBM, 12500 km), – Agni – IV (India, IRBM, 4000km), – Agni-V (India, ICBM, 8000km)
Surface-to-Air Missiles (SAM)	Any guided missile launched from a point on the surface of the Earth to destroy a target in the air. *Defensive weapons*	Small in size, moves at high speeds, and/or are capable of executing complicated maneuvers (e.g., fighter aircraft, helicopters, SSMs)	– Weight – Accuracy – Altitude	– Akash (India, 9km), – Igla (CIS, 6 km), – Stinger (USA, 45km), – Patriot (MIM-104)(US, 80 km)
Air-to-Surface Missiles (ASM)	Launched from an aircraft to destroy targets on Earth. *Offensive/defensive weapons*	Moving (low speed) or stationary	– Versatility – Speed – Modularity	– Gabriel MK-III (Israel, 40 km), – HARM AGM-88A (USA, 25 km), – Nirbhay (India, 1,000 km) – Maverick (AGM-65), (US, up to 22.2 km)
Air-to-Air Missiles (AAM)	Both the launch point and the target are aircraft. *Offensive/defensive weapons*	Targets are small and difficult to locate	– Maneuverability – Performance – Weight & Range	– Super 530 (France, 25 km), – R-77 (CIS, 80+ km), – Sidewinder AIM-9 (USA, 15+ km) – Sparrow III (AIM-7)(USA, 50+ km)

Missile may be sub-classified based on the speed, range and target as follows

Based on speed	Based on range	Based on target
– Subsonic Missile, Mach number < 1 – Supersonic Missile, Mach number > 1 – Hypersonic Missile, Mach number > 5	– Short-range Missile (up to 500 km), Examples: Pershing, Sergeant, and Hawk class – Medium-Range Ballistic Missile (MRBM) (up to 1000 km), Examples: Thor, Jupiter and Trident, – Intermediate-range Ballistic Missile (IRBM) (up to 5000 km), Examples: Minuteman I-III, the MX, and Titan missiles – Intercontinental Ballistic Missile (ICBM) (above 5000km), Examples: Minuteman IV (LGM-30 Minuteman), UGM-133 Trident II	– Anti-Aircraft/Helicopter/UAV – Anti-Ship/Submarine – Anti-Tank – Anti-Satellite – Anti-Missile (Air defense) – Anti-Personnel – Anti-Radiation (Against Radar)

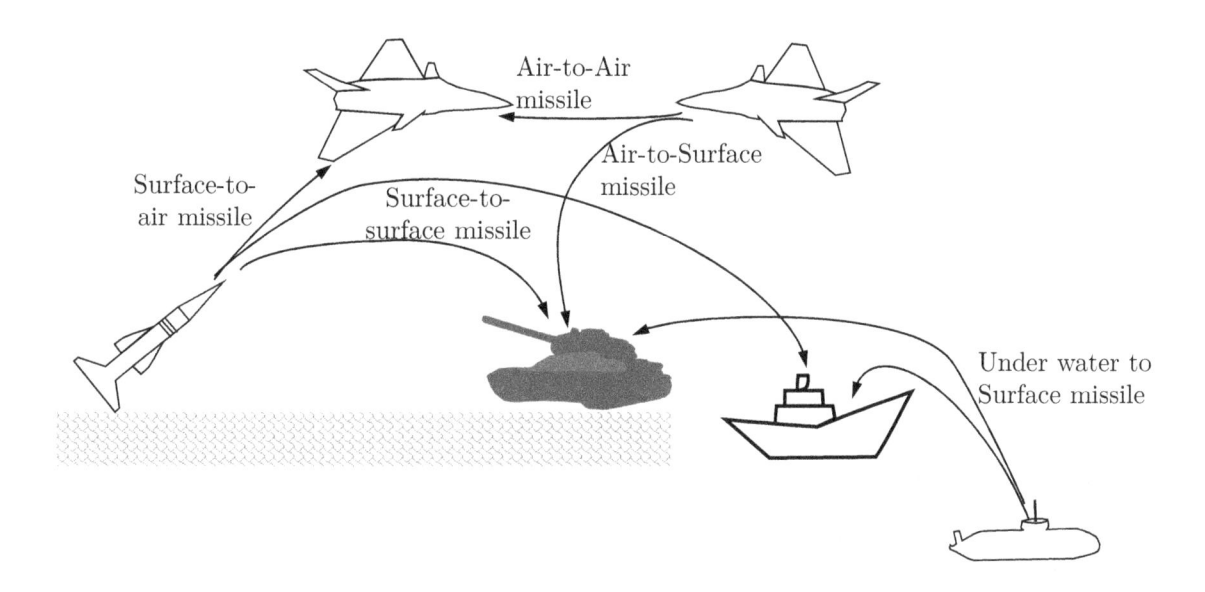

Figure 7.2: Classification of guided missiles based on launch and target locations.

ing) is among the most widely used antitank guided missile. It can fired from helicopters and ground-combat vehicles. The *TOW* was used in Vietnam, Operation Desert Storm, and by the Iranian forces against Iraqi tanks during the 1980–1988 Gulf War.

Hellfire (AGM-114A) (US, Mach 1.1, 8 km range and Laser-guided) is a U.S. Army antitank air-to-ground missile launched from attack helicopters like the *AH-64A Apaches*. *Hellfire II*, developed in 1997 as an anti-ship missile, has a blast fragmentation warhead designed to attack ships, buildings, and bunkers. The warhead penetrates the target before detonation.

Akeron MP (Akeron Moyenne Portée) or MMP (Missile Moyenne Portée - Medium-Range Missile) is a French anti-tank guided missile system, featuring both fire-and-forget and command guidance operating modes, has a range of 4 km.

The MAn Portable Anti-Tank System (MAPATS) is an anti-tank guided missile system, designed for use by infantry, vehicles, and helicopters. It is a laser-guided missile and has a range of 4 km. Known operators of the MAPATS include Israel, Chile, Ecuador, Estonia, and Venezuela.

FGM-148 Javelin (US) is one-man-portable fire-and-forget antitank, guided munition and surveillance weapon system. Fire-and-forget capability means the missile guides itself to the target after launch, allowing the gunner to take cover and avoid counterfire. *Javelin* can be fired from inside buildings or bunkers.

Anti-personnel: *Spike-MR* (Israel) is a lightweight, man-portable, fire and forget, anti-tank guided missile and anti-personnel missile with a range of 2.5 km. The *Spike-MR* has an optional fire, observe and update mode of operation suitable for the modern, multi-faceted battle. The long range version *Spike-LR* has a range of up to 4 km.

Anti-aircraft/helicopter: The *9K38 Igla* man-portable anti-aircraft missile system is designed to engage low-flying air targets. *Roland* is a French / German supersonic short-range surface-to-air missile intended to engage helicopters and aircraft. It is fired from its launch tube using a twin-arm missile launcher. It comes as both armored self-propelled systems for frontline use and as a static system mounted on a truck or shelter. *Crotale* is another French, all-weather, short-range surface-to-air missile system developed to intercept airborne weapons and aircraft.

Anti-ship: *Harpoon (AGM-84)* (Mach 0.85, 139–148 km range and uses a 3-axis attitude reference assembly to monitor the missile's relation to the launching platform) These series of *Harpoons* are long-range sea-skimming anti-ship missiles; they can be launched from bombers, ships, submarines, and coastal defense platforms. Like the French (*Aerospatiale*) *Exocet* and the Norwegian (*Kongsberg*) *AGM-119 Penguin* short-range antiship missiles, the *Harpoon* is a fire-and-forget weapon.

Anti-satellite: Four countries — the USA, China, Russia and India — have destroyed their own satellites in Anti-Satellite (ASAT) missile tests. The *Vought ASM-135A* is a US air-launched, anti-satellite missile developed to destroy a satellite. India's ASAT (*Mission Shakti*) test hit a target satellite at an altitude of 300 kilometers, close to the altitude of the US ASAT test in 2008.

Anti-missile: The *MIM-104 Patriot* is a surface-to-air missile system, the primary of its kind used by the United States Army and several allied states which can target aircraft, cruise missiles and shorter-range ballistic missiles.

7.5.3 Multiple Independently Targetable Reentry Vehicles (MIRVs)

Multiple reentry vehicles are launched by a ballistic missile where each one of them can be directed to a separate target. Such missiles employ a post-boost vehicle or another warhead-dispensing mechanism. The dispensing mechanism maneuvers to achieve successive desired positions and velocities to release each Reentry Vehicle (RV) on a trajectory to attack the desired target. Two characteristics of MIRVs may have significant consequences for military strategy. First, MIRVs can penetrate an anti-ballistic missile system better than missiles with a single warhead. Second, MIRV warheads are numerous and can be guided with great accuracy. MIRVs could lead to an effective counter-force weapon capable of destroying a large part of an adversary's retaliatory forces if used in a surprise attack. MIRVs were initially developed in the early 1960s to permit a missile to deliver multiple nuclear warheads, each capable of aiming to hit a different target. The warheads on these missiles can be released from the missile at different speeds and in different directions, in contrast to a conventional missile, which carries a single warhead. MIRV is an assembly of large missiles, small warheads, accurate guidance, and a complex mechanism for dispensing warheads sequentially during flight. The US was the first to develop MIRV technology, deploying a MIRV Intercontinental Ballistic Missile (ICBM) in 1970 and a MIRVed Submarine-Launched Ballistic Missile (SLBM) in 1971. The USSR followed suit and, by the end of the 1970s, had developed its MIRV-enabled ICBM and SLBM technology. Today, the United States, the United Kingdom, and France use MIRV technology on SLBMs. China has MIRVed ICBMs, while Russia deploys both MIRVed ICBMs and SLBMs. India and Pakistan are also experimenting with MIRV technology. Some of the terminologies used in the MIRV are given below

7.5.4 Missile Defense System (MDS)

The *Missile Defense System* is an integrated, layered approach for missile defense that provides multiple opportunities to destroy the enemy missiles and their warheads before they reach their targets. The missile defense system includes: (1) networked sensors (including satellite-based) and radars for target detection and tracking; (2) missiles for intercepting a ballistic missile using either the force of a direct collision, called hit-to-kill technology, or an explosive blast fragmentation warhead in case of a near-miss situation and (3) a command, control and communications network providing the commanders with the required links between the sensors and interceptor missiles. A missile defense system is intended to shield a country against incoming missiles, such as ICBMs or other ballistic missiles. The United States, Russia, India, France, Israel, Italy, the United Kingdom, China and Iran claim to have developed missile defense systems.

7.5. TYPES OF MISSILES

Major MIRVs around the world

Missile Name	Country	Designer/Manufacturer	Range (km)	Missile mass (tons)	Warhead	First flight	Launch
R-29RMU Sineva	Russia	Krasnoyarsk Machine-Building Plant	11,547	40.3	4 × 500kt	2004	Delta IV submarine
R-29RMU2 Layner	Russia	Krasnoyarsk Machine-Building Plant	11,000 +	40.0	4 × 500kt	2011	Delta IV submarine
R-36M2 Voevoda	Russia	Yuzhny Machine-Building Plant	11,000	211.4	10 × 800 kt / 8730 kg	1986	Silo
UR-100N UTTKh	Russia	Khrunichev Machine-Building Plant	10,000	105.6	6 × 550 kt / 4350 kg	1977	Silo
RS-24	Russia	Votkinsk Machine Building Plant	12,000	49.0	3-4 × 300 kt	2007	Silo, road-mobile TEL
R-29R	Russia	State Rocket Center Makeyev	6,500	35.3	3 × 500kt	1978	Delta III submarine
RSM-56 Bulava	Russia	Votkinsk Plant State Production Association	8000–9300	36.8	6 × 150 kt	2005	Borei-class submarine
Minuteman III	US	Boeing Corporation	13,000	35.3	3 × 300 kt of TNT	1970	Silo
Trident II	UK & US	Lockheed Martin Space Systems	11,300 +	58.5	8 × 475 kt or 14 × 100 kt	1987	Ohio-class and Vanguard-class submarines
M45	France	Aérospatiale/EADS SPACE Transportation	6,000	35.0	6 × 110 kt TNT	1986	Triomphant-class submarine
M51.1	France	EADS Astrium Space Transportation	>10,000	52.0	6 to 10 × 100 kt of TNT	2006	Triomphant-class submarine
DF-5B	China	China Academy of Launch Vehicle Technology	15,000	183.0	3 to 8	2015	Silo
DF-5C	China	China Academy of Launch Vehicle Technology	15,000	183.0	10 × 1 Mt	2015	Silo
DF-31A	China	Academy of Rocket Motors Technology (ARMT)	12,000	42.0	3 × 20 or 90 or 150 kt	2007	Road-mobile TEL
DF-41	China	China Academy of Launch Vehicle Technology	12,000-15,000	80.0	10 × 1 Mt	2012	Road-mobile TEL / Rail-mobile
JL-2	China	Factory 307 (Nanjing Dawn Group)	7200	42.0	1 × 1 Mt		
Agni-V	India	DRDO	5000-6000	50.0	3-6	2012	Road mobile TEL, Rail Mobile

Table 7.2: General designator for missile identification based on launch environment, target environment/mission and type of vehicle

Symbol	Example and Description
Status (Prefix)	Items of information are included as 1. Status prefix 2. Launch environment 3. Primary mission 4. Vehicle type 5. Vehicle design number 6. Vehicle series 7. Manufacturer's code 8. Serial number. Example - JLGM-30BO03 here, J - Status prefix (temporary), L - Launch environment (Silo), G - Mission symbol (Surface attack), M-Vehicle type symbol (missile), 30 - Design number, B - Series symbol, BO - Manufacturer's code, 03 - Serial number
J - Special test, temporary	
N - Special test, permanent	
X - Experimental	
Y - Prototype	
Z - Planning	
Launch Environment (1st Letter)	
A - Air	ADM - 20A (Quail) Air launch, Decoy missile. Vehicles designed or modified to confuse, deceive, or divert enemy defenses by simulating an attack vehicle
B - Multiple	AGM-45A (Shrike) Launched from aircraft while in flight
C - Coffin	LGM-30G (Minuteman III) Vertically stored and launched from below
F - Individual	MIM-23A K (Hawk) Launched from a ground vehicle or movable platform UGM-27C (Polaris) - Launched from a submarine or other underwater device
G - Ground	
H - Silo stored	
L - Silo launched	
M - Mobile	
P - Soft pad	
R - Ship	
U - Underwater	
Target Environment (2nd Letter)	
D - Decoy	XFEM-43B (Redeye) Vehicles designed or modified with electronic equipment for communications, countermeasures, electronic radiation sounding, or other electronic recording or relay missions.
E - Special electronic	RIM-46A (Sea Mauler) Launched from a submarine or other underwater device
G - Surface attack	HGM-25A (Titan) Vertically stored below ground level and launched from the ground
I - Intercept	BQM-34A (Firebee) Capable of being launched from more than one environment and vehicles designed for target, reconnaissance, or surveillance purposes
Q - Drone	AIR-2B (Super Genie) Air launched, unguided air-to-air rocket
T - Training	CGM-13B (Mace) Horizontally stored in a protective enclosure and launched from the ground
V - Underwater attack	XFIM (Redeye) Carried by one man
W - Weather	MIM-23A K (Hawk) Missile launched from a ground vehicle or movable platform PGM-17A (Thor) - Partially or non-protected in storage and launched from the ground RIM-46A (Sea Mauler) - Launched from a surface vessel such as a ship or barge
Type of the Vehicle (3rd Letter)	
M - Guided missile [a]	ATM-12B - (Bullpup) Vehicles designed or permanently modified for training purposes
N - Probe [b]	UUM-44A (SUBROCK) Vehicles designed to destroy enemy submarines or other underwater targets
R - Rocket [c]	PWN-5A Vehicles designed to observe, record, or relay data pertaining to meteorological phenomena

[a] As the third letter in a missile designator, it identifies an unmanned, self-propelled vehicle. Such a vehicle is designed to move in a trajectory or flight path that may be entirely or partially above the Earth's surface. While in motion, this vehicle can be controlled remotely or by homing systems, or by inertial and/or programmed guidance from within. The term "guided missile" does not include space vehicles, space boosters, or naval torpedoes, but it does include target and reconnaissance drones

[b] Vehicle is the nonorbital-instrumented vehicle that is not involved in space missions. These vehicles are used to penetrate the space environment and transmit or report back information. Once launched, the trajectory or flight path of such a vehicle cannot be changed

[c] a self-propelled vehicle without installed or remote control guidance mechanism.

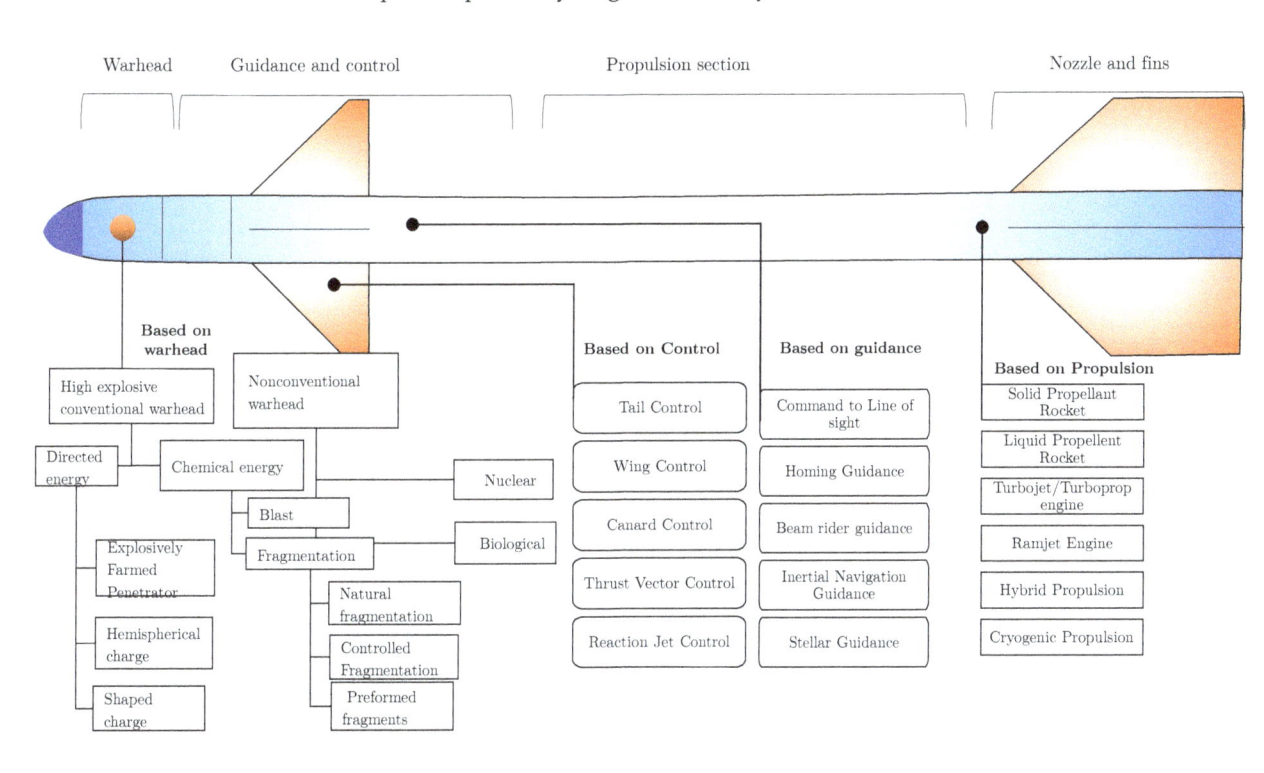

Figure 7.3: Classification based on subsystem configuration

7.6 Some Examples

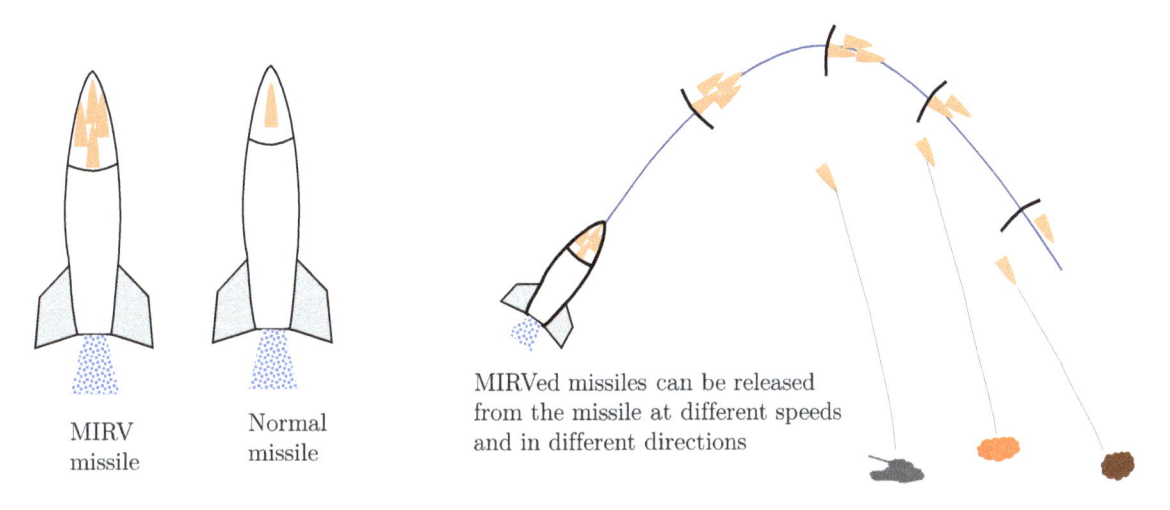

In this section, we will look at the details of some of the missiles. The examples are chosen to represent the different types of missiles, their capabilities and some of their characteristics.

V-2 missile: The *V* stands for *Vergeltungswaffen* (German for Retaliatory weapons). The *V-2* which uses liquid fuel (75% ethanol and 25% water mixture for fuel and liquid oxygen for oxidizer) is the first long-range ballistic missile developed by Germany during World War II with a development program that cost 2 billion in 1944 dollars. *V-2* (14 meters long, 12.5 ton weight) represents an enormous quantum leap in technology, the missile had a maximum range of about 320 km and a one-

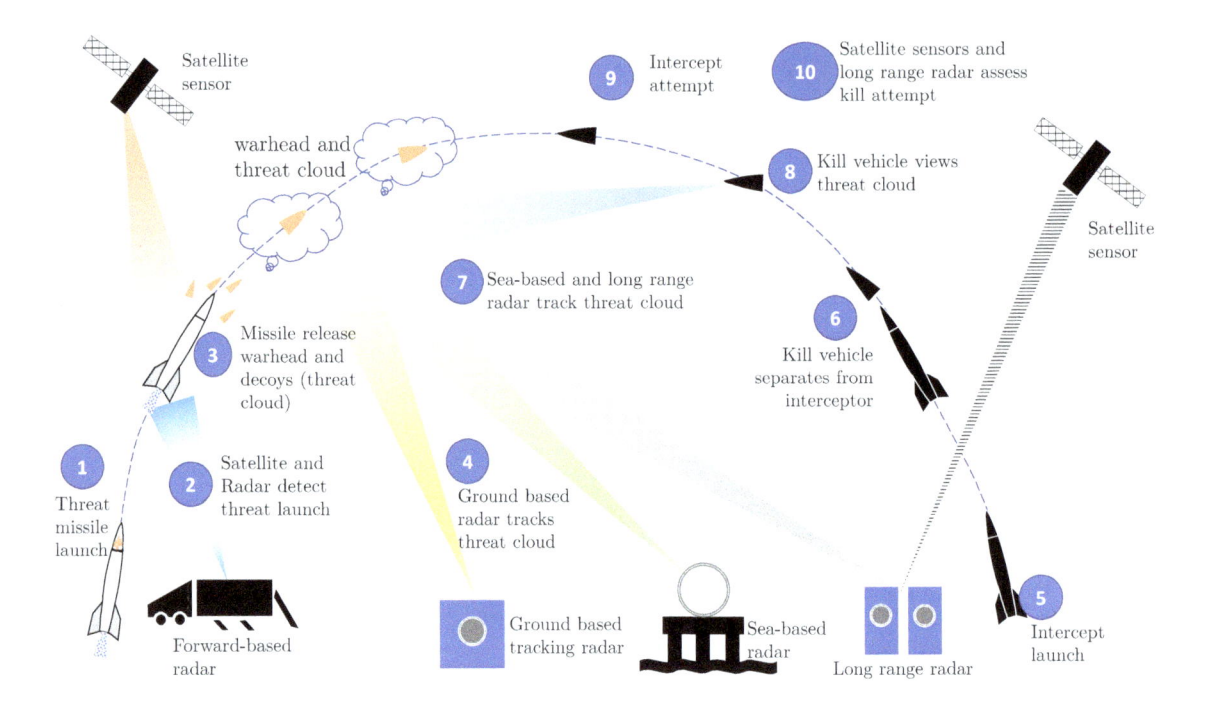

Missile Defense System - capable of intercepting and destroying incoming missiles with nuclear or conventional warheads.

ton warhead. The peak altitude usually reached was roughly 80 km. A total of around 6,000 *V-2* missiles were built. More than three thousand of those missiles were launched in combat, primarily against Antwerp and London, and a further 1,000 to 1,750 were fired in tests and training. *V-2* missile reliability as tested increased from 30% in January 1944 to 70% immediately before combat firings began in September 1944 and it reached nearly 100% when introduced into production in December 1944. The *V-2* killed an estimated five persons per launch. Tests of prototype *V-2*'s in 1943 indicated a 4.5 km CEP (circular error probable - the radius within which 50% of the impact of the shots). 100% of the shots fell within 18 km of the target. A radio beam guidance update system was introduced in December 1944, which produced a 2 km CEP in tests. In reality, in the attack against Britain, 518 rockets were recorded as falling in the Greater London Air defense Zone of 1225 fired, implying an average CEP of 12 km. After the war, both the USA and the USSR captured large numbers of *V-2*s and used them in research that led to the development of their missile and space exploration programs. Personnel and technology from the *V-2* program formed the starting point for post-war rocketry development in America, Russia, and France.

R36M2 Voivode is an intercontinental-range, silo-based, liquid propellant ballistic missile initially developed by the Soviet Union and now the Russian Federation. It is believed that a total of six versions of *R-36* have existed since the program's inception, with only the *R36M2 Voivode* still operationally deployed. In the early 1960's *Vladimir Chelomey*, a Soviet scientist developed the gas dynamic launch method for this *UR-100* missile that used exhaust gases from the rocket to 'pop' the missile out of its silo. In 1969, the idea of a 'cold launch' was conceived. This would use a gas generator - essentially a huge mortar charge – to pop the missile out of the silo. Some of the advantages are that the crucial initial energy impulse would be given to the

7.6. SOME EXAMPLES

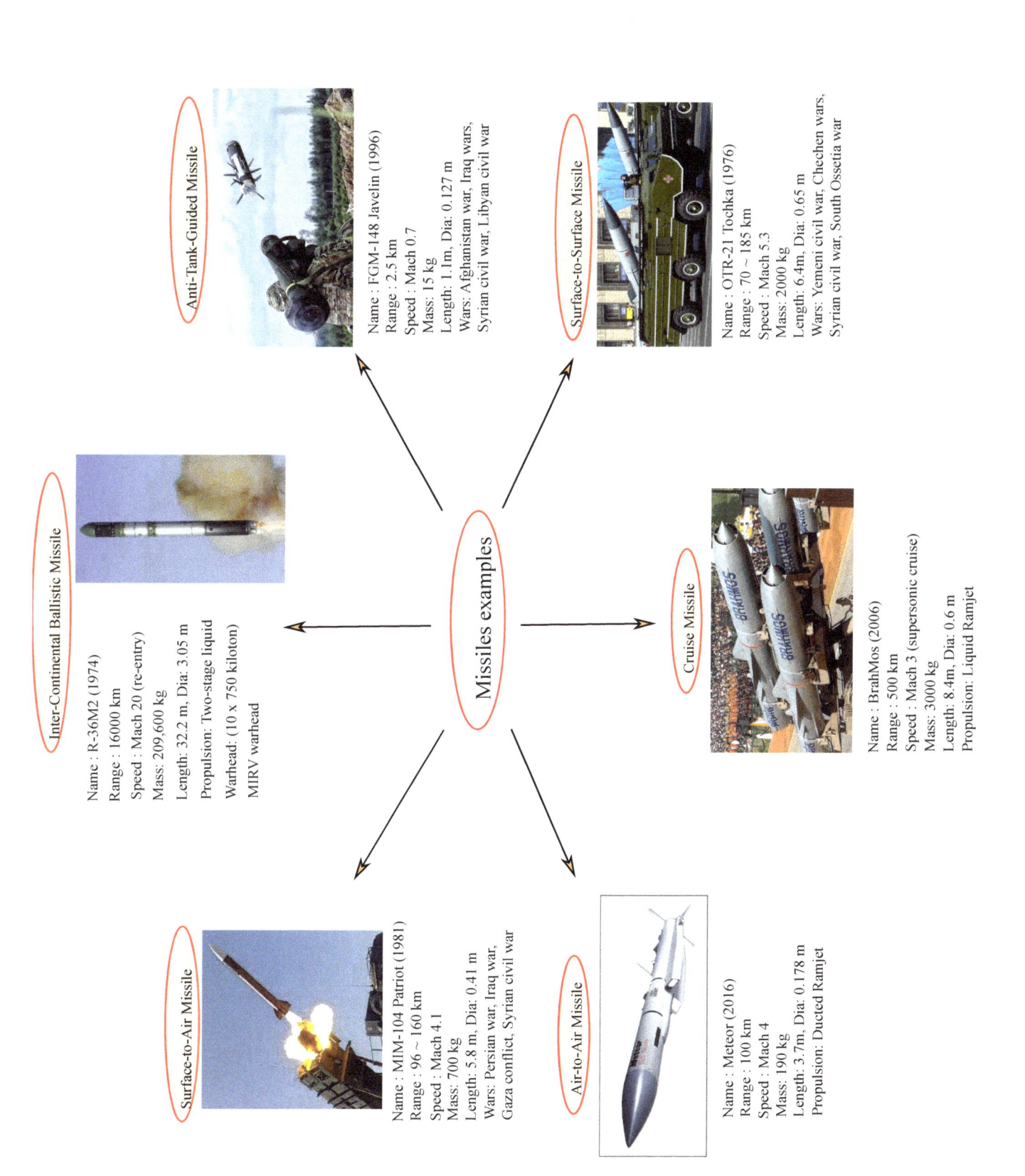

Missiles examples

Anti-Tank-Guided Missile
Name : FGM-148 Javelin (1996)
Range : 2.5 km
Speed : Mach 0.7
Mass: 15 kg
Length: 1.1m, Dia: 0.127 m
Wars: Afghanistan war, Iraq wars, Syrian civil war, Libyan civil war

Surface-to-Surface Missile
Name : OTR-21 Tochka (1976)
Range : 70 ~ 185 km
Speed : Mach 5.3
Mass: 2000 kg
Length: 6.4m, Dia: 0.65 m
Wars: Yemeni civil war, Chechen wars, Syrian civil war, South Ossetia war

Inter-Continental Ballistic Missile
Name : R-36M2 (1974)
Range : 16000 km
Speed : Mach 20 (re-entry)
Mass: 209,600 kg
Length: 32.2 m, Dia: 3.05 m
Propulsion: Two-stage liquid
Warhead: (10 x 750 kiloton) MIRV warhead

Cruise Missile
Name : BrahMos (2006)
Range : 500 km
Speed : Mach 3 (supersonic cruise)
Mass: 3000 kg
Length: 8.4m, Dia: 0.6 m
Propulsion: Liquid Ramjet

Surface-to-Air Missile
Name : MIM-104 Patriot (1981)
Range : 96 ~ 160 km
Speed : Mach 4.1
Mass: 700 kg
Length: 5.8 m, Dia: 0.41 m
Wars: Persian war, Iraq war, Gaza conflict, Syrian civil war

Air-to-Air Missile
Name : Meteor (2016)
Range : 100 km
Speed : Mach 4
Mass: 190 kg
Length: 3.7m, Dia: 0.178 m
Propulsion: Ducted Ramjet

rocket by a device that would not cut into the total missile mass and the silo would not be damaged by the hot gases from the rocket, providing the ability for quick turnaround of the silo for subsequent firing.

The pilot project for the *R-36M* missile was completed in December 1969. Four different warhead variants were proposed – a single light reentry vehicle (RV), a single heavy RV, multiple independently-targeted RVs, and ballistic and maneuverable RVs to overcome enemy anti-ballistic missile defenses. *R36M2 Voivode* has 10 MIRV, with each warhead having 500-750 kiloTons with a CEP of 500 m and a range of 11,000 km.

LGM-30G Minuteman III is a three-stage, solid-fueled ICBM. It is the only land-based component of the US nuclear triad with a fast launch time, near-perfect reliability and backup airborne launch controllers to preserve retaliatory capabilities. *LGM-30G Minuteman III* was the first US missile with MIRV capability. Each missile was fitted with three warheads for a total of 1,500 warheads on 500 launchers. Subsequently as part of the START (Strategic Arms Reduction Treaty) obligations, the U.S. removed 2 warheads from 150 missiles in 2001. Later, the *Obama* administration began de-MIRVing the rest of the *Minuteman* arsenal as part of the New START treaty.

AGM-114 Hellfire is a short-range tactical missile mostly used for land-attack missions and saw large scale deployment during the Global War on Terror. The *Hellfire* is mostly air-launched, although it has also seen limited deployment from sea platforms. The name of the missile is an abbreviation of "HELiborne, Laser, FIRE and forget missile."

Exocet: *MBDA*, a European missile company, builds the *Exocet* missile. Development started in 1967 by the French aircraft manufacturer *Nord Aviation* as a ship-launched missile named *MM 38*. The preliminary design was based on the *Nord AS30* air-to-ground tactical missile. The air-launched *Exocet* was developed in 1974 and entered service with the French Navy. Work on the *AM39*, the air-launched variant, began in 1974 and entered service in 1979. The development of an improved version of called *MM40* began in 1976. Subsequent upgrades were made, resulting in the development of the *AM39/MM40 Block 2* in 1992. The *MM40 Block 3* missile began development in 2004 and was first tested via ship launch in 2010. All the variants are in service and all but the *MM 38* remain in production. There is a submarine-launched variant of the *AM 39* called *SM39*, which began development in 1979 and was ready for deployment in 1984. Coastal defense modifications made to the *MM38* and *MM40* have resulted in the *BC38* and *BC40*, which can be fitted to various platforms like helicopters, destroyers, and frigates.

The missile is designed for attacking small- to medium-size warships, although multiple hits can destroy larger vessels. It uses inertial guidance in mid-flight and turns on active radar later find and hit the target. For reducing the detectability, it maintains a low altitude during ingress, staying 1–2 m above the sea surface. Due to the radar horizon's effect, the target may not detect an incoming missile. It use solid propellant and has a range of 70 kilometers. Later, it was replaced on the *Block 3 MM40* ship-launched version of the missile with a solid-propellant booster and a turbojet sustainer motor which extended the range to 180 kilometers.

7.7 Major Missile Manufacturers

For several decades, the trade in weapons has been among the world's most lucrative businesses, with predictable increases yearly. Sales of arms and military services by the world's 100 largest defense companies rose by 1.9% in 2021 to reach $592 billion. Almost one third of the total sales from the top 100 companies lies with the top five companies on the list, viz., *Lockheed Martin*, *Raytheon Technologies*, *Boeing*, *Northrop Grumman* and *General Dynamics*, among them they had $192 billion sales in 2021. The next five on the list include *BAE Systems* which is based in UK, and four Chinese firms – *Norinco*,

7.8 Treaties & Agreements

AVIC, *CASC* and *CETC* – between them, these five reported arms sales of almost 102 billion.

Manufacturer	Origin	Key Products
Lockheed Martin	USA	THAAD, Patriot, Javelin, Hellfire, Trident
Raytheon Technologies	USA	Tomahawk, AMRAAM, AGM-154, AIM-7, AIM-9, TOW, Stinger, Sea Sparrow
Boeing	USA	Minuteman, Peacekeeper, Harpoon, AGM-86
MBDA	France	Meteor, Milan, MMP, Rapier, Mistral Brimstone, Sea Eagle, Storm Shadow, ASRAAM
IAI	Israel	Arrow, Barak-8, Iron Dome, Gabriel, LAHAT, Nimrod
(Russia co.)	Russia	Kh-31, Kh-35, R-77, Kh-38, Kh-59
Rafael	Israel	Python, Spike, Popeye, Iron Dome, Sparrow
Diehl	Germany	IRIS-T, PARS 3, RIM-116, Hope, HOSBO
Bharat Dynamics Ltd.	India	Akash, Nag, Prithvi, Agni, Astra - (Developed by DRDO) LRSAM, MRSAM, INVAR, MILAN - (Foreign ToT)
Thales	UK	Starstreak, Starburst, Javelin, Blowpipe, Hellfire
BAE Systems	UK	M2/M3 Bradley, M109 Paladin, Archer

Figure 7.4: Major missile producing companies with key products.

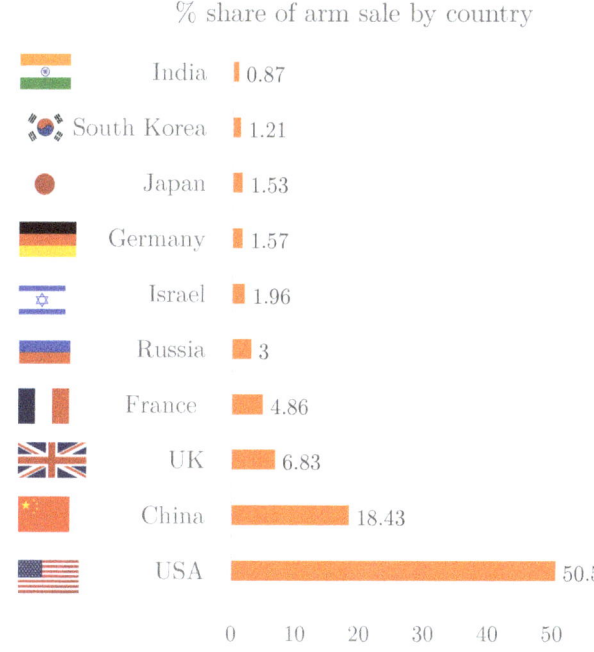

% share of arm sale by country

- India: 0.87
- South Korea: 1.21
- Japan: 1.53
- Germany: 1.57
- Israel: 1.96
- Russia: 3
- France: 4.86
- UK: 6.83
- China: 18.43
- USA: 50.53

Figure 7.5: Country-wise share in the arms sale around the world.

7.8 Treaties & Agreements

The Anti-Ballistic Missile (ABM) Treaty is a treaty signed in 1972 by the USA and the USSR to limit the deployment of missile systems that could be used to destroy incoming ICBMs launched by the other superpower. Negotiations to prohibit ballistic missile defenses were put forward by the USA during the Strategic Arms Limitation Talks (SALT) in 1966. The ABM Treaty was signed by then US President *Richard Nixon* and Soviet leader *Leonid Brezhnev* in Moscow in 1972.

The treaty limited each side to only two ABM deployment sites, one to protect the national capital and the second one to protect an ICBM launch site, with each deployment area limited to 100 launch systems and 100 interceptor missiles. Further in 1974, a protocol reduced the agreement to one ABM site for each country. The Soviet Union opted to maintain a site in Moscow and the USA decided to protect an ICBM site at Grand Forks in North Dakota. To prevent the deployment of a nationwide battle management system, the treaty required all early-warning radars to be placed near the boundary of the country and facing outward. In 1984, the USA claimed that a Soviet radar system was placed near the city of Krasnoyarsk, 800 km from the nearest border, violated this provision. In 1989, the USSR acknowledged this and agreed to dismantle the radar. In addition to conventional interceptor missiles, launchers, and radars, the treaty also covers systems based on other principles, such as lasers.

The **Strategic Arms Reduction Treaty (START) I** was signed between the United States and the Soviet Union on the limitation of strategic weapons. It came into effect in 1994. The treaty limited the two countries from deploying more than 6,000 nuclear warheads and 1,600 ICBMs and bombers.

7.9 Missile Programs in India

India's strategic missile capabilities have significantly advanced over the years. The country has developed the ability to deploy a range of ballistic missiles, including short-range, medium-range, and long-range variants. This progress has been driven by four decades of investment in domestic missile-related research, development, and manufacturing capabilities. As a result, the Indian missile industry has become largely resilient to potential disruptions caused by export control measures.

India has demonstrated its commitment to the development of comprehensive ballistic and cruise missile capabilities through the advancement of the *Agni-V* land-based intercontinental ballistic missile (ICBM), *K series* sea-based ballistic missile, and *BrahMos* cruise missile. In particular, Indian defense strategists are transitioning their emphasis from innovating new missile technologies to the mass production of current missiles. Departing from past self-reliance-centric approaches, India is progressively seeking international partnerships, indicating trust in its native missile capabilities.

India's Ministry of Defense (MOD) brought together all defense-related research and development efforts in the country by combining the Indian Army's Technical Development Establishment and the Directorate of Technical Development & Production with the Defense Science Organization. This merger led to the establishment of a new entity called the Defense Research & Development Organization (DRDO). Initially, in 1958, DRDO comprised only 10 laboratories under its purview.

The government of India formed a Special Weapons Development Team in Delhi to research guided missile weapons development. Later, it shifted to Hyderabad, which marked the inception of the Defence Research and Development Laboratory (DRDL), dedicated solely to the research and development of missile technology. Their first endeavor, an indigenous anti-tank missile, proved successful upon its test firing, laying the groundwork for India's missile program. Later, Bharat Dynamics Limited (BDL), India's premier missile production agency, was established. In the 1970s, the BDL commenced the licensed manufacture of *SS-11B anti-tank missiles*.

In May 1964, India received a two-stage solid propellant sounding rocket, *Centaure* from France and began licensed production with modifications. Before India launched a dedicated missile program in the 1980s, Indian engineers gained good rocketry experience, encompassing satellite launches and recovery operations.

During the 1970s, Indian focused on two significant programs: Project *Devil* and *Valiant*. Project *Valiant* aimed to develop a long-range ballistic missile. However, technological challenges in propulsion led to the discontinuation of the project. The *Project Devil*, initiated in 1972, aimed to gain comprehensive knowledge of an operational missile by reverse-engineering the Soviet-designed *SA-2* a guided medium to high-altitude surface-to-air missile (SAM). Despite facing significant technological and capacity challenges, the project fell short of replicating the *SA-2* entirely but succeeded in producing essential components: two solid-fuel boosters and a three-ton liquid sustainer engine. These components later formed the foundation for the *Prithvi* missile series.

By the early 1980s, DRDL had established expertise in propulsion, navigation, and material manufacturing. This laid the foundation for the initiation of the Integrated Guided Missile Development Program (IGMDP). *APJ Abdul Kalam*, who previously served as the project director for the *Satellite Launch Vehicle (SLV)-3* program at Indian Space Research Organisation (ISRO), assumed the role of DRDL Director in 1983 to conceive and oversee this program. Four projects were launched under IGMDP:

- Short-range surface-to-surface missile (known as *Prithvi* series)

- Short-range low-level surface-to-air missile (known as *Trishul* series)

- Medium-range surface-to-air missile (known as *Akash*)

- Anti-tank missile (known as *Nag*)

The *Agni* missile began its journey within the IGMDP as a technology demonstrator project focusing on re-entry vehicle technology. Over time, it evolved into a ballistic missile with various range capabilities. In addition, the development of the Interim Test Range in *Balasore*, Odisha, was integral to this program to carry out missile tests.

In the face of technology-denial sanctions enforced by the international community following India's 1974 nuclear test, India pushed for increased research in technologies like the *SLV*. As an example, some regarded the motor of *SLV-3* was an inaugural *Agni* technology demonstrator.

Following India's test-firing of the first *Prithvi* missile in 1988 and the *Agni* missile in 1989, the Missile Technology Control Regime (MTCR), a coalition by Canada, France, Germany, Italy, Japan, the United Kingdom, and the United States, opted to restrict access to technologies aiding India's missile development efforts. In response to MTCR restrictions, the IGMDP team formed a consortium comprising DRDO laboratories, industries, and academic institutions to domestically produce these restricted sub-systems, components, and materials. While this approach slowed down program progress, India successfully developed all restricted components indigenously, despite MTCR limitations.

Prithvi-I offered India a basic short-range (150 km) capability for potentially carrying out a restricted nuclear strike. By 1994, two successful flight tests of the 1,400 km-range *Agni-1* missile validated India's atmospheric entry vehicle technology and showcased proficiency in staging. Consequently, the *Agni* program laid the groundwork for designing and advancing longer-range ballistic missile systems, whereas *Prithvi* continued to be India's sole operational strategic missile at the time.

The late 1990s and early 2000s witnessed ongoing technological advancements in the *Prithvi* and *Agni* missiles, along with efforts to explore more advanced missile delivery capabilities. The DRDO initiated programs aimed at developing extended-range variants of the *Agni* series: *Agni-II* (with a range of 3000 km) and A*gni-III* (with a range of 5000 km), as well as the *Prithvi* series: *Prithvi-II* (with a range of 350 km) and *Prithvi-III* (with a range of 600 km). Additionally, in 2001, India successfully conducted the maiden test of its first supersonic cruise missile, named *BrahMos*, developed through collaboration with Russia.

The major missile programs in India are

- *Prithvi Missiles* - tactical surface-to-surface short-range ballistic missiles (SRBM)

- *Agni* missile series started as a Re-Entry Vehicle project (later known as Agni Technology Demonstrator) in the IGMDP

- *K family* of missiles (named after APJ Abdul Kalam) - a family of submarine-launched ballistic missiles (SLBM)

- *Shaurya* missile – a short-medium range hypersonic surface-to-surface hybrid ballistic-cruise missile

- *Prahaar* is a solid-fuelled Surface-to-surface guided short-range tactical ballistic missile

- *Nirbhay* - Long range sub-sonic Cruise Missile SURFACE TO AIR MISSILES

- *Trishul* - a short range, surface-to-air missile.

- *Akash* is a medium-range, surface-to-air missile Air-To-Air Missiles

- *Astra* – a Beyond Visual Range Air-to-Air Missile (BVRAAM)

- *Dhanush* - Antiship missile which is a variant of the surface-to-surface *Prithvi III* missile

- *Nag* - Fire-and-forget anti-tank missile

- The *Rudram-1* - new generation anti-radiation missile (NGARM).

India's ballistic missile defense capabilities received a significant boost in 2006 with the successful testing of the first tier of the country's ballistic missile defense system, *Prithvi Air Defence (PAD) or Pradyumna*. PAD was tested with a maximum interception altitude of 80 km and is designed to intercept missiles within a range of 300-2000 km traveling at speeds up to Mach number 5.0. Subsequently, in 2007, the second tier known as the

Advanced Air Defense (AAD) was also tested. Furthermore, in 2008, DRDO initiated the development of a sea-launched ballistic missile named *Sagarika*, which underwent testing from submersible pontoons.

India and Israel entered into an agreement to collaboratively develop the *Barak-8* surface-to-air missile (formerly known as LR-SAM and MR-SAM). This missile is designed to counter airborne threats including aircraft, helicopters, anti-ship missiles, and UAVs. Both sea-based and land-based variants of the system are available. The inaugural successful test of the sea-based version occurred in 2010.

In 2012, India achieved its first successful test of the *Dhanush* missile. This variant, derived from the surface-to-surface or ship-to-ship *Prithvi-3* missile, was specifically designed for deployment by the Indian Navy. With a range of 350 km, it possesses the capability to carry both conventional and nuclear warheads.

Recent advances have significantly enhanced the *Agni* ballistic missile family, both in terms of range and sophistication. *Agni-III*, boasting a range exceeding 3,200 km, was formally integrated into the armed forces in 2011 following a series of successful tests and user trials. Moreover, India conducted multiple successful flight tests of *Agni-IV*, with a declared range of 4000 km, starting from 2011 before officially inducting it into the armed forces in 2014. The much-anticipated *Agni-V* has also undergone successful testing on numerous occasions since 2012. Recently, *Agni-V* equipped with nuclear Multiple Independently Targetable Reentry Vehicle (MIRV) capabilities with a range surpassing 7,000 km was tested.

Bibliography

E. L. Fleeman, J. A. Schetz (2012) *Missile Design and System Engineering*, American Institute of Aeronautics and Astronautics.

J. N. Nielsen (1988) *Missile Aerodynamics*, American Institute of Aeronautics and Astronautics.

G. M. Siouris (2004) *Missile Guidance and Control Systems*, Springer-Verlag.

R. M. Lloyd (1998) *Conventional Warhead Systems Physics and Engineering Design*, American Institute of Aeronautics and Astronautics.

P. Zarchan (2002) *Tactical and Strategic Missile Guidance*, American Institute of Aeronautics and Astronautics.

R. Alderliesten (2018) *Introduction to aerospace structures and materials*, Delft University of Technology.

V. Sundaram (1996) *Missile materials—Current status and challenges*, Bulletin of Materials Science

P. Garnell and D. Jo East (1977) *Guided weapon control systems*, Pergamon Press

G. M. Siouris (2004) *Missile guidance and control systems*, Springer Science & Business Media

S. S. Chin (1961) *Missile configuration design*. McGraw-Hill

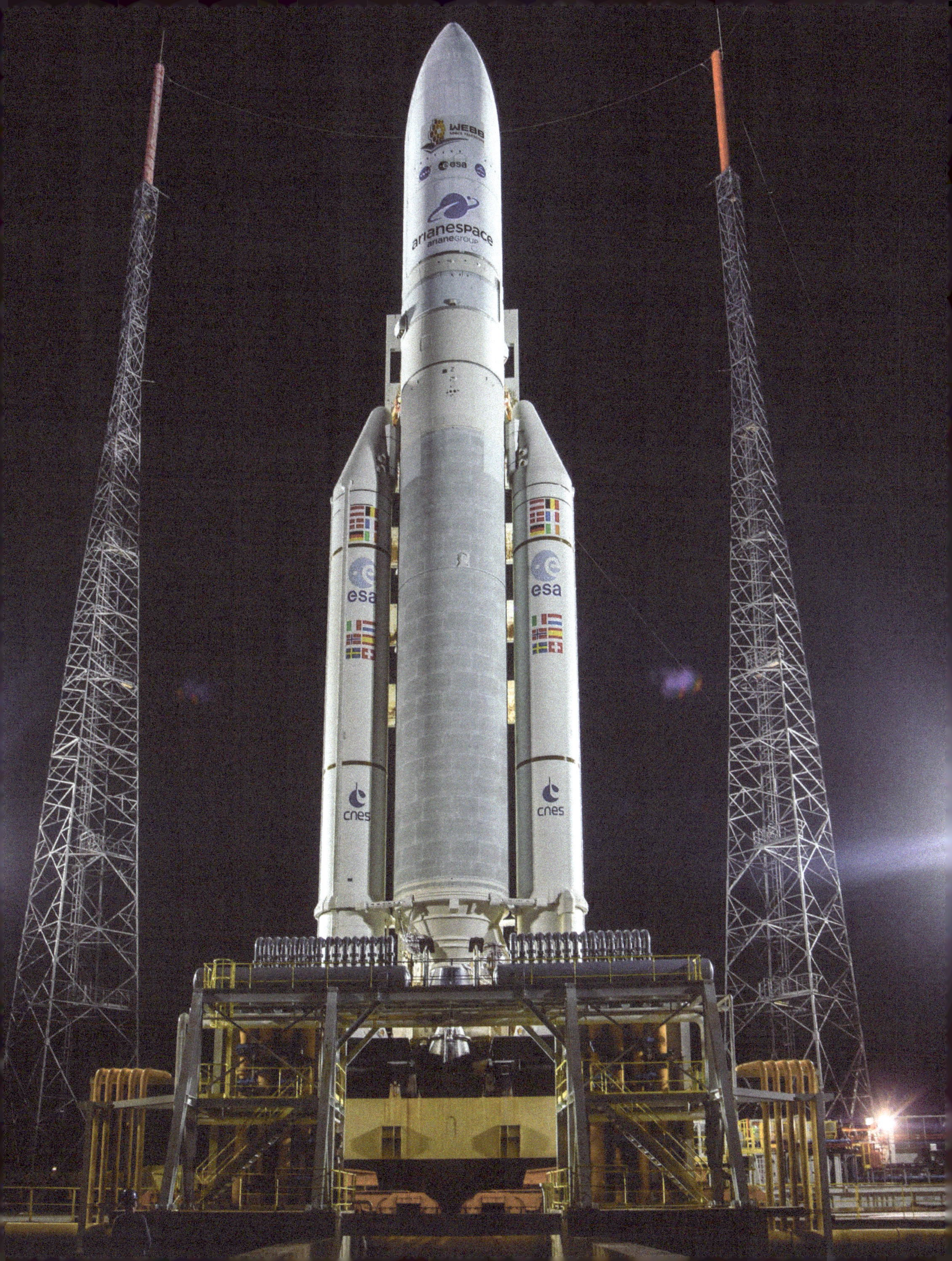
WEBB
esa
arianespace
arianeGROUP
esa
esa
cnes
cnes

Chapter 8

Space Vehicles

Contents

8.1	**Introduction**		**363**
	8.1.1	What is Space	363
8.2	**Types of Space Vehicles**		**364**
	8.2.1	Satellites	364
	8.2.2	Lunar and Interplanetary Vehicles	367
	8.2.3	Launch Vehicles	367
	8.2.4	Space Stations and Space Shuttles	370
8.3	**Satellite Launching Stations**		**370**
8.4	**Types of Trajectories**		**373**
	8.4.1	Orbital Parameters	374
8.5	**Spacecraft Maneuvers and Trajectories**		**376**
8.6	**Space Debris**		**382**
8.7	**Space Agencies**		**382**
8.8	**Space Laws**		**384**
8.9	**India's Space Program**		**384**

8.1 Introduction

The fascination for outer space has often captured human attention. From very early days, there were many myths and beliefs about the Sun, Moon, stars and the planets and their influence on the daily lives of human beings. *Claudius Ptolemy*, who lived in Alexandria during the second century AD, wrote a treatise on astronomy where a geocentric planetary model was presented. Many centuries later, *Nicolaus Copernicus* (1473 - 1543) gave the heliocentric model of the solar system, where the Sun is placed motionless at the center and the planets are moving in circular orbits around the Sun. *Galileo Galilei* (1564 - 1642), the famous Italian astronomer who supported Copernican heliocentrism, extensively used telescopes to make many systematic observations of celestial objects. *Tycho Brahe*, a Danish astronomer who lived in the sixteenth century, made a significant number of accurate astronomical observations using the instruments he made, which included quadrants and sextants. *Johannes Kepler* (1571 - 1630), who was an assistant of *Tycho Brahe* and had access to his astronomical observations, proposed the three classical laws of planetary motion. In these laws, the elliptic motion of the planets with the Sun at one of the foci of the ellipse was propounded. These laws were the most convincing model of the solar system. Subsequently, *Isaac Newton* formulated the law of universal gravitation, which can be used to derive *Kepler*'s laws of motion. Interestingly, the trajectories of modern-day artificial satellites and space missions can be calculated using the law of universal gravitation and Newtonian dynamics.

Since the launch of *Sputnik* by the USSR in 1957, many applications have been found for satellites. They have changed human life in substantial ways. For example, the use of satellite systems like the GPS for navigation has become very much part of our daily lives. Weather predictions have become more accurate due to data collected using weather satellites. Satellites can be used for a wide range of telecommunication applications including TV transmission. Through the *Starlink* system of satellites, *Elon Musk* aims to provide internet services for humanity. Launch of space vehicles for scientific exploration like the *James Webb* telescope and the recent *Chandrayaan* mission capture the imagination of the general public. The number of satellite launches have steadily increased over the years and the technology and services provided by the satellites is becoming more accessible to the general public. From the 133 successful launches in 2021, 27 humans as well as 1,778 spacecraft and satellites were put into orbit. Additionally, 16,675kg of cargo was sent to the *International Space Station*. According to the "Index of Objects Launched into Outer Space" (https://www.unoosa.org/), maintained by the United Nations Office for Outer Space Affairs, there are more than 8,500 individual satellites in space as of April 22, 2022. The database also shows that since inception more than 12,500 satellites have been launched. *Elon Musk*'s *SpaceX* surpassed *ISRO*'s record by successfully launching 143 satellites aboard *Falcon 9* into space on January 24, 2022.

8.1.1 What is Space

Space is the vast, three-dimensional expanse that exists beyond the well-known environmental condition on Earth. It is an almost perfect vacuum, devoid of air and matter, and extends indefinitely in all directions. Space is not completely empty; it contains various entities and phenomena. These include cosmic radiation, short-wave solar radiation (electromagnetic waves), high-energy particles (electrons, protons, neutrons and alpha particles), ultraviolet X-rays and gamma radiation from the galactic background, and dust. Additionally, space is home to numerous celestial objects, such as stars, planets, asteroids, comets, and black holes, each with its own gravitational influence and physical properties.

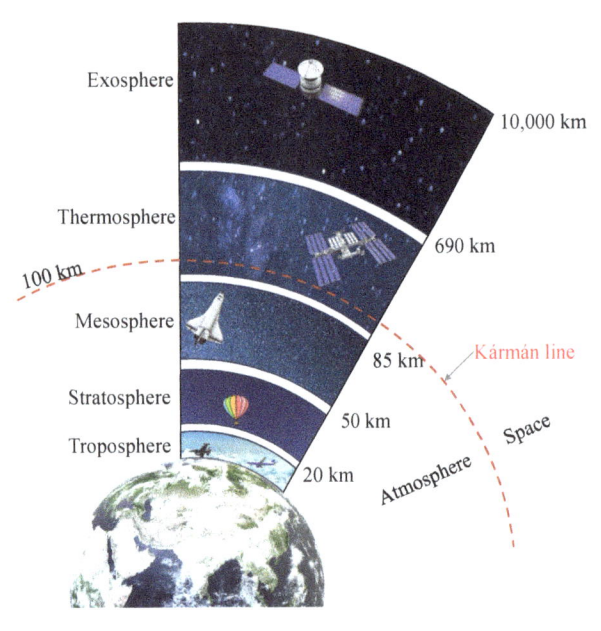

Sketch of Earth's atmosphere with the Kármán line

The Kármán line: The question arise at what altitude from the earth the space starts. However, defining exactly where space begins can be rather tricky as Earth's atmosphere doesn't end abruptly but instead gets thinner and thinner at higher altitudes, which means there's no definitive upper boundary for the Earth's atmosphere. Broadly, it is accepted that space starts at the point where *orbital dynamic forces* become more important than *aerodynamic forces*, or where the atmosphere alone is not enough to support a flying vessel at suborbital speeds.

We learned that the aerodynamics lift that keeps the aircraft up in the air by balancing the downward pull of gravity is directly proportional to air density and the square of the velocity. As the aircraft fly at higher altitude, the velocity has to increase in order for its wings to generate the necessary lift. As we increase the speed (at higher altitude) beyond orbital speed (the speed that an object must maintain to remain in orbit around a planet), the aircraft goes all the way around the Earth without ever hitting the ground.

However, Hungarian-American aerospace engineer, *Theodore von Kármán* (1881 - 1963) did the necessary calculations to find the altitude at which the speed required to generate necessary lift to keep an aircraft aloft become higher than the orbital velocity. The calculations yielded a rounded memorable altitude of 100 kilometers above mean sea level. This altitude is now known as the *Kármán line* in his honor which now widely accepted conventional boundary between Earth's atmosphere and outer space. At altitude 100 kilometers above mean sea level, the atmosphere becomes extremely thin, and the conditions are more akin to those of space rather than a typical atmosphere.

8.2 Types of Space Vehicles

Space vehicles are designed for travel and operations in outer space. They come in various types, each tailored for specific purposes and missions. Some of the broad categories of space vehicles are satellites, lunar and inter-planetary vehicles, launch vehicles, space stations and space shuttles.

8.2.1 Satellites

Satellites are artificial unmanned spacecraft (or natural body) that orbit around celestial bodies such as a planet or a star with Earth being the most common target for artificial satellites. These satellite are put into the desired orbit using rockets and have payload depending upon the intended application. These artificial satellites serve a wide range of purposes, including communication, weather monitoring, navigation, scientific research, and surveillance. Satellites can be further categorized into different types, such as communication satellites, weather satellites, and remote sensing satellites.

Satellites composed of various components such as payload, power system, communication system, Attitude and Orbit Control Systems (AOCS), and onboard computers. The payload can be instruments, sensors, or equipment carried by the satellite to fulfill its mission objectives. It can include cameras, antennas, transponders, or scientific instruments, depending on the satellite's pur-

8.2. TYPES OF SPACE VEHICLES

pose. Satellites use solar panels to generate electrical power from the Sun which is stored in batteries and used to operate the satellite's systems and payloads. Communication systems allow the satellite to transmit and receive signals to and from Earth. They involve antennas, transmitters, receivers, and signal processing equipment. AOCS help the satellite maintain its desired orientation, stability, and orbit. They use thrusters, reaction wheels, gyroscopes, and star trackers to adjust its position and attitude. They have onboard computers that control its operations, process data, and execute commands. These computers manage data storage, perform calculations, and handle communication between different subsystems.

Satellites come in many shapes and sizes. But most have at least two parts in common - an antenna and a power source. *Antennas:* Satellite antenna systems are used to receive and transmit signals to and from Earth. The *power source* can be a solar panel or battery. Solar panels make power by turning sunlight into electricity. Many satellites carry cameras and scientific sensors. Sometimes these instruments point toward Earth to gather information about its land, air and water. Other times they face toward space to collect data from the solar system and universe. Other important components are *Command and Data Handling:* The operational center of a satellite, the command and control systems monitor every aspect of the satellite and receive commands from Earth for various operations. *Guidance and Stabilization:* Sensors monitor the satellite's position and orientation to ensure that it remains in the desired orbit and is oriented towards the target. If needed, thrusters and other maneuvers are used to fine-tune the satellite's position and orientation. *Housing:* Constructed from strong materials that can withstand the harsh space environment. *Thermal Control:* Guards satellite equipment against extreme temperature changes. *Transponders:* Uplink and downlink signals arrive and depart at different frequencies. Transponders convert uplinked frequencies to downlink frequencies and then amplify the converted transmission for sending to Earth.

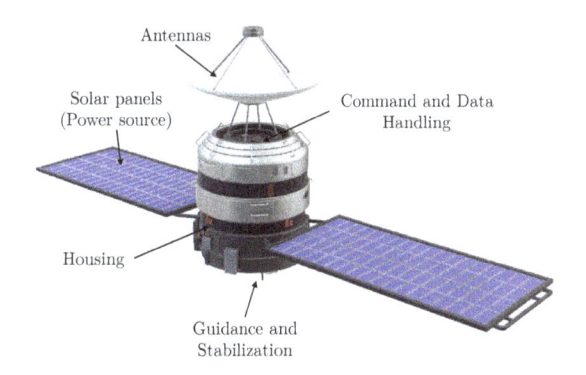

Components of a satellite

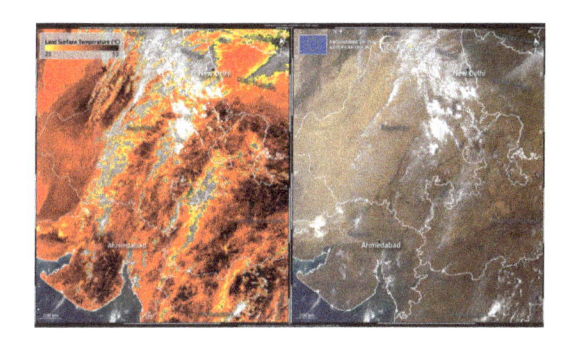

The satellite image shows a cloudless view (right) and land surface temperature (left) over northwest India on April 12, 2022. Credit: European Union, *Copernicus Sentinel-3A* imagery.

Molniya-1, a military communications satellite, the first successful launch occurred on April 23, 1965. Photo: Captured at Air and Space Museum, Paris

Table 8.1: Types of satellites

Types of satellites	Purposes	Examples
Communication Satellites	Facilitate communication by transmitting and receiving signals for television, telephone, internet, and other forms of communication. They operate in geostationary or geosynchronous orbits, allowing them to remain fixed above a specific location on Earth.	*Intelsat* Provides global communication services, including television broadcasting, telephony, and internet connectivity. *Iridium* Enables global voice and data communication, particularly for satellite phones and other mobile devices. *SES* offers direct-to-home broadcasting, data connectivity, and other communication services worldwide.
Weather Satellites	Monitor Earth's weather patterns, atmospheric conditions, and climate changes. They provide valuable data for meteorologists and help in forecasting weather phenomena like storms, hurricanes, and cyclones	*NOAA GOES* provides continuous monitoring of weather conditions, severe storms, hurricanes, and other meteorological phenomena over the United States. *Meteosat* offers weather data for Europe, Africa, and the Atlantic Ocean region, operated by *EUMETSAT*. *Himawari* provides real-time imagery and data for weather forecasting and disaster monitoring over the Asia-Pacific region.
Navigation Satellites	The global navigation satellite system (GNSS), provide precise positioning, navigation, and timing services worldwide. These satellites transmit signals that enable devices on Earth to determine their geographic location with high accuracy	*Global Positioning System (GPS)* provides accurate positioning, navigation, and timing services worldwide. *GLONASS* is Russia's satellite navigation system and *BeiDou* is China's satellite navigation system
Earth Observation Satellites	Capture images and collect data about Earth's surface, atmosphere, oceans, and vegetation. They are used for environmental monitoring, disaster management, urban planning, agriculture, and scientific research.	*Landsat* Monitors Earth's land surfaces, vegetation, and environmental changes, operated by NASA and the USGS. *Sentinel-2* Part of the European Union's Copernicus program, it captures high-resolution imagery for land and coastal areas monitoring. *Terra and Aqua* Part of NASA's Earth Observing System, they study various aspects of Earth's climate and environment.
Scientific Satellites	Designed to conduct research and gather data for various scientific disciplines. They study the universe, celestial bodies, and space phenomena.	*Hubble Space Telescope* is a renowned space telescope operated by NASA and ESA for astronomical observations. *Chandra X-ray Observatory* detects X-ray emissions from celestial objects, providing insights into high-energy astrophysical phenomena. *JWST (James Webb Space Telescope)* is the infrared space telescope designed to study distant galaxies and planetary systems.
Spy Satellites	Also known as reconnaissance satellites, are used for military and intelligence purposes. They gather classified information and perform surveillance activities, providing governments with valuable data for security and strategic purposes	*KH-11 Kennen* are classified reconnaissance satellites used by US for intelligence gathering and national security purposes. *Lacrosse* Radar imaging satellites designed for military reconnaissance and surveillance applications.
Space Stations	Space Station is a habitable space station that orbits Earth and serves as a research laboratory and living space for astronauts from multiple countries. Space stations could potentially deploy smaller satellites (CubeSats or nanosatellites) to their designated orbits.	*International Space Station (ISS)* A habitable space station and research laboratory in low Earth orbit, hosting astronauts from various nations. *Tiangong Space Station* is China's space station that is under construction and expected to be operational for scientific research

8.2.2 Lunar and Interplanetary Vehicles

Space probes are robotic spacecraft designed to explore celestial bodies beyond Earth. These autonomous or remotely controlled spacecraft are sent on missions to gather scientific data, conduct research, and provide insight into the mysteries of our solar system and beyond. Space probes are launched with specific objectives in mind. Probes are sent to study planets, moons, asteroids, and comets, providing valuable information about their composition, geology, atmosphere, and magnetic fields. Examples include Mars rovers (such as *Spirit*, *Opportunity*, *Curiosity*, and *Perseverance*) and the *Voyager* spacecraft (*Voyager 1* and *Voyager 2*) that have explored multiple planets and are now venturing into interstellar space. Probes such as *Hayabusa2* (Japan), *OSIRIS-REx* (NASA), and *Rosetta* (ESA) have been sent to study asteroids and comets up close, collecting samples and conducting detailed observations to learn more about the origin and evolution of these celestial bodies. Probes like *Voyager 1* and *Voyager 2*, currently the farthest human-made objects from Earth, are on a mission to explore interstellar space and provide data about the boundaries of our solar system.

Crewed spacecraft are vehicles designed to carry astronauts or cosmonauts into space. They are specifically built to sustain human life in the harsh conditions of space. Crewed spacecraft can be further classified into two categories: orbital spacecraft and interplanetary spacecraft. Orbital spacecraft, like the *Space Shuttle* and the Russian *Soyuz* spacecraft, are designed for orbital missions, such as transporting astronauts to and from space stations. Interplanetary spacecraft, such as the *Apollo* spacecraft and the upcoming *Artemis* program, are built for missions beyond Earth's orbit, such as lunar landings and future crewed missions to Mars.

8.2.3 Launch Vehicles

Rockets are the primary means of propulsion for space vehicles. They are used to launch satellites,

A self-portrait of NASA's Curiosity Mars rover in 2015

space probes, and crewed spacecraft into space. Rockets come in various sizes and configurations, from small solid-fuel rockets to massive multistage liquid-fueled rockets like the *Saturn V*, which was used in the *Apollo* program.

Launch vehicles, also known as rockets or space launchers, are complex vehicles designed to carry payloads such as satellites, spacecraft, or scientific instruments into space. They consist of several key components, each serving a specific purpose in the launch process. The main parts of a typical launch vehicle include

Payload Fairing: The payload fairing is the protective nose cone at the top of the rocket that encapsulates the payload during the launch and ascent phase. It shields the payload from aerodynamic forces and the harsh conditions of the atmosphere.

Upper Stage: The upper stage is the part of the rocket that ignites after the lower stages have expended their fuel. It provides the final push to reach the desired orbit or trajectory for the payload. Some rockets have multiple upper stages for more complex missions.

Lower Stages: The lower stages are the initial stages of the rocket that ignite at liftoff. They provide the primary thrust needed to overcome Earth's gravity and carry the vehicle through the lower atmosphere.

Rocket Engines: Rocket engines are the propulsion systems that produce thrust by expelling high-speed exhaust gases in the opposite direction. Different stages of the rocket may have different types of engines, such as liquid-fueled or solid-fueled engines.

Viking 5 rocket engine, a member of a series of bipropellant engines operated on storable, hypergolic propellants for the first and second stages of the *Ariane 1* launch vehicles. Image: Captured at Air and Space Museum, Paris

Guidance and Control System: The guidance and control system includes various sensors, computers, and actuators that steer and stabilize the rocket during its flight. It ensures the rocket follows its intended trajectory accurately.

Avionics: Avionics refers to the electronic systems on board the rocket, including communication equipment, navigation systems, and telemetry devices. These systems collect and transmit data during the launch and flight.

Telemetry System: The telemetry system collects data from various sensors on the rocket and transmits it back to ground control. This data is crucial for monitoring the rocket's performance and ensuring a safe flight.

Propellant Tanks: Propellant tanks store the rocket's fuel and oxidizer. In liquid-fueled rockets, the propellants are stored in separate tanks and mixed in the rocket's engines during combustion.

Ignition System: The ignition system initiates the rocket's engines at liftoff. It ensures a controlled and synchronized start of the engines to achieve proper thrust.

Fins and Aerodynamic Surfaces: Some rockets have fins or aerodynamic surfaces attached to their lower stages to provide stability and control during atmospheric flight.

Separation Systems: Separation systems are mechanisms that separate the different stages of the rocket once its fuel is depleted. This allows the upper stages to continue the ascent with the payload while shedding the mass of the empty stages.

The Diamant, the first exclusive French expendable launch system and a key predecessor for all subsequent European launcher projects. Photo: captured at Air and Space Museum, Paris

These components work together to enable the launch vehicle to carry its payload into space, placing it into the desired orbit or trajectory for its mission. The design and configuration of launch vehicle parts can vary based on the rocket's intended mission, payload requirements, and desired destination in space.

Maiden launch of *SpaceX's Falcon Heavy* launch vehicle, a super heavy-lift launch vehicle with partial reusability capable of transporting cargo into Earth orbit and beyond.

Arianespace's Ariane 5 rocket, carrying NASA's James Webb Space Telescope, stands ready for launch at Europe's Spaceport in Kourou, French Guiana, in December 2021.

The trajectory of a satellite launch vehicle refers to the path it follows during its ascent from the Earth's surface into space. The trajectory is carefully planned to achieve the desired orbit or trajectory for the satellite it is carrying. The launch vehicle's trajectory can be divided into several key phases:

Liftoff and Vertical Ascent: The trajectory begins with liftoff from the launch pad. Initially, the launch vehicle follows a vertical ascent, rising directly upwards while gaining altitude. During this phase, the rocket's lower stages provide the primary thrust to overcome Earth's gravity.

Pitch and Roll Maneuvers: As the launch vehicle gains altitude, it starts to perform pitch and roll maneuvers. Pitch refers to the rotation of the vehicle around its lateral axis, and roll refers to the rotation around its longitudinal axis. These maneuvers help steer the rocket on the desired trajectory.

Gravity Turn: As the launch vehicle continues to ascend, it performs a gravity turn. This is a crucial maneuver where the rocket starts tilting its trajectory towards the horizontal to take advantage of the Earth's rotation and conserve energy. The gravity turn helps the rocket achieve higher horizontal velocity and reach the desired orbit more efficiently.

Staging: Multi-stage launch vehicles have multiple rocket stages. As each stage depletes its fuel, it is jettisoned, and the next stage ignites. This process is known as staging. Each stage contributes to the rocket's velocity and altitude, with the upper stage eventually carrying the payload into space.

Orbital Insertion: The final phase of the trajectory is orbital insertion. Once the upper stage reaches the desired velocity and altitude, it shuts down its engines, and the satellite or payload is deployed into its intended orbit. This completes the launch vehicle's mission.

The trajectory of a launch vehicle is carefully planned and calculated by aerospace engineers to optimize fuel consumption, achieve the desired orbital parameters, and ensure a successful satellite deployment. Precise control over the trajectory is critical to achieving mission objectives, such as placing satellites into specific orbits for communication, Earth observation, scientific research, or interplanetary exploration.

Table 8.2 lists some of the launch vehicles from

ISRO's GSLV Mk III Launch vehicle lifting off with Chandrayan-3

around the world along with their specifications.

8.2.4 Space Stations and Space Shuttles

A space station is similar to a satellite orbiting the Earth and is capable of supporting a human crew for extended periods of time, sometimes for months. It has docking ports to allow the space shuttles dock to transfer crew and supplies. The purpose of maintaining an outpost in space is to conduct experiments and scientific studies related to gravity-free environment and human response to the unique environment. As of January 2024, there are two operational space stations: the *International Space Station*, a collaboration of *NASA*, *Roscosmos*, *JAXA*, *ESA* and *CSA*, and China's *Tiangong Space Station*. *Salyut I* was the first space station put into orbit by the USSR in 1971.

Spaceplanes are hybrid vehicles that combine the characteristics of airplanes and spacecraft. They are designed to take off and land like conventional aircraft but have the capability to reach space and operate in the vacuum environment. A notable example of a spaceplane is the experimental *X-37B*, an unmanned vehicle used for various classified missions. Space shuttles, like NASA's *Space Shuttle* program, were a unique type of spacecraft that could be launched into space and return to Earth for reuse. They were used to deploy satellites, conduct research experiments, and perform crewed missions.

Soviet Union's *Vostok-1*, a single-pilot crewed spacecraft placed at Air and Space Museum, Paris. The inaugural human spaceflight was achieved on April 12, 1961, piloted by Soviet cosmonaut Yuri Gagarin.

8.3 Satellite Launching Stations

A launch station is a specialized facility designed to support the preparation, fueling, and launching of space launch vehicles, and spacecraft. It serves as the central location for conducting space missions and plays a crucial role in ensuring the safe and successful liftoff of the vehicles into space. They are strategically located to take advantage of specific launch azimuths and to provide safe trajectories for the launched vehicles. These facilities are staffed with highly skilled engineers, technicians, and support personnel who work together to ensure a successful and safe launch.

Here is a description of a typical launch station:

A launch pad is a specially designed platform or structure from which space launch vehicles, and spacecraft are launched into space. It is a crucial infrastructure component of any launch site and provides a stable and controlled environment for the safe liftoff of the vehicle. Launch pads are complex structures equipped with various systems and facilities to support pre-launch preparations, fueling, and the launch itself. The launch pad is the main area of the launch station where the space launch vehicle is positioned and secured before liftoff. It

Table 8.2: Examples of Launch Vehicles and Their Specifications

Launch Vehicle	Payload to LEO (kg)	Payload to GTO (kg)	Reusability	Manufacturer	Notable Missions
Falcon 9 (SpaceX)	Up to 22,800	Up to 8,300	Yes	SpaceX	Dragon CRS missions to ISS, Starlink satellite launches
Atlas V (ULA)	Up to 18,810	Up to 8,900	No	United Launch Alliance	Mars rovers (e.g., Curiosity, Perseverance), numerous communication satellites
Delta IV Heavy (ULA)	Up to 28,370	Up to 14,220	No	United Launch Alliance	Orion spacecraft test flights, Parker Solar Probe
Soyuz (Roscosmos)	Up to 7,900	-	No	TsSKB-Progress (Russia)	Manned missions to the ISS, Luna and Mars missions
Long March 5 (CZ-5)	Up to 25,000	Up to 14,000	No	China Aerospace Science and Technology Corporation (CASC)	Chang'e lunar missions, Tianwen-1 Mars mission
Ariane 5 (Arianespace)	Up to 21,000	Up to 10,500	No	ArianeGroup	Galileo navigation satellites, James Webb Space Telescope
H-IIA (Mitsubishi Heavy Industries)	Up to 6,000	Up to 6,000	No	Mitsubishi Heavy Industries	Hayabusa missions to asteroids, Akatsuki Venus mission
GSLV Mk III (ISRO)	Up to 10,000	Up to 4,000	No	Indian Space Research Organisation (ISRO)	Chandrayaan missions to the Moon, Gaganyaan human spaceflight mission
Proton-M (Roscosmos)	Up to 22,800	Up to 6,300	No	Khrunichev State Research and Production Space Center (Russia)	Launches of communication and navigation satellites
Falcon Heavy (SpaceX)	Up to 63,800	Up to 26,700	Yes	SpaceX	Launch of the Tesla Roadster, Arabsat 6A satellite

is a large, flat platform equipped with systems to handle the massive thrust and heat generated during liftoff. Launch pads are constructed with materials that can withstand the high forces and extreme conditions associated with rocket launches.

Flame Trench or Deflector: To protect the launch vehicle from the intense heat and exhaust gases during liftoff, the launch pad may have a flame trench or deflector. This structure redirects the hot gases away from the rocket, ensuring its safety.

Service Structures: Launch pads are equipped with service structures that provide access to the launch vehicle for maintenance and pre-launch preparations. These structures also house umbilicals that connect to the rocket for fueling and communication.

Sound Suppression Systems: The noise generated during liftoff can be incredibly loud and may damage the vehicle or payload. Sound suppression systems, such as water deluge systems, release large volumes of water around the launch vehicle to dampen the noise and protect the hardware.

Fueling Systems: Launch stations have fueling systems to handle the rocket's propellants, whether they are liquid or solid fuels. These systems ensure precise fueling of the rocket before launch.

Telemetry and Communication Systems: The launch station is equipped with telemetry and communication systems to monitor the launch vehicle's health and transmit data to the mission control center. It also facilitates real-time communication with the vehicle during the launch.

Environmental Control Systems: Launch pads have environmental control systems to regulate temperature and humidity in and around the launch vehicle, ensuring that it remains within the specified operating conditions.

Lightning Protection: Lightning strikes can pose a risk to the launch vehicle and its sensitive electronics. To mitigate this risk, launch stations are equipped with lightning protection systems that safely conduct electricity away from the rocket.

Safety and Emergency Systems: Launch stations have safety systems and emergency protocols to handle contingencies and ensure the safety of personnel and the surrounding environment.

A Mission Control Station, often referred to simply as Mission Control, is a dedicated facility where space missions are planned, monitored, and controlled. It serves as the nerve center for space agencies or organizations to oversee the operations of satellites, spacecraft, and crewed missions, ensuring their successful execution and safety. Mission Control Stations play a crucial role in space exploration, scientific research, satellite operations, and human spaceflight missions. Mission Control Stations are highly secure and specialized facilities equipped with advanced technology, communication systems, monitoring tools, and mission-specific software. They often operate around the clock, monitoring and controlling space missions, regardless of day or night. These facilities are staffed by teams of highly skilled engineers, scientists, and operators who work collaboratively to ensure the success and safety of space missions. Key functions and features of a Mission Control Station include commanding and controlling the satellites, spacecraft, or crewed missions during various phases of their missions. It communicates with the vehicles in space via ground stations to send commands and receive telemetry data. Mission Control prepares detailed flight plans for each mission, outlining the sequence of operations and maneuvers that the satellite or spacecraft needs to follow. These plans are constantly updated based on mission requirements and changing conditions. Mission Control experts analyze the satellite's or spacecraft's orbit and trajectory to ensure they remain on the desired path and achieve their mission objectives. Mission Control develops safety protocols and contingency plans to address potential emergencies or unexpected events during missions. This includes measures to safeguard the spacecraft and its mission in case of critical failures.

Assembly and Integration Buildings: Launch stations have assembly and integration buildings where rockets or spacecraft are assembled, integrated with their payloads, and prepared for launch. These facilities ensure that all components are properly integrated and tested before being transported to the launch pad.

Payload Processing Facilities: Launch stations have facilities for handling and processing satellites, scientific instruments, and other payloads before integration with the launch vehicle. These facilities ensure that the payloads are prepared and tested for space missions.

Support Infrastructure: Launch stations include a range of support infrastructure, such as fuel storage facilities, environmental control systems, communications networks, safety and security measures, and ground support equipment to facilitate space mission operations.

Range Safety Systems: Launch stations are equipped with range safety systems to ensure public safety and protect against potential accidents or off-nominal events during rocket launches. These systems can include destruct mechanisms for rockets in case they deviate from their planned trajectories.

Landing Facilities: Some launch stations have landing facilities for reusable rockets or crewed spacecraft returning from space missions. These facilities support the recovery and refurbishment of reusable vehicles.

Environmental Considerations: Launch stations are chosen or designed with consideration of various environmental factors, such as proximity to populated areas, local weather conditions, and environmental impact assessments.

Infrastructure for Crewed Missions: Launch stations that support crewed missions have additional facilities for astronaut training, crew accommodations, and medical support.

Ground Support Personnel: Launch stations are staffed with a team of engineers, technicians, and support personnel responsible for mission planning, launch operations, and mission execution.

Launch stations are located in various regions around the world and are often associated with national space agencies, commercial space companies, or international collaborations. Each launch station may have unique features and capabilities, tailored to its specific missions and goals. These facilities are critical for space exploration, scientific research, Earth observation, telecommunications, and a wide range of other applications that rely on space-based technologies.

Table 8.3: Some of the satellite launching stations around the world

Name	Location	Country	Coordinates
Baikonur Cosmodrome	Kazakhstan	Kazakhstan	45.9640° N, 63.3050° E
Plesetsk Cosmodrome	Arkhangelsk Oblast	Russia	62.9275° N, 40.5772° E
Guiana Space Centre (CSG)	Kourou, French Guiana	France	5.2360° N, 52.7696° W
Tanegashima Space Center	Tanegashima Island, Kagoshima	Japan	30.3852° N, 130.9700° E
Kennedy Space Center	Merritt Island, Florida	USA	28.6083° N, 80.6045° W
Vandenberg Space Force Base	California	USA	34.7554° N, 120.580° W
Xichang Satellite Launch Center	Sichuan Province	China	28.2466° N, 102.0263° E
Satish Dhawan Space Centre	Sriharikota, Andhra Pradesh	India	13.7337° N, 80.2350° E
Wenchang Space Launch Site	Hainan Island	China	19.6147° N, 110.9515° E
Jiuquan Satellite Launch Center	Gansu Province	China	40.9845° N, 100.2864° E

8.4 Types of Trajectories

In space exploration, various types of trajectories are used to guide spacecraft to their destinations efficiently. These trajectories are carefully planned to take advantage of celestial bodies' gravitational forces and to minimize fuel consumption.

In many instances like the motion of a planet around the Sun or the motion of a Moon around a planet or the motion of an artificial satellite around a planet, the motion of the lighter body can be described by the gravitational attraction between the two bodies. Newton's law of gravitation states that the gravitational force between two masses, M and m, having a distance, r, between their centers is

$$F = \frac{GMm}{r^2} \qquad (8.1)$$

which is attractive in nature and is directed along the vector joining the two centers. Here, G is the universal gravitational constant and has a value $6.6742 \times 10^{-11}\ Nm^2/kg^2$. In other words, the lighter body will "feel" that it is being pulled to the heavier body. Using the Newton's law of gravitation, it can be be shown that the path of the lighter mass with respect to the heavier mass is a conic section (circle, ellipse, parabola or hyperbola). Recall that the conic section is obtained when a plane cuts a circular cone at an angle. The general equation of a conic section is given by

$$r = \frac{h^2}{\mu} \frac{1}{1 + e\cos\theta} \qquad (8.2)$$

where h is the angular momentum per unit mass for the lighter body, $\mu = GM$ and e is the eccentricity of the conic section.

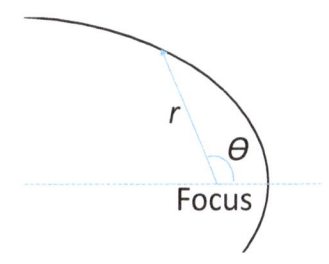

Conic section.

Circular Orbit:

$e = 0$ corresponds to a circular orbit. Then, $r = \frac{h^2}{\mu}$ and the time period required for one orbit is $T_{circular} = \frac{2\pi}{\sqrt{\mu}} r^{\frac{3}{2}}$. The total energy (potential energy + kinetic energy) per unit mass associated with a circular orbit is $-\frac{\mu}{2r}$. In other words, larger the radius of the orbit, the greater is the energy associated with the orbit. Raising a satellite from the Earth's surface to an orbit is associated with increase in energy. This energy is supplied by the work done by the rocket motors of the satellite launch vehicle. Typical remote sensing, imaging and navigation satellites occupy circular Low Earth Orbits (LEO). A LEO lies between below $2000 km$ above sea level. If a satellite appears stationary above the same point on the Earth's equator, then it is in a circular Geostationary Equatorial Orbit. Communication satellites are placed in the GEOs. The GEO satellites are at an altitude of $35,786 km$ above sea level.

Elliptic ($0 < e < 1$), Parabolic ($e = 1$) and Hyperbolic ($e > 1$) Orbits

If $0 < e < 1$, the relative position of the satellite remains bounded. Such orbits are called elliptic orbits. If $e = 1$, the trajectory becomes unbounded. Parabolic trajectories help us to calculate escape velocity. The escape velocity can be calculated to be $v = \sqrt{\frac{2\mu}{r}}$. If $e > 1$, the trajectory is a hyperbola. The turn angle is defined as $\delta = 2\sin^{-1}(1/e)$. At the perigee of a hyperbolic trajectory $v^2 = v_{esc}^2 + v_\infty^2$. v_∞^2 represents the excess kinetic energy over that which is required to simply escape from the gravitational pull and is a measure of the energy required for interplanetary mission.

8.4.1 Orbital Parameters

The parameters of an orbit describe its size, shape, orientation, and position in space. These parameters define the motion of an object, such as a spacecraft or a satellite, as it moves around another celestial body, such as a planet or a moon. The key parameters of an orbit are as follows

Semi-Major Axis (a) is half of the longest diameter of the elliptical orbit. For circular orbits, the semi-major axis is equal to the radius of the circle.

Eccentricity describes the shape of the orbit. For circular orbits, the eccentricity is 0, while for highly elongated (elliptical) orbits, the eccentricity approaches 1. An eccentricity of 1 corresponds to a parabolic trajectory (escape orbit), and values greater than 1 indicate hyperbolic trajectories.

Inclination is the angle between the orbital plane and the reference plane, which is usually the equatorial plane of the celestial body being orbited. Inclination is measured in degrees and ranges from 0° (equatorial orbit) to 90° (polar orbit).

Argument of Periapsis is the angle between the ascending node and the periapsis (closest point to the central body) measured in the orbital plane. It defines the orientation of the orbit within the plane.

Longitude of the Ascending Node is the angle between the reference direction (usually the vernal equinox) and the ascending node, which is the point where the orbit crosses the reference plane moving from the southern hemisphere to the northern hemisphere.

True Anomaly is the angle between the periapsis and the current position of the orbiting object, measured in the orbital plane. It represents the position of the object along its orbit at a specific time.

Period is the time taken for the orbiting object to complete one full revolution around the central body. It is related to the semi-major axis and the gravitational parameter of the central body.

These parameters fully characterize an orbit and are used to precisely define the motion of spacecraft, satellites, and celestial bodies as they move through space. The combination of these parameters uniquely describes the specific trajectory and behavior of an object in its orbit.

Lagrange Points

Lagrange points, also known as *libration points*, are specific locations in space where the gravitational forces of two large celestial bodies (such as a planet and its moon or the Earth and the Sun) create regions of gravitational equilibrium for a body moving under the influence of the two celestial bodies. In these points, the gravitational forces and the cen-

8.4. TYPES OF TRAJECTORIES

Table 8.4: Orbit of Satellites

Satellite Orbits	Definition
Low Earth Orbit (LEO)	LEO lies below 2,000 kilometers above Earth's surface. Satellites in LEO complete orbits relatively quickly, typically within a couple of hours. This orbit is commonly used for Earth observation, remote sensing, and some communication satellites.
Medium Earth Orbit (MEO)	MEO is an orbit located between LEO and geostationary orbit. Satellites in MEO are primarily used for navigation systems like GPS, offering a good balance between coverage area and orbital lifetime.
Geosynchronous orbit	Geosynchronous orbit is an Earth-centered orbit with time period that matches the Earth's rotation about its axis. This means that for an observer on Earth's surface, a satellite in geosynchronous orbit returns to exactly the same position in the sky after a period of one sidereal day.
Geostationary Orbit (GEO)	GEO, a special class of geosynchronous orbit, is an orbit approximately 35,786 kilometers above Earth's equator. Satellites in GEO have an orbital period matching Earth's rotation, allowing them to appear stationary relative to the ground. This orbit is ideal for communication and weather satellites that require continuous coverage.
Polar Orbit	Polar orbit satellites pass over Earth's poles in each orbit, providing global coverage over time. They are often used for Earth observation and scientific missions, as they can scan the entire planet.

trifugal force experienced by a smaller object, like a satellite or spacecraft, balance out, allowing the object to attain equilibrium position relative to the two larger bodies. These can be used by spacecraft as "parking spots" in space to remain in a fixed position with minimal fuel consumption. The stability of Lagrange points makes them useful for various space missions and satellite operations. For the Sun-Earth system, there are five Lagrange points – three are unstable denoted as *L1, L2, L3* and two are stable labeled *L4,L5* – are situated along the line connecting the two large bodies. The five Lagrange points have different stability characteristics:

- *L1 (Lagrange Point 1):* Located between the two large bodies, along the line connecting them and closer to the smaller of the two, L1 is an unstable point. A spacecraft at L1 requires continuous thruster adjustments to maintain its position because any small perturbation will cause it to drift away from the point.

- *L2 (Lagrange Point 2):* Also located between the two large bodies but farther from the larger body, L2 is an unstable point, similar to L1. A

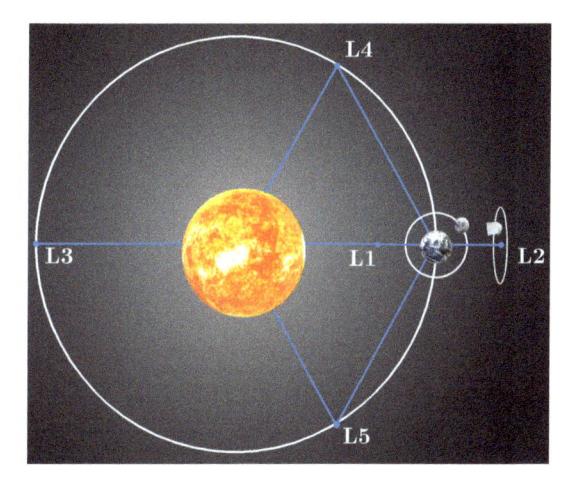

The Lagrange Points: parking spots for spacecraft in space with minimal fuel consumption.

spacecraft at L2 needs continuous corrections to prevent it from drifting away.

- *L3 (Lagrange Point 3):* L3 is located on the opposite side of the larger body from the smaller body, forming an equilateral triangle with the two bodies. It is also an unstable point.

- *L4 (Lagrange Point 4)*: Situated 60 degrees ahead of the smaller body in its orbit around the larger body, L4 and L5 are stable points. They form an equilateral triangle with the two large bodies, and objects placed at L4 and L5 tend to remain near these points with minimal correction.

- L5 (Lagrange Point 5): Located 60 degrees behind the smaller body in its orbit around the larger body, L5 is also a stable point, like L4.

Lagrange points are crucial in space exploration and provide unique opportunities for scientific research, satellite operations, and future space missions. Some space telescopes, like the James Webb Space Telescope (JWST), are positioned at L2 position relative to Earth and the Sun, allowing for continuous observations without the Earth or Moon obstructing their view. Spacecraft have been placed at L1 and L2 for solar and Earth observations, space weather monitoring, and communication relay purposes. The Sun-Earth L4 and L5 points have been identified as locations for potential future space habitats known as "Lagrange point colonies".

8.5 Spacecraft Maneuvers and Trajectories

Hohmann transfer orbit, named after the German space scientist *Walter Hohmann*, is an energy-efficient trajectory used to transfer a spacecraft between two circular orbits at different altitudes or inclinations. It involves two engine burns: one to raise the spacecraft's orbit and another to lower it into the target orbit. This trajectory is often used for interplanetary missions, such as traveling from Earth to Mars. The *Hohmann* transfer has been extensively used in space missions, including missions to other planets within the solar system. For instance, robotic missions to Mars, such as NASA's *Mars rovers*, have employed the *Hohmann* transfer. The *Hohmann* transfer is also commonly used for satellite deployment and spacecraft rendezvous operations.

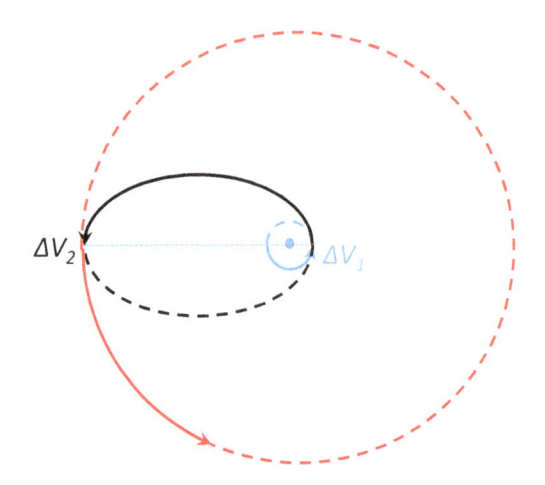

The *Hohmann* transfer.

Lunar trajectories are the paths spacecraft follow when traveling to and from the Moon. The spacecraft is launched from Earth using a rocket. It initially enters a temporary parking orbit around Earth to reach the necessary altitude and velocity for the lunar journey. Once in the Earth parking orbit, the spacecraft performs a critical maneuver called *Translunar Injection(TLI)*. During TLI, the spacecraft's propulsion system is fired to increase its velocity and escape Earth's gravity. This maneuver places the spacecraft on a trajectory headed towards the Moon. Then the spacecraft enters a trajectory called the *Lunar Transfer Orbit*, which carries it from Earth towards the Moon. Upon reaching the vicinity of the Moon, the spacecraft performs another critical maneuver called *Lunar Orbit Insertion (LOI)*. During LOI, the spacecraft's propulsion system is fired again to slow down the spacecraft and allow it to be captured by the Moon's gravity. This maneuver puts the spacecraft into a lunar orbit. Depending on the mission's objectives, the spacecraft may deploy landers or rovers to study and explore the lunar surface. Human missions may involve landing astronauts on the Moon to conduct extravehicular activities (EVAs) or moonwalks. After completing the mission objectives on the Moon, the spacecraft leaves the lunar surface through a process called Lunar Ascent. Once in lunar orbit, the spacecraft performs a Trans-Earth Injection burn, which propels it back toward Earth. As the spacecraft approaches Earth, it performs a re-entry burn to slow down and enter the Earth's atmosphere safely. This burn ensures that the spacecraft enters the atmosphere at the correct angle to avoid excessive heating during re-entry. Depending on the mission type, the spacecraft will either splash down in an ocean or make a controlled landing on land.

Phasing maneuvers are orbital maneuvers performed by spacecraft to synchronize their orbits with other objects, such as satellites, space stations, or celestial bodies. These maneuvers are essential for spacecraft to rendezvous with their targets or maintain specific relative positions during joint operations. When two spacecraft aim to rendezvous in space, they must align their position and velocity vectors simultaneously. Examples of rendezvous include resupply missions for the ISS and the docking of the Lunar Module (LM) with the Command and Service Module (CSM) during the Apollo missions. Phasing maneuvers are also used in space missions for satellite deployment, docking, space station rendezvous, and planetary missions.

There are two primary types of phasing maneuvers: *(1) Relative Orbital Phasing*: In this type of phasing maneuver, a spacecraft adjusts its orbit to align with or catch up to another object in space. The goal is to match the position, velocity, and relative orbital phase with the target object. A common method used for relative phasing, as described earlier, to transfer a spacecraft from one circular orbit to another with the same orbital plane is *"Hohmann Transfer"*. A more complex transfer trajectory that uses two elliptical orbits to adjust the phasing with the target object is the "Bi-elliptic Transfer". There is a "Phasing Burn" where a specific engine burn executed to adjust the spacecraft's orbit, enabling it to achieve the desired relative position with the target.

(2) Rendezvous and Docking Phasing: These maneuvers are used when one spacecraft needs to meet and dock with another spacecraft or a space station. Rendezvous and docking phasing involves fine-tuning the spacecraft's approach and relative velocities for a successful docking operation. In coelliptic rendezvous, the spacecraft performs a series of small burns to gradually reduce its distance to the target while remaining in the same elliptical orbit. In close-approach rendezvous, the spacecraft follows a trajectory that brings it close to the target before performing a final burn to match the target's velocity and initiate docking. In non-coelliptic rendezvous, the spacecraft uses different orbits from the target and performs multiple phasing burns to catch up to the target.

Phasing maneuvers require precise calculations and planning to ensure safe and accurate rendezvous and docking operations. These maneuvers are crucial for missions like satellite constellation deployment, crewed spaceflights, resupply missions to the *International Space Station* (ISS), and deep-space missions involving spacecraft rendezvous with asteroids or other planets. Successful phasing maneuvers are critical for achieving mission objectives and maximizing the efficiency of space missions.

Chase maneuver in spaceflight refers to a series of orbital adjustments performed by one spacecraft to approach and match the relative velocity and position of another spacecraft or object in space. The goal of a chase maneuver is to achieve a close distance and synchronized motion with the target, often for purposes such as inspection, rendezvous, or docking.

Chase maneuvers are particularly important for crewed space missions, satellite servicing, and space missions involving multiple spacecraft working together. They require precise calculations, real-time monitoring, and coordinated control to ensure a safe and successful approach.

Chase maneuvers require careful planning and execution, as any unintended or abrupt movements during the approach could lead to dangerous situations or mission failure. Advanced sensors, guidance systems, and communication between the spacecraft and mission control are essential for safe and precise execution of chase maneuvers in space.

Plane change maneuver in spaceflight refers to a change in the orbital plane of a spacecraft, altering the inclination or orientation of its orbit with respect to a reference plane. This maneuver is necessary when a spacecraft needs to transition from one orbital plane to another, typically to reach a different target or align with the desired orbital plane for a specific mission objective. One way to perform a plane change is to give a normal component of velocity at the crossing point of the initial and final orbital planes. The δV required is the vector change in velocity between the two planes at that point.

Plane change maneuvers are energetically expensive and require a significant amount of propellant. Plane change maneuvers are commonly encountered in various space missions, including interplanetary missions, satellite launches, and rendezvous and docking operations. For example, interplanetary missions often require plane change maneuvers to align the spacecraft's trajectory with the target planet's orbital plane. Satellite launches from Earth frequently include plane change maneuvers to adjust the spacecraft's inclination to match the desired orbit. Spacecraft rendezvous and docking missions may involve plane change maneuvers to align the spacecraft's orbit with that of the target space station or another satellite.

Due to their fuel-intensive nature, efficient planning and execution of plane change maneuvers are critical to ensure the success of space missions and to maximize the spacecraft's operational lifetime.

Interplanetary Trajectories: *Hohmann* trajectories are the simplest way to analyze interplanetary trajectories. Such trajectories assume that the planets involved lie on the same plane and the planets must be positioned just right for such trajectories to happen. Another method called the method of patched conics divides the trajectory into three parts, viz., the hyperbolic departure trajectory from the parent planet, the cruise ellipse trajectory and the hyperbolic arrival trajectory towards the target planet.

Flyby maneuvers in space exploration refer to the use of a celestial body's gravitational pull to alter a spacecraft's trajectory, speed, or direction without going into orbit around the body. Instead of entering orbit, the spacecraft uses the gravitational slingshot effect to gain or lose energy, allowing it to achieve specific mission objectives more efficiently. Flyby maneuvers are also known as gravity assists or gravitational slingshots.

Flyby maneuvers have been widely used in space exploration and have enabled ambitious missions to explore distant regions of the solar system. Some well-known examples of flyby maneuvers include

- *Voyager* Missions: The *Voyager 1* and *Voyager 2* spacecraft used multiple gravity assists to perform flybys of Jupiter, Saturn, Uranus, and Neptune, providing valuable data about these planets and their moons.

- *New Horizons* spacecraft conducted a flyby of Pluto and its moons, providing the first detailed images and data of the dwarf planet.

- *Cassini-Huygens* spacecraft used multiple flybys of Saturn's moons to study their geology, atmosphere, and potential for hosting life.

- *Parker Solar Probe* uses Venusian gravity assists to gradually decrease its orbital distance to the Sun, enabling close observations of its corona.

Flyby maneuvers continue to play a crucial role in space missions, allowing scientists and engineers to explore distant and challenging locations in our solar system while conserving propellant and optimizing mission efficiency.

Gravity assist, also known as gravitational slingshot or gravitational assist, is a spaceflight maneuver that utilizes the gravitational pull of a celestial body, usually a planet or a moon, to alter the trajectory and speed of a spacecraft. By executing a carefully planned flyby of the celestial body, the spacecraft gains or loses energy, changing its velocity and direction without expending additional propellant. This technique allows spacecraft to achieve complex and ambitious missions with minimal fuel consumption. This technique has been used in missions such as *Voyager 1* and *Voyager 2*, which employed multiple gravity assists to explore the outer planets. Early NASA missions, *Pioneer 10 and 11*, were the first to use gravity assist to accelerate toward Jupiter and Saturn, respectively, before heading to the outer solar system. The *New Horizons* spacecraft executed a gravity assist with Jupiter to accelerate its journey to Pluto, where it conducted the first flyby of the dwarf planet. NASA's *Cassini spacecraft* used multiple gravity assists from Venus and Jupiter to reach Saturn and explore its moons and rings.

Aerobraking is a spaceflight maneuver used to reduce their velocity and adjust their orbits by interacting with a planet or moon's atmosphere. During an aerobraking maneuver, the spacecraft passes through the upper layers of a celestial body's atmosphere, which generates drag that slows down the spacecraft. By using the planet or moon's atmosphere as a braking mechanism, the spacecraft can reduce its velocity and lower its orbital altitude without expending significant amounts of propellant.

NASA's *Mars Reconnaissance Orbiter* used aerobraking to adjust its orbit around Mars. The spacecraft conducted multiple passes through Mars' atmosphere to circularize its polar orbit and achieve the desired orbit. NASA's *Galileo spacecraft* also used aerobraking maneuvers during its mission to Jupiter to adjust its trajectory.

Aerobraking is a valuable technique in space exploration, enabling spacecraft to reach their intended destinations efficiently and conduct extended missions with limited propellant resources.

Solar escape trajectories, also known as escape trajectories or escape orbits, refer to the paths followed by spacecraft or celestial objects that leave the gravitational influence of the Sun and enter interstellar space. These trajectories are used when a spacecraft or object gains enough energy to overcome the gravitational pull of the Sun, enabling it to journey beyond the solar system.

The primary conditions for a solar escape trajectory are sufficient velocity and minimal gravitational encounter. The spacecraft must attain a velocity, known as the escape velocity, that allows it to break free from the Sun's gravitational field. The escape velocity from the surface of the Sun is about 617.5 km/s, which is much higher than the typical velocities achieved by spacecraft in the solar system. To minimize the energy required for the escape, spacecraft often execute gravitational assists with planets, moons, or other celestial bodies. These assists use the gravity of these bodies to increase the spacecraft's velocity while decreasing its relative velocity to the Sun, making it easier to escape the Sun's gravitational pull.

Solar escape trajectories are used in specific interplanetary missions and deep space exploration. Some famous examples of missions that employed solar escape trajectories are NASA's *Voyager 1 and Voyager 2* spacecraft with the primary mission of exploring the outer planets in our solar system. After completing their encounters with Jupiter, Saturn, Uranus, and Neptune, both *Voyager* spacecraft executed gravitational assists to gain sufficient velocity to escape the solar system. *Voyager 1* crossed the heliopause (the boundary between the solar system and interstellar space) in 2012, becoming the first human-made object to reach interstellar space. NASA's *New Horizons* spacecraft, launched in 2006, performed a flyby of Pluto and its moons before executing a gravitational assist with Jupiter to gain the velocity needed for a solar escape trajectory. After its Pluto encounter in 2015, *New Horizons* continued its journey into the Kuiper Belt, where it conducted flybys of other Kuiper Belt objects.

Solar escape trajectories allow spacecraft to explore the vast regions beyond our solar system, providing valuable insights into interstellar space and potentially reaching other star systems in the distant future. These missions have significantly expanded our understanding of the cosmos and paved the way for future deep space exploration.

Ballistic trajectories are used for specific maneuvers or mission phases that do not involve continuous propulsion. These trajectories rely solely on the initial velocity and the force of gravity to guide the spacecraft or its components to their intended destinations or orbits. Ballistic trajectories involve a spacecraft following a free-falling path with minimal propulsion. They are typically used for flyby missions where the spacecraft only needs to pass by a planet or moon to collect data

Station-keeping is a critical orbital maneuver used by spacecraft, satellites, and space stations to maintain a stable position or orbit relative to a specific point or object in space. This maneuver involves periodic adjustments to counteract perturbations and external forces that can cause the spacecraft to drift away from its intended position. The goal of station-keeping is to ensure that the spacecraft remains within a designated operational region and performs its intended mission effectively. To counteract these influences and maintain the desired position or orbit, station-keeping maneuvers are performed. These maneuvers typically involve firing thrusters or using reaction control systems (RCS) to apply small amounts of propulsion at specific intervals. By adjusting the spacecraft's velocity and direction strategically, it can counteract the perturbing forces and maintain its position relative to the target point or object.

Station-keeping is essential for various space missions and applications, including: *Geostationary Satellites:* Communications satellites in geostationary orbit need to maintain a fixed position above a specific point on Earth to provide continuous coverage to a specific region. *Space stations* like the International Space Station (ISS) require constant station-keeping to maintain their orbits and relative positions for rendezvous and docking operations. *Satellite Constellations:* Groups of satellites such as GPS satellites, maintain their relative positions and orbits to provide continuous global coverage. Orbiting *scientific Observatories* such as space telescopes and scientific satellites need precise positioning to observe specific celestial objects or regions of interest.

Station-keeping requires precise orbital analysis, accurate propulsion systems, and continuous monitoring to ensure the spacecraft remains on its intended path and achieves its mission objectives effectively. It is an essential aspect of space operations, especially for long-duration missions and critical applications.

Impulse maneuvers are those in which brief firings of on-board rocket motors change the magnitude and direction of velocity vector.

$$\frac{\Delta m}{m} = 1 - e^{-\frac{\Delta v}{gI_{sp}}} \qquad (8.3)$$

where Δm is the mass of propellant consumed, m is the mass of the spacecraft, g is the standard acceleration due to gravity and I_{sp} is the specific impulse of the propellant and Δv is the change in velocity.

Attitude correction of satellites is a critical aspect of space operations that involves adjusting the orientation and attitude of a satellite in space to ensure it maintains the desired position and orientation relative to the Earth or a target celestial body. Attitude control is crucial for various space missions, including Earth observation, communications, weather monitoring, and scientific research. The main objectives of attitude correction are:

Pointing Accuracy: Satellites often carry instruments, antennas, or sensors that require precise pointing towards specific targets on the Earth's surface or in space. Attitude correction ensures the satellite accurately points its instruments in the desired direction.

Stability: To achieve stable operation and data acquisition, satellites need to maintain a fixed attitude or a specific orientation relative to the Earth, the Sun, or other celestial objects.

Power Management: Satellite solar panels are designed to receive maximum sunlight for power generation. Correcting the satellite's attitude optimizes solar panel alignment with the Sun.

Communication: For effective communication with ground stations or other satellites, the satellite's antennas need to be accurately pointed toward the target.

Several methods and devices are employed for attitude correction of satellites:

Reaction Wheels: Reaction wheels are spinning flywheels mounted inside the satellite. By changing the speed of these wheels, the satellite can control its angular momentum and orientation.

Thrusters: Small onboard thrusters, often called reaction control system (RCS) thrusters, provide short bursts of propulsion to adjust the satellite's orientation.

Magnetic Torquers: Magnetic coils interact with Earth's magnetic field, generating torque to control the satellite's orientation.

Control Moment Gyroscopes (CMGs): CMGs use gyroscopic principles to control attitude by changing the direction of angular momentum.

Sun and Earth Sensors: Sun sensors detect the position of the Sun relative to the satellite, while Earth sensors determine the satellite's position relative to the Earth's surface. These sensors provide crucial information for attitude control algorithms.

Star Trackers identify stars in the sky to determine the satellite's orientation relative to known celestial reference points.

Satellite attitude correction involves continuous monitoring and adjustments based on telemetry data and ground commands. Advanced algorithms and software calculate the necessary corrections and implement them in real-time. Attitude control systems play a fundamental role in the overall success and operational efficiency of satellite missions.

Atmospheric entry is a critical phase in space missions, during which a spacecraft descends from space into the atmosphere of Earth or another planet. This phase poses several significant challenges that must be overcome to ensure the safety and success of the mission. Some of the key challenges during atmospheric entry are:

Aerodynamic Heating: As a spacecraft enters the atmosphere at high speeds, it encounters atmospheric molecules that cause friction and generate intense heat. This heat can reach extremely high temperatures and can damage or destroy the spacecraft if not properly managed. To withstand the extreme heat of atmospheric entry, spacecraft require advanced thermal protection systems (heat shields). These shields dissipate the heat and keep the spacecraft's interior at a safe operating temperature.

Rapid Deceleration: Atmospheric entry causes rapid deceleration of the spacecraft due to air resistance. This deceleration can subject the spacecraft to intense forces, which must be managed to prevent structural damage. Determining the appropriate entry angle and speed is crucial to manage heating and deceleration forces effectively. A steep entry angle can increase heat loads, while a shallow angle may prolong the descent.

Entry Trajectory Control: Precise control of the spacecraft's entry trajectory is essential to ensure it reaches the desired landing site or target. Small deviations in trajectory can lead to significant differences in landing location. Also, the spacecraft must enter the atmosphere at the correct entry corridor to ensure it safely reaches its intended landing site or entry point.

Communication Blackout: During certain portions of atmospheric entry, the spacecraft may experience a communication blackout due to ionized plasma surrounding the vehicle, blocking radio signals.

Reusability: For reusable spacecraft like the *Space Shuttle* or *SpaceX's Falcon 9*, additional challenges include handling the stresses of multiple missions and refurbishment between the missions.

Entry Environment Uncertainties: Variability in atmospheric conditions and density during entry can introduce uncertainties in the spacecraft's behavior, requiring robust control systems and algorithms.

Successfully addressing these challenges requires

INITIAL CONCEPT BLUNT BODY CONCEPT 1953

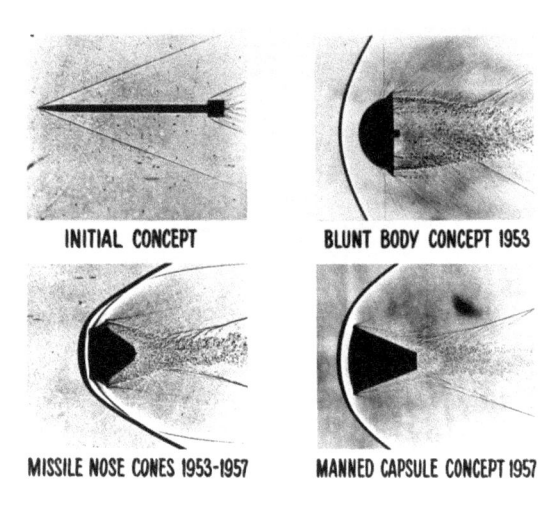

MISSILE NOSE CONES 1953-1957 MANNED CAPSULE CONCEPT 1957

Reentry-vehicle concepts visualized in shadowgraphs of high-speed wind tunnel tests. *H. Julian Allen* developed the Blunt Body Theory, which revolutionized heat shield designs crucial for the Mercury, Gemini, and Apollo space capsules. These innovations facilitated astronauts' survival during fiery re-entry into Earth's atmosphere. A blunt body generates a protective shockwave, as depicted in the accompanying photo, effectively shielding the vehicle from excessive heating. Consequently, blunt-body vehicles maintain lower temperatures compared to sleek, low-drag counterparts.

extensive engineering, testing, and simulations to ensure the spacecraft can endure the harsh conditions of atmospheric entry. Precise navigation, advanced thermal protection systems, and reliable communication systems are essential for a safe and successful entry, especially for crewed missions returning to Earth or missions that require precise landings. Atmospheric entry remains one of the most critical and high-stress phases in any space mission, and careful planning and execution are essential for mission success.

8.6 Space Debris

Space debris also known as space junk refers to defunct satellites, spent rocket stages, fragments from collisions, discarded hardware from past space missions and other celestial bodies. Space debris poses significant challenges in space exploration and has become a growing concern for space agencies and companies operating in Earth's orbit. The long-term sustainability of space activities is at risk due to the accumulation of space debris. It may reach a point where it becomes challenging to conduct space missions safely and efficiently. The challenges it presents in space exploration include:

- With thousands of trackable objects and countless smaller, untrackable fragments, the risk of collisions with space debris is a constant concern. Even small pieces of debris can cause significant damage to operational satellites, spacecraft, and the *International Space Station* (ISS).

- Active satellites must regularly adjust their orbits to avoid potential collisions with space debris. These collision avoidance maneuvers consume valuable fuel and limit the operational lifetime of satellites.

- Collisions between space debris can create even more fragments, leading to the "Kessler Syndrome" where cascading collisions generate an increasing amount of debris, making space increasingly hazardous.

- Large pieces of space debris can interfere with astronomical observations, affecting the quality of data collected by ground-based and space-based observatories.

Addressing the challenges posed by space debris requires global cooperation and efforts from the international space community. Efforts to address space debris are essential to maintaining a safe and sustainable space environment for future space exploration endeavors and ensuring the long-term viability of space activities. As the number of satellites and missions in space continues to increase, effective space debris management becomes increasingly crucial to safeguarding our use of space.

8.7 Space Agencies

Space agencies play pivotal roles in advancing space exploration, conducting scientific research,

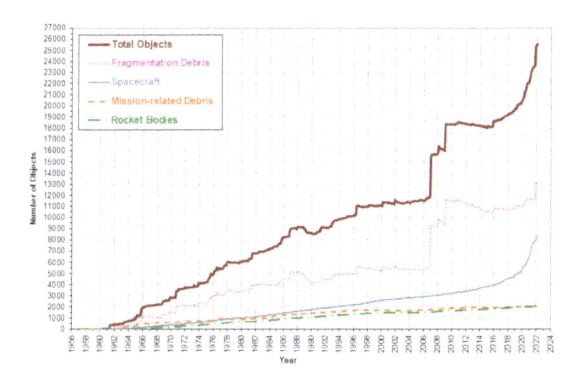

Monthly number of objects in Earth orbit by object type. The data represent the number of objects > 10 cm in LEO. *Source: NASA orbital debris program office.*

launching satellites for communication and observation, and pushing the boundaries of human knowledge about space and our universe.

China National Space Administration (CNSA): The CNSA oversees China's space program formed in 1993, conducting a wide array of missions encompassing crewed spaceflight, lunar exploration, Mars missions, satellite deployment, and space station construction. Notable achievements include the *Chang'e lunar missions*, the *Tianzhou* cargo spacecraft, the *Tiangong* space station, and the successful deployment of the *Chang'e-4* rover on the far side of the Moon. Continued lunar exploration, Mars exploration, space station assembly, and developing advanced launch vehicles.

European Space Agency (ESA): Comprising 22 member states formed in 1975, the ESA focuses on collaborative space endeavors, encompassing scientific research, satellite launches, human spaceflight, Earth observation, and interplanetary exploration. ESA's significant projects involve the Ariane launch vehicles, the Mars Express mission, the Hubble Space Telescope, and the Rosetta mission which successfully landed a probe on a comet. Ongoing missions include Earth observation (Copernicus), Mars exploration (ExoMars), and developing next-generation launch vehicles.

Indian Space Research Organisation (ISRO): India's premier space agency established in 1969, ISRO, conducts a diverse range of missions, including satellite launches, lunar exploration, Mars exploration, Earth observation, and navigation satellite systems. ISRO's achievements include the Chandrayaan and Mangalyaan missions to the Moon and Mars, the development of the PSLV and GSLV launch vehicles, and numerous successful satellite deployments. Continued lunar exploration, satellite launches for various applications, human spaceflight program (Gaganyaan), and Mars and Venus missions.

Japan Aerospace Exploration Agency (JAXA): JAXA (formed in 2003) leads Japan's space exploration efforts, conducting missions in planetary exploration, Earth observation, satellite launches, and space science research. Noteworthy missions encompass the Hayabusa asteroid missions, the Akatsuki Venus orbiter, the Kibo laboratory module on the ISS, and contributions to global satellite navigation systems. Mars moon exploration (MMX), lunar exploration, asteroid missions, and satellite technology development.

National Aeronautics and Space Administration (NASA): NASA (Established in 1958), the United States' space agency, conducts a broad spectrum of space missions encompassing space exploration, Earth and planetary science, astrophysics, human spaceflight, and aeronautics research. Notable missions include the Apollo Moon landings, Mars rovers (e.g., Curiosity, Perseverance), the Hubble Space Telescope, the ISS, and upcoming Artemis missions for lunar exploration. Mars exploration (Artemis program), return to the Moon, deep space exploration, Earth science missions, and studying exoplanets are the current ongoing and future Programs.

Roscosmos: Russia's state space corporation, Roscosmos, oversees the country's space exploration efforts, including crewed space missions, satellite launches, lunar and planetary exploration, and space station operations. Roscosmos has been involved in pioneering space missions such as the Sputnik satellite, the Vostok missions, the Soyuz spacecraft, and collaborations with international partners in the ISS project.

8.8 Space Laws

Space laws or space legislation are a set of legal principles, treaties, and agreements that govern human activities in outer space. As the exploration and utilization of space have advanced, international cooperation has been necessary to address the legal issues that arise. The following are some significant space laws and agreements:

Outer Space Treaty (OST) (1967): Also known as the Treaty on Principles Governing the Activities of States in the Exploration and Use of Outer Space, including the Moon and Other Celestial Bodies, this treaty is one of the fundamental pillars of space law. It establishes space as the province of all humankind, prohibits the placement of nuclear weapons in space, and restricts the use of the Moon and other celestial bodies exclusively for peaceful purposes. It also emphasizes international cooperation and responsibility for space activities.

Rescue Agreement (1968): The Agreement on the Rescue of Astronauts, the Return of Astronauts, and the Return of Objects Launched into Outer Space deals with the prompt return of astronauts to their respective countries and the return of space objects in case of accidents or emergencies.

Liability Convention (1972): The Convention on International Liability for Damage Caused by Space Objects establishes rules for liability in case a space object causes damage to another country or its space objects on Earth or in space.

Registration Convention (1976): The Convention on Registration of Objects Launched into Outer Space requires countries to provide information on space objects they launch to an international registry maintained by the United Nations.

Moon Agreement (1984): The Agreement Governing the Activities of States on the Moon and Other Celestial Bodies outlines specific guidelines for the exploration and use of the Moon and its resources. It aims to ensure the Moon's use for the benefit of all countries and prevent the Moon's ownership or territorial claims.

Space Debris Mitigation Guidelines: Although not legally binding, the Inter-Agency Space Debris Coordination Committee (IADC) has developed guidelines and best practices for space agencies to minimize space debris creation and ensure the sustainable use of space.

Commercial Space Launch Competitiveness Act (2015): This U.S. law addresses various aspects of commercial space activities, including the establishment of a legal framework for commercial space mining and property rights for extracted resources from celestial bodies.

Space Activities Act (Australia) (1998): This Australian law regulates space activities launched from Australia and establishes the framework for granting licenses for space launches and space objects.

Space laws are continuously evolving to address emerging challenges and advancements in space technology. As commercial space activities increase, international cooperation and clear legal frameworks become even more crucial for ensuring peaceful and responsible exploration and use of outer space.

8.9 India's Space Program

India is striving to become a key player in the global space market through advancements in its space program. The Indian Space Research Organisation (ISRO), the country's space agency based in Bengaluru, oversees India's space initiatives. Established in 1969, ISRO operates under the Department of Space (DoS) and reports directly to the Prime Minister of India. The organization's primary goal is to leverage space technology for national progress. ISRO's missions showcase its commitment to global cooperation and technological advancement. With a total launch of 417 foreign satellites from 34 countries, alongside 116 spacecraft missions and 86 launch missions, ISRO has demonstrated its capability and reliability in space exploration. Moreover, ISRO's proficiency extends to re-entry missions, with a couple of of successful endeavors. These statistics underscore ISRO's significant contributions to the global space community and its ongoing pursuit of excellence in space exploration and technology. India's space program is advancing significantly, with missions targeting the moon, Mars, and further into space.

The origins of ISRO can be traced back to the Indian National Committee for Space Research (INCOSPAR), which was established in 1962 under the leadership of *Vikram Sarabhai*, the father of the Indian space program. INCOSPAR was responsible for setting up the *Thumba Equatorial Rocket Launching Station* (TERLS) in Kerala, which became the first launch site in India. In 1969, INCOSPAR was reorganized and renamed the ISRO, with *Vikram Sarabhai* as its first chairman. He was instrumental in establishing ISRO as a premier space agency, overseeing the successful launch of several satellites and the development of launch vehicles. *Satish Dhawan* was known for his leadership, vision, and commitment to scientific excellence, which helped ISRO achieve significant milestones during his tenure. Over the years, ISRO has grown to become one of the most advanced space agencies in the world, with a wide range of capabilities in satellite technology, launch vehicles, and space exploration.

ISRO has a vast network of research and development laboratories spread across different locations in India. *Vikram Sarabhai Space Centre (VSSC), Thiruvananthapuram*, established in 1962, is the largest ISRO center and the primary site for the development of launch vehicles. It is responsible for the design, development, and testing of rockets, including the launchers like *Rohini* and *Menaka*, as well as the SLV, ASLV, PSLV, GSLV, and LVM3 families of launch vehicles. The center also houses facilities for the assembly, integration, and testing of satellites. *Satish Dhawan Space Centre (SDSC)* (formerly Sriharikota Range – SHAR)), Sriharikota in Andhra Pradesh is ISRO's primary launch center which is responsible for the launch of various satellites and rockets, including the PSLV and GSLV. At present, the Centre operates two active launch pads for deploying sounding rockets, polar satellites, and geosynchronous satellites. Notable missions launched from here include India's lunar exploration probes Chandrayaan-1, Chandrayaan-2, and Chandrayaan-3, along with the Mars Orbiter Mission, solar research mission Aditya-L1, and space observatory XPoSat. *U R Rao Satellite Centre (URSC)*, Bengaluru (formerly ISRO Satellite Centre (ISAC)) is responsible for the design, development, and testing of satellites. It houses facilities for the assembly, integration, and testing of various satellite systems, including communication, earth observation, and scientific satellites. ISAC also has a dedicated facility for the development of satellite subsystems and payloads. Another lab, *Space Applications Centre (SAC)*, Ahmedabad focuses on the development of satellite-based applications and technologies. It works on the design and development of payloads for earth observation, communication, and meteorological satellites. SAC also conducts research and development in areas such as remote sensing, satellite communication, and navigation. *Liquid Propulsion Systems Centre (LPSC)*, Thiruvananthapuram is responsible for the design, development, and testing of liquid propulsion systems for launch vehicles and spacecraft. It works on the development of cryogenic and storable liquid propulsion systems, including engines, propellant tanks, and associated subsystems. *ISRO Propulsion Complex (IPRC)*, Mahendragiri in Tamilnadu focuses on the development of solid and hybrid propulsion systems for launch vehicles. It is also responsible for testing, assembling, and integrating propulsion systems and stages that are developed at LPSC.

ISRO begins its jounery in space exploration, launching India's first satellite, *Aryabhata*, in 1975. Aryabhata, named after the renowned astronomer, was notable for being entirely designed and built in India. Following *Aryabhata*'s launch, numerous satellite series and constellations have been launched and ISRO manages one of the most extensive constellations of active communication and earth imaging satellites, catering to both military and civilian needs. ISRO has executed numerous triumphant missions, such as the *Mars Orbiter Mission* and the *Chandrayaan missions* to the Moon, showcasing India's prowess in space technology. Successful Mars Orbiter Mission (Mangalyaan), which made India the first country to successfully reach Mars on its first attempt. Chandrayaan-1, India's first lunar mission, which discovered the presence of water on the Moon. Other missions include launching satellites, exploring other planets, and creating launch vehicles for communication, TV broadcasting, weather forecasting, resource management, and more. India's *Chandrayaan-3* mission marks the country's successful landing on the moon, making India the fourth nation to achieve this milestone. ISRO has successfully developed and mastered the technology for cryogenic engines, which are crucial for the launch of heavier satellites. The indigenous development of cryogenic engines has reduced India's dependence on foreign technology and has been a significant achievement.

Wing Commander *Rakesh Sharma*, who joined the Indian Air Force as a test pilot in 1970, is a celebrated Indian former Air Force pilot who was selected on 20 September 1982 to become a cosmonaut and go into space as part of a joint programme between the Indian Air Force and the Soviet *Interkosmos* space programme. He embarked on a historic spaceflight aboard the Soyuz T-11 on 3 April 1984, and holds the distinction of being the first and only Indian citizen to have the privilege of traveling to space.

In the 1960s and 1970s, India embarked on the development of its own launch vehicles, driven by geopolitical and economic factors. During this period, India engineered a sounding rocket, and by the 1980s, the efforts were fruitful with the development of the *Satellite Launch Vehicle-3* (SLV-3) and the more sophisticated *Augmented Satellite Launch Vehicle* (ASLV), accompanied by essential operational infrastructure. The *Polar Satellite Launch Vehicle (PSLV)* is India's initial medium-lift launch vehicle, facilitating the launch of all remote-sensing satellites into Sun-synchronous orbit. Despite a setback in its inaugural launch in 1993 and two partial failures, PSLV has emerged as ISRO's mainstay, with over 50 successful launches, deploying numerous Indian and foreign satellites into orbit. Conceived in the 1990s, the *Geosynchronous Satellite Launch Vehicle (GSLV)* aimed to transport substantial payloads to geostationary orbit. ISRO encountered significant challenges initially in realizing GSLV due to the decade-long development of a cryogenic rocket engine within India. Hindered by the U.S. block on India's access to cryogenic technology from Russia, India was compelled to independently develop its own cryogenic engines. The *Launch Vehicle Mark-3 (LVM3)*, formerly recognized as GSLV Mk III, stands as ISRO's most robust operational rocket. Enhanced with a more potent cryogenic engine and boosters compared to GSLV, it boasts a substantially increased payload capacity, enabling India to deploy all its communication satellites.

Bibliography

H. D. Curtis (2015) *Orbital Mechanics: For Engineering Students*, Elsevier Science.

A. H. Julian, and A. J. Eggers Jr. (1958) *A Study of the Motion and Aerodynamic Heating of Ballistic Missiles Entering the Earth's Atmosphere at High Supersonic Speeds*. NACA Rep. 1381

D. Edberg and W. Costa (2022) *Design of Rockets and Space Launch Vehicles*, AIAA, Incorporated.

R. K. Yedavalli (2020) *Flight dynamics and control of aero and space vehicles*. John Wiley & Sons

Table 8.5: ISRO's space missions till 2024.
[1] New Space India Limited (NSIL) is a Public Sector Undertaking (PSU) of the Government of India and under Department of Space responsible for producing, assembling and integrating the launch vehicle with the help of industry consortium

Year	Mission Name	Mission Description
1975	Aryabhata	First satellite.
1980	Rohini Satellite	First satellite launched by own launch vehicle, the SLV-3.
1983	INSAT-1B	First satellite launched as a part of the Indian National Satellite System (INSAT) for telecommunications, broadcasting, meteorology and search & rescue operations
1987	SROSS Series (SROSS-1)	Stretched Rohini Satellite Series (SROSS) deployed for conducting astrophysics, Earth Remote Sensing, and upper atmospheric monitoring experiments.
1993	IRS-1E	Earth observation mission launched under the National Natural Resources Management System program.
1999	INSAT-2E	A multi - purpose satellite for telecommunication, television broadcasting and meteorological services.
2001	GSAT-1	experimental satellite launched aboard the maiden flight of the GSLV rocket and used for conducting communication experiments like digital audio broadcast, internet services, and compressed digital TV transmission.
2005	CARTOSAT-1	An Earth observation satellite in a Sun-synchronous orbit, the first Indian Remote Sensing Satellite capable of providing in-orbit stereo images
2008	Chandrayaan-1	India's first mission to Moon under the Chandrayaan programme which discovered water molecules on the Moon.
2013	Mars Orbiter Mission	Mangalyaan, India's first interplanetary mission launched bya Polar Satellite Launch Vehicle (PSLV) rocket, making India the first Asian nation and the first country in the world to reach Mars on its maiden attempt.
2014	IRNSS-1C	The third out of seven in the Indian Regional Navigation Satellite System (IRNSS) series of satellites for providing accurate position information.
2015	Astrosat	India's first dedicated multi-wavelength space observatory telescope for astronomical observations.
2016	GSAT-18	Advanced communication satellite operated by INSAT.
2017	CARTOSAT-2	An Earth observation satellite in a Sun-synchronous orbit and the second of the Cartosat series of satellites. It is used for high-resolution earth observation satellite for cartographic and military applications.
2018	GSAT-29	High-throughput communication satellite to provide broadband connectivity to rural and remote areas.
2019	Chandrayaan-2	India's second lunar mission to explore south pole of the Moon launched by LMV3-M1 rocket and it was partially successful.
2020	GSAT-30	Replacement satellite for INSAT-4A, providing enhanced communication services
2021	PSLV-C51	launch of Brazil's Amazonia-1 satellite and 18 co-passenger payloads by PSLV. The first dedicated commercial launch executed by NSIL[1].
2022	GSAT-24	A communication satellite launched by NSIL to provide DTH television services.
2022	LVM3-M3	Launch of 36 OneWeb broadband communication satellites (OneWeb India-1) aboard the LVM3 rocket.
2023	Aditya-L1	India's first solar mission launched by PSLV-XL/C-57 to study the Sun's corona and chromosphere.
2023	Chandrayaan-3	The third lunar-exploration missions in the Chandrayaan programme
2024	XPoSat	X-Ray Polarimeter Satellite(XPoSat) Launched to investigate the polarization of cosmic X-rays and examine the 50 brightest known celestial objects in the universe.

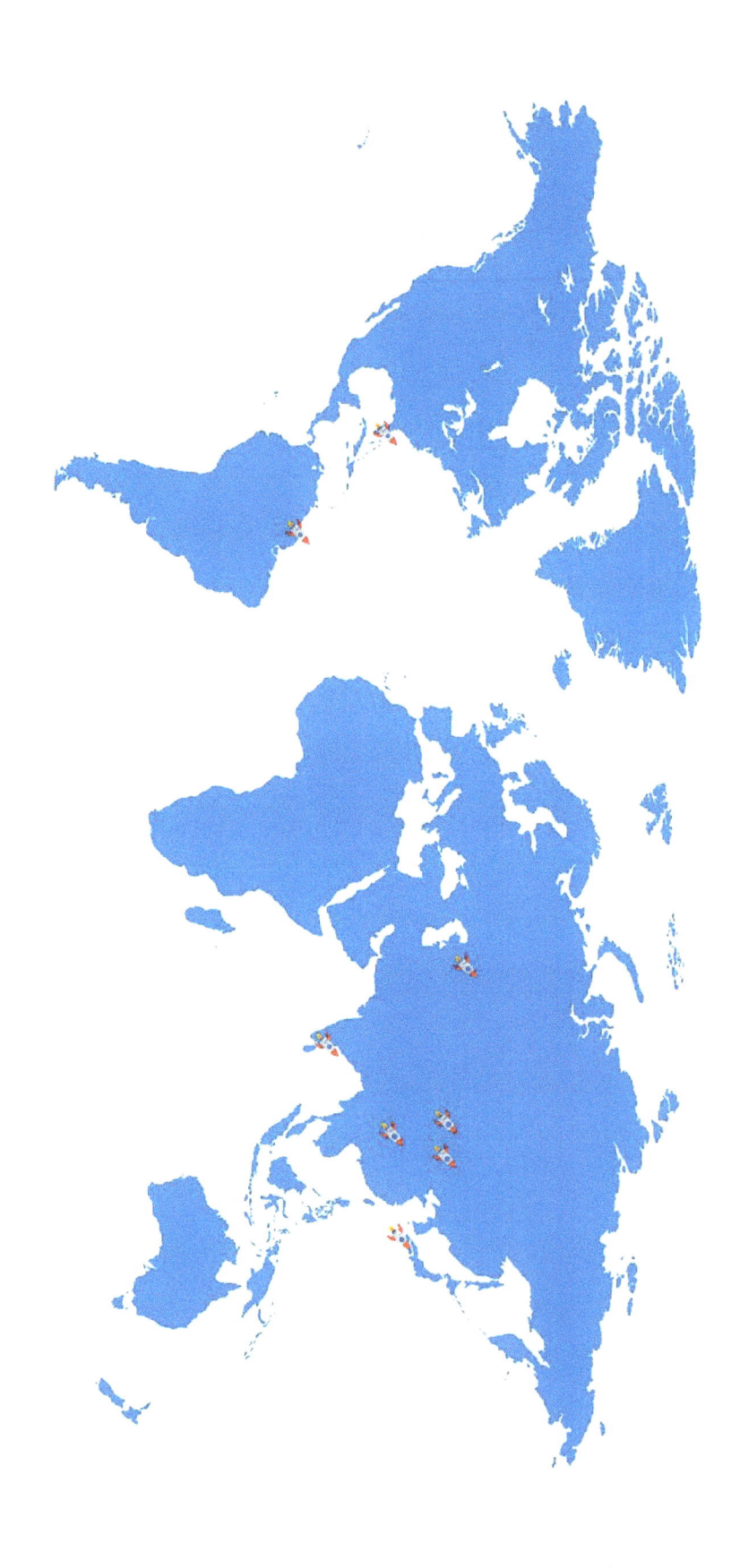

Figure 8.1: World map showing space stations. Attribution: Images by rawpixel.com on Freepik.

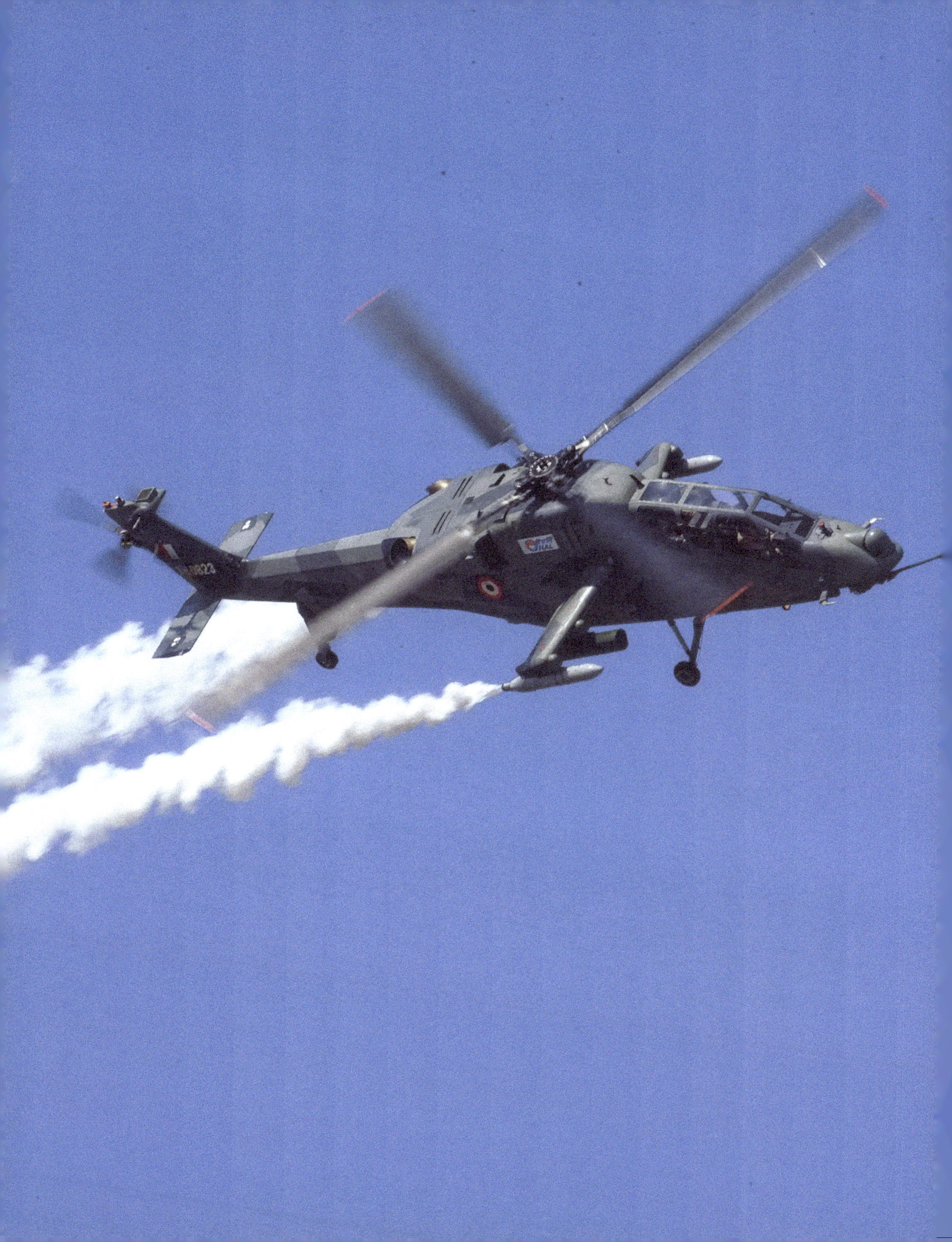

Chapter 9

Helicopters

Contents

9.1	**Introduction**	**391**
9.2	**Working Principle**	**392**
	9.2.1 Components of Helicopter	392
	9.2.2 Maneuvering	394
9.3	**Helicopter Rotor Configurations**	**398**
	9.3.1 Single Main Rotor with Tail Rotor	399
	9.3.2 Tandem Rotors	400
	9.3.3 Coaxial Rotors	400
	9.3.4 Intermeshing Rotors	400
	9.3.5 NOTAR (No Tail Rotor)	401
	9.3.6 Tilt Rotor	401
	9.3.7 Compound Helicopters	401
9.4	**Autogyros**	**402**
9.5	**Examples of Helicopters**	**404**
	9.5.1 Older Helicopters	404
	9.5.2 Advanced Helicopters	406
	9.5.3 Future Vertical Lift (FVL) Programs	414
9.6	**Helicopters in India**	**415**

9.1 Introduction

Let's open the chapter by recalling two renowned incidents that have captured widespread attention in the news.

Incident 1: *On May 1, 2023, a Cessna 206 aircraft, under the control of two pilots, went down due to engine failure over the verdant expanse of the Amazon rainforest in south Colombia. Onboard were a mother and her four children. After a span of two weeks following the crash, precisely on May 16, a search and rescue team successfully located the wreckage of the aircraft and recovered the bodies of the two pilots and the mother. However, four children (the oldest was only 13 years and the youngest would mark his first birthday soon when plane crashed) were nowhere to be found.*

A massive search through the virgin, inhospitable forest began encompassing 150 soldiers and 200 volunteers from local Indigenous communities and a team of 10 Belgian shepherd dogs, covering an area of more than 323 sq km. Helicopters hovered over the area around the crash, broadcasting messages from the children's grandmother, telling them they hadn't been forgotten, urging them to stay in one place, and dropping packets of food that may have helped them survive. Days melded into weeks, weeks into a month, and doubts cast shadows over hopes.

A few tantalising clues, the remains of fruit with bitemarks made by small human teeth, a pair of scissors and nappies in the rainforest mud, kept the rescuers going. Then, on June 9, 2023, after 40 days of the plane crash, a breakthrough emerged as army radios crackled to life "miracle, miracle, miracle, miracle." It echoed through the air, an invocation of hope. "Miracle" was the army code for a childr found alive; repeated four times meant all four had survived, in a remarkable feat of resilience.

Wrapped in rags to shield their feet from the forest floor's muck, the children were discovered. The thick jungle posed such an obstacle that the rescue helicopter had no place to touch down, leading to the children being lifted up one by one using a rope. Colombia's president, Gustavo Petro, calling it a joy for the whole country said "They have given us an example of total survival that will go down in history".

Incident 2: *In the shadowy realm of covert operations, a daring mission unfolded that would forever alter the course of history. It was the early hours of May 2, 2011, when a team of elite Navy SEALs embarked on a mission codenamed "Operation Neptune Spear". Their target: Osama bin Laden, the elusive mastermind behind the September 11, 2001, terrorist attacks.*

Under the cover of night, two stealthy helicopters–a modified MH-60 Black Hawk coated with special radar evading paint and panels as well as noise suppression devices–swept through the dark skies over Abbottabad, Pakistan that is closely followed by two other helicopters, most likely Chinooks. Their mission was to infiltrate the compound where intelligence had indicated bin Laden might be hiding.

As the SEALs descended, the unexpected happened. A controlled crash-landing resulted in the loss of one helicopter, the MH-60 Stealth Black Hawk. Yet, undeterred, the team continued its mission. Moving swiftly, they breached the compound's walls, navigating the labyrinthine structure with precision.

In the heart of the compound, a climactic encounter unfolded. The SEALs confronted bin Laden in a tense firefight that ended with the terrorist leader's demise. Amidst the chaos, the SEALs seized vital intelligence materials, and with the information secured, they swiftly retreated.

In a final dramatic twist, as dawn's first light touched the horizon, the surviving SEALs embarked on a daring escape. Despite the loss of one helicopter, the mission's success was undeniable. Laden with crucial intelligence and carrying the legacy of bin Laden's end, they vanished into the Pakistani night.

Helicopters have proven to be indispensable assets in critical moments of history. Whether rescuing stranded children from the Amazon rainforest or executing daring missions like the elimination of *Osama bin Laden*, helicopters exemplify agility, versatility, and resilience. Their ability to navigate challenging terrains and swiftly respond to emergent situations underscores their pivotal role in shaping outcomes and delivering success in complex operations. From life-saving missions to high-stakes operations, helicopters stand as steadfast symbols of precision, determination, and achievement.

Helicopters are rotary-wing aircraft and remarkable air vehicles that have revolutionized various aspects of modern life since their inception. Helicopters do things that fixed wing aircraft cannot but it must operate according to the same aerodynamic laws. Unlike fixed-wing aircraft that rely on forward motion to generate lift, helicopters are capable of vertical takeoff and landing (VTOL) due to their rotating blades. They can move backward, forward, sideways, or even straight up. They have a unique feature of hovering in the air and can be maneuvered with exceptional agility, and access areas that are inaccessible to other types of aircraft.

This aspect makes helicopters invaluable tools in emergency situations and disaster relief efforts. They can transport personnel, supplies, and equipment directly to disaster sites, saving precious time and potentially lives. Additionally, helicopters are essential in providing medical assistance in emergencies, enabling quick access to critical care for patients in remote areas or congested urban settings. In the military realm, helicopters serve various roles, including troop transportation, reconnaissance and surveillance, and combat support.

Despite their numerous advantages, helicopters also face challenges. Helicopters with spinning rotors are mechanically more complex than airplanes with static wings, more prone to failure, need more maintenance, and are expensive to operate. The complexity of their mechanical components and the need for regular inspections and upkeep can be financially demanding. Noise pollution is another issue associated with helicopters, especially in densely populated urban areas. Efforts are being made to develop quieter rotor systems and explore alternative propulsion technologies to mitigate this problem. Unlike propeller in the fixed wing aircraft that always operates in an axially symmetric environment means each blade experiences a constant situation as it rotates, helicopter rotor operate in asymmetric environment while in motion. During forward flight, the airflow through the main rotor disc differs between the advancing and retreating sides. The advancing side encounters greater relative airflow due to the helicopter's forward speed, while the retreating side experiences reduced relative airflow. This leads to asymmetry of lift, presenting specific design challenges.

The future of helicopters is exciting, with ongoing research and development aiming to improve their efficiency, safety, and environmental impact. Advancements in materials, propulsion systems, and autonomous flight technology will likely play a significant role in shaping the next generation of helicopters. Electric and hybrid-electric propulsion systems are being explored to reduce emissions and operating costs. Additionally, advances in autonomous flight technology could lead to more efficient and safer helicopter operations, especially in challenging conditions.

9.2 Working Principle

The movement of air over the helicopter rotor blades generate lift. Unlike the airfoils that are part of the fixed wing in an airplane, the helicopters have them built into their rotor blades, which spin at high speed (typically about 400–500 RPM on a small helicopter or about 225 RPM on a huge *Chinook*. With skillful piloting, a helicopter, unlike a fixed wing airplane, can take off or land vertically, hover or spin on the spot, or move in any direction.

9.2.1 Components of Helicopter

A helicopter is a complex machine with various components that work together to achieve flight and control. The main parts of a typical helicopter are shown in the figure.

Fuselage and Cockpit

The fuselage is the main body of the helicopter, housing the crew, passengers (if applicable), and

9.2. WORKING PRINCIPLE

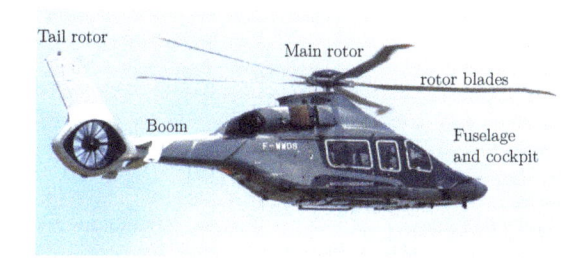

Parts of a helicopter. Picture shows the *H160M, Joint Light Helicopter*, developed by *Airbus* for the French Armed Forces.

cargo. It provides structural support for all other components. It is typically made from strong but relatively lightweight composite materials. The cockpit is the front section of the fuselage where the pilot and co-pilot (if present) sit. The cockpit is equipped with various instruments and avionics, including navigation systems, communication equipment, altimeters, airspeed indicators, and engine gauges, to provide critical information to the pilot during flight.

Big and heavy aerofoil shaped helicopter's rotor blades. As the main rotor blades rotate, they create lift through the principles of aerodynamics. The shape of the blades and their angle of attack create areas of high and low air pressure, lifting the helicopter off the ground.

Main Rotor

The main rotor is the most critical component of a helicopter. It consists of two or more large blades mounted on the top of the helicopter. These blades are connected to the main rotor hub, which is driven by the engine through the transmission system.

Each blade can swivel about a feathering hinge as it spins. When the engine powers the main rotor, it starts to spin rapidly. As the main rotor spins, it creates lift through the process of aerodynamic forces. This aerodynamic lift enables vertical takeoff and landing. The shape of the rotor blades is designed

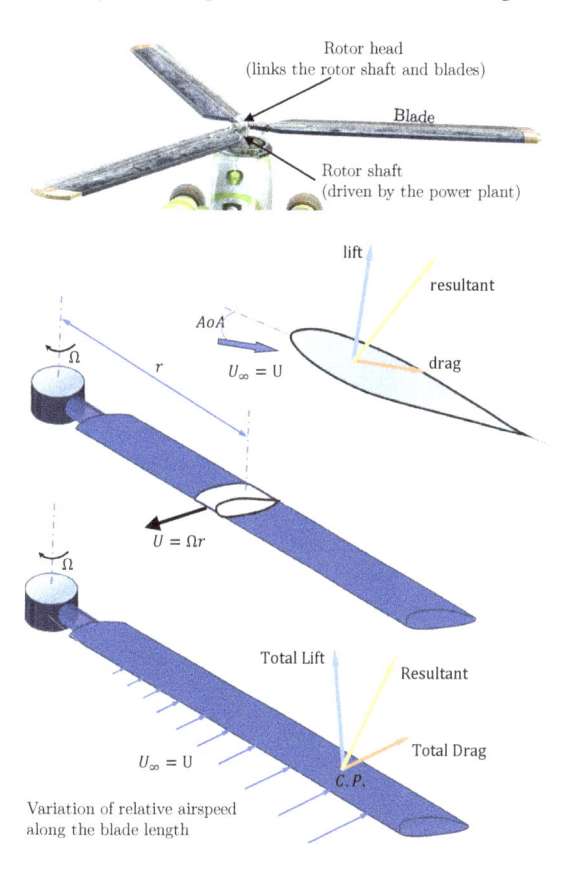

Total lift and drag on the rotor blade and variation of the relative air speed along the blade length

in such a way that air flows faster over the curved upper surface of the blades than the flat lower surface. This creates lower pressure on the upper surface and higher pressure on the lower surface, causing the helicopter to rise into the air.

Helicopter blade is twisted to counteract lift disparity resulting from varying rotational wind speeds across the blade. The twist design mitigates internal blade stress and ensures an even distribution of lift along the blade. It involves higher pitch angles at the root, where velocity is lower, and lower pitch angles near the tip, where velocity is higher. This mechanism boosts induced air velocity and enhances blade loading closer to the inboard section of the blade. A pitch link that is used to at-

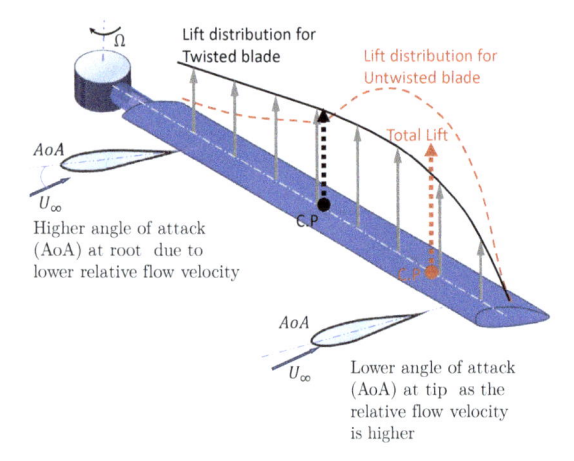

Higher angle of attack (AoA) at root due to lower relative flow velocity

Lift distribution for Twisted blade

Lift distribution for Untwisted blade

Total Lift

Lower angle of attack (AoA) at tip as the relative flow velocity is higher

Rotor blade twist – the change in blade incidence from the root (attachment to the hub) to the tip (outer blade). The twist in the blade helps in achieving better load distribution by operating at a higher angle of attach (AoA) near the rotor head (lower relative flow velocity) and at a lower AoA towards the blade tip where relative velocity is higher. It also shifts center of pressure (C.P.)– point at which aerodynamic forces act– towards the root, hence reduces the moment at the blade root. Although, sometime, the even lift distribution along the blade length is achieved by the trapezoidal shape blade

tach each blade to the rotor hub can tilt the blade to a steeper or shallower angle. That's how a helicopter hovers and steers. The two swash plates can be moved up and down or tilted to the side by the pilot's cyclic and collective cockpit controls. The rotor is powered by a driveshaft connected through a transmission and gearbox from the engine.

Tail Rotor (or Fenestron)

The tail rotor is a small vertical rotor mounted at the tail end of the helicopter. Its main purpose is to counteract the torque produced by the main rotor's rotation. Without the tail rotor, the helicopter's body would rotate in the opposite direction of the main rotor. The pilot uses the anti-torque pedals to control the tail rotor's thrust, allowing the helicopter to rotate or yaw. For instance, *Airbus Eurocopter Tiger* shown in the figure has a small, sideways-pointing tail rotor– powered by a drive-

shaft from the engine that runs through the tail end of the craft– to counteract the torque. There are other configuration where helicopter have two counter rotating main rotors. For examples, *Boeing CH-47 Chinook* military helicopter has one rotor at the front and one at the back and they spin in opposite directions to cancel one another's torque.

Engine

The engine provides the power necessary to turn the main rotor and, in some cases, the tail rotor. Helicopter engines are usually gas turbine engines that run on aviation fuel. Some helicopters (usually smaller) have a single engine mounted horizontally, underneath and just behind the rotor. Others (bigger military helicopter such as *Seahawk* and *Apache*) have one engine mounted either side of the rotor mast. Most modern helicopters have turboshaft engines, which have similarities to normal jet engines on airplanes. However, instead of having the hot jet of exhaust gas that thrusts them forward, they use the energy from the burning gas to spin a central turbine and driveshaft that powers the transmission. The throttle controls the engine's power output, which, in turn, determines the speed of the main rotor. By adjusting the throttle, the pilot can control the altitude and overall performance of the helicopter.

The *transmission system* transfers power from the engine to the main rotor and tail rotor. It adjusts the speed of the rotors and allows the pilot to control the helicopter's flight characteristics. The *landing gear* supports the helicopter during takeoff, landing, and when it is on the ground. The landing gear can vary in design, from skids for some models to wheels or floats for others.

9.2.2 Maneuvering

As mentioned, the rotor blades of a helicopter are similar to the wings of an airplane but unlike plane, the air flowing over spinning rotor blades produces lift. The direction and magnitude of the lift can be

9.2. WORKING PRINCIPLE

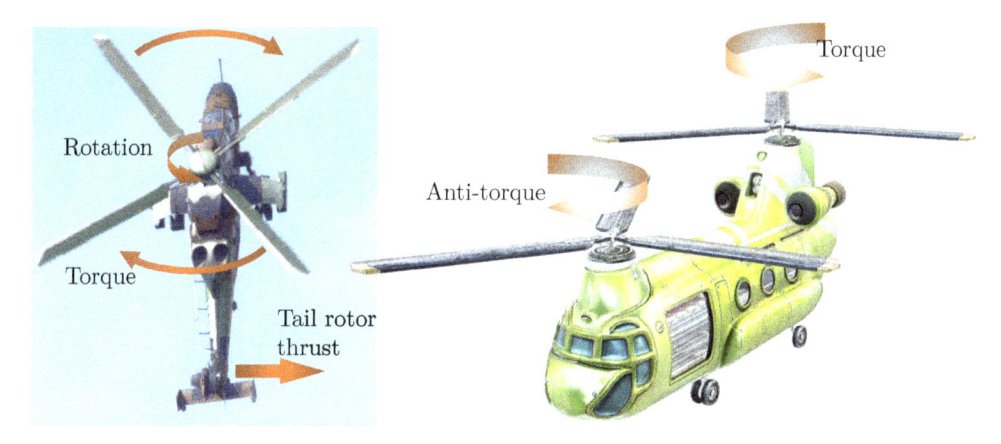

Counteracting the torque of the main rotor either using tail rotor (*Airbus Eurocopter Tiger* is shown) or two counter-rotating main rotors (Ex. *Boeing CH-47 Chinook*).

altered in order to control the helicopter.

For a helicopter to generate lift, the rotor blades need to be turning. The rotation of the rotor system propels the blades into the air, creating a relative wind without necessitating movement of the airframe through the atmosphere, unlike airplanes or gliders. The relative wind's direction varies due to factors such as blade motion and helicopter airframe movement. The centrifugal force due to rotation motion pulls the blades outward from the main rotor hub. The strength and rigidity of the blades stem from this force, enabling them to support the helicopter's weight. The maximum centrifugal force depends on the maximum operating rotation speed in revolutions per minute (RPM).

During takeoff, the lift is generated. The centrifugal force and lift both act simultaneously. This leads the blades to follow a conical path rather than staying in a plane perpendicular to the mast. This conical path becomes evident during takeoff when the rotor disk shifts from a flat orientation to a slight cone shape. If the rotor RPM drops, centrifugal force decreases while the coning angle increases significantly. Consequently, if RPM is too low, the rotor blades can fold up without recovery.

As the helicopter moves forward through the air, the airflow relative to the main rotor disk varies between its advancing and retreating sides. The advancing blade encounters heightened relative wind

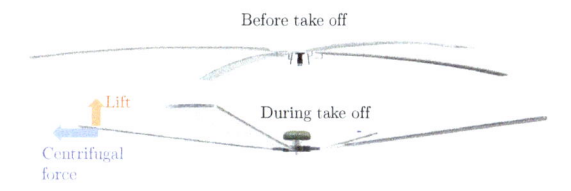

During takeoff, the rotor disk inclines upward due to the combined influence of centrifugal force and lift.

due to the helicopter's forward motion, whereas the retreating blade experiences decreased relative wind due to the aircraft's forward airspeed. Consequently, this difference in relative wind results in the advancing blade side of the rotor disk generating more lift compared to the retreating blade side. This lift discrepancy would render the helicopter uncontrollable in scenarios other than calm wind hovering. To achieve a balanced lift, a mechanism is required to compensate for, correct, or eliminate this unequal lift effect.

The main rotor is connected to the hub at the top of the mast, a central axle connected to the engine by the transmission, by a feathering hinge that allows each blade to swivel as it spins. This enables each blade to make a steeper or shallower angle to the oncoming air. There are two flat disc like structures called swash plates: upper one is rotating on ball bearings around lower non-rotating plate. The swashplate is a critical mechanical component found in helicopter rotor configuration which plays

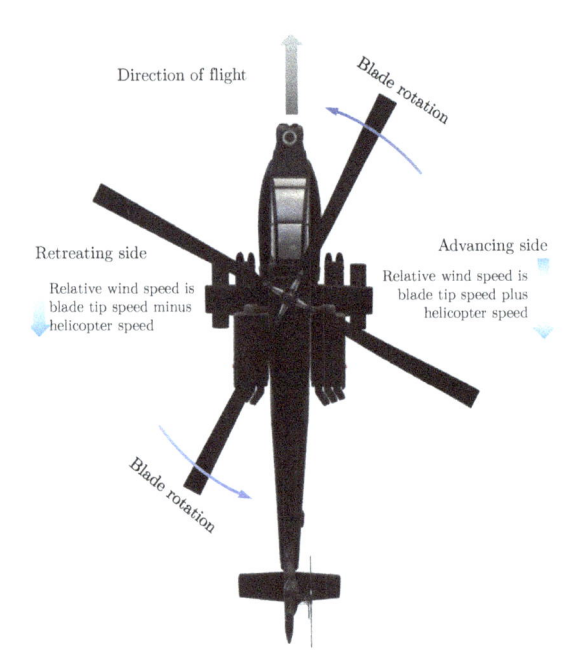

Asymmetry of lift between the advancing and retreating halves of the rotor disk, resulting from varying wind flow velocities across each section.

the critical role in converting the control inputs from the pilot into the required movements of the main rotor blades for maneuvering the helicopter. Pitch links connect the upper swash plate to the rotor blades. These vertical pitch links push the blades up and down, making them swivel as they rotate according to the angle of the swash plates, and thus able to change the direction of the force produced by the rotors. For instance, when swashplate is tilted left, the rotor blades are forced to pitch up when they're on the right and down when they're on the left through the pitch links. This make each rotor blade tilt to a steeper angle (more lift) when it's on the right and a shallower angle (less lift) when it's on the left which creates a net force for steering the chopper to the left.

Movement and Steering Controls

The pilot has the following basic movement and steering controls in a helicopter.

Collective pitch control hand lever – Usually located on the left side of the cockpit, collective pitch can increase or decrease the pitch angle that all the blades

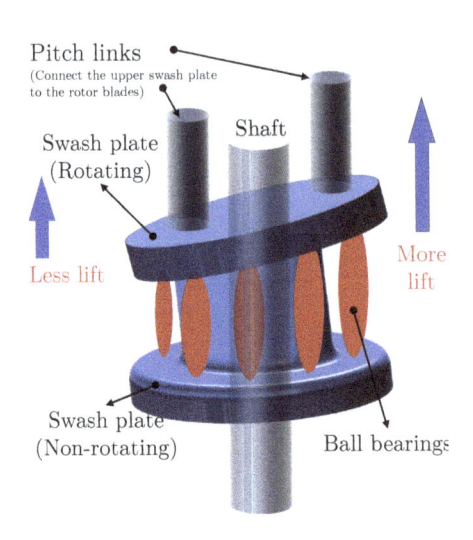

A simplified view of the swash plate mechanism that steers a helicopter.

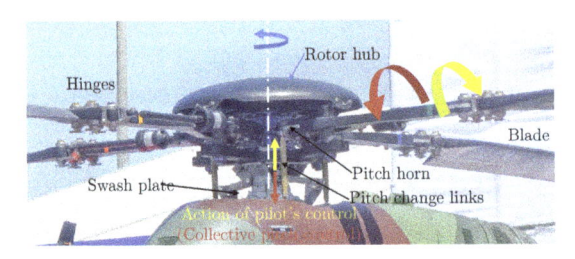

Swash plate and main rotor of the *HAL's Light Combat Helicopter (LCH)*. The thrust from the main rotor in a helicopter is controlled by adjusting the pitch angle of all the rotor blades at the same time.

make to the oncoming air as they spin around. The collective control changes the pitch angle of each of the rotor blades by the same amount at the same time. If the blades make a steeper pitch angle, the lift generated by the blades increases.

Cyclic pitch control hand lever – Usually a stick or handle in the cockpit, allows the pilot to tilt the main rotor disc (swash plate) in different directions. By tilting the rotor disc, the pilot can move the helicopter forward, backward, left, and right while hovering. This means the cyclic pitch control changes the angle of selective rotor blades as they spin.

Throttle control The pedal connected by a cable to the engine is like the accelerator used for increasing or decreasing the engine speed so the rotor makes more or less lift.

Two anti-torque pedals The tail rotor at the rear end

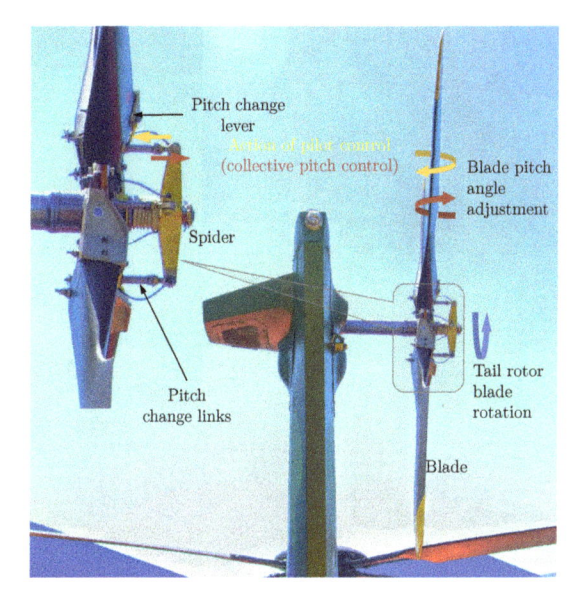

The tail rotor thrust —which counterbalances the main rotor's reaction torque — is controlled by the collective variation of blade pitch angle. The pitch change is provided by a spider which is connected to the blade pitch change levers and to the pilot control. The blade pitch change lever is connected to the blade via links. The pilot control is operated by the control rod movements from the pilot station (rudder bar). When the pilot moves a specific control in the cockpit, it moves the control rod, which then adjusts pitch angles of the rotor blades through the spider mechanism.

counteract the torque produced by the main rotor's rotation. The pilot uses the anti-torque pedals, located on the helicopter's floor, to control the tail rotor's thrust and to change the pitch of the tail rotor blades so they make more or less sideways thrust than in normal straight flight. By adjusting the tail rotor's thrust, the pilot can prevent the helicopter from rotating uncontrollably and maintain stable hovering. Pressing on one of the anti-torque pedals increases the tail rotor's thrust on that side, causing the helicopter to yaw or rotate in the opposite direction. By adjusting the anti-torque pedal positions, the pilot can control the helicopter's heading or direction. Some configurations have two rotors that spin in opposite directions, as in the *Chinook*, and more recently, designers have experimented with blowing air out of ducts onto the tail boom.

Helicopters can have control surfaces that allow the pilot to maneuver the aircraft in different directions. The cyclic control allows the helicopter to move forward, backward, left, and right. The collective control changes the pitch of all the main rotor blades simultaneously, controlling the overall lift of the helicopter. The anti-torque pedals control the tail rotor to counteract the torque generated by the main rotor's rotation, preventing the helicopter from spinning uncontrollably. Most maneuvers of a helicopter that a pilot executes involve an interplay between these different controls, which is why flying a helicopter requires skill and concentration. By skillfully manipulating the cyclic, collective, and anti-torque controls, the pilot can control the helicopter's movement in all directions and perform various maneuvers, including vertical take-offs, landings, and hovering. The intricate coordination of these controls allows the helicopter to navigate through the air with impressive agility and precision. The working principle of a helicopter is a fascinating combination of aerodynamics, engineering, and pilot skill, enabling this remarkable flying machine to perform a wide range of tasks and missions.

Hovering, Steering and Controlling in Flight

A helicopter can hover and steer due to its unique design and control systems. Hovering is one of the most remarkable abilities of a helicopter. To hover, the main rotor blades spin at a constant speed, generating lift that exactly balances the helicopter's weight.

For *hovering*, the pilot operates throttle control to generate lift that exactly balances the helicopter's weight, collective control correct any asymmetry so that all the blade make same pitch angle and use anti-torque pedals to control the yaw motion of the helicopter. However, collective control increases the pitch angle and hence lift, which means that there is more drag. The pilot needs to open the throttle further to produce more engine power in order to prevent the rotor blades from slowing down. In most modern helicopters there is an electronic governor that monitors the rotor RPM and adjusts the engine power as required. As main rotor speed modified, the anti-torque pedals has to be corrected to avoid the yaw motion.

For *steering*, the cyclic control and the anti-torque pedals play critical roles. Tilting the cyclic control forward causes the helicopter to move forward, tilting it backward moves the helicopter backward, tilting it left or right moves the helicopter in those respective directions. By adjusting the anti-torque pedal positions, the pilot can control the helicopter's heading or direction.

Controlling the helicopter in flight is done by altering the direction of the aerodynamics force which is achieved by altering the tilt of rotor system through the cyclic control. When the pilot moves the cyclic control forward, the rotor disc tilts in the same direction. The lift, instead of being vertical, now has a horizontal component, so the helicopter moves forward. This increases the drag, if nothing is done, the helicopter will start to descend. Therefore, in order to stay level, the pilot needs to raise the collective and also adequately apply anti-torque pedal to prevent the helicopter from yawing. Thus all the controls affect each other and their use needs to be coordinated.

Rearward/Sideward and Turning Flight

During forward flight, the rotor disk tilts forward, causing the entire lift-thrust force of the rotor disk to also tilt forward. During backward flight, the tip-path plane is inclined rearward, consequently tilting the lift-thrust vector in the same direction. In this scenario, drag acts forward, while the lift component remains vertically upward and weight vertically downward.

When a helicopter is banked, the rotor disk tilts sideways, resulting in the division of lift into two components. The vertical component counteracts weight, while the horizontal component counters centrifugal force. With an increase in angle of bank, the total lift force tilts more horizontally, leading to a increased rate of turn. This shift reduces the ver-

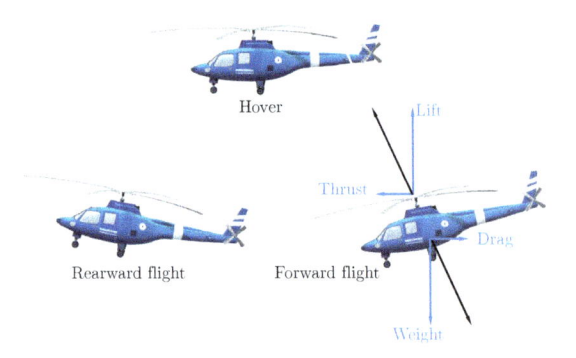

Figure 9.1: Forces on the helicopter.

tical component of the lift. To counteract reduced vertical lift, the rotor blade's angle of attack (AOA) must increase to maintain the altitude. Greater bank angles necessitate a more pronounced AOA of rotor blades to maintain height. Consequently, increasing bank and AOA boosts the resultant lifting force, elevating the rate of turn. In simple terms, raising collective pitch is crucial for maintaining altitude and airspeed during turns. During sideward flight, the tip-path plane[1] is inclined towards the desired flight direction. This adjustment causes the total lift-thrust vector to shift sideways. While the vertical lift component remains directed upwards and counteracts weight downwards, the horizontal thrust component now operates sideways, while drag opposes it on the opposite side.

9.3 Helicopter Rotor Configurations

Helicopters can have various rotor configurations, each designed to suit specific purposes and operating conditions. The main rotor configuration primarily determines the helicopter's flight characteristics, performance, and capabilities.

---

[1] The tip-path plane is the imaginary plane traced out by the tips of the helicopter's rotor blades as they rotate through their circular path during flight. The tip-path plane is perpendicular to the axis of rotation of the helicopter's rotor. This means that it is oriented parallel to the Earth's surface when the helicopter is in level flight.

9.3. HELICOPTER ROTOR CONFIGURATIONS

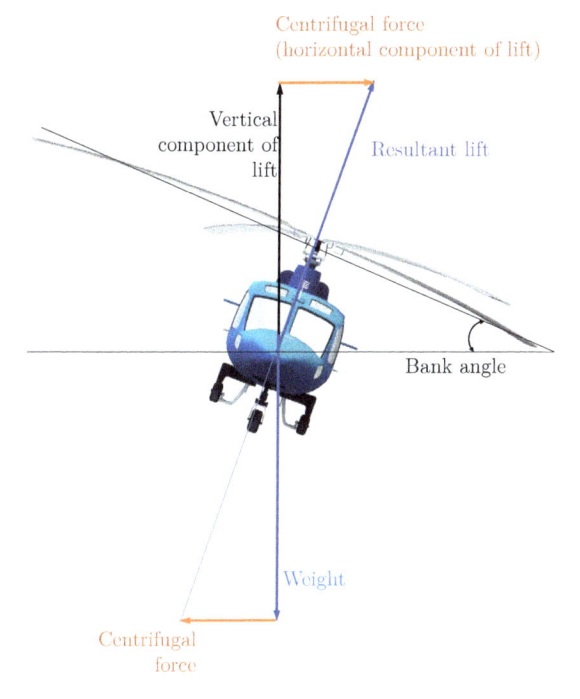

Forces acting on the helicopter during turning flight.

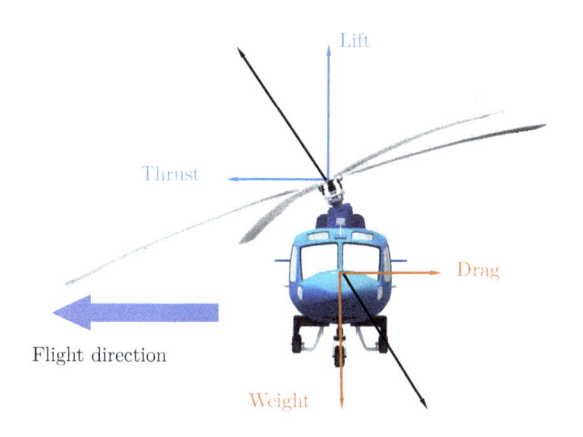

Forces acting on the helicopter during sideward flight.

9.3.1 Single Main Rotor with Tail Rotor

This is the most common rotor configuration used in helicopters. It consists of a large main rotor mounted on top of the helicopter's fuselage and a smaller tail rotor at the rear. The main rotor provides lift and thrust for vertical takeoff and landing, as well as forward flight. The tail rotor counteracts the torque produced by the main rotor's rotation, allowing the helicopter to maintain its heading during flight. A few examples of these configurations are *Bell UH-1 Iroquois* (commonly known as the Huey), *Eurocopter* (now *Airbus Helicopters*) *AS350 Ecureuil (AStar)* or *Sikorsky SH-60 Seahawk*.

Bell's UH-1 Iroquois, utility helicopter and the first turbine-powered helicopter.

Airbus Helicopters H125, light utility helicopter.

Sikorsky SH-60B Seahawk, a twin turboshaft engine, multi-mission helicopter.

Single main rotor with tail rotor (conventional) configuration helicopter.

9.3.2 Tandem Rotors

In tandem rotor helicopters, there are two main rotors, one mounted in front of the other on the same axis and no tail rotor. Usually the rear rotor is mounted at a higher position than the front rotor, and the two are designed to avoid the blades colliding. The rotor discs are slightly tilted toward each other to provide control along the vertical axis during the hover. Both rotors provide lift and thrust, and they rotate in opposite directions to counteract each other's torque. Tandem rotor helicopters have increased lifting capacity with shorter blades and stability, making them suitable for heavy-lift operations and transporting bulky cargo. The design of the drive and control system are more complicated than the ones of a single main rotor helicopter. Typical examples of tandem rotor helicopters are *Boeing CH-47 Chinook*, *Piasecki H-21 Shawnee* and *Boeing Vertol CH-46 Sea Knight*.

9.3.3 Coaxial Rotors

Coaxial rotor helicopters have two main rotors mounted on the same axis, one on top of the other. The rotors rotate in opposite directions, canceling out each other's torque. Coaxial rotor configurations eliminate the need for a tail rotor, making the helicopter more compact and reducing the risk of tail rotor-related accidents. They also offer improved lifting capacity and increased maneuverability. The control along the vertical axis is produced as a result of dissimilar lifts, thus differential torque, of the two rotor discs. The helicopter will yaw to the left if the clockwise rotor produces more lift, and vice versa. The drag due to the rotors is relatively larger because of the interference of airflows by the two rotor discs. So, these helicopters do not normally have a high cruising speed. Example: The *Kamov Ka-32* is a utility helicopter that features a coaxial rotor configuration. It is known for its maneuverability and lifting capabilities, making it suitable for firefighting, construction work, and offshore operations.

Boeing CH-47 Chinook, the heavy lifting helicopter with tandem rotor configuration.

Piasecki H-21 Shawnee (the flying banana), an American multi-mission helicopter.

Boeing-Vertol CH-46 Sea Knight, an American medium-lift transport helicopter.

Tandem-rotor helicopters.

9.3.4 Intermeshing Rotors

Intermeshing rotor helicopters have two main rotors mounted on separate masts slightly inclined towards each other, and the rotor blades intermesh with each other (generally called as synchropter). The two rotors mesh with one another, like a gear-

The *Kamov Ka-226* – a small, twin-engine Russian utility helicopter.

The *Kamov Ka-50, Black Shark* is a Soviet/Russian single-seat attack helicopter.

The *Kamov Ka-27* is a military helicopter developed for the Soviet Navy.

Coaxial rotor helicopters.

wheel. As there are two rotors moving in opposite directions, this configuration does not require a tail rotor. They have high stability and powerful lifting capabilities. Intermeshing rotor configurations offer enhanced lifting capacity, agility, and control, making them suitable for various military and civilian applications. Example: The *Kaman K-MAX* is a unique intermeshing rotor helicopter designed for external load operations. It is often used in firefighting, logging, and other missions that require precise and heavy lift capabilities.

9.3.5 NOTAR (No Tail Rotor)

The NOTAR system is a unique rotor configuration that eliminates the need for a tail rotor. Instead, it uses a system of high-pressure air directed through slots in the tail boom to counteract torque and provide anti-torque control. NOTAR-equipped helicopters offer reduced noise levels and increased safety due to the absence of a tail rotor. Example: *MD Helicopters MD 520N* is a lightweight, single-engine helicopter that utilizes the NOTAR system instead of a tail rotor. The NOTAR system reduces noise and vibration, making it suitable for law enforcement and civilian applications.

9.3.6 Tilt Rotor

In this configuration, the rotors are mounted at the edge of the wings, on nacelles that can rotate in order to transition the rotors from the vertical position to the horizontal position. A tiltrotor helicopter is a unique type of aircraft that combines the vertical takeoff and landing (VTOL) capabilities of a helicopter with the high-speed and range capabilities of a fixed-wing aircraft. Both lift and propulsion are generated by the rotors, which act as helicopter main rotors in the vertical position, and like airplane propellers when in the horizontal position. Lift is generated similar to that like fixed wing airplanes.

9.3.7 Compound Helicopters

Compound helicopters effectively combine the best features of both helicopters and fixed-wing aircraft. Like a traditional helicopter, they have a main rotor for lift and a tail rotor for stability and utilize

Kaman HH-43 Huskie.

MD Helicopters MD 500

Kellett XR-10 was a military transport helicopter developed in the United States.

MD Helicopters MD 600N, a light utility civilian helicopter designed in the United States in 1994

NOTAR (No Tail Rotor) helicopter

Kaman K-MAX, produced by the American manufacturer Kaman Corporation

Intermeshing Rotors.

an additional propulsion method–typically either a tail-mounted propeller or a jet engine–to achieve increased air speeds. In addition, many models are equipped with a small set of wings to help sustain forward flight. They are designed to overcome some of the limitations of conventional helicopters and offer improved performance in terms of speed, range, and efficiency. Compound helicopters achieve this by incorporating additional propulsion methods or aerodynamic enhancements to complement their primary rotor-based lift system.

Each rotor configuration has its advantages and disadvantages, and they are chosen based on the specific requirements of the helicopter's intended use. The selection of the rotor configuration can significantly impact the helicopter's performance, stability, and overall capabilities.

9.4 Autogyros

Autogyros, also known as gyroplanes or gyrocopters –the precursors to helicopters– are unique

9.4. AUTOGYROS

Bell Boeing V-22 Osprey

AgustaWestland AW609

Examples of tiltrotor helicopters.

Lockheed AH-56 Cheyenne, a compound helicopter.

rotary-wing aircraft that operate on the principle of autorotation to generate lift. They combine elements of fixed-wing airplanes and helicopters, offering a safe and stable flying experience with relatively simple mechanical systems. Autogyros were first developed in the early 20th century and have since evolved into popular recreational and utility aircraft.

Autogyros and helicopters are both types of rotary-wing aircraft, but they have significant differences in their design, flight characteristics, and capabilities. Autogyros generate lift by an unpowered, freely rotating main rotor, also known as an autorotation. The rotor spins due to the forward motion of the aircraft, creating lift without the need for engine-driven power to turn the rotor. However, they rely on a separate engine-driven propeller to provide forward thrust during level flight and climb. The engine also powers other essential systems, such as the controls and instrumentation. They require a short takeoff roll to build up forward airspeed before they can achieve lift for takeoff. They can land vertically, but they generally require more space for landing compared to helicopters due to their slower descent rate. Autogyros have a limited maximum forward speed, typically lower compared to helicopters, due to the autorotation of their main rotor.

Autogyros can maintain a controlled descent rate in a state of autorotation, but they are not capable of hovering in a stationary position like helicopters. They are generally more maneuverable and agile than fixed-wing aircraft. Autogyros are mechanically simpler than helicopters, as they lack complex pitch controls for the main rotor blades.

Autogyros find applications in various fields. Autogyros are popular for recreational flying and aviation enthusiasts due to their unique flight characteristics and open-cockpit designs. Their slow flight capabilities and stability make autogyros suitable for aerial photography, environmental surveying, and mapping tasks. Also, they can be used for crop dusting and agricultural applications, where slow, low-level flight is required. In some regions, they are utilized for search and rescue missions, especially in remote or inaccessible areas. While not as common as helicopters, autogyros have been explored for surveillance and reconnaissance purposes in the military.

Some examples of autogyros are *AutoGyro Calidus, Magni M24 Orion, Carpenter P-1 Gyroplane, ELA Aviacion ELA-07S, Sportcopter II, Rotorvox C2, ArrowCopter AC10 and Gyros-2 Smartflie*.

The *Magni M16* autogyro is potentially a tandem-seat model designed for flight training purposes by *Magni Gyro*, a reputable Italian autogyro manufacturer. It emphasizes safety, comfortable tandem seating, stable flight characteristics, and modern avionics to facilitate effective flight training experiences for both instructors and students.

Dimension	4.70 m (Length)
Weight	261 kg (Empty), 450 kg (MTOW)
Capacity	72 L (Fuel)
Power plant	1 x Rotax 914UL turbocharged flat four providing 85 kW
Propellers	3-bladed Arplast, ground adjustable pitch, 1.70 m diameter
Maximum speed	185 km/h
Cruise speed	145 km/h
Range	480 km
Endurance	3 hr
Service ceiling	3,500 m
Rate of climb	5.0 m/s

In summary, autogyros and helicopters have different lift-generation mechanisms, vertical takeoff and landing capabilities, forward flight speeds, and hovering capabilities. Autogyros are mechanically simpler and typically have a lower forward speed but can glide safely to the ground in case of engine failure. Helicopters, on the other hand, offer true VTOL capabilities, hovering, and higher forward speeds, making them suitable for a wider range of missions, especially those that require precise and flexible operations.

9.5 Examples of Helicopters

Within the realm of helicopters, there exist several widely recognized and celebrated models. These helicopters have been meticulously designed to cater to diverse applications, each possessing unique configurations tailored to their intended purposes. Through a combination of engineering prowess and innovation, the following helicopters have become emblematic examples in their respective roles.

9.5.1 Older Helicopters

Focke-Wulf FW-61 was the world's first fully operational helicopter. It was designed and produced in Germany by *Focke-Wulf Flugzeugbau AG*. The development began in the early 1930s, and the first flight took place in 1936. The *Focke-Wulf FW-61* was a groundbreaking achievement in aviation history, being the first helicopter to demonstrate controlled vertical flight and successful forward flight. The helicopter was used primarily for flight testing and development, contributing valuable insights into the challenges and possibilities of helicopter flight. The successful flight of the *FW-61* had a significant influence on subsequent helicopter designs, inspiring further research and development in the field of rotorcraft.

The *FW-61* featured a unique dual-rotor configuration with the rotors rotating in opposite directions. This arrangement provided stability and eliminated the need for a tail rotor. The helicopter had a lightweight airframe made of tubular steel and covered with fabric. Its compact design allowed for ease of handling and transportation. The *FW-61* had an open cockpit, and the pilot and instructor sat side by side during flight. This design allowed for direct visibility and communication during flight tests.

Bell 47 (*Bell H-13 Sioux* is its military variant) was developed in the United States by *Bell Helicopter*.

9.5. EXAMPLES OF HELICOPTERS

The *Focke-Wulf Fw 61* – Considered the first practical, functional helicopter, first flown in 1936.

The development of the *Bell 47* began in the late 1940s, and it entered production in 1946. The *Bell 47* was the first helicopter ever certified for use by civilians in March 1946. The *Bell 47* was a highly successful helicopter, and it became one of the most-produced helicopters in history. Over 5,600 *Bell 47*s were manufactured during its production run, making it a widely used and iconic rotorcraft.

It had a classic helicopter design, featuring a single main rotor and a tail rotor for anti-torque control. Its familiar bubble-shaped cockpit became one of its most recognizable features. The *Bell 47* was a light and versatile helicopter, suitable for various applications such as civilian use, military roles, agriculture, and aerial work. The helicopter was equipped with a two-blade main rotor that provided lift and stability during flight.

The *Bell 47* was widely used in civilian applications, including as a general-purpose utility helicopter, for aerial surveying, crop dusting, and as a platform for various commercial tasks. The military variant, the *Bell H-13 Sioux*, served as a reconnaissance and light observation helicopter during the Korean War. It played a vital role in medical evacuation missions, earning the nickname *Angel of Mercy* for its lifesaving missions on the battlefield. It was also used as a training aircraft for pilots learning helicopter flight, due to its relatively straightforward controls and stable flight characteristics. The *Bell 47* gained fame in popular culture and media, thanks to its appearances in movies, television shows and other forms of entertainment. Its distinctive design and role in aviation history contributed to its lasting legacy.

The *Bell 47* was used in various countries around the world due to its versatility and widespread production. It served in both civilian and military capacities, not only in the United States but also in many other nations. The helicopter's long and successful production run ensured that it remained in service for several decades, making it an essential part of helicopter history and a beloved aircraft among aviators and enthusiasts alike.

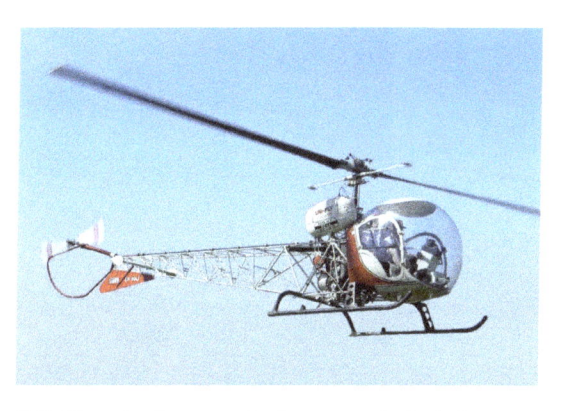

Bell 47, a single-rotor single-engine light helicopter and the first one certified for civilian use.

Aerospatiale SA-313 Alouette II is a light utility helicopter that was developed in France. It was designed by *Sud Aviation*, which later became part of *Aerospatiale*. The *Alouette II* first flew in 1955, and production continued from the late 1950s until the 1970s. It was the first production helicopter powered by a gas turbine engine instead of the heavier conventional piston powerplant. The *Aerospatiale SA-313 Alouette II* was a highly successful helicopter, and it became one of the most-produced helicopters in its class. Over 1,300 *Alouette II* helicopters were manufactured during its production run.

The *Alouette II* had a compact design, making it well-suited for various applications, including civilian, military, and aerial work. Its small size and agility made it ideal for operations in tight or confined spaces. The helicopter was powered by a single turbine engine, providing it with adequate power for its size and capabilities. It was equipped with a three-blade main rotor, which provided lift and stability during flight.

The *Alouette II* was widely used in civilian roles, including as a general-purpose utility helicopter, for aerial surveying, crop spraying, and search and rescue missions. It served in various military roles, such as reconnaissance, liaison, and light transport. It was also utilized for training purposes in some armed forces. The helicopter was used for various aerial work tasks, including external load lifting, pipeline and power line inspections, and geological surveying. Its simple and robust design made it relatively easy to maintain, contributing to its popularity among operators.

The *Aerospatiale SA-313 Alouette II* was used in many countries around the world, making it a widely deployed and successful helicopter. It was operated by both civilian and military operators in various regions, including Europe, Asia, Africa, and South America. Its reliability, versatility, and relatively low operating costs contributed to its popularity and enduring service life. The *Alouette II* played a crucial role in civilian and military operations, and its legacy continues in some regions, where it remains in service even to this day.

Aerospatiale SA-313, the first jet-powered helicopter.

9.5.2 Advanced Helicopters

Bell AH-1 Cobra is an attack helicopter developed in the United States. It was designed by *Bell Helicopter* and first entered production in 1967. The *AH-1 Cobra* has seen various production variants over the years, and the total number produced across all versions is over 1,100 helicopters. A member of the prolific *Huey* family, the *AH-1* is also referred to as the *HueyCobra* or *Snake*. Other members of the larger *Huey* family includes *AH-1 SuperCobra*, the *AH-1J SeaCobra*, the *AH-1T Improved SeaCobra*, and the *AH-1W SuperCobra*.

The *Bell AH-1 Cobra* first flew in 1965, and it played a pioneering role in the development of attack helicopters, setting the standard for many subsequent designs. As the world's first dedicated attack helicopter, the *AH-1 Cobra* introduced new concepts and capabilities that became fundamental to future generations of attack helicopters.

The *AH-1 Cobra* features a tandem seating arrangement, with the pilot in the rear seat and the co-pilot/gunner in the front seat. This design provides the front-seater with an unobstructed view for engaging targets. The *Cobra* has a relatively narrow fuselage, which reduces its frontal area and enhances its agility. The sleek design contributes to the helicopter's speed and maneuverability. The *AH-1 Cobra* is armed with various weapons systems, typically including a chin-mounted *M197* 20mm three-barrel cannon, rocket pods, and missile launchers (such as the *TOW* or H*ellfire* missiles).

The primary role of the *AH-1 Cobra* is as an attack helicopter, providing close air support to ground forces. Its powerful armament allows it to engage and destroy enemy vehicles, structures, and personnel on the battlefield. The *Cobra* is highly effective in anti-tank operations, capable of employing wire-guided anti-tank missiles, such as the *BGM-71 TOW* missile, to engage and neutralize armored threats. The *AH-1 Cobra* can serve as an armed escort for transport helicopters or other vulnerable aircraft, providing protective cover during troop transport or other critical missions. In some configurations, it is equipped with sensors and targeting systems, enabling it to conduct reconnaissance and target acquisition missions.

The *AH-1 Cobra* has been in service with several

Table 9.1: Important helicopters of all time

Helicopter	Importance
Focke-Wulf Fw-61 (Germany, 1936)	World's first helicopter
Sikorsky R-4 (USA, 1942)	First mass-produced helicopter
The Bell 47 (USA, 1945)	First helicopter certified for civilian use
Aerospatiale SA-313 (France, 1949)	First jet-powered helicopter
Bell UH-1 Iroquois (USA, 1959)	Significant use in Vietnam War
Mil Mi-8 (Soviet Union, 1961)	World's most produced helicopter
Boeing CH-47 Chinook (USA, 1962)	Tandem rotor helicopter
Bell 206 JetRanger (USA, 1966)	Popular civilian helicopter
Bell AH-1 Cobra (USA, 1967)	First dedicated attack helicopter
Westland Lynx (UK, 1971)	First fully aerobatic helicopter
Sikorsky H-60 Black Hawk (USA, 1974)	Modern Huey
Robinson R-22 (USA, 1979)	Best-selling, low-cost helicopter
Mil Mi-26 (Soviet Union, 1980)	Largest series production helicopter
Northrop-Grumman MQ-8 (USA, 2002)	First operational autonomous helicopter
Eurocopter X3 (France, 2010)	World's fastest helicopter

countries around the world. It has been operated by the USA and has been widely exported to numerous allied nations and other foreign customers. The *AH-1 Cobra* continues to be used in various roles in several armed forces. Its reputation as an effective attack helicopter and its extensive service history have solidified its place as one of the iconic and influential helicopters in military aviation.

Bell UH-1 Iroquois (Huey), commonly known as the *Huey*, is a utility military helicopter developed in the United States by *Bell Helicopter*. It first entered production in the late 1950s. The *Bell UH-1 Iroquois* has an extensive production history, with over 16,000 units manufactured in various models and configurations. It is one of the most produced helicopters in history, probably one of the world's best-known helicopters. Cementing its place in history during the Vietnam War, when people think of helicopters, the *Huey* is probably the first to spring to mind.

The *Huey* is one of the most recognizable helicopters globally, with its distinctive shape and large windows. Its design has become synonymous with military helicopter operations. The *UH-1* was one of the first military helicopters to feature a turbine engine, which provided increased power, reliability, and performance compared to piston-powered helicopters. The *Huey* has a spacious cabin capable of accommodating troops, cargo, or stretchers for medical evacuation missions. The rear ramp facilitates rapid loading and unloading of personnel and supplies.

The *UH-1* is primarily utilized for troop transport, capable of carrying infantry troops to and from the battlefield. Some variants of the *UH-1 Iroquois* have been armed with door-mounted machine guns and rockets, allowing them to provide close air support to ground forces during combat operations. Its versatility has made it adaptable to various roles, including search and rescue, reconnaissance, aerial firefighting, and disaster relief missions.

The *Huey* has been used by numerous countries worldwide, both by the United States military and through foreign military sales. It has seen exten-

(a) *Bell AH-1G Cobra*, the first dedicated attack helicopter

(b) *Bell AH-1W Super Cobra*

Members of the prolific *Huey* family.

The *Bell UH-1 Iroquois (Huey)*, a helicopter often associated with the Vietnam War.

sive service in various armed conflicts and peacekeeping operations globally. Its reliability, versatility, and ability to operate in diverse environments have made it a trusted workhorse for military forces and humanitarian organizations around the world.

The *UH-1 Iroquois* holds a prominent place in military history and remains in service with many armed forces, even as more modern helicopters have been introduced. Its enduring legacy and widespread use reflect its significant contributions to military aviation and its impact on rotorcraft design.

The Boeing CH-47 Chinook is a heavy-lift, tandem rotor helicopter developed in the United States. It was designed by the American company *Boeing Vertol* (later part of Boeing Rotorcraft Systems). The *CH-47* first entered service with the United States Army in 1962, making it one of the longest-serving and most iconic military helicopters in the world.

One of the most distinctive features of the *CH-47 Chinook* is its tandem rotor design, with two large rotors mounted on top of the aircraft in a fore-and-aft configuration. This setup provides exceptional lifting capabilities, stability, and agility, making it suitable for a wide range of missions. The *Chinook* is specially designed to be able to independently adjust each rotor to enable it to adapt to the weight of different cargos. It is renowned for its impressive cargo capacity. It can transport a variety of payloads, both internally and externally. The external cargo hook allows the helicopter to carry heavy loads, such as vehicles, artillery, or supplies, slung beneath the aircraft. The *CH-47* features a large rear-loading ramp, which enhances its versatility by allowing for quick and efficient loading and unloading of troops and cargo. This feature is particularly valuable in military operations, disaster relief efforts, and humanitarian missions.

The *CH-47 Chinook* is primarily known for its heavy-lift capabilities, enabling it to transport large and heavy loads over significant distances. This includes military equipment, troops, humanitarian supplies, and construction materials. The helicopter can carry a substantial number of fully equipped troops, making it an essential asset for rapid troop deployment and battlefield mobility. Some variants of the *CH-47* are specifically designed for special operations forces, equipped with advanced avionics, defensive systems, and terrain-following radar for low-level flight operations. With its excellent

9.5. EXAMPLES OF HELICOPTERS

lifting capability, endurance, and all-weather operation, the *Chinook* can be configured for medical evacuation missions, providing life-saving medical support and transportation for wounded soldiers or civilians.

The *CH-47 Chinook* is in service with numerous countries around the world, including but not limited to the United States, the United Kingdom, Canada, Australia, Japan, various European nations, and several Middle Eastern and Asian countries. It is widely used by both military forces and civilian operators for various missions, highlighting its global importance and versatility.

CH-47 Chinook Helicopter, one of the most iconic helicopters of all time.

Sikorsky H-60 *Black Hawk* is a versatile medium-lift utility helicopter developed in the United States. It is part of the larger *Sikorsky S-70* family of helicopters. The *Black Hawk* was designed by *Sikorsky Aircraft Corporation* (now a part of *Lockheed Martin Corporation*). The U.S. Army first introduced the *Black Hawk* into service in 1979, and it has since become one of the most widely used and recognized helicopters in the world.

The *Black Hawk* is designed as a medium-lift utility helicopter, capable of performing a wide range of missions. Its design strikes a balance between payload capacity, speed, and agility, making it suitable for various military and civilian applications. The *Black Hawk* features a four-blade main rotor that provides lift and stability. The rotor system is designed for efficient lift and maneuverability, allowing the helicopter to operate in tight spaces and challenging environments. The *Black Hawk* has a tail rotor, which is essential for counteracting the torque generated by the main rotor's rotation. The tail rotor allows the helicopter to yaw and maintain directional control during flight.

The *Black Hawk* is widely used for troop transportation, capable of carrying up to 11 fully equipped combat troops, in addition to the crew. It allows for rapid deployment and extraction of soldiers during military operations. The *Black Hawk* can be configured for medical evacuation missions, with the capacity to carry stretchers and medical personnel to provide critical care and transport injured personnel to medical facilities. Variants of the *Black Hawk* have been developed specifically for special operations forces, equipped with advanced avionics, defensive systems, and enhanced capabilities for covert operations. The *Black Hawk* is employed in search and rescue operations, often operated by both military and civilian agencies, to locate and extract individuals in distress from challenging or remote locations. The *Black Hawk* can provide combat support, such as tactical airlift, reconnaissance, and resupply missions in support of ground forces.

The *Sikorsky H-60 Black Hawk* helicopter is in service with a wide range of countries across the globe. Besides the United States, it is used by numerous nations, including various NATO allies and partner countries. Many armed forces, law enforcement agencies, and civilian operators rely on the *Black Hawk*'s versatility and reliability for their missions, making it one of the most prevalent and widely used helicopters in the world.

Mil Mi-24 *Hind*, commonly known as the *Hind*, is an iconic and heavily armed attack helicopter developed in the Soviet Union (now Russia). The development of the *Mi-24* began in the late 1960s, and it first entered service with the Soviet military in 1972 (more than 4,000 units have been produced). It was designed by the *Mil Moscow Helicopter Plant*.

The *Mil Mi-24* is a unique combination of a heavily armed attack helicopter and a troop-carrying

The *Sikorsky UH-60 Black Hawk*, a four-blade, twin-engine, medium-lift utility military helicopter manufactured by *Sikorsky Aircraft*.

transport helicopter. It is often referred to as an assault helicopter due to its primary role in supporting ground forces with its firepower and troop transport capabilities. The helicopter has a tandem cockpit configuration, with the pilot sitting in the rear seat and the co-pilot/gunner in the front seat. This design allows for better visibility and coordination during combat operations. The *Hind* features a large, extensively glazed canopy for the pilot and co-pilot, providing excellent situational awareness and visibility during flight and combat. The *Mi-24* is equipped with stub wings on each side of the fuselage, allowing it to carry a wide array of weapons, including rockets, missiles, and heavy machine guns. This extensive armament gives it substantial firepower to engage both ground and air targets.

One of the primary roles of the *Mi-24 Hind* is anti-armor warfare. It can carry a variety of guided and unguided anti-tank missiles and rockets, making it highly effective against armored vehicles. The *Hind* is capable of providing close air support to ground troops, delivering accurate and devastating firepower to suppress enemy positions and provide cover for friendly forces. The *Mi-24* can carry up to eight fully equipped troops in its troop compartment. This capability allows it to conduct transport and insertion/extraction missions for special forces or infantry units in combat zones. The helicopter can conduct reconnaissance and battlefield surveillance, gathering valuable intelligence on enemy positions and movements. Some versions of the *Mi-24* have been adapted as airborne command posts, providing communication and coordination capabilities for ground forces during operations.

Typically a *Hind* is equipped with a four-barreled 12.7mm *Yakushev-Borzov Yak-B gatling* gun improved through the installation of a 30mm *GSh-30K* twin-barrel, fixed cannon. It can also be armed with machine gun pods, anti-tank missiles, and rocket pods.

The *Mil Mi-24 Hind* has been widely exported to numerous countries and has seen service in various armed conflicts around the world. It was originally developed for the Soviet military and has been in service with the Russian Armed Forces and several other former Soviet states. Additionally, it has been acquired and operated by many countries in Africa, Asia, the Middle East, and South America.

The *Mil Mi-24 (Hind)*, a large helicopter gunship, attack helicopter.

Bell 222 is a twin-engine light helicopter developed in the United States by *Bell Helicopter*, a division of *Textron*. The development of the *Bell 222A* began in the late 1970s, and it first entered production in 1979. Around 200 units of *Bell 222* were manufactured during its production run, which lasted until the early 1990s.

The *Bell 222A* features a twin-engine configuration, with two *Allison 250-C30* turboshaft engines providing power to the main rotors. The use of two engines enhances safety and redundancy during flight. The helicopter's design includes a relatively spacious cabin capable of accommodating up to eight passengers or various mission-specific equipment. The *Bell 222A* is recognized for its sleek and streamlined exterior design, which was considered modern and stylish for its time.

The *Bell 222A* is well-suited for corporate and VIP transport due to its comfortable cabin and relatively high cruising speed. The helicopter has been utilized by law enforcement agencies for tasks such as surveillance, search and rescue, and aerial law enforcement operations. The *Bell 222A* has been used for aerial photography and cinematography due to its stable flight characteristics and spacious cabin for camera equipment.

It's worth noting that the *Bell 222A* has been succeeded by other models in *Bell Helicopter*'s product line, such as the *Bell 430* and the *Bell 429*, which feature further improvements in performance and capabilities.

The *Bell 222*, the first light commercial helicopter with twin engines constructed in the United States.

Mil V-12, also known as the *Mil Mi-12*, was a Soviet experimental heavy-lift helicopter. It was developed by the *Mil Design Bureau* in the Soviet Union during the 1960s and was one of the largest helicopters ever built. The helicopter's first flight took place in 1968, and it underwent testing and evaluation during the 1970s. The *Mil V-12* was an experimental and one-of-a-kind helicopter. Only two prototypes were built, and it did not enter serial production.

The most significant and unique feature of the *Mil V-12* was its impressive heavy-lift capability. It was designed to lift extremely heavy loads, including military equipment, large artillery pieces, and even other helicopters. It was intended for military transport and logistics purposes. It featured a twin-rotor configuration with an eight-blade rotor system on each rotor. The two rotors turned in opposite directions to counteract torque and provide stability. The cockpit had a tandem seating arrangement for the pilot and co-pilot, allowing for good visibility during flight.

During its testing phase, the *Mil V-12* set several world records, including the record for the heaviest lift achieved by a helicopter. It demonstrated its ability to carry extraordinarily heavy loads, far beyond the capacity of most other helicopters of its time.

The *Mil V-12*, the largest helicopter ever built.

Bell Boeing V-22 Osprey is a tiltrotor aircraft developed jointly by *Bell Helicopter, Textron* company, and *Boeing Defense, Space & Security* in the United States. The development of the *V-22 Osprey* began in the 1980s, and it entered production in the 1990s. Over 400 *V-22 Osprey* aircraft have been produced. It is primarily operated by the United States military and has also been exported to other countries.

The *V-22 Osprey* features a unique tiltrotor configuration, combining the capabilities of a helicopter and a fixed-wing aircraft. The tiltrotor design allows the aircraft to take off and land vertically like a helicopter and transition to horizontal flight like a turboprop airplane. It has large wings with engine nacelles mounted on the wingtips. During vertical flight, the engine nacelles are directed upward to provide lift, and during horizontal flight, they are tilted forward to provide thrust. The wings and rotor blades of the *Osprey* are foldable, enabling the aircraft to be stored more efficiently on naval vessels or in tight spaces.

The *V-22 Osprey*'s VTOL capability allows it to operate from a wide range of locations, including ships, remote areas, and unprepared landing zones. The tiltrotor design gives the *Osprey* enhanced speed and range compared to traditional helicopters, allowing it to cover longer distances more rapidly. The *Osprey* can transport troops, equipment, and cargo, making it valuable for military operations and humanitarian missions. The *Osprey* is used in special operations missions, such as inserting and extracting special forces teams. It also serves in search and rescue roles due to its versatile capabilities.

The *Osprey* is primarily operated by the United States military, including the U.S. Marine Corps and the U.S. Air Force. It has been extensively used by the U.S. Marine Corps for various missions, including troop transport, assault support, and special operations. The U.S. Air Force employs the *CV-22* variant for special operations and search and rescue operations.

Additionally, the *V-22 Osprey* has been exported to other countries, including Japan and Israel, where it serves in their respective armed forces.

Northrop Grumman MQ-8 Fire Scout is an uninhabited autonomous helicopter developed in the USA. It was developed by *Northrop Grumman*'s Aerospace Systems division. The *MQ-8 Fire Scout* program began in the early 2000s, and the first production models were introduced in the mid-2000s. Several dozen *MQ-8 Fire Scout* helicopters have been produced and deployed by the U.S. Navy.

The *MQ-8 Fire Scout* can operate without a pilot on board. It is remotely controlled from the ground or operated in autonomous mode using pre-programmed flight paths. The *Fire Scout* is based on the *Schweizer 330*, a commercial helicopter. It was modified and integrated with advanced avionics and control systems to transform it into a capable military uninhabited aerial vehicle (UAV). It is also equipped with a modular payload system, allowing it to carry a variety of mission-specific sensors and equipment. This flexibility enables it to perform a wide range of roles and missions.

A KC-130J Hercules refuels an Osprey.

The *MQ-8 Fire Scout* excels in providing real-time intelligence, surveillance, and reconnaissance capabilities. Its sensors and cameras can gather valuable data and intelligence from the battlefield. The helicopter is capable of acquiring and tracking targets, assisting ground forces with target designation and enemy movement analysis. The *Fire Scout* can act as a communications relay platform, extending the range of communications between ground units or other assets. Some variants of the *MQ-8 Fire Scout* are equipped with specialized sensors and equipment for anti-submarine warfare missions, supporting naval operations.

The primary operator of the *Northrop Grumman MQ-8 Fire Scout* is the United States Navy. It is utilized by the U.S. Navy for various missions, including intelligence gathering, surveillance, reconnaissance, and supporting naval operations. The *Fire Scout* has been deployed on various naval vessels to extend their capabilities and enhance situational awareness during missions at sea.

Northrop Grumman MQ-8 Fire Scout, the first operational autonomous helicopter.

AH-64 Apache is a family of attack helicopters developed in the USA by *McDonnell Douglas* (later acquired by *Boeing*). The *AH-64 Apache* was first introduced in 1984. Over 2,000 *AH-64 Apache* helicopters have been produced as of 2021, and the aircraft remains in active production to meet global demand.

The *Apache* features a tandem seating configuration, with the pilot in the rear seat and the co-pilot/gunner in the front seat. This setup allows for optimal visibility and coordination during missions. The *Apache* is equipped with sophisticated avionics and sensor systems, including a Longbow fire control radar, electro-optical and infrared sensors, and a helmet-mounted display system for the crew. The helicopter is armed with a 30mm M230 Chain Gun mounted under the nose, as well as a variety of rockets, missiles, and guided munitions on its wing pylons.

The *Apache* is a dedicated attack helicopter designed for engaging and destroying enemy targets on the battlefield, including armored vehicles, bunkers, and infantry positions. The helicopter's primary anti-tank missile is the *AGM-114 Hellfire*, which is highly effective against armored threats. The *Apache*'s advanced sensors and targeting systems enable it to conduct reconnaissance missions and acquire targets for engagement. It is capable of network-centric warfare, allowing it to receive and share real-time battlefield information with other friendly forces.

The *AH-64 Apache* is operated by various countries around the world. It serves as the primary attack helicopter for the United States Army, where it is used in a variety of combat roles. Additionally, it has been exported to numerous allied nations, making it one of the most widely used attack helicopters globally.

The *Apache* has seen extensive service in various conflicts, including *Operation Desert Storm*, the Iraq War, and the Afghanistan War. Its combat-proven capabilities and adaptability have solidified its position as one of the most effective and lethal attack helicopters in modern military aviation.

The *Boeing AH-64 Apache*, one of the most advanced combat helicopters.

Sikorsky-Boeing SB-1 Defiant is a technologically advanced rotorcraft developed through a joint effort between *Sikorsky Aircraft Corporation* (a subsidiary of *Lockheed Martin*) and *Boeing*. The *SB-1 Defiant* was designed as part of the Joint Multi-Role Technology Demonstrator (JMR-TD) program. Development of the aircraft began in the early 2010s, and it first took to the skies for testing in 2019.

The *SB-1 Defiant* is a compound helicopter, a hybrid design that combines the benefits of traditional rotary-wing helicopters and fixed-wing aircraft. It features coaxial rotors for lift, like a conventional helicopter, and a rear pusher propeller for additional thrust, similar to a fixed-wing aircraft. The incorporation of the coaxial rotors and rear pusher propeller allows it to achieve higher speeds than conventional helicopters. This design feature is intended to address the speed limitations often associated with traditional rotorcraft. The coaxial rotor system on the *SB-1 Defiant* offers increased lift efficiency and improved maneuverability. Additionally, the absence of a tail rotor reduces complexity and enhances safety. The helicopter is equipped with advanced integrated avionics and fly-by-wire technology, providing precise and sophisticated control over the aircraft's flight dynamics.

The *SB-1 Defiant* is part of the U.S. Army's Future Vertical Lift program, which aims to develop next-generation rotorcraft to replace existing helicopters. It serves as a technology demonstrator to validate and refine concepts for future military helicopter platforms. The advanced rotor system and fly-by-wire controls contribute to the *Defiant*'s agility and versatility, enabling it to adapt to a diverse range of missions, from troop transport to combat operations. As a technology demonstrator, the *SB-1 Defiant* plays a critical role in evaluating and showcasing innovative technologies and designs that may shape the future of military helicopter development.

The *Sikorsky-Boeing SB-1 Defiant*, a compound helicopter with rigid coaxial rotors helicopter developed as Future Long-Range Assault Aircraft program.

9.5.3 Future Vertical Lift (FVL) Programs

The Future Vertical Lift (FVL) programs are a series of initiatives by the United States Department of Defense (DoD) aimed at developing next-generation vertical lift aircraft. These programs seek to replace aging military helicopters with advanced rotorcraft that offer improved performance, agility, range, and capabilities. The FVL programs are collaborative efforts involving multiple branches of the U.S. military and various aerospace companies.

- **Future Long-Range Assault Aircraft (FLRAA)**: The FLRAA program aims to develop a new advanced assault helicopter to replace the *UH-60 Black Hawk* in the U.S. Army's fleet. The FLRAA helicopter is expected to have enhanced speed, range, and survivability, enabling it to conduct rapid troop transport and insertion in contested environments. The goal is to improve operational efficiency and effectiveness during joint military operations.

Prototype of Future Long Range Assault Aircraft (FLRAA) program aim to replace the Black Hawk by 2030.

- **Future Attack Reconnaissance Aircraft (FARA)**: The FARA program focuses on developing a new armed reconnaissance helicopter to replace the aging *OH-58D Kiowa Warrior*. The FARA aircraft will have improved sensors, weapons, and avionics, providing advanced reconnaissance and attack capabilities to support ground forces. The FARA platform is designed to complement the capabilities of the larger *AH-64 Apache* attack helicopters.

Prototype of Future Attack Reconnaissance Aircraft (FARA) program.

- **Future Unmanned Aircraft System (F-UAS)**: The F-UAS program aims to develop next-generation uninhabited aerial systems with advanced capabilities for intelligence, surveillance, reconnaissance, and other specialized missions. These platforms are expected to operate alongside traditional helicopters to enhance overall mission effectiveness and flexibility.

9.6 Helicopters in India

The existing combat helicopter fleets within both the Indian Air Force (IAF) and the Army Aviation Corps are:

AH-64E Apache Attack Helicopter: The Indian government procured 22 *AH-64E Apache* attack helicopters through a contract with *Boeing*, replacing the aging *Mi-35* fleet. These helicopters, tailored to IAF standards, are also being supplied to the Indian Army. *Tata Boeing Aerospace Limited* in Hyderabad manufactures the aero-structures for these helicopters. The *Apache*, known as the world's most advanced combat helicopter, equips the IAF with potent offensive capabilities, including various weapons and advanced systems. Deployed amid tensions with China, the *Apache* enhances the IAF's operational capacity in challenging terrains like Ladakh. This acquisition bolsters India's military strength, akin to the *Rafale* in the fixed-wing domain, preparing the IAF for future challenges.

MI-35 Attack Helicopter: The IAF's attack helicopter squadron, established in 1983 with *Mi-25s* from the Soviet Union, later upgraded to *Mi-35s* in 1990. Currently, the IAF operates one operational squadron of *Mi-35s* while the second awaits overhaul. These helicopters are twin-engine turbo-shaft, armed with a 12.7mm rotary gun and capable of carrying anti-tank missiles. With a history of active participation in operations, including support in Sri Lanka, the *Mi-35* fleet has served the IAF for three decades. However, with only one squadron operational, the IAF faces significant operational limitations in meeting challenges.

HAL Rudra The *Rudra Mk-IV*, a variant of the *Dhruv ALH*, is India's first indigenous combat helicopter developed by HAL. It obtained initial operational clearance in 2013 and is designed for various missions such as reconnaissance, troop transport, and anti-tank warfare. Armed with advanced weaponry like anti-tank guided missiles and air-to-air missiles, it features a defensive aids suite and HAL-made engines. HAL is set to deliver 76 *Rudras* to the Indian Army and IAF, with 58 currently in use by the Army and 12 by the IAF. The platform's operational capabilities remain untested.

Light Combat Helicopter: HAL developed the *Light Combat Helicopter (LCH)* in response to the need for a high-altitude combat aircraft during the Kargil war. Launched in 2006, the *LCH* is a multi-role attack helicopter with impressive specifications, including a low weight and high operational flight ceiling. Powered by *HAL/Turbomeca* engines, it has been tested at high altitudes and is set to be equipped with advanced weaponry. While still in development, the *LCH* shows promise for anti-infantry and anti-armour operations, with plans for enhanced electronic warfare systems and weapon capabilities. The current prototypes lack certain weapons but are expected to be fully operational in the future.

HAL Dhruv, a utility helicopter developed by

Table 9.2: Specifications of some representative helicopters
Speed-Maximum Speed, Range(F)- Ferry Range, Ceiling-Service Ceiling, DL-Disk Loading (Ratio of the total weight to the total rotor disk area, circular area swept by the rotating rotor blades), P/M-Power-to-Mass Ratio, MTD-Main Rotor Diameter, Payload- Maximum Payload

Helicopter	First Flight	Speed km/h	Range (F)km	Ceiling m	DL kg/m²	P/M (hp/kg)	MTD (m)	MTOW kg	Payload kg	Powerplant	Armament
Attack military helicopters											
Ka-52 Alligator	1997	350	1,220	5,500	245	0.23	14.5	10,800	2,000	2x Klimov VK-2500 turboshaft engines	Cannon, rockets, and missiles
AH-64 Apache	1975	293	1,900	6,400	275	0.24	14.6	10,433	1,814	2x General Electric T700 turboshaft engines	Various anti-tank missiles, rockets, and guns
Mi-28N Havoc	1982	324	1,100	5,700	234	0.19	17.2	11,500	1,700	2x Klimov TV3-117 turboshaft engines	Cannon, rockets, and anti-tank guided missiles
Eurocopter (Airbus) Tiger	1991	800	500	4,000	221	0.21	13	6,000	1,800	2x Rolls-Royce/Turbomeca MTR390 turboshaft engines	Cannon, rockets, and missiles
CAIC Z-10	2003	300	800	6,400	245	0.20	11.0	7,000	1,500	2x WZ-9 turboshaft engines	Cannon, rockets, and missiles
AgustaWestland T129 ATAK	2009	281	537	4,572	240	0.22	11.9	5,000	1,350	2x LHTEC T800 turboshaft engines	Various guided missiles, rockets, and gun pods
Mi-24 Hind	1969	335	450	4,500	214	0.15	17.3	11,500	2,400	2x Isotov TV3-117 turboshaft engines	Machine guns, rockets, and anti-tank guided missiles
AH-1Z Viper	2000	420	724	6,100	267	0.24	14.6	8,391	2,682	2x General Electric T700-GE-401C turboshaft engines	Various anti-tank missiles, rockets, and guns
Sikorsky UH-60 Black Hawk	1974	295	1,275	5,790	245	0.22	16.4	10,660	2,268	2x General Electric T700 turboshaft engines	Door-mounted machine guns, rockets, and missiles
Bell AH-1 Cobra	1965	282	463	4,270	142	0.23	13.41	4,536	363	2x General Electric T700 turboshaft engines	20mm M197 Gatling gun, rockets, and missiles
HAL Light Combat Helicopter	2010	255	700	6,500	210	0.25	11.9	5,800	1,500	2x HAL/Turbomeca Shakti turboshaft engines	8x MBDA Helina anti-tank missiles, 4x 68mm rockets,1x 20mm turret gun
Civilian Transport Helicopters											
Airbus H135	1994	259	696	6,096	87	0.20	10.2	2,950	1,452	2x Turbomeca Arrius 2B2 turboshaft engines	-
Leonardo AW139	2001	310	1,250	6,096	148	0.22	13.8	6,800	2,300	2x Pratt & Whitney Canada PT6C-67C turboshaft engines	-
Heavy-Lift Helicopters											
Mil Mi-26	1977	295	1,920	4,600	200	0.18	32	56,000	20,000	2x Lotarev D-136 turboshaft engines	Various cargo configurations
Sikorsky CH-53 Sea Stallion	1964	315	1,890	4,572	201	0.18	22.0	33,300	14,515	3x General Electric T64 turboshaft engines	Door-mounted machine guns, rockets, and missiles
Boeing MH-47 Chinook	1961	315	1,852	6,096	274	0.25	18.3	22,680	11,340	2x Honeywell T55-GA-714A turboshaft engines	Various configurations for troops and cargo
Search and Rescue Helicopters											
Eurocopter AS365 Dauphin	1975	306	1,440	6,096	136	0.23	11.94	4,300	1,720	2x Turbomeca Arriel 2C turboshaft engines	-
Sikorsky S-92	1998	312	1,340	5,791	218	0.24	17.17	12,100	4,536	2x General Electric CT7-8A turboshaft engines	-
AgustaWestland AW101 (EH101)	1987	309	833	4,575	238	0.20	18.6	15,600	5,443	3x Rolls-Royce, Turbomeca RTM322 turboshaft engines	Machine guns, rockets, torpedoes, and depth charges
MIL Mi-8	1961	250	800	4,500	198	0.14	21.3	12,000	4,000	2x Klimov TV3-117 turboshaft engines	Machine guns, rockets, and bombs
Bell UH-1	1956	204	435	4,265	112	0.11	13.4	4,536	2,268	1x Lycoming T53 turboshaft engine	Machine guns, rockets, and grenade launchers

Hindustan Aeronautics Limited (HAL) in November 1984, took its first flight in 1992 and was commissioned in 2002. The helicopter serves military and civilian needs, with variants tailored for the Indian Armed Forces and commercial use. Various military versions like transport, utility, reconnaissance, and medical evacuation variants are in production. Over 400 *Dhruvs* have been produced as of January 2024 for both domestic and export markets, accounting for more than 340,000 flying hours.

Indian Multi-Role Helicopter (IMRH), currently under development by Hindustan Aeronautics Limited (HAL) for the Indian Armed Forces, is designed to serve various roles including air assault, air-attack, anti-submarine, anti-surface, military transport, and VIP transport. It aims to replace the existing *Mil Mi-17* and *Mil Mi-8* helicopters across the Indian Armed Forces. HAL estimates a need for over 314 rotorcraft of the same class across the Indian Armed Forces to replace existing helicopters. Scaled model tests have been ongoing, and the first flight of a full prototype is expected in 2025–26, with introduction into the armed forces anticipated in 2028 after two years of testing.

HAL Chetak is a light utility helicopter and is used primarily for training, rescue and light transport roles in the IAF.

HAL Cheetah, a light utility helicopter utilized for high-altitude operations, serves in transport and search-and-rescue missions within the IAF.

Bibliography

S. Newman (1994) *Foundations of Helicopter Flight*, Elsevier Science.

J. M. Seddon and S. Newman (2011) *Basic Helicopter Aerodynamics*, John Wiley & Sons.

C. Venkatesan (2014) *Fundamentals of Helicopter Dynamics*, CRC Press.

Federal Aviation Administration (2007) *Rotorcraft Flying Handbook*, Skyhorse Publishing Inc..

J. G. Leishman (2007) *The Helicopter*, College Park Press College Park, Maryland.

A. R. S. Bramwell, D. Balmford, and G. Done (2001) *Bramwell's Helicopter Dynamics*, Elsevier.

G. D. Padfield (2008) *Helicopter Flight Dynamics: The theory and application of flying qualities and simulation modelling*, John Wiley & Sons.

W. Boyne (2011) *How the Helicopter Changed Modern Warfare*, Pelican Publishing Company, Inc.

W. Johnson (2012) *Helicopter Theory*, Courier Corporation.

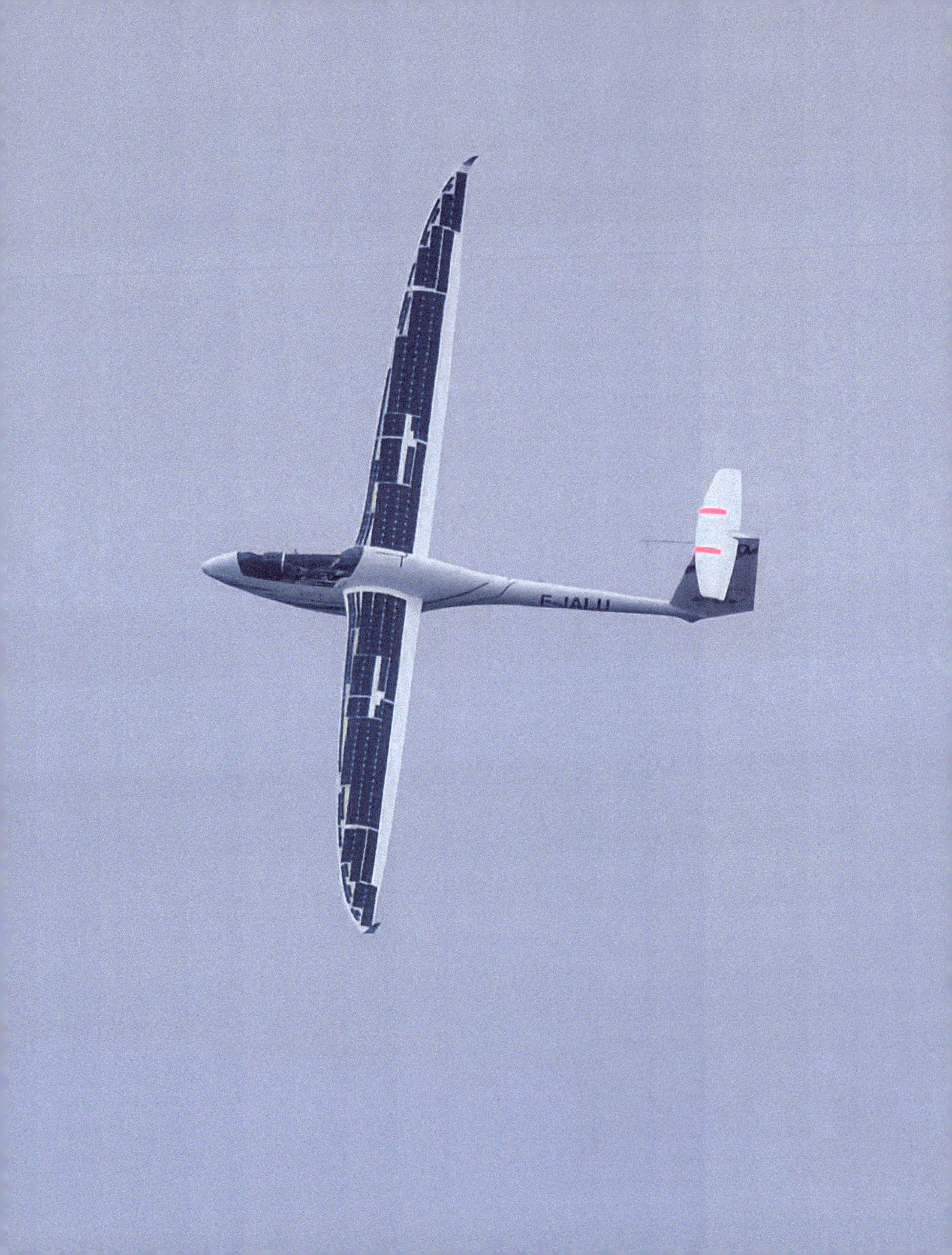
F-JALU

Chapter 10

Miscellaneous Flight Vehicles

Contents

10.1	**Aerostats**		**420**
	10.1.1	Air ships	421
	10.1.2	Balloons	423
10.2	**Autogyros**		**423**
10.3	**Convertiplanes**		**425**
	10.3.1	Tiltrotors	426
	10.3.2	Tiltwings	426
	10.3.3	Stopped Rotors	426
10.4	**Gliders**		**426**
10.5	**Ground-Effect Vehicles**		**428**
10.6	**Jet Packs**		**429**
10.7	**Seaplanes**		**430**
	10.7.1	Flying Boats	430
	10.7.2	Floatplanes	430
	10.7.3	Amphibious Aircraft	431
10.8	**Uninhabited Aerial Vehicles**		**431**
	10.8.1	Micro Aerial Vehicles	432
	10.8.2	Quadcopters	434

Covering a wide range of flight vehicles beyond traditional aircraft, this chapter explores a variety of unconventional flying machines. The chapter covers gliders, ground-effect vehicles, jet packs, and seaplanes, from aerostats like airships and balloons to uninhabited aerial vehicles (UAVs), autogyros, and convertiplanes. Each of these flight vehicles offers a distinct perspective on the boundless possibilities of flight, expanding our understanding of aviation's potential.

10.1 Aerostats

An aerostat is a type of flight vehicle that remains aloft using buoyancy, which is the upward force exerted on an object immersed in a fluid (in this case, air) due to the difference in densities. It consists of a large, enclosed bag, often called an envelope, filled with a gas that is lighter than air. The most commonly used gases are helium and hydrogen and, in some cases, hot air. The difference in density between the gas inside the envelope and the surrounding air allows the aerostat to float in the sky. This allows aerostats to stay aloft without the need for constant propulsion, making them more energy efficient for certain applications.

Aerostats are known for their inherent stability in the air and long-duration flights, making them useful for various applications. Unlike traditional aircraft, aerostats do not have propulsion systems, and they rely on external means for movement and control. Aerostats remain stationary in the sky once they reach a stable altitude. Aerostats can stay airborne for extended periods, ranging from hours to days or even weeks, depending on the design and purpose. This makes them ideal for prolonged missions, such as surveillance, where continuous monitoring is required.

Aerostats come in two main configurations: tethered and free-flying. *Tethered Aerostats* are anchored to the ground with a strong cable, which provides both support and control. The tether allows for easy retrieval and control of the aerostat's altitude and position. Tethered aerostats are commonly used for surveillance and communication purposes, as they can be stationed at strategic locations and operated for extended periods. *Free-Flying Aerostats* are not tethered to the ground and are designed to move with the wind currents. They are often used for scientific research and environmental monitoring, as they can cover large areas and access remote locations.

Tethered Aerostat Radar System, an American low-level airborne ground surveillance system.

Aerostats find numerous applications across different sectors. These are large tethered balloons equipped with surveillance equipment such as radars, cameras, and communication systems. They are used for military and border security purposes to monitor large areas from a high vantage point. Surveillance aerostats are particularly effective for detecting and tracking low-flying aircraft, surface vessels, and ground activities. One of the examples is the *Tethered Aerostat Radar System* (TARS) used by the U.S. military for long-range surveillance along its borders and coasts.

Stratospheric balloons are specialized aerostats that operate at very high altitudes, often in the stratosphere. They carry scientific instruments, communication equipment, or observational payloads for research and commercial purposes. Stratospheric balloons can stay aloft for days, weeks, or even months, providing an affordable alternative to satellites for certain applications. *Google*'s *Project Loon*, aimed to provide internet connectivity to re-

Aerostats used as *Energizer Bunny* hot air balloon.

mote and underserved regions using a network of stratospheric balloons.

In some locations, aerostats are used for tourist activities, offering scenic rides and aerial views of landscapes and cities. These leisure aerostats are typically larger than traditional hot air balloons and can carry more passengers.

Some companies use aerostats as giant floating billboards for advertising and branding purposes. These aerostats are often customized with colorful designs and logos to catch the attention of people on the ground. The *Energizer Bunny* hot air balloon is a famous promotional aerostat used by the *Energizer Battery Company* for branding and marketing purposes.

Aerostats are used for scientific research and data collection in the atmosphere. They can carry instruments for measuring meteorological conditions, air quality, and other environmental parameters. The National Oceanic and Atmospheric Administration (NOAA) uses aerostats to conduct weather and atmospheric research, including the study of hurricanes and other severe weather events.

Aerostats have unique advantages due to their ability to stay aloft for extended periods, stability, and payload capacity. However, they are limited in terms of speed and maneuverability compared to traditional aircraft. Despite this, their diverse applications make them valuable assets in fields ranging from military and surveillance to scientific research and tourism. While aerostats have numerous advantages, they also have some limitations. Their speed and maneuverability are limited compared to traditional aircraft. Weather conditions, especially strong winds, can affect their flight and require careful management. Furthermore, their payload capacity may be constrained due to the need to lift both the envelope and the payload using buoyancy.

10.1.1 Air ships

An airship, also known as a dirigible, is a type of lighter-than-air aircraft that can be steered and propelled through the air. Unlike aerostats, airships have an onboard propulsion system that allows them to move in any direction, providing greater control and maneuverability. Airships have a long history, and while they were once widely used for various purposes, their popularity declined with the advent of more advanced fixed-wing aircraft. There are two main types of airships: Rigid and Non-Rigid (Blimps) Airships.

Rigid airships have a rigid framework or structure that maintains the shape of the airship's envelope and provides support to the internal gas cells. The framework is usually made of lightweight materials such as aluminum or duralumin. The envelope is filled with a lighter-than-air gas, typically helium, which provides buoyancy.

Rigid airships have a streamlined, elongated shape with a gondola or cabin suspended beneath the envelope to accommodate passengers, crew, and cargo. They are steered and propelled by engines and propellers mounted on the airship's framework. Rigid airships were popular in the early 20th century and were used for passenger travel, military reconnaissance, and exploration.

Famous examples of rigid airships include the *Zeppelin LZ 127 Graf Zeppelin*, which conducted the first commercial transatlantic passenger flights, and the ill-fated *Hindenburg*, which tragically caught fire during a landing in 1937.

Non-rigid airships (or blimps) have a flexible en-

The *LZ 129 Hindenburg* was the largest airship ever built and was destroyed in 1937.

Spirit of Innovation (N4A), *Goodyear*'s blimp, was retired in 2017.

LZ 127 Graf Zeppelin, a German passenger-carrying hydrogen-filled rigid airship that flew from 1928 to 1937.

velope without a rigid internal structure. The shape of the envelope is maintained by the pressure of the lifting gas. Blimps are generally smaller and more maneuverable than rigid airships. They are often used for advertising, aerial photography, surveillance, and recreational purposes. Blimps are filled with helium and have a distinctive shape, usually with a cylindrical or cigar-like body. The gondola or cabin is attached directly to the bottom of the envelope. Blimps use propellers and engines for propulsion and steering. Examples of non-rigid airships include the *Goodyear Blimps*, which are well-known for their advertising and promotional appearances at various events.

Airships offer several advantages over fixed-wing aircraft and other forms of transportation. Airships require relatively little power to stay aloft since they rely on buoyancy to offset their weight. This results in lower fuel consumption compared to traditional aircraft. Airships fly at low speeds and offer stable flight characteristics. This stability makes them suitable for certain applications, such as aerial photography or surveillance. Airships can stay aloft for extended periods, making them useful for missions that require prolonged aerial presence, such as surveillance or research. Some airships can hover in place, which is particularly advantageous for tasks that demand precise positioning or observations.

Despite these advantages, airships also have notable limitations, such as their relatively slow speed and susceptibility to adverse weather conditions, particularly strong winds. Additionally, the availability of helium, which is used to provide lift, can sometimes be a limiting factor.

While airships were once widely used and considered a promising mode of air travel and exploration, the development of fixed-wing aircraft and improvements in aviation technology eventually surpassed their capabilities. Today, airships are primarily used for specific applications such as advertising, aerial photography, surveillance, and research, but they continue to capture the imagination and fascination of many people around the world.

10.1.2 Balloons

Balloons are one of the oldest forms of human flight, and they continue to be popular for recreational activities, aerial photography, and as a unique way to experience the joy of flying. There are two main types of balloons: Hot air balloons and Gas balloons.

Hot air balloons use the principle that heated air is less dense than the surrounding cooler air, and therefore able to provide lift by buoyancy. The balloon envelope is made of lightweight and durable materials, such as nylon or polyester, and it is filled with air using a large burner that produces an open flame. As the air inside the envelope is heated, the balloon becomes buoyant and rises. To control the balloon's ascent and descent, the pilot can regulate the burner's intensity to increase or decrease the temperature of the air inside the envelope. By finding different wind layers at various altitudes, pilots can navigate hot air balloons horizontally to some extent, but their direction is largely determined by the wind's course. Hot air balloons are widely used for recreational purposes, offering a serene and peaceful flight experience with breathtaking views of landscapes and natural wonders.

Gas balloons use a lighter-than-air gas, such as helium or hydrogen, to provide lift. Unlike hot air balloons, gas balloons do not rely on heat to achieve buoyancy. Gas balloons are typically used for long-distance and high-altitude flights. They are often employed for record-setting attempts or scientific research missions in the upper atmosphere. The ability to stay aloft for extended periods makes gas balloons valuable for various atmospheric studies and research endeavors.

Both types of balloons have a gondola or basket suspended from the envelope, where passengers, crew, and equipment can be accommodated. The pilot operates the balloon and is responsible for managing the burner (in the case of hot air balloons) and determining the best altitude and wind layers for navigation.

Ballooning is considered one of the safest forms of aviation, with fewer mechanical components and less reliance on technology. It offers a unique and peaceful flying experience, allowing passengers to soar above the ground, almost feeling one with the wind and nature.

As a recreational activity, hot air ballooning has become a popular attraction in many tourist destinations worldwide, drawing people seeking unforgettable experiences and breathtaking aerial views. Additionally, balloons are often used for special events, advertising, and scientific research purposes.

However, it's essential to ensure safety precautions are taken when operating balloons, as weather conditions can significantly affect flight operations, and proper training and equipment maintenance are crucial for a successful and safe ballooning experience.

10.2 Autogyros

Autogyros, also known as gyroplanes or gyrocopters –the precursors to helicopters– are unique rotary-wing aircraft that operate on the principle of autorotation to generate lift. They combine elements of fixed-wing airplanes and helicopters, offering a safe and stable flying experience with relatively simple mechanical systems. Autogyros were

The main differences between airship, balloon, and aerostats. Airships, balloons, and aerostats are all types of lighter-than-air aircraft, but they differ in their design, structure, propulsion, and applications. Airships are large, rigid aircraft with onboard propulsion systems, balloons are simple non-rigid aircraft that depend on the wind for movement, and aerostats encompass both airships and balloons

Airship	Balloon	Aerostats
• Rigid internal structure for shape and support • Onboard propulsion system for maneuverability • Commonly used for passenger travel, military reconnaissance, and exploration. • Generally larger and more sophisticated than other lighter-than-air aircraft	• Relies solely on buoyancy to stay afloat in the air • Single, non-rigid envelope for buoyancy on which the balloons relies to stay afloat in the air • No onboard propulsion; movement is mostly passive and dependent on wind	• Refers to any lighter-than-air aircraft that remains aloft using buoyancy. • Broad category of lighter-than-air aircraft that includes both airships and balloons. • May refer more specifically to tethered balloons used for specialized applications.

first developed in the early 20th century and have since evolved into popular recreational and utility aircraft.

Autogyros and helicopters are both types of rotary-wing aircraft, but they have significant differences in their design, flight characteristics, and capabilities. Autogyros generate lift by an unpowered, freely rotating main rotor, also known as an autorotation. The rotor spins due to the forward motion of the aircraft, creating lift without the need for engine-driven power to turn the rotor. However, they rely on a separate engine-driven propeller to provide forward thrust during level flight and climb. The engine also powers other essential systems, such as the controls and instrumentation. They require a short takeoff roll to build up forward airspeed before they can achieve lift for takeoff. They can land vertically, but they generally require more space for landing compared to helicopters due to their slower descent rate. Autogyros have a limited maximum forward speed, typically lower compared to helicopters, due to the autorotation of their main rotor.

Autogyros can maintain a controlled descent rate in a state of autorotation, but they are not capable of hovering in a stationary position like helicopters. They are generally more maneuverable and agile than fixed-wing aircraft. They can perform sharp turns and fly at lower speeds, making them well-suited for aerial photography, agricultural tasks, and recreational flying. Autogyros are mechanically simpler than helicopters, as they lack complex pitch controls for the main rotor blades.

Typical Gyrocopter

Autogyros find applications in various fields. Autogyros are popular for recreational flying and aviation enthusiasts due to their unique flight characteristics and open-cockpit designs. Their slow flight capabilities and stability make autogyros suitable for aerial photography, environmental surveying, and mapping tasks. Also, they can be used for crop dusting and agricultural applications, where slow, low-level flight is required. In some regions, they are utilized for search and rescue missions, especially in remote or inaccessible areas. While not as common as helicopters, autogyros have been explored for surveillance and reconnaissance purposes in the military.

Examples of autogyros are:

- *AutoGyro Calidus* is a popular two-seat autogyro known for its stability and comfortable flying experience. It features a fully enclosed cabin, a tandem seating arrangement, and a wide range of options for personalization.

- *Magni M24 Orion* is an Italian-made autogyro with a sleek design and excellent performance. It offers an open cockpit with side-by-side seating and is widely used for recreational flying and flight training.

- *Carpenter P-1 Gyroplane* is an experimental gyroplane designed and built by engineer *Igor Bensen*. It is one of the earliest modern autogyros, playing a significant role in popularizing this type of aircraft.

- *ELA Aviacion ELA-07S* is a Spanish autogyro designed for sport and recreational flying. It features a fully enclosed cabin with a bubble canopy, making it suitable for various weather conditions.

Autogyros and helicopters have different lift-generation mechanisms, vertical takeoff and landing capabilities, forward flight speeds, and hovering capabilities. Autogyros are mechanically simpler and typically have a lower forward speed but can glide safely to the ground in case of engine failure. Helicopters, on the other hand, offer true VTOL capabilities, hovering, and higher forward speeds, making them suitable for a wider range of missions, especially those that require precise and flexible operations.

10.3 Convertiplanes

A convertiplane is an aircraft that uses rotor power for vertical takeoff and landing (VTOL) and converts to fixed-wing lift in normal flight. It is a unique type of aircraft that combines features of both helicopters and fixed-wing airplanes. It is designed to take off, land, and hover like a helicopter while also having the capability to fly forward at high speeds like a conventional airplane. This versatile design allows convertiplanes to efficiently operate in a wide range of scenarios and mission profiles.

The key feature of a convertiplane is its tilting rotors or propellers. These rotors are attached to the wingtips or other parts of the aircraft's structure, and they can be mechanically tilted to transition between vertical and horizontal flight modes. In vertical flight mode, the rotors are tilted vertically, allowing the aircraft to take off and land like a helicopter. In horizontal flight mode, the rotors are tilted horizontally, and the aircraft relies on its fixed wings and forward propulsion for efficient, high-speed flight.

The ability to transition between vertical and horizontal flight modes gives convertiplanes several advantages over traditional aircraft. Convertiplanes can take off and land vertically (VTOL), eliminating the need for long runways, which is particularly useful in urban areas or locations with limited infrastructure. Operation in short runways or unprepared surfaces provides them flexibility in deployment. Once in horizontal flight mode, convertiplanes can achieve higher speeds and longer ranges than conventional helicopters, making them suitable for rapid transportation of both passengers and cargo. They can hover in place, offering enhanced maneuverability and the ability to perform tasks like search and rescue, surveillance, and precision landings. Convertiplanes are well-suited for a variety of missions, including military operations, emergency response, aerial reconnaissance, and commercial applications.

Some of the most well-known examples of convertiplanes include *Bell Boeing V-22 Osprey*, *AgustaWestland AW609* or *Leonardo Helicopters AW609*.

Tiltrotor, tilt-wing, and stopped rotor are three distinct types of convertiplanes, each with its own unique design and operational characteristics. While they share the ability to transition between vertical and horizontal flight modes, they differ in how they achieve this transition and the specific applications they are suited for.

10.3.1 Tiltrotors

Tiltrotor aircraft, like the *Bell Boeing V-22 Osprey*, have rotors that are mounted on the wingtips or other parts of the aircraft's structure. These rotors can be mechanically tilted between vertical and horizontal positions, allowing the aircraft to transition between helicopter-like VTOL operations and efficient airplane-like forward flight. Tiltrotors are well-suited for a variety of missions, including military transport, search and rescue, special operations, and other scenarios that require both VTOL capability and fast, long-range flight.

10.3.2 Tiltwings

Tiltwing aircraft, like the *Canadair CL-84* and the *NASA XV-15*, have wings that can tilt relative to the fuselage. The tiltwing is similar to the tiltrotor, except that the rotor mountings are fixed to the wing and the whole assembly tilts between vertical and horizontal positions.

Tiltwing aircraft are less common than tilt rotors and have been primarily explored for experimental and research purposes. The complexity of the wing tilt mechanism and the transition between flight modes present engineering challenges that have limited their practical use in operational aircraft.

10.3.3 Stopped Rotors

Stopped rotor aircraft with variable-incidence rotors, have a different approach to transitioning between vertical and horizontal flight modes. These aircraft have rotors that are capable of stopping and tilting to redirect their thrust during the transition. In VTOL mode, the rotors are used for lift, similar to a helicopter. Once the aircraft is airborne, the rotors can be slowed down or stopped, and their pitch angle is adjusted to provide forward thrust for horizontal flight. The main advantage of this approach is that it eliminates the need for complex wing tilt mechanisms. It works like a compound helicopter where the main rotor is always active and provides lift during all phases of flight but incorporates additional forward propulsion systems as stopped rotor aircraft stops or slows down its rotor blades during the transition from vertical to horizontal flight modes.

The *Boeing X-50 Dragonfly* had a two-bladed rotor driven by the engine for takeoff. In horizontal flight, the rotor stopped to act like a wing. Fixed canard and tail surfaces provided lift during transition, and also stability and control in forward flight. The *Sikorsky X-Wing* is another aircraft in this category.

While convertiplanes offer significant advantages, they also come with some challenges, including complex engineering and control systems, which can make them more expensive to develop and maintain than traditional aircraft. However, ongoing advancements in aerospace technology continue to improve the performance and capabilities of these unique and versatile aircraft.

10.4 Gliders

Gliders are a special kind of fixed-wing aircraft that fly without an engine or any form of onboard propulsion. They achieve flight by relying on the principles of aerodynamics to generate lift and maintain their altitude and forward movement. Gliders are often used for recreational flying, train-

10.4. GLIDERS

Bell Boeing V-22 Osprey, tiltrotor aircraft.

Canadair CL-84 (Dynavert), a V/STOL turbine tiltwing monoplane.

Sikorsky S-72 (X-Wing) a stopped rotor convertiplane.

Three distinct types of convertiplanes: tiltrotor, tiltwing and stopped rotor.

ing, competitive soaring, research, and scientific purposes.

Paper airplanes represent the simplest form of gliders, easy to construct and fly. For educational purposes, balsa wood or Styrofoam toy gliders serve as affordable tools for students to enjoy themselves while learning the basics of aerodynamics. Hang-gliders, on the other hand, are piloted aircraft with cloth wings and minimal structure. Some hang-gliders resemble piloted kites, while others boast maneuverable parachute-like designs. Sailplanes, also piloted gliders, utilize standard aircraft components, construction, and flight control systems but are devoid of engines. They gracefully soar through the skies relying solely on the principles of aerodynamics. Fascinatingly, the *Space Shuttle* exhibits glider-like characteristics during its return to Earth; its rocket engines are exclusively employed during liftoff. Even the pioneering *Wright brothers* gained invaluable piloting experience through a series of glider flights from 1900 to 1903, setting the stage for the future of human-powered flight.

Gliders are given an initial velocity by hand throwing (paper planes) or catapult (balsa gliders), a tow line to provide velocity and some initial altitude or towed aloft by a powered aircraft and then cut loose to begin the glide. The glider can trade the potential energy difference from a higher altitude to a lower altitude to produce kinetic energy, which means velocity.

The design of gliders focuses on maximizing lift and minimizing drag to achieve efficient and sustained flight without the need for engine power. Glider wings are long and slender, often with a high aspect ratio (span divided by chord). This design reduces induced drag and improves lift efficiency. The fuselage of a glider is streamlined and lightweight, reducing drag and improving the glider's overall aerodynamic performance.

When the pilot skillfully identifies an ascending pocket of air moving faster than the glider's descent, the glider gains altitude, accumulating potential energy in the process. These upward-moving air pockets are referred to as *updrafts*, playing a vital role in glider flight. Updrafts are commonly encountered when winds encounter hills or mountains, compelling them to ascend while overcoming the terrain. Additionally, updrafts can be found above dark land masses that readily absorb heat from the sun. As the ground warms up, the surrounding air is heated, causing it to rise. These rising pockets of warm air are known as *thermals*.

In a captivating display of flight mechanics, large gliding birds like owls and hawks are frequently spotted circling within thermals, attaining altitude without the need for wing flapping. Gliders expertly employ the same technique, capitalizing on thermals or updrafts to sustain and extend their flights, even in the absence of an engine. By harnessing these natural phenomena, gliders can remain airborne for prolonged periods, conserving energy and reaching impressive heights in the skies.

In flight, a glider has three forces acting on it– lift, drag, and weight – as compared to the four forces (additional thrust) that act on a powered aircraft. If a glider is in a steady (constant velocity and no acceleration) descent, it loses altitude as it travels. The glider's flight path is a simple straight line with a negative slope equal to the tangent of the descent path angle. The descent path angle, (γ), of a glider refers to the angle at which the glider descends relative to the horizontal plane while in flight. It represents the rate at which the glider loses altitude as it moves forward through the air. For a steady and level flight, $\tan \gamma = D/L$, where D is the drag and L is the lift, the descent path angle is purely dependent on the aerodynamics. The descent velocity of a glider can be defined as the loss of altitude with time, $V_d = V \sin \gamma$, where V is the flight velocity.

There are various types of gliders designed for specific purposes. *Standard gliders* are general-purpose gliders used for basic flight training and recreational flying. Performance gliders are High-performance gliders designed for competitive soaring and long-distance flights. Sailplanes are motorless gliders capable of soaring on thermals and ridge lift for extended periods. Hang gliders are a type of glider where the pilot is suspended from a wingframe, often foot-launching from elevated locations.

Gliders are widely used in the sport of gliding, where pilots compete in various tasks such as cross-country flights, duration flights, and precision landings. They are excellent training aircraft for aspiring pilots to learn fundamental flight skills and aerodynamics principles. Some gliders are used for aerial surveying, photography, and scientific research due to their silent and non-intrusive flight. Gliders equipped with instruments are used for studying atmospheric conditions, weather patterns, and climate research.

Examples of Gliders:

- *Schleicher ASK 21*: A popular two-seat training glider used worldwide for flight training.

- *Schempp-Hirth Ventus-2*: A high-performance glider used in competitive soaring and long-distance flights.

- *Wills Wing Falcon*: A well-known hang glider used for recreational flying and competitions.

- *NASA's X-38*: An experimental crew return vehicle glider used for testing spacecraft's atmospheric entry and landing.

10.5 Ground-Effect Vehicles

A ground-effect vehicle (GEV), also known as a wing-in-ground-effect (WIG) vehicle, is a type of craft that takes advantage of the aerodynamic phenomenon known as ground effect to achieve lift and efficient flight near the surface of the Earth. Ground effect occurs when an aircraft flies close to the ground, within one wingspan's distance or less. In this situation, the ground creates a cushion of high-pressure air beneath the wings, reducing induced drag and increasing lift.

What is the ground effect? The ground effect arises from the interaction between a lifting wing and a fixed surface positioned beneath it. As the wing directs air downward, creating pressure, the fixed surface acts as a boundary, confining the air and forming an air cushion. This phenomenon also avoids the formation of wingtip vortices, which typically reduce wing efficiency. By mitigating these vortices, the ground effect reduces induced drag, theoretically enabling an aircraft to travel with greater efficiency compared to flying above ground effect. It's this efficiency that makes ground-effect vehicles particularly appealing.

GEVs typically have large wings or lifting surfaces that span a considerable distance, allowing them to utilize ground effect efficiently. They are designed to fly at relatively low altitudes, typically a few meters above the surface of the water or land, although some GEVs are capable of flying higher.

One of the primary advantages of ground-effect vehicles is their ability to travel at high speeds while consuming less fuel compared to conventional aircraft flying at similar speeds. This makes them particularly suitable for applications such as maritime patrol, search and rescue operations, coastal surveillance, and military transport.

Ground-effect vehicles come in various sizes and configurations, ranging from small recreational craft to large military transports capable of carrying heavy loads over long distances. Some GEVs are designed for civilian use, offering fast and efficient transportation over water or flat terrain.

While ground-effect vehicles offer several advantages, they also have limitations. They are generally restricted to flying over flat surfaces, such as bodies of water, coastal regions, or flat terrain, due to their reliance on ground effect for lift. Additionally, they may be susceptible to rough weather conditions and are less maneuverable than traditional aircraft.

Overall, ground-effect vehicles represent an intriguing intersection of aviation and maritime technology, offering unique capabilities for a variety of applications where speed, efficiency, and access to coastal and shallow water areas are essential.

10.6 Jet Packs

A jet pack, also referred to as a rocket belt, rocket pack, or flight pack, is a compact personal propulsion device worn on the back, allowing individuals to fly short distances using low rocket power

The *A-90 Orlyonok*, a Soviet GEV designed by *Rostislav Alexeyev* of the Central Hydrofoil Design Bureau. Utilizing the ground effect, the *A-90* gracefully glides a mere few meters above the surface and also can reach altitudes of up to 3,000 meters.

generated by a non-combusting gas. Typically, a jet pack consists of propulsion thrusters that generate enough thrust to lift the wearer off the ground and propel them forward.

Wendell Moore, from *Bell Aerosystems*, was the pioneer behind jet pack development in the mid-1950s. During the 1960s, the U.S. military conducted thorough research on the device, considering its potential as an aid for combat soldiers. However, its brief duration of only a few seconds proved too restrictive.

Jet packs typically operate on a non-explosive hydrogen peroxide fuel which makes jet packs marginally safer. When hydrogen peroxide combines with pressurized liquid nitrogen and a silver catalyst, it undergoes a chemical reaction, producing super-heated steam that propels the wearer. Although there is no flame, the process remains inherently hazardous. While hydrogen peroxide serves as a dependable fuel source, generating only water as a by-product, it is expensive.

A standard rocket belt weighs approximately 56 kilograms), requiring the pilot to weigh around 80 kilograms or less to achieve adequate lift. Control mechanisms entail the throttle operated by the right hand and yaw control, facilitating lateral movement, managed by the left hand. Despite their brief flight durations, rocket belts can achieve speeds of up to 130 km/h, boasting remarkable acceleration

Rocketbelt pilot *Dan Schlund* at the 2007 Rose Parade.

capabilities. Landing is executed by gradually reducing throttle pressure.

Although still primarily a niche technology, jet packs represent a thrilling glimpse into the possibilities of personal flight and continue to inspire innovators and enthusiasts alike with their futuristic appeal and potential for exploration and adventure in the skies.

A Thrilling Jet Pack Display at 1984 Olympics:

At the grand opening of the 1984 Olympics in the iconic Los Angeles Memorial Coliseum, amid the mesmerizing tunes of *John Williams* and a spectacular display of dancers and performers, a moment of true marvel unfolded. *Bill Suitor*, a distinguished *Bell* test pilot, took to the skies with a rocket belt strapped to his back. With a roar of engines, he soared from one end of the stadium, swiftly crossing the vast expanse to gracefully touch down at the center of the field. Witnessed by nearly 100,000 spectators and broadcast to billions worldwide, this remarkable display marked perhaps the most memorable and widely viewed rocket belt flight in history.

10.7 Seaplanes

A seaplane is a powered fixed-wing aircraft capable of taking off and landing on water. Seaplanes can operate from water as well as conventional runways on land, making them versatile and suitable for various applications.

10.7.1 Flying Boats

A flying boat is a type of fixed-winged seaplane that has one or more slender floats mounted under the fuselage to provide buoyancy which makes it capable of floating on water. The main source of buoyancy is the fuselage, which acts like a ship's hull in the water because the fuselage's underside has been hydrodynamically shaped to allow water to flow around it. This hull design enables the aircraft to perform water takeoffs and landings. It is equipped with wings and usually has one or more engines for powered flight. Most flying boats have small floats mounted under the wing or wing-like hull projections (called sponsons) to keep them stable. Not all small seaplanes have been floatplanes, but most large seaplanes have been flying boats, with their great weight supported by their hulls.

Some interesting examples of flying boats from history include the *Felixstowe F.2* (UK, 1917), *Supermarine Southampton* (UK, 1919), *Consolidated PBY Catalina* (US, 1935), the *Short S.25 Sunderland* (UK, 1937), and the *Sikorsky S-42* (US, 1934). These aircraft played crucial roles in aviation history and continue to be remembered as innovative and significant contributions to the development of international air travel.

10.7.2 Floatplanes

A floatplane is also a type of seaplane, but instead of a boat-like hull, it has one or more buoyant floats mounted underneath the fuselage or wings. These floats provide buoyancy and stability on water, allowing the aircraft to take off and land on water surfaces.

Floatplanes have a long history in aviation and played a crucial role in the early development of commercial air travel, especially in regions with vast water bodies and limited infrastructure. Today, floatplanes continue to be used for recreational flying, tourism, and specialized missions in remote or rugged environments.

Some well-known examples of floatplanes include the de *Havilland Canada DHC-2 Beaver*, the *DHC-3 Otter*, and the *Cessna 208 Caravan* on float modifications. These aircraft are valued for their ability to access both water and land, making them valuable assets in various regions and applications.

10.7.3 Amphibious Aircraft

An amphibious aircraft or amphibian is an aircraft, typically fixed-wing equipped with retractable wheels, that can take off and land on both ground and water. Amphibious helicopters are used for a variety of purposes including air-sea rescue, marine salvage and oceanography, in addition to other tasks that can be accomplished with a conventional helicopter. It can be designed with a waterproof hull like a flying boat or it can be attached with utility floats similar to that in floatplanes. Amphibious aircraft offer the flexibility of being able to access both land-based airports and water bodies, making them well-suited for a wide range of applications, including search and rescue, aerial surveying, and tourism.

Typical examples of amphibious aircraft are *Vickers Viking*, *Canadair CL-215T*, Italian Air Force *Piaggio P.136*, Japanese *ShinMaywa US-2*.

Overall, *flying boats* have a boat-like hull and are designed exclusively for water-based operations. *Floatplanes* have buoyant floats attached to the fuselage or wings and can operate from water as well as land. *Amphibious aircraft* have either floats or retractable landing gear, allowing them to take off and land on both water and conventional runways.

10.8 Uninhabited Aerial Vehicles

An Uninhabited Aerial Vehicle (UAV) is an aircraft that operates without a human pilot on board. UAVs are controlled remotely by a human operator or can follow pre-programmed flight paths autonomously.

Supermarine Southampton, a flying boat.

de Havilland Otter floatplane.

A *Canadair CL-215T* amphibian with retractable wheels.

Three distinct types of seaplanes: flying boat, floatplane and amphibian.

UAVs can operate autonomously using onboard sensors, GPS, and sophisticated flight control systems, enabling them to fly predetermined routes and execute specific tasks without continuous human input. Many UAVs are remotely controlled

Table 10.1: Comparison of flying boats, floatplanes, and amphibious aircraft

Flying Boat	Floatplane	Amphibious Aircraft
Design: Boat-like hull for water landings (Hydrodynamic hull design)	Buoyant floats for water landings (Floats mounted under fuselage/wings)	Buoyant floats & retractable landing gear (Floats + retractable wheels)
Operation: Water takeoffs & landings only (Limited to water surfaces)	Water takeoffs & landings only (Limited to water surfaces)	Water & conventional runway operations (Both water & land operations)
Suitable for long-range overwater flights (Long endurance)	Suitable for short to medium-range overwater flights (Limited endurance)	Versatile for overwater & land missions (Flexible mission profiles)
Application: Early long-distance air travel, maritime patrol, search & rescue, military reconnaissance	Sightseeing, transport to remote locations, charter flights	Search & rescue, firefighting, transport, tourism, maritime surveillance, aerial surveying
Advantages: Excellent stability on water, large payload capacity,	Good water handling characteristics, relatively simple design	Access to water and land destinations, greater operational flexibility
Limitations: Slower speeds compared to some land-based aircraft, limited ground mobility without specialized facilities	Limited range for overwater flights	Higher complexity than non-amphibious aircraft, Additional weight from landing gear
Examples: Consolidated PBY Catalina, Short Sunderland, Boeing 314 Clipper	de Havilland Canada DHC-2 Beaver, Cessna 208 Caravan on floats	ShinMaywa US-2, Beriev Be-200, Canadair CL-415, Grumman G-21 Goose

by human operators using ground control stations. The operator communicates with the UAV through radio signals or satellite links. UAVs come in various sizes and configurations, from small hand-launched drones to large, high-altitude, long-endurance (HALE) aircraft. This versatility allows them to serve a wide range of applications. UAVs can carry different payloads, including cameras, sensors, communication equipment, and even weapons, depending on their intended use.

10.8.1 Micro Aerial Vehicles

Micro Aerial Vehicles (MAVs) represent a category of UAVs that are characterized by their small size (wings of length less than 20 cm), lightweight, and maneuverability. MAVs are engineered to mimic the flight capabilities of insects and birds, making them suitable for various applications in both civilian and military domains. Due to their compact size, MAVs often feature innovative engineering designs to ensure stability, control, and efficient propulsion. It is to be noted that UAVs can be categorized based on their weight which is the norm followed by the regulatory agencies in India.

The working principle of MAVs is based on aerodynamics and advanced control systems. These miniature aircraft generate lift through the interaction of their wings or rotors with the air. The flight control systems, often employing microelectromechanical systems (MEMS) and sophisticated algorithms, enable precise control of MAVs in confined spaces and complex environments.

MAVs are designed to be lightweight, agile, and versatile. MAVs typically have fixed-wing or rotary-wing configurations. Fixed-wing MAVs are efficient for longer endurance and higher speed flights while rotary-wing MAVs offer vertical takeoff and landing capabilities, hover, and agility. Electric motors or miniature internal combustion engines are commonly used for propulsion, providing sufficient power for MAV flight while maintaining low weight. Advanced materials, such as carbon fiber composites and lightweight alloys, are used to ensure strength and durability while minimizing overall weight.

Types of MAVs are

- Fixed-wing MAVs are essentially miniature aircraft, typically measuring less than 200 mm in size, and are operated remotely. Some of these MAVs can be as tiny as the palm of a human hand, as illustrated in the figure. These micro aerial vehicles consist of wings and propellers that enable them to achieve flight. Controlling these MAVs is achieved using methods similar to those employed in traditional aircraft. They are ideal for mapping, surveillance, and environmental monitoring.

- Rotary-wing MAVs (also called as quadcopters or multirotors), resembling miniature helicopters, are remotely controlled aircraft. Unlike fixed-wing MAVs, they lack traditional wings and instead rely on their rotors for flight. They use multiple rotors to achieve vertical takeoff and landing and stable hovering. These compact and versatile MAVs have become widely prevalent today, finding applications ranging from aerial photography, inspection, and surveillance to military surveillance and disaster management. Furthermore, innovative ideas propose their utilization for commercial purposes, such as order deliveries by companies like *Amazon*.

- Flapping wing MAVs represent the latest trend in MAV development. They are the most intricate of all MAVs, as they imitate the flight mechanism used by birds. Birds achieve flight by flapping their wings in specific patterns, with variations depending on the bird's size and species. Similarly, flying insects exhibit unique wing-flapping patterns. The advanced MAVs promise exciting possibilities for aerial maneuverability and applications in various fields.

MAVs continue to evolve, driven by advancements in engineering, materials, and miniaturization technologies, unlocking new possibilities and applications in various industries. Nowadays, the increasing number of MAV applications demands

(a) *Honeywell RQ-16A T-Hawk*, a ducted fan VTOL miniature UAV
Gross weight: 8.39 kg *Fuel capacity*: 0.95 kg
Endurance: 40 minutes *Service ceiling*: 3,200 m
Powerplant: 1 x 3W-56 56 cc Boxer Twin-piston engine, 4 hp (3 kW) *Maximum speed*: 130 km/h

(b) *Prox Dynamics's Black Hornet* (Norway, 2012), . Over 3,000 Black Hornets had been delivered as of 2014.
Size: 16 x 2.5 cm, *Endurance* : 20 minutes, *Top speed*: 21 km/h, *weighs* : 18 g

Examples of MAVs, UAV with size less than 200 mm.

enhanced MAV efficiency. Achieving this efficiency boost involves improvements in aerodynamics, size reduction, swarming (simultaneous use of multiple MAVs), increased autonomy, and other innovative approaches. These technological advancements are currently in the development phase and are yet to be fully realized.

Ongoing aerodynamics research focuses on comprehending the intricate physics behind wing flight in MAVs. For instance, examining the rough-winged dragonfly's ability to generate lift poses intriguing questions. Unlike conventional smooth wings on full-size aircraft that reduce energy consumption for flight, dragonfly's complex aerodynamics offer superior efficiency for its size. Consequently, exploring the dragonfly's wing shape and flapping motion is crucial for MAVs of similar size.

Another area of research centers on swarming, wherein MAVs move in synchronized groups. This requires extensive investigation into real-time data transfer and analysis for coordinated movement. MAVs must interact, sharing their relative positions and considering the aerodynamic impact on each other. Each MAV must then process this data in real-time and make adjustments, necessitating vast computational capabilities in a compact space.

Additionally, the diminutive size of MAVs presents unique challenges. The limited space within MAVs must accommodate batteries, electric motors, and navigation electronics. Research strives to miniaturize these components, as even a slight efficiency improvement can substantially enhance MAV capabilities.

10.8.2 Quadcopters

A quadcopter (or quadrotor or quadrotor) is a type of UAV that is powered by four rotors. It is one of the most common and popular configurations for consumer and commercial drones due to its simplicity, stability, and maneuverability. The term quadcopter is derived from quad meaning four and copter referring to a rotary-wing aircraft.

Quadcopters operate on the principle of multirotor flight, where the lift and control are achieved by varying the speed of the rotors. By individually adjusting the rotational speed of each rotor, the quadcopter can generate thrust and control its movement in all directions.

A typical quadcopter consists of a frame, motors, propeller, Electronic Speed Controllers (ESCs), flight controller and battery. The frame is the central structure that holds all the components together. It is usually made of lightweight materials like carbon fiber or plastic to keep the overall weight low. Four electric motors are mounted on the frame, with two rotating clockwise (CW) and the other two counter-

(a) Israeli *IAI Malat's Mosquito*
Size: 35 cm long and 35 cm in total span, *Power:* electric motor with batteries, *Endurance:* 40 minutes, *Radius of action:* about 1.2 km

(b) USA *Aurora Flight Science's Skate*
Size: wingspan of 60 cm and length of 33 cm, *Power:* twin electric motors, *Control:* by tilting the motor/propeller assemblies (no control surfaces), *Payload:* 227 g, *Weight:* 1.1 kg

(c) Australian *Cyber Technology CyberQuad Mini*
Size: $42 \times 42 \times 20$ cm^3 and four ducted fans, each with a diameter < 20 cm,

Examples of very small UAVs (dimensions of the order of a 30-50 cm).

clockwise (CCW). This counter-rotating configuration ensures stability and eliminates the need for a tail rotor found in traditional helicopters. Each motor is attached to a propeller, which generates the lift and thrust needed for flight. The propellers come in various sizes and designs, and their rotation creates the upward force to keep the quadcopter airborne. ESCs are connected to the motors and control the speed of each motor independently. By adjusting the speed of the motors, the quadcopter

10.8. UNINHABITED AERIAL VEHICLES

Table 10.2: Classes of UAVs based on size

UAV type and description	Typical examples
Micro Aerial Vehicle (MAVs) Size of a large insect to around 20 cm wing span. It can have rotary-wing, flapping wing or fixed-wing. The smallest category, with a wingspan of a few cm is called *Nano UAVs*.	*Honeywell RQ-16A T-Hawk*, a ducted fan VTOL miniature UAV developed by Honeywell Aerospace. It is also known as the *Tarantula Hawk*. The T-Hawk is designed to perform various military and civilian tasks in challenging environments, providing intelligence, surveillance, reconnaissance, and other capabilities. *DJI Mavic Mini*, a popular consumer-grade quadcopter MAV for recreational aerial photography. *Parrot Anafi*, a foldable, lightweight quadcopter MAV suitable for aerial photography and videography. *Black Hornet Nano*, a fixed-wing MAV used for military reconnaissance and surveillance. In 2012, the British Army deployed the sixteen-gram *Black Hornet Nano* in Afghanistan to support infantry operations.
Very Small UAVs size 30-50 cm. They are either flapping wings or fixed wings	*Israeli IAI Malat's Mosquito* an oval flying wing with a single tractor propeller, It is hand or bungee-launched and can deploy a parachute for recovery *USA Aurora Flight Science's Skate:* a rectangular flying wing with twin tractor engine/propeller combinations that can be tilted to provide thrust vectoring for control. It folds in half along its centerline for transport and storage. It allows vertical takeoff and landing (VTOL) and transition to efficient horizontal flight. *Australian Cyber Technology CyberQuad Mini* has four ducted fans in a square arrangement
Small UAVs: Size: 50 cm - 200 cm. Mostly fixed-wing model airplanes, they are hand-launched by the operator	*US AeroVironment Raven:* A fixed-wing vehicles. The Raven and its control "station" can be carried around by its operator on his/her back and can carry visible, near-infrared (NIR), and thermal imaging systems for reconnaissance as well as a "laser illuminator" to point out targets to personnel on the ground. The *Turkish Byraktar Mini:* A conventional fixed-wing, electrically powered vehicle, launched by being thrown into the air by its operator. The Bayraktar Mini has a gimbaled day or night camera. It offers waypoint navigation with GPS or other radio navigation systems.
Medium UAVs have typical wingspans of the order of 5-10 m and carry payloads of from 100 - 200 kg	The *Israeli-US Hunter* and the UK *Watchkeeper* are more recent examples of medium-sized, fixed-wing UAVs. US Boeing *Eagle Eye*, which is a medium-sized VTOL system that is notable for using tilt-wing technology. The *RQ-2 Pioneer* is smaller than a light manned aircraft
Large UAVs They are large enough to carry weapons in significant quantities and can fly long distances from their bases, loiter for extended periods to perform surveillance functions	The *US General Atomics Predator A & B*, which has a significant range and endurance, but can carry only two missiles of the weight presently being used. The *US Northrop Grumman Global Hawk*, larger UAV designed for very long range and endurance and capable of flying anywhere in the world on its own.
Expendable UAVs (similar to guided missiles) are not designed to return after accomplishing their mission.	It can be recovered if possible but can have a very high loss rate. They contain an internal warhead and are intended to be crashed into a target destroying it and themselves

(a) US *AeroVironment Raven*
Size: 1.4-m wingspan and about 1 m long, *Weight:* about 2 kg, *propulsion:* electrical and *Endurance:* an hour and a half.

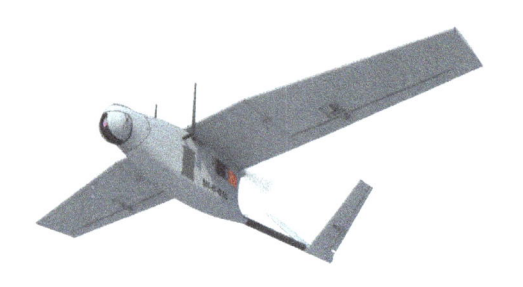

(a) *RQ-2 Pioneer*
Length: 4.3 m, *Wingspan:* 5.151 m *Height:* 1.006 m, *Gross weight:* 205 kg, *Fuel capacity:* 44 to 47 L, *Powerplant:* 1 x ZF Sachs 2-stroke 2-cylinder horizontally-opposed piston engine, 19 kW (26 hp) or UEL AR-741 rotary engine; 28.3 kW (38.0 hp), *Range:* 185 km, *Service ceiling:* 4,600 m (15,100 ft)

(b) Turkish *Bayraktar Mini*
Size: length of 1.2 m, *wingspan* 2 m, *Weight* 5 kg, The data link has a range of 20 km

(b) *Thales Watchkeeper WK450*
Wingspan: Approximately 10.5 meters, *Length:* Approximately 6.2 meters, *Maximum Takeoff Weight:* Around 450 kilograms, *Payload Capacity:* Up to 150 kilograms, *Maximum Altitude:* Over 5,486 meters, *Maximum Endurance:* Over 16 hours,

Examples of Small UAVs (dimensions of the order of a 50-200 cm).

Examples of medium UAVs (dimensions of the order of a 5-10 m).

can be controlled to move in different directions. The flight controller is the brain of the quadcopter. It contains sensors like gyroscopes, accelerometers, and sometimes GPS, which provide real-time data on the quadcopter's orientation and position. The flight controller processes this data and sends appropriate signals to the ESCs to stabilize and control the quadcopter's flight. A rechargeable battery powers the quadcopter's motors and electronics.

Quadcopters are highly maneuverable and can perform various flight movements, including hovering, forward/backward flight, sideways flight (yaw), and rotation (roll and pitch). Their ability to hover and fly in confined spaces makes them suitable for a wide range of applications, including aerial photography, videography, surveillance, inspections, and recreational flying. Quadcopters are inherently stable due to their design, making them easier to fly compared to other types of UAVs. Additionally, many modern quadcopters have built-in safety features, such as altitude hold, return-to-home functions, and obstacle avoidance systems, to prevent collisions and enhance the user experience.

Index

A-10, 307
A-10 Thunderbolt, 307
A400M Atlas, 318
Absolute Ceiling, 219
Aerial refueling, 56
Aerodynamic Center, 190
Aeronautical Society of India, 8
Aeronautics, 19
Aerospace engineering, 19
Aerospace sector, 2
Aerostat, 420
AEW&C or AWACS, 326
AIAA, 8
Aileron, 223
Air Traffic Control (ATC), 284
Airborne Collision Avoidance Systems (ACAS), 275
Airbus, 261
Aircraft, 19
Aircraft Noise, 276
Aircraft Simulator, 81
Airfoil, 189
Airliner, 245
Airmail, 55
Airplane, 19
Airship, 19, 421
Alberto Santos-Dumont, 31
Amelia Earhart, 57
Amphibious Aircraft, 431
AN-124, 319
Anthony Fokker, 47
Anti-Ballistic Missile Treaty, 357
Anti-Satellite missile, 350
Antonov An-124 Ruslan, 319
Apollo Space Program, 146
Area Rule, 194

Ariane 5, 369
Artemis Program, 165
Arthur Charles Hubert Latham, 34
Arthur Clarke, 128
Astronautics, 19
Autogyro, 423
Autonomous Runway Incursion Warning System (ARIWS), 276
Autopilot, 82
Auxiliary Power Unit (APU), 253
Aviation, 19
Aviation Meteorology, 183
Avionics, 19

B-1 Lancer, 311
B-52 Stratofortress, 310
Balloon, 19
Battle of Britain, 63
Bell X-1, 78
Boeing, 261
Boeing 737, 262
Boeing 787, 264
Boeing CH-47 Chinook, 400
Bumper Program, 131

C-130J Super Hercules, 321
C-17 Globemaster, 320
C-27J Spartan, 318
C-390 Millennium, 319, 320
C-5M Super Galaxy, 319
Carrier Takeoff and Landing, 300
Center of Pressure, 190
Chandrayaan, 162
Charles Lindbergh, 57
Chicago Convention, 287
Chord Line, 189
Chuck Yeager, 77

CIRA, 7
Circular Error Probable (CEP), 346
Claude-Louis Navier, 36
Climb Performance, 299
CNSA, 383
Concorde, 86
Controller-Pilot Data Link Communications (CPDLC), 275
Convertiplane, 425
Critical Mach number, 192

D-Day, 64
Daniel Bernoulli, 35
Dassault Rafale, 322, 323
Delta Wing, 79
DGCA, 8
Dirigible, 19
DLR, 7
Drag, 190
Drag Divergence Mach number, 192
Drag Polar, 215
DRDO, 7

EASA, 7
Elevator, 223, 250
Embraer EMB 314 Super Tucano, 308
Emergency Locator Transmitter (ELT), 253, 274
Empennage, 250
Endurance, 299
Environmental Control Systems (ECS), 254
ESA, 383
Eugene Burton Ely, 34
EUROCONTROL, 8
Eurofighter Typhoon, 324, 325
Exocet, 356
Extended-range Twin Operational Performance Standards (ETOPS), 269

F-117 Nighthawk, 84
F-15, 314
F-16, 312, 324
F-16 Fighting Falcon, 324
F-22, 296
F-35 Lightning II, 325

FAA, 8
Flight deck or cockpit, 250
Floatplane, 430
Fly by Wire, 82
Flying Ace, 45
Flying Boat, 430
Flying Boats, 62
Frank Whittle, 54
Fuselage, 248

G-suit, 76
Gaetano Arturo Crocco, 33
George Cayley, 27
George Stokes, 36
Glenn Curtiss, 33
Glider, 19, 426
Global Aeronautical Distress and Safety System (GADSS), 276
Global Hawk, 314
GNSS, 275
GPWS, 274
Ground Effect Vehicle, 428
Guidance and Control System, 343
Gyroplane, 19

HAL, 7
HAL Tejas, 323
Hans von Ohain, 53
Harriet Quimby, 33
Helicopter, 19
Hellfire, 356
Henri Farman, 31
Hermann Oberth, 127
High Lift Device, 217
Hiroshima and Nagasaki bombings, 64
Horizontal Stabilizer, 250
Hubble telescope, 144
Hugo Junkers, 50

IATA, 7
ICAO, 7
Il-78, 315
Ilyushin Il-76, 321
Instrument Landing System (ILS), 272

INTA, 7
International Air Transport Association (IATA), 287
International Civil Aviation Organization (ICAO), 287
International Space Station, 138
ISRO, 383

James Webb telescope, 144
Javelin Misile, 349
JAXA, 383
Jet fuel, 251
Jet Pack, 429
John Montgomery, 29
Junkers Ju 88, 312

Kalpana Chawla, 142
KC-46 Pegasus, 320
kite balloon, 19
Konstantin Tsiolkovsky, 126

Landing Gear, 249
Launch Vehicle Mark-3, 163
LCA Tejas, 323
Leonardo Da Vinci, 27
Leonhard Euler, 36
Lift, 190
Lockheed U-2, 313
Louis Blériot, 31
Ludwig Prandtl, 37
Luna program, 145

Mach Number, 188
Mangalyaan, 153
Mars Odyssey, 153
Martin Wilhelm Kutta, 37
MAV, 432
Maximum Speed, 298
Mean Camber Line, 189
Mean Sea Level, 179
mid-air refueling, 315
MiG-15, 293
Military Airplane Generations, 292
Military Transport Airplanes, 317
Minimum Speed, 299

Minuteman, 356
Mir Space Station, 136
MIRV, 350
Missile, 19
Missile Defense System, 350
Montgolfier brothers, 27
MQ-9 Reaper, 314

NAL, 7
Narrow-body Airliner, 248
NASA, 383
Neil Armstrong, 129
Nikolay Zhukovsky, 36

Octave Chanute, 28
ONERA, 7
Operation Overlord, 64
Ornithopter, 19, 27
Otto Lilienthal, 29

P-51 Mustang, 66
Patrouille de France, 302
Pearl Harbor, 64
Pilatus PC-9, 316
Pioneer Space Vehicles, 155
Pitch, 223
Precision Approach Path Indicator (PAPI), 272
Pressurized Cabins, 74
Pylons, 249

Quadcopter, 434

Radar Cross-Section (RCS), 301
Range, 248, 299
Regulatory bodies, 7
RESA, 274
Retractable Landing Gear, 74
Reynolds Number, 188
Richard Whitcomb, 80
Robert Goddard, 126
Roll, 223
Roscosmos, 383
Rotorcraft, 19
Royal Aeronautical Society, 8
Rudder, 223, 250

Safety Management System (SMS), 271
Samuel Cowdery, 30
Samuel Langley, 28
Seaplane, 430
Sergei Korolëv, 127
Sikorsky UH-60 Black Hawk, 409
Soyuz, 142
Space agencies, 5
Space Planes, 370
Space Shuttle, 139
Space stations, 136
Spacecraft, 19
SpaceX Falcon Heavy launch vehicle, 369
Spitfire, 65
Sputnik satellite, 133
Stall Velocity, 217
Standard Atmosphere, 174, 180
Stealth Technology, 84, 301
Stopped rotor Aircraft, 426
Strategic Arms Reduction Treaty, 357
Stuka bombings, 68
Su-25, 308
Su-34, 307
Sukhoi Su-27, 312
Sukhoi Su-34, 307
Supercritical airfoils, 192
Surya Kiran, 303

Technical societies, 8
Terrain Awareness and Warning Systems (TAWS), 274
Theodore von Kármán, 37
Thunderbirds, 303
Tiangong Space Station, 138
Tiltrotor, 426
Tiltwing, 426
Trainer Airplanes, 316
Trim Tabs, 250
TsAGI, 7
Tsiolkovsky rocket equation, 126
Tu-144, 88
Tupolev Tu-160, 310
Turning Performance, 300

UAV, 431
Underwater Locator Beacon (ULB), 253
US Navy Blue Angels, 303

V-2 Rocket, 131
V-22 Osprey, 426
Venera Program, 154
Vertical Stabilizer, 250
Vikram Lander, 163
Voivode, 354
Voyager Space Programs, 155

Warheads, 346
Wernher von Braun, 128
Whitcomb Area Rule, 80
Wide-body Airliner, 248
Wing Dihedral, 194
Wing-tail configurations, 10
Winglets, 249
Wings, 249
World War I Flying Aces, 43
World War I Reconnaissance, 38
World War II Flying Aces, 71
Wright brothers, 30
Wright Flyer, 30

X-43, 142
X-planes, 157

Yaw, 223
Yuri Gagarin, 129

Zeppelin, 42

About the Authors

Mahesh M. Sucheendran attended the *Indian Institute of Technology, Madras* and obtained the Bachelor of Technology and Master of Technology degrees in Aerospace Engineering in 2004. Subsequent to that, he worked as a Scientist in the Aerodynamics Research and Development Division of *Vikram Sarabhai Space Centre, Indian Space Research Organisation* till 2007. Further, he attended the *University of Illinois at Urbana-Champaign*, graduating with a Ph.D. in Aerospace Engineering in 2013. The Ph.D. was funded by the *Air Force Research Laboratory, Ohio* through the *MidWest Structural Sciences Center* and was guided by Professor *Daniel Bodony* and Professor *Philippe Geubelle*. On returning to India, he joined as an Assistant Professor in the Aerospace Engineering Department of *Defence Institute of Advanced Technology, Pune*. Since 2016, he has been a faculty member in the Department of Mechanical and Aerospace Engineering at the *Indian Institute of Technology, Hyderabad*. He has done various projects related to Ballistics, Aerodynamics, Aeroacoustics and Aeroelasticity for various aerospace and defence agencies like the *Armament Research & Development Establishment, Armament Research Board, Advanced Naval Systems Program, Aeronautical Development Agency* and *Advanced Systems Laboratory*. He is a lifetime member of the International Ballistics Society.

Lokanna Hoskoti has been working as a Project Scientist in the Department of Mechanical and Aerospace Engineering at the *Indian Institute of Technology Hyderabad* since the year 2020.

Prior to this role, he pursued his academic journey at the *Defence Institute of Advanced Technology, Pune*, where he earned his Ph.D. in Aerospace Engineering and a Master's degree in Mechanical Engineering with a specialization in Gas Turbine Technology. He holds a Bachelor's degree in Mechanical Engineering from *Visvesvaraya Technological University* in Belagavi, Karnataka.

During the years 2011 to 2015, Lokanna worked as an Assistant Professor in the Department of Mechanical Engineering at *Sinhagad College of Engineering, Pune*. With approximately eight years of experience in Mechanical and Aerospace Engineering, including four years of postdoctoral research experience, Dr. Lokanna was involved in several sponsored research projects funded by various defense agencies such as the *Aeronautical Development Agency (ADA)* and *Advanced Systems Laboratory (ASL)*.